AQA A-level

Mathematics

Year 2

Authors

Sophie Goldie

Val Hanrahan

Cath Moore

Jean-Paul Muscat

Susan Whitehouse

Series editors

Roger Porkess

Catherine Berry

Consultant Editor

Heather Davis

Approval message from AQA

This textbook has been approved by AQA for use with our qualification. This means that we have checked that it broadly covers the specification and we are satisfied with the overall quality. Full details of our approval process can be found on our website.

We approve textbooks because we know how important it is for teachers and students to have the right resources to support their teaching and learning. However, the publisher is ultimately responsible for the editorial control and quality of this book.

Please note that when teaching the *AQA A-level Mathematics* course, you must refer to AQA's specification as your definitive source of information. While this book has been written to match the specification, it cannot provide complete coverage of every aspect of the course.

A wide range of other useful resources can be found on the relevant subject pages of our website: www.aqa.org.uk.

Acknowledgements

The Publishers would like to thank the following for permission to reproduce copyright material.

Questions from past AS and A Level Mathematics papers are reproduced by permission of MEI and OCR.

Practice questions have been provided by Chris Little (p319–320), Neil Sheldon (p410–413), Rose Jewell (p518–520) and MEI (p127–129 and p237–239).

p35 Figure 3.1 data from United Nations Department of Economics and Social Affairs, Population Division. World Population prospects: The 2015 Revisions, New York, 2015.

p350 Table source: adapted from Table Q1.6(i), Executive summary tables: June 2013, Criminal justice statistics quarterly: June 2013

p354 Data from Table NTS0905 published by gov.uk, reproduced under the Open Government Licence www.nationalarchives.gov.uk/doc/open-government-licence/version/3/

p403 Figure 17.13 data from *The World Factbook* 2013–14. Washington, DC: Central Intelligence Agency, 2013. https://www.cia.gov/library/publications/the-world-factbook/index.html.

Every effort has been made to trace all copyright holders, but if any have been inadvertently overlooked, the Publishers will be pleased to make the necessary arrangements at the first opportunity.

Although every effort has been made to ensure that website addresses are correct at time of going to press, Hodder Education cannot be held responsible for the content of any website mentioned in this book. It is sometimes possible to find a relocated web page by typing in the address of the home page for a website in the URL window of your browser.

Hachette UK's policy is to use papers that are natural, renewable and recyclable products and made from wood grown in sustainable forests. The logging and manufacturing processes are expected to conform to the environmental regulations of the country of origin.

Orders: please contact Bookpoint Ltd, 130 Park Drive, Milton Park, Abingdon, Oxon OX14 4SE. Telephone: (44) 01235 827720. Fax: (44) 01235 400454. Email education@bookpoint.co.uk Lines are open from 9 a.m. to 5 p.m., Monday to Saturday, with a 24-hour message answering service. You can also order through our website: www.hoddereducation.co.uk

ISBN: 978 1 4718 52893

© Sophie Goldie, Val Hanrahan, Jean-Paul Muscat, Roger Porkess, Susan Whitehouse and MEI 2017

First published in 2017 by

Hodder Education,

An Hachette UK Company

Carmelite House

50 Victoria Embankment

London EC4Y 0DZ

www.hoddereducation.co.uk

Impression number 10 9 8 7 6 5 4 3 2 1

Year   2021   2020   2019   2018   2017

Cover photo © Natinkabu/iStock/Thinkstock/Getty Images

Typeset in Bembo Std, 11/13 pts. by Aptara®, Inc.

Printed in Italy

A catalogue record for this title is available from the British Library.

Contents

Getting the most from this book

Mathematics is not only a beautiful and exciting subject in its own right but also one that underpins many other branches of learning. It is consequently fundamental to our national wellbeing.

This book covers the remaining content of A Level Mathematics and so provides a complete course for the second of the two years of Advanced Level study. The requirements of the first year are met in the first book.

Between 2014 and 2016 A level Mathematics and Further Mathematics were very substantially revised, for first teaching in 2017. Major changes include increased emphasis on

■ Problem solving

■ Proof

■ Use of ICT

■ Modelling

■ Working with large data sets in statistics.

This book embraces these ideas. The first section of Chapter 1 is on **problem solving** and this theme is continued throughout the book with several spreads based on the problem solving cycle. In addition a large number of exercise questions involve elements of problem solving; these are identified by the PS icon beside them. The ideas of **mathematical proof** and rigorous logical argument are also introduced in Chapter 1 and are then involved in suitable exercise questions throughout the book. The same is true of **modelling**; the modelling cycle is introduced in the first chapter and the ideas are reinforced through the rest of the book.

The use of **technology**, including graphing software, spreadsheets and high specification calculators, is encouraged wherever possible, for example in the Activities used to introduce some of the topics in Pure mathematics, and particularly in the analysis and processing of **large data sets** in Statistics. Places where ICT can be used are highlighted by a T icon. A large data set is provided at the end of the book but this is essentially only for reference. It is also available online as a spreadsheet (www.hoddereducation.co.uk/AQAMathsYear2) and it is in this form that readers are expected to store and work on this data set, including answering the exercise questions that are based on it. These are found at the end of each exercise in the Statistics chapters and identified with a purple bar. They illustrate, for each topic, how a large data set can be used to provide the background information.

Throughout the book the emphasis is on understanding and interpretation rather than mere routine calculations, but the various exercises do nonetheless provide plenty of scope for practising basic techniques. The exercise questions are split into three bands. Band 1 questions (indicated by a green bar) are designed to reinforce basic understanding. Band 2 questions (yellow bar) are broadly typical of what might be expected in an examination: some of them cover routine techniques; others are designed to provide some stretch and challenge for readers. Band 3 questions (red bar) explore round the topic and some of them are rather more demanding. Questions in the Statistics chapters that are based on the large data set are identified with a purple bar. In addition, extensive online support, including further questions, is available by subscription to MEI's Integral website, http://integralmaths.org.

In addition to the exercise questions, there are five sets of questions, called Practice questions, covering groups of chapters. All of these sets include identified questions requiring **problem solving** PS, **mathematical proof** MP, **use of ICT** T and **modelling** M. There are some multiple choice questions preceding each of these sets of practice questions to reflect those in the AQA papers.

This book follows on from *A Level Mathematics for Year 1 (AS)* and most readers will be familiar with the material covered in it. However, there may be occasions when they want to check on topics in the earlier book; the parts entitled Review allow them to do this without having to look elsewhere. The five short Review chapters provide a condensed summary of the work that was covered in the earlier book,

including one or more exercises; in addition there are nine chapters that begin with a Review section and exercise, and then go on to new work based on it. Confident readers may choose to miss out the Review material, and just refer to these parts of the book when they are uncertain about particular topics. Others, however, will find it helpful to work through some or all of the Review material to consolidate their understanding of the first year work.

There are places where the work depends on knowledge from earlier in the book and this is flagged up in the margin in Prior knowledge boxes. This should be seen as an invitation to those who have problems with the particular topic to revisit it earlier in the book. At the end of each chapter there is a summary of the new knowledge that readers should have gained.

Two common features of the book are Activities and Discussion points. These serve rather different purposes. The Activities are designed to help readers get into the thought processes of the new work that they are about to meet; having done an Activity, what follows will seem much easier. The Discussion points invite readers to talk about particular points with their fellow students and their teacher and so enhance their understanding. Callout boxes and Note boxes are two other common features. Callout boxes provide explanations for the current work. Note boxes set the work in a broader or deeper context. Another feature is a Caution icon ❗, highlighting points where it is easy to go wrong.

The authors have taken considerable care to ensure that the mathematical vocabulary and notation are used correctly in this book, including those for variance and standard deviation, as defined in the AQA specification for A-level Mathematics. In the paragraph on notation for sample variance and sample standard deviation (page 327), it explains that the meanings of 'sample variance', denoted by s^2, and 'sample standard deviation', denoted by s, are defined to be calculated with divisor $(n-1)$. In early work in statistics it is common practice to introduce these concepts with divisor n rather than $(n-1)$. However there is no recognised notation to denote the quantities so derived. Students should be aware of the variations in notation used by manufacturers on calculators and know what the symbols on their particular models represent.

When answering questions, students are expected to match the level of accuracy of the given information. However, there are times when this can be ambiguous. For example 'The mass of the block is 5 kg' could be taken to be an exact statement or to be true to just 1 significant figure. In many of the worked examples in this book such statements are taken to be exact. A particular issue arises with the value of g, the acceleration due to gravity. This varies from place to place around the world. Unless stated otherwise questions in this book are taken to be at a place where, to 3 significant figures, it is $9.80 \, \mathrm{m\,s^{-2}}$. So, providing that other information in the question is either exact or given to at least 3 significant figures, answers based on this value are usually given to 3 significant figures. However, in the solutions to worked examples it is usually written as 9.8 rather than 9.80. Examination questions often include a statement of the value of g to be used and candidates should not give their answers to a greater number of significant figures; typically this will be 3 figures for values of g of $9.80 \, \mathrm{m\,s^{-2}}$ and $9.81 \, \mathrm{m\,s^{-2}}$, and 2 figures for $9.8 \, \mathrm{m\,s^{-2}}$ and $10 \, \mathrm{m\,s^{-2}}$.

Answers to all exercise questions and practice questions are provided at the back of the book, and also online at www.hoddereducation.co.uk/AQAMathsYear2. Full step-by-step worked solutions to all of the practice questions are available online at www.hoddereducation.co.uk/AQAMathsYear2. All answers are also available on Hodder Education's Dynamic Learning platform.

Finally a word of caution. This book covers the content of Year 2 of A Level Mathematics and is designed to provide readers with the skills and knowledge they will need for the examination. However, it is not the same as the specification, which is where the detailed examination requirements are set out. So, for example, the book uses the data set of cycling accidents to give readers experience of working with a large data set, but this is not the data set that will form the basis of any examination questions. Similarly, in the book cumulative binomial tables are used in the explanation of the output from a calculator, but such tables will not be available in examinations. Individual specifications will also make it clear how standard deviation is expected to be calculated. So, when preparing for the examination, it is essential to check the specification.

Catherine Berry and Roger Porkess

★Please note that the marks stated on the example questions are to be used as a guideline only, AQA have not reviewed and approved the marks.

Prior knowledge

This book builds on work from AS/Year 1 A level Mathematics. AS work is reviewed either in **sections at the start of chapters, or in separate review chapters in this Year 2 A level Mathematics book.**

The order of the chapters has been designed to allow later ones to use and build on work in earlier chapters. The list below identifies cases where the dependency is particularly strong.

The Statistics and Mechanics chapters are placed in separate sections of the book for easy reference, but it is expected that these will be studied alongside the Pure mathematics work rather than after it.

- The work in **Chapter 1: Proof** pervades the whole book. It builds on the work on problem solving and proof covered in Chapter 1 of AS/Year 1 Mathematics.

- **Chapter 2: Trigonometry** builds on the trigonometry work in Chapter 6 of AS/Year 1 Mathematics.

- **Review: Algebra 1** reviews the work on surds, indices, exponentials and logarithms from Chapters 2 and 13 of AS/Year 1 Mathematics.

- **Chapter 3: Sequences and series** requires some use of logarithms, covered in **Review: Algebra 1**.

- **Review: Algebra 2** reviews the work on equations, inequalities and polynomials from Chapters 3, 4 and 7 of AS/Year 1 Mathematics.

- **Chapter 4: Functions** begins with a review of the work on transformations covered in Chapter 8 of AS/Year 1 Mathematics.

- **Chapter 5: Differentiation** begins with a review of the work on differentiation covered in Chapter 10 of AS/Year 1 Mathematics.

- **Review: The sine and cosine rules** reviews the work on triangles covered in part of Chapter 6 of AS/Year 1 Mathematics.

- **Chapter 6: Trigonometric functions** builds on the work in **Chapter 2**, and uses ideas about functions from **Chapter 4**.

- **Chapter 7: Further algebra** starts with a review of the work on the binomial expansion from Chapter 9 of AS/Year 1 Mathematics. It also builds on work on the factor theorem and algebraic division, covered in **Review: Algebra 2**.

- **Chapter 8: Trigonometric identities** builds on the work in **Chapter 2** and **Chapter 6**.

- **Chapter 9: Further differentiation** builds on the work in **Chapter 5**. It also requires the use of radians, covered in **Chapter 2**.

- **Chapter 10: Integration** starts with a review of the work on integration covered in Chapter 11 of AS/Year 1 Mathematics. It follows on from the differentiation work in **Chapter 9**, and also requires the use of radians, covered in **Chapter 2**, and partial fractions, covered in **Chapter 7**.

- **Review: Coordinate geometry** reviews the work in Chapter 5 of AS/Year 1 Mathematics.

- **Chapter 11: Parametric equations** uses trigonometric identities covered in **Chapter 6** and **Chapter 8**. You should also recall the equation of a circle, covered in **Review: Coordinate geometry**, and be confident in the differentiation techniques covered in **Chapter 5** and **Chapter 9**.

- **Chapter 12: Vectors** builds on the vectors work in Chapter 12 of AS/Year 1 Mathematics.

- **Chapter 13: Differential equations** uses integration work covered in **Chapter 10**.

- **Chapter 14: Numerical methods** requires some simple differentiation and knowledge of how integration relates to the area under a graph.

- **Review: Working with data** reviews the work in **Chapters 14 and 15 of AS/Year 1 Mathematics**.

- **Chapter 15: Probability** starts with a review of the probability work in **Chapter 16 of AS/Year 1 Mathematics**.

- **Chapter 16: Statistical distributions** starts with a review of the work on the binomial distribution covered in **Chapter 17 of AS/Year 1 Mathematics**. It involves use of probability covered in **Chapter 15**.

- **Chapter 17: Statistical hypothesis testing** starts with a review of the work on hypothesis testing covered in **Chapter 18 of AS/Year 1 Mathematics**. It requires use of the Normal distribution covered in **Chapter 16**.

- **Chapter 18: Kinematics** starts with a review of the work on kinematics covered in **Chapters 19 and 21 of AS/Year 1 Mathematics**. You should be confident in working with vectors in two dimensions (reviewed in **Chapter 12**) and in working with parametric equations (**Chapter 11**).

- **Chapter 19: Forces and motion** starts with a review of the work on force covered in **Chapter 20 of AS/Year 1 Mathematics**. It requires the use of vectors in two dimensions (reviewed in **Chapter 12**).

- **Chapter 20: Moments of forces** uses work on force covered in **Chapter 19**, and the use of vectors in two dimensions (reviewed in **Chapter 12**).

- **Chapter 21: Projectiles** uses trigonometric identities from **Chapter 6** and **Chapter 8**, and work on parametric equations from **Chapter 11**. It also requires use of vectors in two dimensions (reviewed in **Chapter 12**).

- **Chapter 22: A model for friction** uses work on force and moments covered in **Chapters 19 and 20**, as well as vectors in two dimensions (reviewed in **Chapter 12**).

Photo credits

p1 © Tatiana Popova – Shutterstock; **p12** © Adam Majchrzak – Shutterstock; **p35** © Dmitry Nikolaev – Fotolia; **p44** © Granger Historical Picture Archive/Alamy Stock Photo; **p64** © Gianfranco Bella – Fotolia; **p96** © andreadonetti – 123RF; **p117** © focal point – Shutterstock; **p137** © Design Pics Inc – Getty Images; **p150** © REUTERS/Alamy Stock Photo; **p169** © EpicStockMedia – Fotolia; **p184** © hanapon1002/Istock via Thinkstock; **p197** © Alberto Loyo – Shutterstock; **p248** © Ken Hewitt/iStockphoto.com; **p250** © Sinibomb Images/Alamy Stock Photo; **p266** © Matt Tilghman/123RF; **p278** © volff – Fotolia; **p291** © Imagestate Media (John Foxx)/Vol 08 Modern Lifestyles; **p332** © www.hollandfoto.net – Shutterstock; **p334** © Stephen Finn – Shutterstock; **p353** © Matt Cardy/Getty Images; **p355** © sixpixx –Shutterstock; **p362** © RTimages/ Alamy Stock Photo; **p378** © Marc Darkin – Fotolia; **p384** © wavebreakmedia – Shutterstock; **p402** (upper) © Topham/Fotomas; **p402** (lower) © Bibliothèque nationale de France via Wikipedia Commons (Public Domain); **p414** © NASA; **p441** (left) © Georgios Kollidas/Shutterstock; **p441** (right) © Georgios Kollidas/Alamy Stock Photo; **p438** © Ambient Ideas – Shutterstock; **p470** © almgren – Shutterstock; **p478** © Simon Whaley/Alamy Stock Photo; **p478** (left) Public Domain; **p482** © Lano Lan – Shutterstock; **p504** © Nieuwland Photography – Shutterstock; **p506** © Anucha Saorong – 123RF.

Every effort has been made to trace all copyright holders, but if any have been inadvertently overlooked, the Publishers will be pleased to make the necessary arrangments at the first opportunity.

1 Proof

Figure 1.1 shows a square of side c inside a square of side $a + b$.

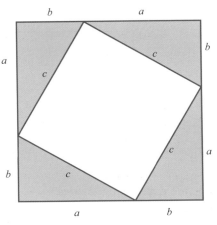

Figure 1.1

→ How can you deduce Pythagoras' theorem ($c^2 = a^2 + b^2$) by finding two ways of expressing the area of the central square?

1 Problem solving

Mathematical problem solving sometimes involves solving purely mathematical problems, and sometimes involves using mathematics to find a solution to a 'real-life' problem.

The **problem solving cycle** in Figure 1.2 shows the processes involved in solving a problem.

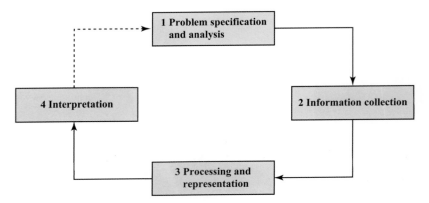

Figure 1.2

In purely mathematical problems, the same cycle can often be expressed using different words, as in Figure 1.3.

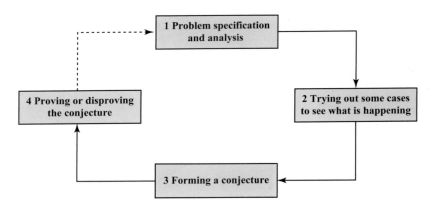

Figure 1.3

Forming a conjecture

Rob is investigating what happens when he adds the terms of the sequence

$$2, 1, \frac{1}{2}, \frac{1}{4}, \frac{1}{8}, \ldots$$

He says:

- 'However many terms I add, the answer is always less than 4.'
- 'If I add enough terms, I can get as close to 4 as I like.'

These statements are **conjectures**. They are Rob's theories. ← A statement that has not yet been proved is called a conjecture.

A conjecture may or may not be true, but once you have made a conjecture you obviously want to know whether it is true. If it is, you will want to convince other people, and that means you must prove it.

Rob draws the diagram in Figure 1.4.

Discussion points

→ How can Rob use the diagram to prove his conjectures?

→ How would he make the arguments watertight?

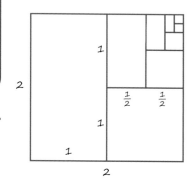

Figure 1.4

Here is a well-known conjecture that no one has managed to prove yet.

Discussion point

→ Explain why Goldbach's conjecture does not apply to all integers.

Goldbach's conjecture

Every even integer greater than 2 can be written as the sum of two prime numbers.

Most people believe it is true and many have tried to prove it, but so far without success.

Rob's conjectures can be shown geometrically, but many other conjectures need algebra. Proving a conjecture always requires rigorous, logical argument.

Here are some symbols and words that are very useful in this:

Discussion point

→ Explain why the **converse** statement 'n is a prime number $\Rightarrow n = 5$' is not true.

■ The symbol $\Rightarrow$ means 'leads to' or 'implies' and is very helpful when you want to present an argument logically, step by step.

$$n = 5 \Rightarrow n \text{ is a prime number.}$$

> You can say that the statement '$n = 5$' is a **sufficient** condition for the statement 'n is a prime number'.

Discussion point

→ Explain why the converse statement '$2n$ is even $\Rightarrow n$ is even' is not true.

■ You can write the symbol $\Rightarrow$ the other way round, as $\Leftarrow$. In that case, it means 'is implied by' or 'follows from'.

$$2n \text{ is even} \Leftarrow n \text{ is even.}$$

This statement could also be written in the form

$$n \text{ is even} \Rightarrow 2n \text{ is even.}$$

> You can say that the statement '$2n$ is even' is a **necessary** condition for the statement 'n is even'.

■ In situations where both the symbols $\Rightarrow$ and $\Leftarrow$ give true statements, the two symbols are written together as $\Leftrightarrow$. You can read this as 'implies and is implied by' or 'is equivalent to'.

$$n \text{ is an even number} \Leftrightarrow n^2 \text{ is an even number.}$$

> You can say that 'n is an even number' is a **necessary and sufficient** condition for the statement 'n^2 is an even number'.

Problem solving

① Is the statement 'n is odd $\Leftarrow n^3$ is odd' true or false?

② In each case, write one of the symbols $\Rightarrow$, $\Leftarrow$ or $\Leftrightarrow$ between the two statements A and B.

(i) A: PQRS is a rectangle.

B: PQRS has two pairs of equal sides.

(ii) A: The point P is inside a circle centre O, radius 3.

B: The distance OP is less than 3.

(iii) A: p is a prime number greater than 2.

B: p is odd.

(iv) A: $(x - 3)(x - 4) > 0$

B: $x > 4$

③ Samir writes:

AB is parallel to CD $\Rightarrow$ ABCD is a parallelogram.

(i) Is Samir correct? Explain your answer.

(ii) Write down the converse of Samir's statement. Is the converse true?

④ Winnie lives in a village in rural Africa; it is marked P on the diagram in Figure 1.5.

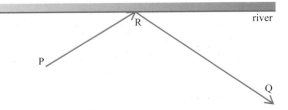

Figure 1.5

Each day she goes to a river which flows due east. She fills a bucket with water at R and takes it to her grandmother who lives in a nearby village, Q. Winnie wants to know where to fill the bucket so that she has the shortest distance to walk.

Referred to a coordinate system with axes east and north, P is the point (2, 3), Q is (8, 1) and the equation of the river is $y = 5$.

(i) Draw the diagram accurately on graph paper, using equal scales for both axes.

(ii) Winnie thinks that the best point for R is (5, 5). Show that her conjecture is wrong.

(iii) Find the coordinates for the best position of R. Explain carefully how you know that this is indeed the case.

⑤ Place the numbers from 1 to 8 in a copy of the grid in Figure 1.6 so that consecutive numbers are not in adjacent cells (i.e. cells that have a common edge or vertex).

Figure 1.6

If you can't do it, explain why not.

If you can do it, state in how many ways it can be done, justifying your answer.

⑥ Figure 1.7 shows a square of side 1 m and four circles.

The small red circle fits in the gap in the middle.

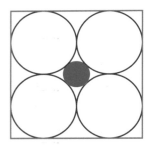

Figure 1.7

Show that the diameter of the red circle is $\left(\frac{1}{2}\sqrt{2} - \frac{1}{2}\right)$ m.

⑦ A game is played using a standard dartboard.

Figure 1.8

In this game, the 'doubles' count as squares and the 'trebles' count as cubes.

A player has three darts and must score one 'single', one 'double' and one 'treble' to make a total of 501.

(i)   Find two ways in which a player can finish (ignoring the order in which the darts are thrown).

(ii) Prove that there are no other possible ways.

2 Methods of proof

> **Discussion point**
>
> Sarah challenges her classmates to find two consecutive numbers such that the difference between their squares is even.
>
> ➜ What answers do you think she will get?

You have probably found that Sarah's challenge in the discussion point appears to be impossible. You have formed the conjecture that the difference between the squares of two consecutive numbers is always odd. The next step is to prove that your conjecture is true, and then you will know for certain that Sarah's challenge is impossible.

Two of Sarah's classmates decide to prove that her challenge is impossible.

Jamie writes:

> For two consecutive numbers, one must be even and one must be odd.
> An even number squared is even.
> An odd number squared is odd.
> The difference between an even number and an odd number is always odd, so the difference between the square of an even number and the square of an odd number must be odd.
> So the difference between the squares of consecutive numbers must be odd.

Zarah writes:

> Let the first number be n.
> So the next number is n + 1.
> The difference between their squares = $(n + 1)^2 - n^2$
> $$= n^2 + 2n + 1 - n^2$$
> $$= 2n + 1$$
> 2n + 1 is an odd number, so the difference between the squares of consecutive numbers is always odd.

> 2n is a multiple of 2, so it is an even number. So 2n + 1 must be an odd number.

Discussion point

→ Which proof do you prefer?

Jamie and Zarah have both proved the conjecture, in different ways.

You have now reached the stage where it is no longer always satisfactory to assume that a fact is true without proving it, since one fact is often used to deduce another.

There are a number of different techniques that you can use.

Proof by direct argument

Both Jamie's proof and Zarah's proof are examples of proof by direct argument, or deductive proof. You start from known facts and deduce further facts, step by step, until you reach the statement that you wanted to prove.

Example 1.1

Prove that the opposite angles of a cyclic quadrilateral are supplementary (add up to 180°).

You may assume the result that the angle subtended by an arc at the centre of a circle is twice the angle subtended by the same arc at the circumference.

Solution

Figure 1.9 shows a circle centre O and a cyclic quadrilateral ABCD.

$\angle ADC = x$ and $\angle ABC = y$.

The minor arc AC subtends angle x at the circumference of the circle, and angle p at the centre of the circle.

So $p = 2x$.

The major arc AC subtends angle y at the circumference of the circle, and angle q at the centre of the circle.

So $q = 2y$.

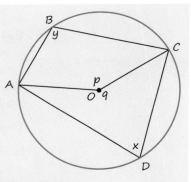

Figure 1.9

These two statements use the result that you may assume, given in the question.

Adding the two angles at O gives $p + q = 360°$

$$\Rightarrow 2x + 2y = 360°$$

$$\Rightarrow \quad x + y = 180°.$$

The sum of the four angles of any quadrilateral is 360°, so the sum of each pair of opposite angles of a cyclic quadrilateral is 180°.

Proof by exhaustion

For some conjectures it is possible to test all possible cases, as in Example 1.2.

Example 1.2

Prove that when a two-digit number is divisible by 9, reversing its digits also gives a number that is divisible by 9.

Solution

There are only 10 two–digit numbers divisible by 9:

$$18, 27, 36, 45, 54, 63, 72, 81, 90, 99.$$

Reversing each of these gives the following:

$$81, 72, 63, 54, 45, 36, 27, 18, 09, 99.$$

These numbers are also divisible by 9, so the conjecture has been proved.

ACTIVITY 1.1

Prove the result from Example 1.2 using direct proof.

Prove the corresponding result for a three-digit number.

Discussion points

→ Is it true that reversing the digits of a two-digit number that is divisible by 9 always gives a two-digit number that is divisible by 9?

→ How is this question different from the one in Example 1.2? ←

> It is important to be precise about wording.

Proof by contradiction

In some cases it is possible to deduce a result by showing that the opposite is impossible, as in the following examples.

Example 1.3

Prove that the sum of the interior angles x and y for a pair of parallel lines, as shown in Figure 1.10, is 180°.

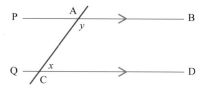

Figure 1.10

Solution

If the conjecture is false, then either $x + y < 180°$ or $x + y > 180°$. Look at these two cases separately.

→ Assume that $x + y < 180°$ as shown in Figure 1.11.

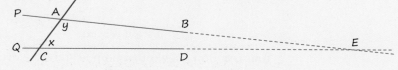

Figure 1.11

→

In this case the lines AB and CD, when extended, will meet at a point E, where

$$\angle BED = 180° - x - y.$$ ← Using the sum of the angles in a triangle.

This means that AB and CD are not parallel.

Similarly, assuming that $x + y > 180°$, as shown in Figure 1.12, will give angles $(180° - x)$ and $(180° - y)$, with a sum of $(360° - (x + y))$.

Figure 1.12

> It has now been shown that assuming that either $x + y < 180°$ or $x + y > 180°$ leads to a contradiction, so the only remaining possibility is that $x + y = 180°$.

$360° - (x + y) < 180°$, so now AP and CQ when extended will meet at a point R, showing that AP and CQ are not parallel.

Consequently, $x + y = 180°$.

Example 1.4

Prove that $\sqrt{2}$ is irrational.

Solution

Assume that $\sqrt{2}$ is rational, so $\sqrt{2} = \dfrac{m}{n}$, where m and n have no common factor.

Squaring $\Rightarrow 2 = \dfrac{m^2}{n^2}$

$\Rightarrow 2n^2 = m^2$ ①

$2n^2$ is a multiple of $2 \Rightarrow 2n^2$ is even

$\Rightarrow m^2$ is even

$\Rightarrow m$ is even.

So let $m = 2p$. ← As m is even, it can be expressed as $2p$, where p is an integer.

In equation ① this gives

$$2n^2 = (2p)^2 = 4p^2$$

$$\Rightarrow n^2 = 2p^2$$

$2p^2$ is a multiple of $2 \Rightarrow 2p^2$ is even

$\Rightarrow n^2$ is even

$\Rightarrow n$ is even.

You have now shown that both m and n are even numbers, which contradicts the assumption that m and n have no common factor.

Consequently, $\sqrt{2}$ is not rational, so it must be irrational.

Example 1.5

Prove that there are an infinite number of prime numbers.

Solution

Suppose there are a finite number of prime numbers: $2, 3, 5, \ldots, p_n$.

Let $q = (2 \times 3 \times 5 \times \ldots \times p_n) + 1.$ ← q is formed by multiplying together all the prime numbers in the list and then adding 1.

Is q prime?

- If q is prime, then it is a new prime number, not in the original list.

- If q is not prime, then it has a prime factor.
2 cannot be a factor of q, because q is one more than a multiple of 2.
3 cannot be a factor of q, because q is one more than a multiple of 3.
Similarly, none of the primes in the list can be a factor of q.
So if q is not prime, then it must have a prime factor which is not in the list.
So there is another prime number that is not in the list.

So, whether q is prime or not, there is another prime number not in the list. This contradicts the original assertion that there are a finite number of prime numbers.

Disproof by the use of a counter-example

Sometimes you may come across a conjecture that looks as if it might be true, but is in fact false. Always start by checking the result for a few particular values, to try to get a 'feel' for what is happening. Next, if you think that it is true, you could try to prove it using any of the methods discussed earlier. If you seem to be getting nowhere, then finding just one case, a **counter-example**, when it fails is sufficient to disprove it.

Example 1.6

Hassan says that 1003 is a prime number.

Is Hassan correct? Either prove his conjecture, or find a counter-example.

Solution

Checking for prime factors of 1003:

2 is not a factor of 1003

3 is not a factor of 1003

5 is not a factor of 1003.

...

However, it turns out that 17 is a factor of 1003:

$17 \times 59 = 1003$

Hassan is wrong.

Exercise 1.2

In questions 1–12 a conjecture is given. Decide whether it is true or false. If it is true, prove it using a suitable method and name the method. If it is false, give a counter-example.

① Numbers raised to the power 4 only end in the digits 0, 1, 4, 5 and 6.

② $2^n + 3$ is prime for $n > 1$.

③ $(a + b)^2 - (a - b)^2 = 4ab$, where a and b are real numbers.

④ The triangle with sides of length $\sqrt{2n + 1}$, n and $(n + 1)$ is right-angled.

⑤ No square number ends in 8.

⑥ The number of diagonals of a regular polygon with n sides is $\leqslant n$.

⑦ The sum of the squares of any two consecutive integers is an odd number.

⑧ $\sqrt{3}$ is irrational.

⑨ If T is a triangular number (given by $T = \frac{1}{2}n(n + 1)$ where n is an integer), then

   (i) $9T + 1$ is a triangular number

   (ii) $8T + 1$ is a square number.

⑩ (i) A four-digit number formed by writing down two digits and then repeating them is divisible by 101.

   (ii) A four-digit number formed by writing down two digits and then reversing them is divisible by 11.

⑪ The value of $(n^2 + n + 11)$ is a prime number for all positive integer values of n.

⑫ The tangent to a circle at a point P is perpendicular to the radius at P.

⑬ (i) The sum of the squares of any five consecutive integers is divisible by 5.

   (ii) The sum of the squares of any four consecutive integers is divisible by 4.

⑭ For any pair of numbers x and y, $2(x^2 + y^2)$ is the sum of two squares.

⑮ (i) Prove that $n^3 - n$ is a multiple of 6 for all positive integers n.

   (ii) Hence prove that $n^3 + 11n$ is a multiple of 6 for all positive integers n.

⑯ Prove that no number in the infinite sequence

   10, 110, 210, 310, 410, …

can be written in the form a^n where a is an integer and n is an integer $\geqslant 2$.

⑰ Prove that if (a, b, c) and (A, B, C) are Pythagorean triples then so is $(aA - bB, aB + bA, cC)$.

⑱ Which positive integers cannot be written as the sum of two or more consecutive numbers? Prove your conjecture.

⑲ An integer N is the sum of the squares of two different integers.

   (i)   Prove that N^2 is also the sum of the squares of two integers.

   (ii) State the converse of this result and either prove it is true or provide a counter-example to disprove it.

2 Trigonometry

This compass has two scales: the inner scale shows degrees and the outer
scale shows angular mils. There are 6400 angular mils in 360 degrees.

Mils are used by the military, in navigation and in mapping because they are
more accurate than degrees.

→ A pilot flies one degree off course. How far from the intended position is
the aeroplane after it has flown 10 km?

1 Radians

Have you ever wondered why angles are measured in degrees, and why there are
360° in one revolution?

There are various legends to support the choice of 360, most of them based in
astronomy. One of these is that the shepherd-astronomers of Sumeria thought
that the solar year was 360 days long; this number was then used by the ancient
Babylonian mathematicians to divide one revolution into 360 equal parts.

Degrees are not the only way in which you can measure angles. Some calculators have modes which are called 'rad' and 'gra' (or 'grad'); if yours is one of these, you have probably noticed that these give different answers when you are using the sin, cos or tan keys. These answers are only wrong when the calculator mode is different from the angular measure used in the calculation.

The **gradian** (mode 'gra' or 'grad') is a unit which was introduced to give a means of angle measurement which was compatible with the metric system. There are 100 gradians in a right angle, so when you are in the gradian mode, $\sin 100 = 1$, just as when you are in the degree mode, $\sin 90 = 1$. Gradians are largely of historical interest and are only mentioned here to remove any mystery surrounding this calculator mode.

By contrast, radians are used extensively in mathematics because they simplify many calculations. The **radian** (mode 'rad') is sometimes referred to as the natural unit of angular measure. If, as in Figure 2.1, the arc AB of a circle centre O is drawn so that it is equal in length to the radius of the circle, then the angle AOB is 1 radian, about 57.3°.

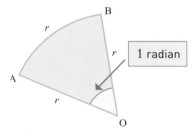

Figure 2.1

You will sometimes see 1 radian written as 1^c, just as 1 degree is written $1°$.

Since the circumference of a circle is given by $2\pi r$, it follows that the angle of a complete turn is 2π radians.

$$360° = 2\pi \text{ radians}$$

Consequently

$$180° = \pi \text{ radians}$$

$$90° = \frac{\pi}{2} \text{ radians}$$

$$60° = \frac{\pi}{3} \text{ radians}$$

$$45° = \frac{\pi}{4} \text{ radians}$$

$$30° = \frac{\pi}{6} \text{ radians}$$

To convert degrees into radians you multiply by $\frac{\pi}{180}$.

To convert radians into degrees you multiply by $\frac{180}{\pi}$.

Example 2.1

(i) Express in radians   (a)   30°   (b)   315°   (c)   29°.

(ii) Express in degrees   (a) $\dfrac{\pi}{12}$ (b) $\dfrac{8\pi}{3}$ (c)   1.2 radians.

Notes

1 If an angle is a simple fraction or multiple of 180° and you wish to give its value in radians, it is usual to leave the answer as a fraction of π, for example $\dfrac{\pi}{6}$.

2 When an angle is given as a multiple of π it is assumed to be in radians.

Solution

(i) (a) $30° = 30 \times \dfrac{\pi}{180} = \dfrac{\pi}{6}$

   (b) $315° = 315 \times \dfrac{\pi}{180} = \dfrac{7\pi}{4}$

   (c) $29° = 29 \times \dfrac{\pi}{180} = 0.506$ radians (to 3 s.f.).

(ii) (a) $\dfrac{\pi}{12} = \dfrac{\pi}{12} \times \dfrac{180}{\pi} = 15°$

   (b) $\dfrac{8\pi}{3} = \dfrac{8\pi}{3} \times \dfrac{180}{\pi} = 480°$

   (c)   1.2 radians $= 1.2 \times \dfrac{180}{\pi} = 68.8°$ (to 3 s.f.).

Note

Angular mils (see page 12) are derived from the milliradian ($\frac{1}{1000}$th of a radian).

There are approximately 6283 milliradians in 360° ($2\pi \times 1000$) and this number is rounded to make 6400 angular mils, which is a more convenient unit for navigation.

Trigonometry and radians

You can use radians when working with trigonometric functions.

Remember that the x–y plane is divided into four quadrants and that angles are measured from the x-axis (see Figure 2.2).

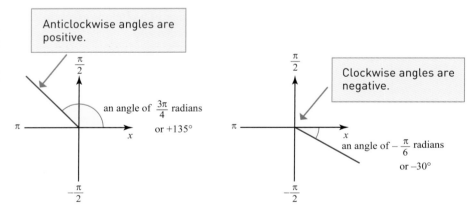

Figure 2.2

You can extend the definitions for sine, cosine and tangent by drawing a unit circle drawn on the x–y plane, as in Figure 2.3.

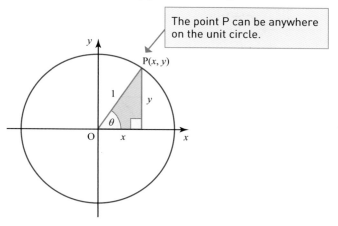

The point P can be anywhere on the unit circle.

Figure 2.3

For any angle (in degrees or radians):

$$\sin\theta = y, \quad \cos\theta = x, \quad \tan\theta = \frac{y}{x} \quad \text{and} \quad \tan\theta = \frac{\sin\theta}{\cos\theta}, \cos\theta \neq 0.$$

Graphs of trigonometric functions

The graphs of the trigonometric functions can be drawn using radians. The graph of $y = \sin\theta$ is shown in Figure 2.4.

- Period is 2π radians.
- Rotational symmetry of order 2 about the origin.
- Oscillates between -1 and 1, so $-1 \leqslant \sin\theta \leqslant 1$.

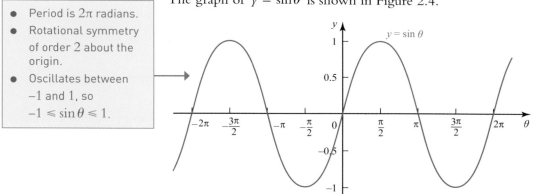

Figure 2.4

The graph of $y = \cos\theta$ is shown in Figure 2.5.

- Period is 2π radians.
- Symmetrical about y-axis.
- Oscillates between -1 and 1, so $-1 \leqslant \cos\theta \leqslant 1$.

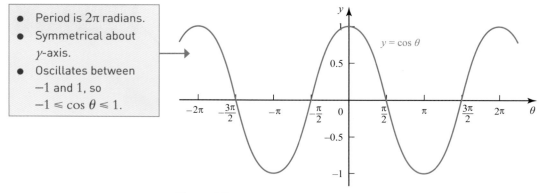

Figure 2.5

- Period is π radians.
- Rotational symmetry of order 2 about the origin.
- Asymptotes at $\pm\frac{\pi}{2}$, $\pm\frac{3\pi}{2}$, ...

The graph of $y = \tan\theta$ is shown in Figure 2.6.

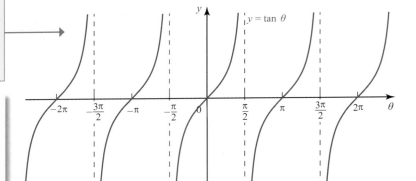

Figure 2.6

Note

If you wish to find the value of, say, $\sin 1.4^{c}$ or $\cos\frac{\pi}{12}$, use the 'rad' mode on your calculator. This will give the answers directly – in these examples, 0.9854... and 0.9659...

You could convert the angles into degrees (by multiplying by $\frac{180}{\pi}$) but this is an inefficient method. It is much better to get into the habit of working in radians.

The diagrams in Figure 2.7 can be used to help you remember the sign of $\sin\theta$, $\cos\theta$ and $\tan\theta$ in each of the four quadrants.

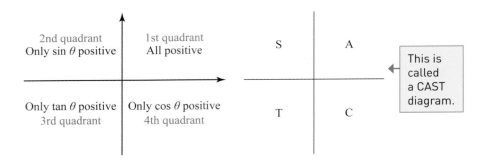

This is called a CAST diagram.

Figure 2.7

Exercise 2.1

① What other angle in the range $0 \le \theta \le 2\pi$ has the same cosine as $\frac{\pi}{6}$?

② Express the following angles in radians, leaving your answers in terms of π where appropriate.
  (i) $45°$
  (ii) $90°$
  (iii) $120°$
  (iv) $75°$
  (v) $300°$
  (vi) $23°$
  (vii) $450°$
  (viii) $209°$
  (ix) $150°$
  (x) $7.2°$

③ Express the following angles in degrees, using a suitable approximation where necessary.
  (i) $\frac{\pi}{10}$
  (ii) $\frac{3\pi}{4}$
  (iii) 2 radians
  (iv) $\frac{4\pi}{9}$
  (v) 3π
  (vi) $\frac{4\pi}{3}$
  (vii) 0.5 radians
  (viii) $\frac{5\pi}{2}$
  (ix) $\frac{7\pi}{3}$
  (x) $\frac{3\pi}{7}$

④ In Figure 2.8, ABC is an equilateral triangle. AB = 2 cm. D is the midpoint of BC.

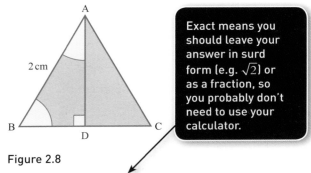

Exact means you should leave your answer in surd form (e.g. $\sqrt{2}$) or as a fraction, so you probably don't need to use your calculator.

Figure 2.8

  (i) Find the exact lengths of
     (a) BD   (b) AD.
  (ii) Write down each of these angles in radians.
     (a) ABD   (b) BAD

(iii) Hence find the exact values of the following.

(a) $\sin \dfrac{\pi}{3}$ (b) $\cos \dfrac{\pi}{3}$

(c) $\tan \dfrac{\pi}{3}$ (d) $\sin \dfrac{\pi}{6}$

(e) $\cos \dfrac{\pi}{6}$ (f) $\tan \dfrac{\pi}{6}$

⑤ By drawing a suitable right-angled triangle, prove that

(i) $\sin \dfrac{\pi}{4} = \dfrac{\sqrt{2}}{2}$ (ii) $\cos \dfrac{\pi}{4} = \dfrac{\sqrt{2}}{2}$

(iii) $\tan \dfrac{\pi}{4} = 1$

⑥ Match together the expressions with the same value.

Do not use your calculator.

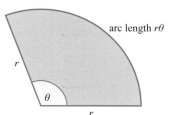

⑦ Draw the graph of $y = \sin x$ for $0 \le x \le 2\pi$.

Use your graph to find two values of x, in radians, for which $\sin x = 0.6$.

> You can use a graphical calculator or graphing software.

⑧ Draw the graphs of $y = \sin x$ and $y = \cos x$ on the same pair of axes for $0 \le x \le 2\pi$.

Use your graphs to solve the equation $\sin x = \cos x$.

⑨ Write down the smallest positive value of k, where k is in radians, to make each of the following statements true.

(i) $\sin (x - k) = -\sin x$

(ii) $\cos (x - k) = \sin x$

(iii) $\tan (x - k) = \tan x$

(iv) $\cos (k - x) = -\cos (k + x)$

⑩ (i) Given that $\sin x = \sin \dfrac{5\pi}{7}$ where $0 < x < \dfrac{\pi}{2}$, find x.

(ii) Given that $\cos y = \cos\left(-\dfrac{2\pi}{5}\right)$ where $\pi < y < 2\pi$, find y.

(iii) Given that $\tan k\pi = \tan\left(-\dfrac{5\pi}{3}\right)$ where $0 < k < \dfrac{1}{2}$, find k.

2 Circular measure

The length of an arc of a circle

From the definition of a radian (see p. 13), an angle of 1 radian at the centre of a circle corresponds to an arc of length r (the radius of the circle).

Similarly, an angle of 2 radians corresponds to an arc length of $2r$ and, in general, an angle of θ radians corresponds to an arc length of θr, which is usually written $r\theta$ (Figure 2.9).

arc length $r\theta$

Figure 2.9

The area of a sector of a circle

A **sector** of a circle is the shape enclosed by an arc of the circle and two radii (Figure 2.10).

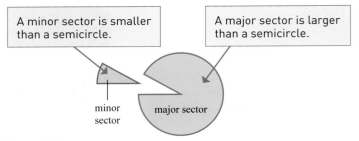

A minor sector is smaller than a semicircle.

A major sector is larger than a semicircle.

minor sector

major sector

Figure 2.10

The area of a sector is a fraction of the area of the whole circle. The fraction is found by writing the angle θ as a fraction of one revolution, i.e. 2π (Figure 2.11). So the area of the shaded sector is $\dfrac{\theta}{2\pi}$ of the area of the whole circle

$$= \frac{\theta}{2\pi} \times \pi r^2 = \frac{1}{2} r^2 \theta$$

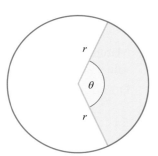

Figure 2.11

ACTIVITY 2.1

You can work out the length of an arc and the area of a sector using degrees instead of radians, but it is much simpler to use radians. Copy and complete Table 2.1 to show the formulae for arc length and sector area using radians and degrees.

Table 2.1

	Radians	Degrees
Angle	θ^c	$\alpha° \left(\alpha = \theta \times \dfrac{180}{\pi} \right)$
Arc length		
Area of sector		

Note

A chord divides a circle into two regions called **segments**.

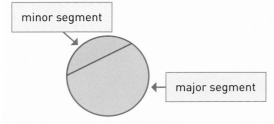

minor segment

major segment

Figure 2.12

Example 2.2

(i) Calculate the exact arc length, perimeter and area of a sector of angle $\frac{2\pi}{3}$ and radius 6 cm.

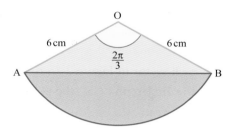

Figure 2.13

(ii) Calculate the area of the segment bounded by the chord AB and the arc AB.

Solution

(i) Arc length $= r\theta$

$$= 6 \times \frac{2\pi}{3}$$

$$= 4\pi \text{ cm}$$

> Draw a sketch if one is not given in the question.

Perimeter $= 4\pi + 6 + 6$

$$= (4\pi + 12) \text{ cm}$$

> Don't forget to add on the two radii.

Area $= \frac{1}{2} r^2 \theta$

$$= \frac{1}{2} \times 6^2 \times \frac{2\pi}{3} = 12\pi \text{ cm}^2$$

(ii)

Figure 2.14

Area of segment = area of sector AOB − area of triangle AOB

Area of a triangle $= \frac{1}{2} \times$ base $\times$ height

Using OA as the base, the height of the triangle is $6\sin\frac{2\pi}{3}$.

Area of triangle AOB $= \frac{1}{2} \times 6 \times 6\sin\frac{2\pi}{3}$

> In general, the area of a triangle is $\frac{1}{2}ab\sin C$, where a and b are two sides and C is the angle between them.

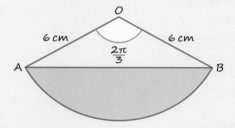

$\sin\frac{2\pi}{3} = \sin\frac{\pi}{3} = \frac{\sqrt{3}}{2}$

$$= 18 \times \frac{\sqrt{3}}{2}$$

$$= 9\sqrt{3}$$

> From part (i), the area of the sector is $12\pi \text{ cm}^2$.

Area of segment $= 12\pi - 9\sqrt{3} = 22 \text{ cm}^2$ to 3 s.f.

Exercise 2.2

① An arc, with angle $\frac{\pi}{2}$, of a circle, has length 2π cm. What is the radius of the circle?

② For each sector in Figure 2.15 find

    (a) the arc length   (b) the perimeter

    (c) the area.

    (i)

    (ii)

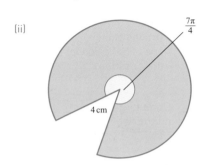

Figure 2.15

③ Each row of Table 2.2 gives dimensions of a sector of a circle of radius r cm.

The angle subtended at the centre of the circle is θ radians, the arc length of the sector is s cm and its area is A cm^2. Copy and complete the table.

Table 2.2

r (cm)	θ (rad)	s (cm)	A (cm²)
4		2	
	$\frac{\pi}{3}$	$\frac{\pi}{2}$	
5			10
	0.8	1.5	
	$\frac{2\pi}{3}$		4π

④ In a cricket match, a particular cricketer generally hits the ball anywhere in a sector of angle 100°. If the boundary (assumed circular) is 80 yards away, find

    (i) the length of boundary which the fielders should patrol

    (ii) the area of the ground which the fielders need to cover.

⑤ The perimeter of the sector in Figure 2.16 is $(5\pi + 12)$ cm.

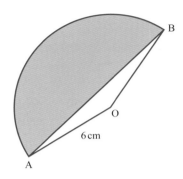

Figure 2.16

Find the exact area of

    (i) the sector AOB

    (ii) the triangle AOB

    (iii) the shaded segment.

⑥ A circle, centre O, has two radii OA and OB. The line AB divides the circle into two regions with areas in the ratio 3 : 1. The angle AOB is θ (radians).

Show that

$$\theta - \sin\theta = \frac{\pi}{2}.$$

⑦ (i) Show that the perimeter of the shaded segment in Figure 2.17 is $r\left(\theta + 2\sin\frac{\theta}{2}\right)$.

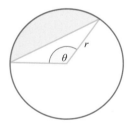

Figure 2.17

    (ii) Show that the area of the shaded segment is $\frac{1}{2}r^2(\theta - \sin\theta)$.

⑧ The silver brooch illustrated in Figure 2.18 is in the shape of an ornamental cross.

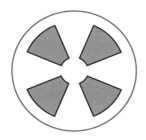

Figure 2.18

The dark shaded areas represent where the metal is cut away. Each is part of a sector of a circle of angle $\frac{\pi}{4}$ and radius 1.8 cm.
The overall diameter of the brooch is 4.4 cm, and the diameter of the centre is 1 cm. The brooch is 1 mm thick.
Find the volume of silver in the brooch.

⑨ In the triangle OAB in Figure 2.19, OA = 3 m, OB = 8 m and angle AOB = $\frac{\pi}{12}$.

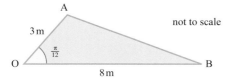

Figure 2.19

Calculate, correct to 2 decimal places

(i) the length of AB

(ii) the area of triangle OAB.

⑩ The plan of an ornamental garden in Figure 2.20 shows two circles, centre O, with radii 3 m and 8 m.

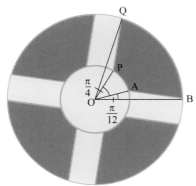

not to scale

Figure 2.20

Grass paths of equal width are cut symmetrically across the circles.

The brown areas represent flower beds.

BQ and AP are arcs of the circles.

Triangle OAB is the same triangle as shown in Figure 2.19.

Given that angle POA = $\frac{\pi}{4}$, calculate the area of

(i) sector OPA

(ii) sector OQB

(iii) the flower bed PABQ.

⑪ (i) Find the area of the shaded segment in Figure 2.21.

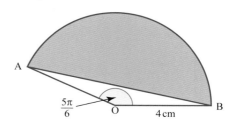

Figure 2.21

(ii) Figure 2.22 shows two circles, each of radius 4 cm, with each one passing through the centre of the other.

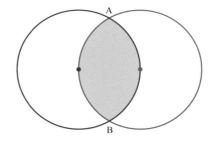

Figure 2.22

Calculate the shaded area.

⑫ Figure 2.23 shows the cross-section of three pencils, each of radius 3.5 mm, held together by a stretched elastic band.
Find

(i) the shaded area

(ii) the stretched length of the band.

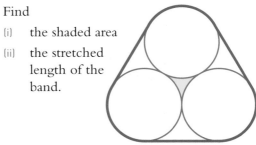

Figure 2.23

3 Small-angle approximations

Figure 2.24 shows the graphs of $y = \theta$, $y = \sin\theta$ and $y = \tan\theta$ on the same axes, for $0 \leq \theta \leq \frac{\pi}{2}$.

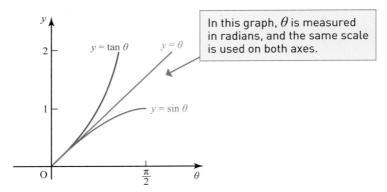

In this graph, θ is measured in radians, and the same scale is used on both axes.

Figure 2.24

From this, you can see that for small values of θ, where θ is measured in radians, both $\sin\theta$ and $\tan\theta$ are approximately equal to θ.

To prove this result, look at Figure 2.25. PT is a tangent to the circle, radius r units and centre O.

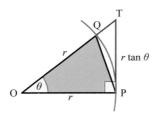

Figure 2.25

Area of sector OPQ $= \frac{1}{2}r^2\theta$

θ must be in radians for this formula to work!

Area of triangle OPQ $= \frac{1}{2}r^2\sin\theta$

Area of triangle is $\frac{1}{2}ab\sin C$.

Area of triangle OPT $= \frac{1}{2}r^2\tan\theta$

Using $\frac{1}{2} \times$ base $\times$ height.

Discussion point

➜ Why does θ need to be in radians?

When θ is very small, these three areas are very close in size.

Figure 2.26

So $\frac{1}{2}r^2\theta \approx \frac{1}{2}r^2\sin\theta \approx \frac{1}{2} \times r \times r\tan\theta$

$\Rightarrow \theta \approx \sin\theta \approx \tan\theta$

The small-angle approximation for $\cos\theta$

The result for $\cos\theta$ can be derived by considering a right-angled triangle drawn on a unit circle (Figure 2.27). The angle θ is small and in radians.

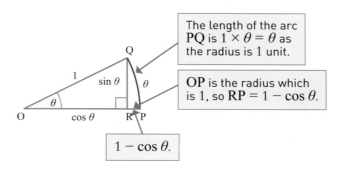

The length of the arc PQ is $1 \times \theta = \theta$ as the radius is 1 unit.

OP is the radius which is 1, so $RP = 1 - \cos\theta$.

$1 - \cos\theta$.

Figure 2.27

In the right-angled triangle PQR, $PQ \approx \theta$ when θ is small. ◄ The length of the arc is approximately the same as the hypotenuse of triangle PQR.

Using right-angled trigonometry

$$\cos\theta = \frac{\text{OR}}{1} \Rightarrow \text{OR} = \cos\theta$$

and $\qquad \sin\theta = \dfrac{\text{RQ}}{1} \Rightarrow \text{RQ} = \sin\theta$

Pythagoras' theorem gives $\quad PQ^2 = PR^2 + RQ^2$ — Expand brackets.

$$\theta^2 \approx (1 - \cos\theta)^2 + \sin^2\theta$$

$$\theta^2 \approx 1 - 2\cos\theta + \cos^2\theta + \sin^2\theta$$

$$\theta^2 \approx 1 - 2\cos\theta + 1 \qquad \blacktriangleleft \boxed{\sin^2\theta + \cos^2\theta \equiv 1, \text{ see p 137.}}$$

$$2\cos\theta \approx 2 - \theta^2$$

$$\cos\theta \approx 1 - \frac{\theta^2}{2} \qquad \blacktriangleleft \boxed{\text{Make } \cos\theta \text{ the subject.}}$$

All of these approximations are very good for $-0.1 \leqslant \theta \leqslant 0.1$ radians.

Discussion points

➜ What do you think is meant by the expression 'very good' here?

➜ Quantify this by calculating the maximum percentage error.

Example 2.3

(i) Find an approximation for $\cos\theta - \cos 2\theta$ when θ and 2θ are both small.

(ii) Hence find

$$\lim_{\theta \to 0} \frac{\cos\theta - \cos 2\theta}{\theta^2}.$$

> This means 'the limit as θ tends to zero'.

Solution

(i) When θ and 2θ are both small

$$\cos\theta \approx 1 - \frac{\theta^2}{2}$$

and

$$\cos 2\theta \approx 1 - \frac{(2\theta)^2}{2}$$
$$\approx 1 - 2\theta^2$$

Using these approximations, when θ is small

$$\cos\theta - \cos 2\theta \approx \left(1 - \frac{\theta^2}{2}\right) - \left(1 - 2\theta^2\right)$$

$$\approx \frac{3\theta^2}{2}$$

(ii)
$$\frac{\cos\theta - \cos 2\theta}{\theta^2} \approx \frac{3\theta^2}{2\theta^2}$$

$$\approx \frac{3}{2}$$

> Check this result by substituting in values of θ (in radians) starting with $\theta = 0.2$ and decreasing in steps of 0.02.

Hence $\lim\limits_{\theta \to 0} \dfrac{\cos\theta - \cos 2\theta}{\theta^2} = \dfrac{3}{2}$

Exercise 2.3

① When θ is small, find approximate expressions for the following.

(i) $\theta \tan\theta$

(ii) $1 - \cos\theta$

(iii) $\cos 2\theta$

(iv) $\sin\theta + \tan\theta$

② When θ is small enough for θ^3 to be ignored, find approximate expressions for the following.

(i) $\dfrac{\theta \sin\theta}{1 - \cos\theta}$

(ii) $\cos\theta \cos 2\theta$

(iii) $\dfrac{\theta \tan\theta}{1 - \cos 2\theta}$

(iv) $\dfrac{\cos 4\theta - \cos 2\theta}{\sin 4\theta - \sin 2\theta}$

③ (i) Find an approximate expression for $\sin 2\theta + \tan 3\theta$ when θ is small enough for 3θ to be considered as small.

(ii) Hence find

$$\lim_{\theta \to 0} \frac{\sin 2\theta + \tan 3\theta}{\theta}$$

④ (i) Find an approximate expression for $1 - \cos\theta$ when θ is small.

(ii) Hence find

$$\lim_{\theta \to 0} \frac{1 - \cos\theta}{4\theta \sin\theta}$$

⑤ (i) Find an approximate expression for $1 - \cos 4\theta$ when θ is small enough for 4θ to be considered as small.

(ii) Find an approximate expression for $\tan^2 2\theta$ when θ is small enough for 2θ to be considered as small.

(iii) Hence find

$$\lim_{\theta \to 0} \frac{1 - \cos 4\theta}{\tan^2 2\theta}$$

⑥ Use a trial and improvement method to find the largest value of θ correct to 2 decimal places such that $\theta = \sin\theta = \tan\theta$ where θ is in radians.

⑦ Use small-angle approximations to find the smallest positive root of

$$\cos x + \sin x + \tan x = 1.2$$

Why can't you use small-angle approximations to find a second root to this equation?

⑧ There are regulations in fencing to ensure that the blades used are not too bent. For épées, the rule states that the blade must not depart by more than 1 cm from the straight line joining the base to the point (see Figure 2.28a). For sabres, the corresponding rule states that the point must not be more than 4 cm out of line, i.e. away from the tangent at the base of the blade (see Figure 2.28b).

(a)

(b)

(c)

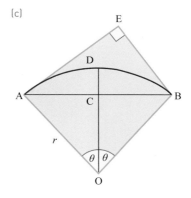

Figure 2.28

Suppose that a blade AB is bent to form an arc of a circle of radius r, and that AB subtends an angle 2θ at the centre O of the circle. Then with the notation of Figure 2.28c, the épée bend is measured by CD, and the sabre bend by BE.

(i) Show that $CD = r(1 - \cos\theta)$.

(ii) Explain why angle $BAE = \theta$.

(iii) Show that $BE = 2r\sin^2\theta$.

(iv) Deduce that if θ is small, $BE \approx 4CD$ and hence that the rules for épée and sabre amount to the same thing.

LEARNING OUTCOMES

When you have completed this chapter, you should be able to:

➤ work with radian measure, including use for arc length and area of sector

➤ understand and use the standard small angle approximations of sine, cosine and tangent

　○ $\sin\theta \approx \theta$

　○ $\cos\theta \approx 1 - \dfrac{\theta^2}{2}$

　○ $\tan\theta \approx \theta$

➤ know and use exact values of sin and cos for $0, \dfrac{\pi}{6}, \dfrac{\pi}{4}, \dfrac{\pi}{3}, \dfrac{\pi}{2}, \pi$ and multiples thereof

➤ know and use exact values of tan for $0, \dfrac{\pi}{6}, \dfrac{\pi}{4}, \dfrac{\pi}{3}, \pi$ and multiples thereof.

KEY POINTS

1 2π radians $= 360°$

To convert degrees into radians you multiply by $\dfrac{\pi}{180}$.

To convert radians into degrees you multiply by $\dfrac{180}{\pi}$.

2 Table 2.3

$\theta°$	0°	30°	45°	60°	90°
θ **radians**	0	$\dfrac{\pi}{6}$	$\dfrac{\pi}{4}$	$\dfrac{\pi}{3}$	$\dfrac{\pi}{2}$
$\sin\theta$	0	$\dfrac{1}{2}$	$\dfrac{\sqrt{2}}{2}$	$\dfrac{\sqrt{3}}{2}$	1
$\cos\theta$	1	$\dfrac{\sqrt{3}}{2}$	$\dfrac{\sqrt{2}}{2}$	$\dfrac{1}{2}$	0
$\tan\theta$	0	$\dfrac{\sqrt{3}}{3}$	1	$\sqrt{3}$	undefined

3

$y = \sin\theta$
Period is 2π radians. Rotational symmetry of order 2 about the origin. Oscillates between -1 and 1, so $-1 \leqslant \sin\theta \leqslant 1$.

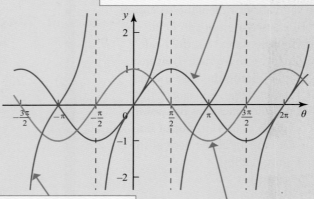

$y = \tan\theta$
Period is $\dfrac{\pi}{2}$ radians. Rotational symmetry of order 2 about the origin. Asymptotes at $\pm\dfrac{\pi}{2}, \pm\dfrac{3\pi}{2} \cdots$

$y = \cos\theta$
Period is 2π radians. Symmetrical about y-axis. Oscillates between -1 and 1, so $-1 \leqslant \cos\theta \leqslant 1$.

Figure 2.29

FUTURE USES

- You will often need to use radians in the trigonometry work in Chapters 6 and 8.
- Radians are also important when you differentiate and integrate trigonometric functions (covered in Chapters 9 and 10).

4 Area of a sector $= \dfrac{1}{2}r^2\theta$
Arc length $= r\theta$

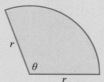

Figure 2.30

5 For small angles where $-0.1 \leqslant \theta \leqslant 0.1$ radians, you can use the following approximations.

$$\sin\theta \approx \theta$$
$$\tan\theta \approx \theta$$
$$\cos\theta \approx 1 - \dfrac{\theta^2}{2}$$

R

Review: Algebra 1

1 Surds and indices

Surds

Sometimes you need to simplify expressions containing **surds**. ←

> A surd is a number involving a root (such as a square root) that cannot be written as a rational number.

Remember

- $(\sqrt{x})^2 = x$
- $\sqrt{xy} = \sqrt{x}\sqrt{y}$

Addition, subtraction and multiplication follow the same rules as any other algebraic manipulation.

- Addition: $(2 + \sqrt{3}) + (3 - 4\sqrt{3}) = 5 - 3\sqrt{3}$
- Subtraction: $(2 + \sqrt{3}) - (3 - 4\sqrt{3}) = -1 + 5\sqrt{3}$
- Multiplication: $(2 + \sqrt{3})(3 - 4\sqrt{3}) = 6 + 3\sqrt{3} - 8\sqrt{3} - 4\sqrt{3}\sqrt{3}$

$$= 6 - 5\sqrt{3} - 12 \quad \longleftarrow$$
$$= -6 - 5\sqrt{3}$$

> Notice that in this case the final term is a rational number.

When dividing by a surd you need to **rationalise the denominator** as shown in Example R.1 overleaf.

Example R.1

Simplify the following by rationalising the denominator.

(i) $\dfrac{5}{2\sqrt{3}}$ (ii) $\dfrac{2 - \sqrt{3}}{1 + \sqrt{3}}$

TECHNOLOGY

Most calculators will simplify expressions involving surds for you. You can use a calculator to check your work.

Solution

(i) $\dfrac{5}{2\sqrt{3}} = \dfrac{5}{2\sqrt{3}} \times \dfrac{\sqrt{3}}{\sqrt{3}}$ ← Multiply top and bottom by $\sqrt{3}$.

      $= \dfrac{5\sqrt{3}}{(2\sqrt{3})\sqrt{3}}$ ← $2\sqrt{3}\sqrt{3} = 2 \times 3 = 6$

      $= \dfrac{5\sqrt{3}}{6}$ ← 6 is a rational number.

(ii)   To rationalise this denominator you can make use of the result $(a + b)(a - b) = a^2 - b^2$.

      $\dfrac{2 - \sqrt{3}}{1 + \sqrt{3}} = \dfrac{2 - \sqrt{3}}{1 + \sqrt{3}} \times \dfrac{1 - \sqrt{3}}{1 - \sqrt{3}}$ ← Multiplying top and bottom by $(1 - \sqrt{3})$.

             $= \dfrac{(2 - \sqrt{3})(1 - \sqrt{3})}{(1 + \sqrt{3})(1 - \sqrt{3})}$

             $= \dfrac{2 - \sqrt{3} - 2\sqrt{3} + 3}{1 - 3}$ ← The denominator is now rational.

             $= \dfrac{5 - 3\sqrt{3}}{-2}$

             $= \dfrac{3\sqrt{3} - 5}{2}$

Indices

The rules for manipulating indices are:

- Multiplication: $a^m \times a^n = a^{m+n}$ ← Add the indices.

- Division: $a^m \div a^n = a^{m-n}$ ← Subtract the indices.

- Power of a power: $(a^m)^n = a^{mn}$ ← Multiply the indices.

- Power zero: $a^0 = 1$

- Negative indices: $a^{-m} = \dfrac{1}{a^m}$

- Fractional indices: $a^{\frac{1}{n}} = \sqrt[n]{a}$

Example R.2

Simplify the following.

(i) $(2x^{-2})^3$ (ii) $\sqrt{16x^4 y^6}$

Solution

(i) $(2x^{-2})^3 = 2^3 \times (x^{-2})^3$ (ii)    $\sqrt{16x^4 y^6} = \sqrt{16} \times \sqrt{x^4 y^6}$

          $= 8x^{-6}$ $= 4x^2 y^3$

① Simplify the following.

   (i) $(4 + 2\sqrt{3}) + (2 - 3\sqrt{3})$

   (ii) $(2 + 3\sqrt{2}) - (3 - 2\sqrt{2})$

   (iii) $(3 - 2\sqrt{2})^2$

   (iiii) $(3 - 2\sqrt{2})(3 + 2\sqrt{2})$

② Simplify the following by rationalising the denominator.

   (i) $\dfrac{9}{\sqrt{3}}$ (ii) $\dfrac{5 - \sqrt{5}}{\sqrt{5}}$

   (iii) $\dfrac{4 + 2\sqrt{2}}{3 - \sqrt{2}}$ (iv) $\dfrac{5}{3\sqrt{5} - 4}$

③ Evaluate the following.

   (i) 2^{-5} (ii) $27^{\frac{1}{3}}$

   (iii) $25^{\frac{3}{2}}$ (iv) $\left(\dfrac{16}{9}\right)^{-\frac{1}{2}}$

④ Simplify the following, writing your answers in the form x^n.

   (i) $x^4 \times x^3$

   (ii) $x^4 \div x^3$

   (iii) $(x^{-3})^2$

   (iv) $\sqrt{x^8}$

   (v) $\sqrt[3]{x^{12}}$

⑤ Simplify the following.

   (i) $2x^2 y^3 \times 3x^3 y^2$

   (ii) $\sqrt{x^2 y^3} \times \sqrt{x^2 y^5}$

   (iii) $12x^6 y^5 \div 4x^2 y^3$

   (iv) $(1 + x)^3 + 2x(1 + x)^2$

2 Exponentials and logarithms

Exponentials

Exponent is another name for power or index: for example, when you write $x = a^n$, a is the base and n can be referred to using any of the names power, index or exponent.

A function of the form $y = a^x$, where a is positive and x is the variable, is called an **exponential function**.

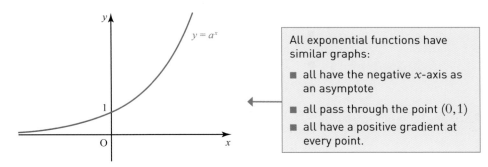

All exponential functions have similar graphs:

- all have the negative x-axis as an asymptote
- all pass through the point $(0, 1)$
- all have a positive gradient at every point.

Figure R.1

Logarithms

A **logarithm** is the inverse of an index:

$$x = a^n \iff n = \log_a x \qquad \text{for } a \text{ (the base)} > 0 \text{ and } x > 0.$$

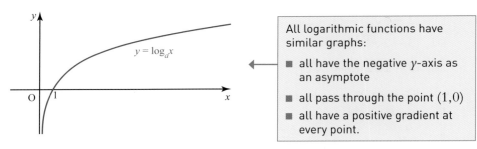

All logarithmic functions have similar graphs:

■ all have the negative y-axis as an asymptote

■ all pass through the point $(1,0)$

■ all have a positive gradient at every point.

Figure R.2

When using the same scale on both axes, the graphs of $y = a^x$ and $y = \log_a x$ are reflections of each other in the line $y = x$. This is because $\log_a x$ and a^x are inverse functions.

For more on inverse functions see page 83.

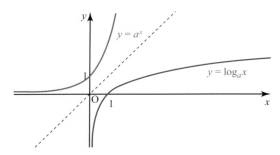

Figure R.3

The rules of logarithms are derived from those for indices:

■ Multiplication: $\qquad \log xy = \log x + \log y$

■ Division: $\qquad \log\left(\dfrac{x}{y}\right) = \log x - \log y$

■ Powers: $\qquad \log x^n = n \log x$

■ Logarithm of 1: $\qquad \log 1 = 0$

■ Reciprocals: $\qquad \log\left(\dfrac{1}{y}\right) = -\log y$

■ Roots: $\qquad \log \sqrt[n]{x} = \dfrac{1}{n}\log x$

■ Logarithm to its own base: $\qquad \log_a a = 1$

Any positive number can be used as the base for a logarithm, but the two most common bases are 10 and the irrational number $2.718\,28\ldots$, which is denoted by the letter e. Logarithms to base e are written as ln and on your calculator you will see that, just as log and 10^x are inverse functions and appear on the same button, so are ln and e^x.

Solve the equation $3^x = 2$, giving your answer correct to 3 significant figures.

Solution

Taking logarithms of both sides:

$$3^x = 2 \implies \log 3^x = \log 2$$

$$\implies x \log 3 = \log 2$$

$$\implies x = \frac{\log 2}{\log 3}$$

$$= 0.631 \ (3 \ \text{s.f.})$$

> You can use logs to any base.

The number of people infected with a disease varies according to the formula

$$N = 200(e^{-0.04t})$$

where N is the number of people infected and t is the time in weeks from the first detection of the disease.

(i)   How many people had the disease when it was first detected?

(ii)   How many months did it take until there were only 10 people infected?

Solution

(i)   Initially, $t = 0$.

Substituting this into $N = 200(e^{-0.04t})$ gives $N = 200$.

(ii)   When $N = 10$,

$$10 = 200(e^{-0.04t})$$

$$\implies 0.05 = e^{-0.04t}$$

$$\implies -0.04t = \ln(0.05)$$

$$\implies t = 74.8933$$

So it took around 18 months for only 10 people to be infected.

Reducing to linear form

Logarithms are particularly useful when trying to find an equation to fit experimental data, since this often involves an exponential relationship, such as $y = kx^n$ or $y = a \times b^x$. Equations of this form can be written in linear form, which makes it easier to plot a graph and estimate its equation.

Example R.5

In an experiment, Yuen obtains the data in Table R.1.

Table R.1

t	1	2	3	4	5	6
y	1.3	3.4	6.2	9.8	13.3	17.5

Yuen thinks that the relationship between y and t is given by an equation of the form $y = kt^n$.

(i) Show that if Yuen is correct, then plotting $\log y$ against $\log t$ will give an approximately straight line graph.

(ii) Plot this graph and use it to estimate the values of n and k.

Solution

(i) Taking logarithms of both sides:

$$\log y = \log kt^n$$
$$= \log k + \log t^n$$
$$= \log k + n \log t$$
$$= n \log t + \log k$$

> Logarithms to base 10 have been used here, but any base can be used.

This is of the form $y = mx + c$, so plotting a graph of $\log y$ against $\log t$ should give a straight line with gradient n and vertical axis intercept $\log k$.

(ii) **Table R.2**

t	1	2	3	4	5	6
y	1.3	3.4	6.2	9.8	13.3	17.5
$\log t$	0	0.301	0.478	0.602	0.699	0.778
$\log y$	0.114	0.531	0.792	0.991	1.124	1.243

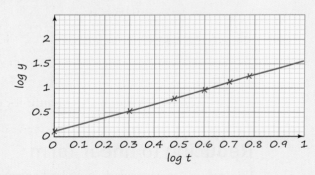

Figure R.4

Gradient of graph ≈ 1.5 so $n \approx 1.5$.

Intercept ≈ 0.1

$$\Rightarrow \quad \log k \approx 0.1$$
$$\Rightarrow \quad k \approx 10^{0.1} \approx 1.26 \text{ (3 s.f.)}$$

TECHNOLOGY

Use a spreadsheet or graphing software to verify that the equation $y = 1.26 \times t^{1.5}$ is a good fit for the original data.

If the relationship is of the form $y = a \times b^x$, then taking logs of both sides gives

$$\log y = \log(a \times b^x)$$
$$= \log a + \log b^x$$
$$= \log a + x \log b \longleftarrow \boxed{\text{Compare this with } y = mx + c.}$$

In this case you would plot $\log y$ against x to obtain a straight line with gradient $\log b$ and intercept $\log a$.

Exercise R.2

① Write the following expressions in the form $\log x$ where x is a number.
  (i) $\log 3 + \log 6 - \log 2$
  (ii) $\frac{1}{2}\log 9 + 2\log 4$
  (iii) $\log 1 + \log 2 - \log 3$

② Write the following expressions in terms of $\log x$.
  (i) $\log x^4 - 2\log x$
  (ii) $3\log x + \log x^3$
  (iii) $\log(\sqrt{x})^3$

③ Solve the following equations.
  (i) $10^{2x+3} = 5$ (ii) $10^{1-x} = 6$
  (iii) $e^{3x-2} = 5$ (iv) $e^{4-2x} = 1$

④ Solve the following equations, giving exact solutions.
  (i) $\log(x + 1) = 5$
  (ii) $\log(x + 2) + \log(x - 2) = 0$
  (iii) $\ln(x + 3) = 4$
  (iv) $\ln(x + 2) + \ln(x - 2) = 0$

⑤ Make x the subject of the following formulae.
  (i) $a = \log_c(x - b)$
  (ii) $e^{px+q} = s$

⑥ The number N of bees in a hive is given by $N = 200\,e^{0.1t}$ where t is the number of days since observations began.
  (i) How many bees were in the hive initially?
  (ii) Sketch the graph of N against t.
  (iii) What is the population of the hive after 30 days?
  (iv) How good do you think this model is? Explain your answer.

⑦ During a chemical reaction, the concentration C ($\mathrm{kg\,m^{-3}}$) of a particular chemical at time t minutes is believed to be given by an equation of the form $C = p \times q^t$. The data in Table R.3 are obtained.

Table R.3

t	0	1	2	3	4	5	6
C	1.32	1.23	0.95	0.76	0.62	0.51	0.38

  (i) Explain why, if this relationship is correct, plotting a graph of $\log C$ against t will give an approximate straight line.
  (ii) Plot the graph and use it to estimate the values of p and q.

⑧ The data in Table R.4 are obtained in an experiment.

Table R.4

r	1.2	2.3	2.8	3.2	3.9
s	5.4	7.6	8.3	9.0	9.8

The relationship between r and s is believed to be of the form $s = kr^n$.
  (i) Explain why, if this relationship is correct, plotting a graph of $\log s$ against $\log r$ will give an approximate straight line.
  (ii) Plot the graph and use it to estimate the values of k and n.

KEY POINTS

1 **The laws of indices**
   - $a^m \times a^n = a^{m+n}$
   - $a^m \div a^n = a^{m-n}$
   - $(a^m)^n = a^{mn}$
   - $a^0 = 1$
   - $a^{-m} = \dfrac{1}{a^m}$
   - $a^{\frac{1}{n}} = \sqrt[n]{a}$

2 - $(\sqrt{x})^2 = x$
     $\sqrt{xy} = \sqrt{x}\sqrt{y}$

3 When simplifying expressions containing square roots you need to
   - make the number under the square root as small as possible
   - rationalise the denominator if necessary.

4 A function of the form a^x is described as exponential.

5 A logarithm is the inverse of an index:
   $$y = \log_a x \quad \Leftrightarrow \quad a^y = x$$

6 **The laws of logarithms to any base**
   - $\log xy = \log x + \log y$
   - $\log\left(\dfrac{x}{y}\right) = \log x - \log y$
   - $\log 1 = 0$
   - $\log x^n = n \log x$
   - $\log\left(\dfrac{1}{y}\right) = -\log y$
   - $\log \sqrt[n]{x} = \log x^{\frac{1}{n}} = \dfrac{1}{n}\log x$
   - $\log_a a = 1$

7 The function e^x, where e is the irrational number 2.718 28..., is called the exponential function.
   $\log_e x$ is called the natural logarithm of x and is denoted by $\ln x$.

8 Logarithms may be used to discover the relationship between variables in two types of situation.
   - $y = kx^n \quad \Leftrightarrow \quad \log y = n \log x + \log k$

     Plotting $\log y$ against $\log x$ gives a straight line where n is the gradient and $\log k$ is the y-axis intercept.
   - $y = ka^x \quad \Leftrightarrow \quad \log y = x \log a + \log k$

     Plotting $\log y$ against x gives a straight line where $\log a$ is the gradient and $\log k$ is the y-axis intercept.

3 Sequences and series

> *Population, when unchecked, increases in a geometrical ratio. Subsistence increases only in an arithmetical ratio. A slight acquaintance with numbers will show the immensity of the first power in comparison with the second.*
> Thomas Malthus (1766–1834)

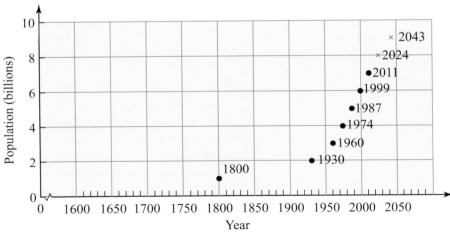

Figure 3.1 Human population, 1600–2050 A.D. (projections shown by red crosses)

→ What was the approximate world population when Malthus made his famous statement that is quoted on the left?

→ Look at the graph and comment on whether things are turning out the way he predicted. What other information would you find helpful in answering this question?

1 Definitions and notation

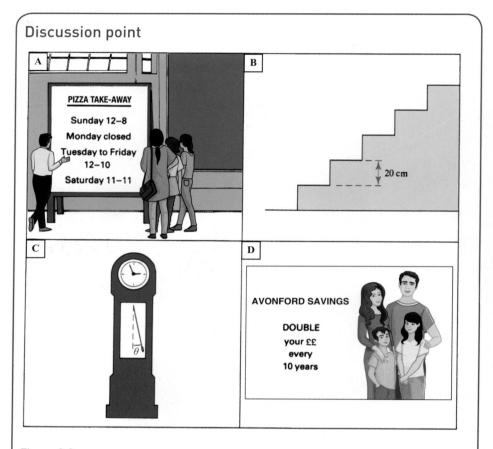
A **sequence** is an ordered set of objects with an underlying rule. Each of the numbers or letters is called a **term** of the sequence. It can be **finite** or **infinite**.

Examples of sequences are:

■   1, 3, 5, 7, ...

■ $\frac{1}{2}, \frac{1}{4}, \frac{1}{8}, \frac{1}{16}, \ldots$

■

■   Monday, Tuesday, Wednesday, ...

For the sequence $1, 3, 5, 7, \ldots$,

$$\text{the 3rd term is } 5 = (2 \times 3) - 1$$

$$\text{the 4th term is } 7 = (2 \times 4) - 1.$$

Continuing this pattern, the general term a_k is written as $2k - 1$.

Series

When the terms of a sequence are added together, like

$$1 + 3 + 5 + 7 + \ldots$$

the resulting sum is called a **series**. The process of adding the terms together is
called **summation** and is indicated by the symbol Σ (the Greek letter sigma),
with the position of the first and last terms involved given as **limits**.

You can write $a_1 + a_2 + a_3 + a_4 + a_5$ as $\displaystyle\sum_{k=1}^{k=5} a_k$ or just as $\displaystyle\sum_{k=1}^{5} a_k$ or even $\displaystyle\sum_{1}^{5} a_k$.

If all the terms of a sequence are to be added, the sum may be written as $\displaystyle\sum a_k$.

A sequence may have an infinite number of terms, in which case it is called an
infinite sequence and the corresponding series is called an **infinite series**.

For example

$$1 + 5x + 10x^2 + 10x^3 + 5x^4 + x^5 = (1 + x)^5$$

> This series has a finite number of terms (6).

$$4\left(1 - \frac{1}{3} + \frac{1}{5} - \frac{1}{7} + \ldots\right) = \pi$$

$$\text{and } x - \frac{x^3}{3!} + \frac{x^5}{5!} - \ldots = \sin x.$$

> Both of these series have an infinite number of terms.

> x is measured in radians.

Defining sequences

Definitions which give the value of a_k directly are called **deductive** definitions
(position to term). Alternatively, there is an **inductive** definition (term to
term), where each term is defined by relating it to the previous one. Look
at the sequence $5, 8, 11, 14, \ldots$. This sequence, and many others, can be
written algebraically in more than one way. Try out the following formulae for
yourself.

- $a_k = 2 + 3k$ for $k = 1, 2, 3, \ldots$
- $a_k = 5 + 3(k - 1)$ for $k = 1, 2, 3, \ldots$
- $a_1 = 5; \ a_{k+1} = a_k + 3$

> This formula has the advantage that it contains both the number 5, which is the first term, and the number 3, which is the difference between the terms.

> In this case, substituting $k = 1$ gives $a_2 = a_1 + 3 = 5 + 3 = 8$,
> substituting $k = 2$ gives $a_3 = a_2 + 3 = 8 + 3 = 11$, etc.

Example 3.1

A sequence is defined deductively by $a_k = 3k - 1$ for $k = 1, 2, 3, \ldots$.

(i) Write down the first six terms of the sequence, and describe the sequence.

(ii) Find the value of the series $\displaystyle\sum_{k=1}^{k=6} a_k$.

Solution

(i) Substituting $k = 1, 2, 3, \ldots 6$ in $a_k = 3k - 1$ gives

$$a_1 = 3(1) - 1 = 2$$
$$a_2 = 3(2) - 1 = 5$$
$$a_3 = 3(3) - 1 = 8$$
$$a_4 = 3(4) - 1 = 11$$
$$a_5 = 3(5) - 1 = 14$$
$$a_6 = 3(6) - 1 = 17$$

So the sequence is $2, 5, 8, 11, 14, 17$, which starts with 2 and then adds 3 each time.

(ii)
$$\sum_{k=1}^{k=6} a_k = a_1 + a_2 + a_3 + a_4 + a_5 + a_6$$
$$= 2 + 5 + 8 + 11 + 14 + 17$$
$$= 57$$

Arithmetic sequences

A sequence in which consecutive terms differ by the addition of a fixed (positive or negative) number is described as **arithmetic**. This number is called the **common difference**.

The general form of an arithmetic sequence is $a, a + d, a + 2d, a + 3d, \ldots$, where the first term is a and the common difference is d.

Thus the sequence $\underbrace{5 \quad 8 \quad 11 \quad 14}_{+3 \quad +3 \quad +3} \ldots$ is arithmetic with $a = 5$ and $d = 3$.

You will look at arithmetic sequences in more detail later in the chapter.

Geometric sequences

A sequence in which you find each term by multiplying the previous one by a fixed number is described as **geometric**; the fixed number is called the **common ratio**.

The general form of a geometric sequence is $a, ar, ar^2, ar^3, \ldots$, where the first term is a and the common ratio is r.

Thus the sequence $\underbrace{10 \quad 20 \quad 40 \quad 80}_{\times 2 \quad \times 2 \quad \times 2} \ldots$ is a geometric sequence with $a = 10$ and $r = 2$.

It may be written algebraically as

$$a_k = 5 \times 2^k \text{ for } k = 1, 2, 3, \ldots \qquad \text{(deductive definition)}$$

or as $\quad a_1 = 10; \; a_{k+1} = 2a_k \text{ for } k = 1, 2, 3, \ldots \qquad$ (inductive definition).

Find the common ratio for the geometric sequence $2, -\frac{1}{2}, \frac{1}{8}, -\frac{1}{32}, \ldots$

Solution

The first term, $a = 2$ and the second term $ar = -\frac{1}{2}$. Hence

$$r = \frac{ar}{a}$$

$$= \frac{-\frac{1}{2}}{2}$$

$\Rightarrow$ common ratio $= -\frac{1}{4}$.

Geometric sequences are also dealt with in more detail later in the chapter.

Periodic sequences

A sequence which repeats itself at regular intervals is called **periodic**. In the case of the pizza take-away in Figure 3.2, the number of hours it is open each day forms the sequence

$$a_1 = 8, \quad a_2 = 0, \quad a_3 = 10, \quad a_4 = 10, \quad a_5 = 10, \quad a_6 = 10, \quad a_7 = 12,$$
$$\text{(Sun)} \quad \text{(Mon)} \quad \text{(Tues)} \quad \text{(Wed)} \quad \text{(Thurs)} \quad \text{(Fri)} \quad \text{(Sat)}$$

$$a_8 = 8, \quad a_9 = 0, \ldots$$
$$\text{(Sun)} \quad \text{(Mon)}$$

There is no neat algebraic formula for the terms of this sequence but you can see that $a_8 = a_1, a_9 = a_2$ and so on.

In general, this sequence can be written as

$$a_{k+7} = a_k \qquad \text{for } k = 1, 2, 3, \ldots$$

This sequence is periodic with period 7 since each term is repeated after seven terms.

A sequence for which

$$a_{k+p} = a_k \qquad \text{for } k = 1, 2, 3, \ldots \text{ (for a fixed integer } p)$$

is periodic. The period is the smallest positive value of p for which this is true.

Increasing and decreasing sequences

A sequence is **increasing** if each term of the sequence is greater than the term immediately preceding it.

For example

- $2, 5, 8, 11, \ldots$ is an increasing arithmetic sequence with a common difference of 3.
- $2, 6, 18, 54, \ldots$ is an increasing geometric sequence with a common ratio of 3.

- 1, 4, 9, 16, ... is the increasing sequence of the squares of positive integers.
- 0.9, 0.909, 0.90909, 0.9090909, ... is an increasing sequence which converges to $\frac{10}{11}$.

Similarly a sequence is **decreasing** if each term of the sequence is less than the term immediately preceding it.

For example

- 9, 5, 1, −3, ... is a decreasing arithmetic sequence with a common difference of −4. This sequence **diverges**. ← The terms become increasingly large negatively so they do not converge.

> The terms get closer and closer to zero without ever getting there. However small a number you can think of, there comes a point when all the subsequent terms are closer to zero.

- $\frac{1}{2}, \frac{1}{4}, \frac{1}{8}, \frac{1}{16}, \ldots$ is a decreasing geometric sequence with a common ratio of $\frac{1}{2}$. This sequence **converges to zero**.

- $\frac{1}{2}, \frac{2}{3}, \frac{3}{4}, \frac{4}{5}, \ldots$ is an increasing sequence with general term $a_k = \frac{k}{k+1}$. This sequence converges to 1.

Example 3.3

A sequence is defined by $a_k = (-1)^k$ for $k = 1, 2, 3, \ldots$

(i)   Write down the first six terms of the sequence and describe its pattern in as many ways as you can.

(ii) Find the value of the series $\sum_{2}^{5} a_k$.

(iii) Describe the sequence defined by $b_k = 5 + (-1)^k \times 2$ for $k = 1, 2, 3, \ldots$

Solution

> **Note**
>
> Multiplying an expression for a_k by $(-1)^k$ ensures that the signs of the terms alternate between positive and negative. It is a very useful device.

(i) $a_1 = (-1)^1 = -1$

$a_2 = (-1)^2 = 1$

$a_3 = (-1)^3 = -1$

$a_4 = (-1)^4 = 1$

$a_5 = (-1)^5 = -1$

$a_6 = (-1)^6 = 1$

The sequence is $-1, +1, -1, +1, -1, +1, \ldots$.

It is periodic with period 2.

It is also geometric with first term −1 and common ratio −1.

(ii) $\sum_{2}^{5} a_k = a_2 + a_3 + a_4 + a_5$

$= (+1) + (-1) + (+1) + (-1)$

$= 0$

Discussion point

Identify some more examples of each of these types of sequence: arithmetic, geometric, periodic, increasing and decreasing. Try to find some examples that fit into more than one category, and for each of your sequences define a_k.

(iii) $\quad b_1 = 5 + (-1) \times 2 = 3$

$b_2 = 5 + (-1)^2 \times 2 = 7$

$b_3 = 5 + (-1)^3 \times 2 = 3$

$b_4 = 5 + (-1)^4 \times 2 = 7$

and so on, giving the sequence $3, 7, 3, 7, \ldots$ which is periodic with period 2.

Sequences with other patterns

There are many other possible patterns in sequences. Figure 3.3 shows a well-known children's toy in which blocks with a square cross-section are stacked to make a tower. The smallest square shape has sides 1 cm long, and the length of the sides increase in steps of 1 cm.

The areas of these squares, in cm², form the sequence

$$1^2, 2^2, 3^2, 4^2, 5^2 \ldots$$
$$\text{or } 1, 4, 9, 16, 25, \ldots$$

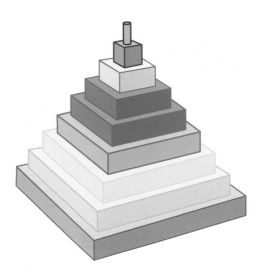

Figure 3.3

This is the sequence of square numbers and it does not fit any of the patterns described so far. If you subtract each term from the next, however, you will find that the differences form a pattern.

Sequence	1		4		9		16		25 …
Difference		3		5		7		9 …	

These differences form an arithmetic sequence with common difference 2. The next difference in the sequence will be $9 + 2 = 11$, and so the next term in the areas sequence will be $25 + 11 = 36$, which is indeed 6^2.

Looking at the differences between the terms often helps you to spot the pattern within a sequence. Sometimes you may need to look at the differences between the differences, or go even further.

Exercise 3.1

① Write down two things that these sequences have in common:

$a_1 = 1, a_{k+1} = -a_k$ and $a_1 = 1, a_{k+1} = 3 - a_k$

② For each of the following sequences, write down the next four terms (assuming the same pattern continues) and describe its pattern as fully as you can.

(i) $7, 10, 13, 16, \ldots$

(ii) $8, 7, 6, 5, \ldots$

(iii) $4.1, 3.9, 3.7, 3.5, \ldots$

(iv) $3, 6, 12, 24, \ldots$

(v) $64, 32, 16, 8, \ldots$

(vi) $1, -2, 4, -8, \ldots$

(vii) $2, 2, 2, 5, 2, 2, 2, 5, 2, 2, 2, 5, \ldots$

(viii) $1, 3, 5, 3, 1, 3, 5, 3, \ldots$

③ Write down the first four terms of each of the sequences defined deductively below. In each case, k takes the values $1, 2, 3, \ldots$.

  (i) $a_k = 2k + 1$

  (ii) $a_k = 3 \times 2^k$

  (iii) $a_k = 2k + 2^k$

  (iv) $a_k = \dfrac{1}{k}$

  (v) $a_k = 5 + (-1)^k$

④ Write down the first four terms of each of the sequences defined inductively below.

  (i) $a_{k+1} = a_k + 3;\ a_1 = 12$

  (ii) $a_{k+1} = -a_k;\ a_1 = -5$

  (iii) $a_{k+1} = \frac{1}{2}a_k;\ a_1 = 72$

  (iv) $a_{k+1} = a_k + 2k + 1;\ a_1 = 1$

  (v) $a_{k+2} = a_k;\ a_1 = 4, a_2 = 6$

⑤ Find the value of the series $\displaystyle\sum_1^4 a_k$ in each of the following cases.

  (i) $a_k = 2 + 5k$

  (ii) $a_k = 3 \times 2^k$

  (iii) $a_k = \dfrac{12}{k}$

  (iv) $a_k = 2 + (-1)^k$

  (v) $a_{k+1} = 3a_k, a_1 = 1$

⑥ Express each of the following series in the form $\displaystyle\sum_1^n a_k$, where n is an integer and a_k is an algebraic expression for the kth term of the series.

  (i) $1 + 2 + 3 + \ldots + 10$

  (ii) $21 + 22 + 23 + \ldots + 30$

  (iii) $210 + 220 + 230 + \ldots + 300$

  (iv) $211 + 222 + 233 + \ldots + 310$

  (v) $190 + 180 + 170 + \ldots + 100$

⑦ Find the value of each of the following.

  (i) $\displaystyle\sum_1^5 k$

  (ii) $\displaystyle\sum_1^{20} k^2 - \sum_1^{19} k^2$

  (iii) $\displaystyle\sum_0^5 (k^2 - 5k)$

  (iv) $\displaystyle\sum_1^{10} (k^2 - (k-1)^2)$

  (v) $\displaystyle\sum_1^{10} ((k+1)^2 - (k-1)^2)$

⑧ The Fibonacci sequence is given by $1, 1, 2, 3, 5, 8, \ldots$.

  (i) Write down the sequence of differences between the terms of this sequence, and comment on what you find.

  (ii) Write down the next three terms of the Fibonacci sequence.

  (iii) Write down the sequence formed by the ratio of one term to the next, $\dfrac{a_{k+1}}{a_k}$, using decimals. What do you notice about it?

⑨ The terms of a sequence are defined by
$$a_k = 4 + (-1)^k \times 2.$$

  (i) Write down the first six terms of this sequence.

  (ii) Describe the sequence.

  (iii) What would be the effect of changing the sequence to

   (a) $a_k = 4 + (-2)^k$

   (b) $a_k = 4 + (-\frac{1}{2})^k$?

⑩ A sequence of numbers $t_1, t_2, t_3, \ldots$ is formed by taking a starting value of t_1 and using the result
$$t_{k+1} = t_k^2 - 2 \quad \text{for } k = 1, 2, 3, \ldots$$

  (i) If $t_1 = \sqrt{2}$, calculate t_2, t_3 and t_4. Show that $t_5 = 2$, and write down the value of t_{100}.

  (ii) If $t_1 = 2$, show that all the terms of the sequence are the same.
  Find the other value of t_1 for which all the terms of the sequence are the same.

  (iii) Determine whether the sequence converges, diverges or is periodic in the cases where

   (a) $t_1 = 3$ (b) $t_1 = 1$

   (c) $t_1 = \dfrac{\sqrt{5} - 1}{2}$. [MEI]

⑪ Throughout time there have been many attempts to find a series that will calculate π, two of which are the following:

$$\frac{\pi}{4} = 1 - \frac{1}{3} + \frac{1}{5} - \frac{1}{7} \dots$$ (James Gregory 1671 and Gottfried Liebnitz 1674)

$$\frac{\pi^2}{6} = \frac{1}{1^2} + \frac{1}{2^2} + \frac{1}{3^2} + \frac{1}{4^2} + \dots$$

(Leonhard Euler 1748)

(i) Which of these would you expect to converge more quickly? Why?

(ii) Use a spreadsheet to find out how many terms of each are needed to give an approximation to π which is correct to

(a) one d.p.

(b) two d.p.

(c) three d.p.

(d) four d.p.

2 Arithmetic sequences and series

When the terms of an arithmetic sequence, or arithmetic progression, are added together, the result is called an **arithmetic series**.

> ### Notation
>
> The following conventions will be used:
>
> ■ first term, $a_1 = a$
> ■ number of terms $= n$
> ■ last term, $a_n = l$
> ■ common difference $= d$
> ■ the general term $= a_k$, i.e. the term in the kth position.

For the arithmetic sequence 5, 8, 11, 14, 17, 20,

$$a = 5, n = 6, l = 20 \text{ and } d = 3.$$

To find an expression for the general term, look at how the terms are formed:

$$
\begin{aligned}
a_1 &= a & &= 5 \\
a_2 &= a + d & &= 5 + 3 & &= 8 \\
a_3 &= a + 2d & &= 5 + 2(3) & &= 11 \\
a_4 &= a + 3d & &= 5 + 3(3) & &= 14 \\
a_5 &= a + 4d & &= 5 + 4(3) & &= 17
\end{aligned}
$$

The 4th term is the first term (5) plus three times the common difference (3).

In each case, the number of differences that are added is one less than the position number of the term you are finding. This gives rise to the formula

$$a_k = a + (k - 1)d .$$

For the last term this becomes

$$l = a + (n - 1)d.$$

These are both general formulae which apply to any arithmetic sequence.

Example 3.4

How many terms are there in the arithmetic sequence $14, 18, 22, \ldots, 162$?

Solution

This is an arithmetic sequence with first term $a = 14$, last term $l = 162$ and common difference $d = 4$.

Using the result $l = a + (n - 1)d$ gives

$$162 = 14 + 4(n - 1)$$
$$\Rightarrow 148 = 4(n - 1)$$
$$\Rightarrow 37 = n - 1$$
$$\Rightarrow n = 38$$

There are 38 terms.

Example 3.5

Find the 25th term in the arithmetic sequence $7, 4, 1, \ldots$.

Solution

In this case $a = 7$ and $d = -3$ (negative since the terms are decreasing).

Using the result $a_k = a + (k - 1)d$ gives

$$a_{25} = 7 + (25 - 1) \times (-3)$$
$$= -65$$

The 25th term is -65.

Historical note

When Carl Friedrich Gauss (1777–1855) was at school he was always quick to answer mathematical questions. One day his teacher, hoping for half an hour of peace and quiet, told his class to add up all of the whole numbers from 1 to 100. Almost at once the 10-year-old Gauss announced that he had done it and that the answer was 5050.

Gauss had not, of course, added the terms one by one. Instead he wrote the series down twice, once in the given order and once backwards, and added the two together:

$$S = 1 + 2 + 3 + \ldots + 98 + 99 + 100$$
$$S = 100 + 99 + 98 + \ldots + 3 + 2 + 1$$

Adding,

$$2S = 101 + 101 + 101 + \ldots + 101 + 101 + 101$$

Since there are 100 terms in the series,

$$2S = 100 \times 101$$
$$\Rightarrow S = 5050.$$

The numbers $1, 2, 3, \ldots, 100$ form an arithmetic sequence with common difference 1. Gauss' method gives rise to a formula which can be used to find the sum of any arithmetic sequence.

The sum of the terms of an arithmetic sequence

It is common to use the letter S to denote the sum of the terms of a sequence. When there is any doubt about the number of terms that are being summed, this is indicated by a subscript: S_4 indicates 4 terms and S_n indicates n terms.

Find the sum of the arithmetic series $10 + 7 + 3 + \ldots + (-20)$.

Solution

This arithmetic series has a first term of 10 and a common difference of -3.

Using the formula $l = a + (n - 1)d$,

$$-20 = 10 + (n - 1)(-3)$$

$$\Rightarrow -30 = (n - 1)(-3)$$

$$\Rightarrow 10 = n - 1$$

so there are 11 terms.

Writing down the series both forwards and backwards gives

$$S = \quad 10 \; + \quad 7 \; + \ldots + (-17) + (-20)$$

and

$$S = (-20) + (-17) + \ldots + \quad 7 \; + \quad 10$$

Adding gives $2S = (-10) + (-10) + \ldots + (-10) + (-10)$

Since there are 11 terms this gives

$$2S = (-10) \times 11$$

$$\Rightarrow S = -55.$$

Obviously it would be rather tedious to have to do this each time, so this method is generalised to give a formula for the sum of n terms of the arithmetic sequence that has a first term a and a common difference d:

TECHNOLOGY

Use a spreadsheet to generate an arithmetic sequence and to calculate the sum of the terms. Verify that the formula gives the same answers.

$$S = \quad [a] \quad + \quad [a + d] \quad + \ldots + \; [a + (n - 2)d] + \; [a + (n - 1)d]$$

$$S = [a + (n - 1)d] \; + \; [a + (n - 2)d] \; + \ldots + \quad [a + d] \; + \quad [a]$$

$$2S = [2a + (n - 1)d] \; + [2a + (n - 1)d] \quad + \ldots + [2a + (n - 1)d] + [2a + (n - 1)d]$$

Since there are n terms it follows that

$$S = \tfrac{1}{2}n[2a + (n - 1)d].$$

This result may also be written as

$l = a + (n - 1)d$ is the last term.

$$\rightarrow S = \tfrac{1}{2}n(a + l).$$

Example 3.7

The arithmetic sequence 2, $1\frac{2}{3}$, $1\frac{1}{3}$, 1, ... has 60 terms. Find the sum of the sequence.

Solution

This is an arithmetic series of 60 terms with first term 2 and common difference $\left(-\frac{1}{3}\right)$.

Using $S = \frac{1}{2}n[2a + (n-1)d]$ gives

$$S = \frac{60}{2}\left[4 + 59 \times \left(-\frac{1}{3}\right)\right]$$
$$= -470.$$

Example 3.8

Catherine has just completed her degree and her starting salary with her first employer is £21 000 per annum. She has been promised an increase of £1200 each year for the first five years and then a salary review at that point.

(i)   What is her salary in her fifth year?

(ii)   What will her total earnings be after five years?

Solution

(i)   Using $a_k = a_0 + (k-1)d$,
$$a_5 = 21\,000 + 4(1200)$$
$$= 25\,800.$$
Catherine's salary is £25 800.

(ii)   Using $S = \frac{1}{2}n(a + l)$,

total salary $= \frac{5}{2}(21\,000 + 25\,800)$
$$= £117\,000.$$
Her total earnings are £117 000.

Exercise 3.2

① Are the following sequences arithmetic? If so, state the common difference and the tenth term.
  (i)   5, 8, 11, 14, ...
  (ii)   1, 4, 8, 12, ...
  (iii)   5, 2, −1, −4, ...
  (iv)   3, 7, 11, 15, ...
  (v)   −1, 2, 3, 4, ...
  (vi)   2, 3.5, 5, 6.5, ...

② The first term of an arithmetic sequence is −10 and the common difference is 4.
  (i)   Find the twelfth term.
  (ii)   The last term is 102. How many terms are there?

③ The first term of an arithmetic sequence is 3, the common difference is −2 and the last term is −71. How many terms are there?

④ The first term of an arithmetic sequence is −5 and the sixth term is 10.
  (i)   Find the common difference.
  (ii)   Find the sum of the first 20 terms.

⑤ The kth term of an arithmetic sequence is given by $a_k = -7 + 4k$.
  (i)   Find the common difference.
  (ii)   Write down the first four terms of the sequence.
  (iii)   Find the sum of the first 15 terms.

⑥ The sixth term of an arithmetic sequence is twice the third term and the first term is 3.

   (i)   Find the common difference.

   (ii)   Find the sum of the first 100 terms.

⑦ In an arithmetic sequence, the third term is 7 and the common difference is 2.

   (i)   Find the sum of the first ten terms.

   (ii)   After how many terms does the sum equal 528?

⑧ (i)   Find the sum of all the odd numbers from 49 to 149 inclusive.

   (ii)   Find the sum of all the even numbers from 50 to 150 inclusive.

   (iii)   Explain how you could deduce the answer to (ii) from the answer to (i).

⑨ A ball rolls down a slope. The distances it travels in successive seconds are 4 cm, 12 cm, 20 cm, 28 cm, … .
How many seconds elapse before it has travelled a total of 36 metres?

⑩ In an arithmetic sequence the 8th term is twice the 4th term and the 20th term is 40.

   (i)   Find the first term and the common difference.

   (ii)   Find the sum of the terms from the 10th to the 20th inclusive.

⑪ 50 m of adhesive tape is wound onto a reel of circumference 12 cm. Owing to the thickness of the tape, each turn takes 0.5 mm more tape than the previous one.

How many complete turns are needed?

⑫ A piece of string 10 m long is to be cut into pieces, so that the lengths of the pieces form an arithmetic sequence. The lengths of the longest and shortest pieces are 1 m and 25 cm respectively.

   (i)   How many pieces are there?

   (ii)   If the same string had been cut into 20 pieces with lengths that formed an arithmetic sequence, and if the length of the second longest had been 92.5 cm, how long would the shortest piece have been?

⑬ The sum of the first n terms of a sequence is S_n where $S_n = 2n^2 - n$.

   (i)   Prove that the sequence is arithmetic, stating the first term and the common difference.

   (ii)   Find the sum of the terms from the 3rd to the 12th inclusive.

⑭ (i)   Rewrite the formula
$S_n = \frac{1}{2}n[2a + (n-1)d]$ in the form
$S_n = pn^2 + qn$, where p and q are constants, stating the values of p and q in terms of a and d.

   (ii)   Explain why the nth term of the sequence is given by $S_n - S_{n-1}$, and hence find a formula for the nth term in terms of p and q.

   (iii)   Show that any sequence where S_n is of the form $pn^2 + qn$ is arithmetic and find the first term and common difference in terms of p and q.

3 Geometric sequences and series

When the terms of a geometric sequence, or geometric progression, are added together, the result is called a **geometric series**.

Prior knowledge

You need to know how to use logarithms. These are covered in Review: Algebra (1) on page 30.

Notation

The following conventions will be used:

- first term $a_1 = a$
- common ratio $= r$
- number of terms $= n$
- the general term $= a_k$, i.e. the term in the kth position.

For the geometric sequence $3, 6, 12, 24, 48$,

$$a = 3, r = 2 \text{ and } n = 5.$$

To find an expression for the general term, look at how the terms are formed:

$$
\begin{aligned}
a_1 &= a & &= 3 \\
a_2 &= ar & = 3 \times 2 & = 6 \\
a_3 &= ar^2 & = 3 \times 2^2 & = 12 \\
a_4 &= ar^3 & = 3 \times 2^3 & = 24 \\
a_5 &= ar^4 & = 3 \times 2^4 & = 48
\end{aligned}
$$

In each case, the power of r is one less than the position number of the term, for example $a_4 = ar^3$. This can be written deductively as

$$a_k = ar^{k-1} \qquad \text{(the general term).}$$

The last term is

$$a_n = ar^{n-1}.$$

These are both general formulae which apply to any geometric sequence.

Example 3.9

Find the eighth term in the geometric sequence $5, 15, 45, 135, \ldots$.

Solution

In the sequence the first term $a = 5$ and the common ratio $r = 3$.
The kth term is given by $a_k = ar^{k-1}$.

$$
\begin{aligned}
\Rightarrow a_8 &= 5 \times 3^7 \\
&= 10\,935
\end{aligned}
$$

Example 3.10

How many terms are there in the geometric sequence

$$0.2, 1, 5, 25 \ldots, 390\,625?$$

Solution

The last (nth) term of a geometric sequence is given by $a_n = a \times r^{n-1}$.

In this case, $a = 0.2$ and $r = 5$, so

$$0.2 \times 5^{n-1} = 390\,625$$

$$\Rightarrow \quad 5^{n-1} = \frac{390\,625}{0.2} = 1\,953\,125$$

$$\Rightarrow \quad \log 5^{n-1} = \log 1\,953\,125 \quad \longleftarrow$$

$$\Rightarrow (n-1)\log 5 = \log 1\,953\,125$$

$$\Rightarrow \quad n - 1 = \frac{\log 1\,953\,125}{\log 5} = 9$$

$$\Rightarrow \quad n = 10$$

> You can solve the equation by taking logarithms of both sides. Alternatively, you can use trial and improvement to solve the equation.

There are 10 terms.

The sum of the terms of a geometric sequence

The next example shows you how to derive a formula for the sum of a geometric sequence.

Example 3.11

Find the value of $2 + 6 + 18 + 54 + \ldots + 1\,062\,882$.

Solution

This is a geometric series with first term 2 and a common ratio of 3.

Let $\quad S = 2 + 6 + 18 + 54 + \ldots + 1\,062\,882$ ①

Multiplying by the common ratio 3 gives

$$3S = 6 + 18 + 54 + \ldots + 1\,062\,882 + 3\,188\,646 \qquad ②$$

Subtracting ① from ②

$$
\begin{aligned}
3S &= \phantom{2 +{}} 6 + 18 + 54 + \ldots + 1\,062\,882 + 3\,188\,646 \\
- \quad S &= 2 + 6 + 18 + 54 + \ldots + 1\,062\,882 \\
\hline
\Rightarrow \quad 2S &= -2 + 0 + 0 + 0 + + 0 + 3\,188\,646 \\
\Rightarrow \quad 2S &= 3\,188\,644 \\
\Rightarrow \quad S &= 1\,594\,322
\end{aligned}
$$

Applying this to the general geometric series to give a formula for the sum:

$$S = a + ar + ar^2 + \ldots + ar^{n-1} \qquad ①$$

Multiplying by the common ratio r gives

$$rS = ar + ar^2 + ar^3 + \ldots + ar^n \qquad ②$$

Subtracting ① from ② as before gives

$$
\begin{aligned}
(r - 1)S &= -a + ar^n \\
&= ar^n - a \\
&= a(r^n - 1) \\
\Rightarrow \quad S &= \frac{a(r^n - 1)}{r - 1}
\end{aligned}
$$

The sum of n terms is usually written as S_n and that for an infinite sum as S or S_∞.

For values of r between -1 and $+1$, the sum of n terms is usually written as:

$$S_n = \frac{a(1 - r^n)}{1 - r}.$$

← This means that you are working with positive numbers inside the brackets.

🖥 TECHNOLOGY

Use a spreadsheet to generate a geometric sequence and to calculate the sum of the terms. Verify that the formula gives the same answers.

Example 3.12

A geometric sequence has terms 4, 8, 16, ... 1024.

(i)   How many terms are there in the sequence?

(ii) Find the sum of the terms in this sequence.

Solution

(i)

$$n\text{th term} = ar^{n-1}$$

$$\Rightarrow \quad 4 \times 2^{n-1} = 1024 \quad \longleftarrow \boxed{a = 4, r = 2 \text{ and } n\text{th term} = 1024.}$$

$$\Rightarrow \quad 2^{n-1} = 256$$

$$\Rightarrow \quad \log 2^{n-1} = \log 256$$

$$\Rightarrow (n-1)\log 2 = \log 256$$

$$\Rightarrow \quad n - 1 = \frac{\log 256}{\log 2}$$

$$\Rightarrow \quad n - 1 = 8$$

There are 9 terms in the sequence.

(ii)
$$S = \frac{a(r^n - 1)}{r - 1}$$

$$= \frac{4(2^9 - 1)}{2 - 1}$$

$$= 2044$$

Note

$-1 < r < 1$ is sometimes written as $|r| < 1$. $|r|$ is the modulus of r (or mod r) and is the absolute value or magnitude of r.

For example, $|-1| = 1$, $\left|-\frac{1}{2}\right| = \frac{1}{2}$ and $|1| = 1$.

So if the magnitude of the common ratio is less than 1 then the sequence converges.

Infinite geometric series

In the general geometric series $a + ar + ar^2 + \ldots$, the terms become progressively smaller in size if the common ratio, r, is between -1 and $+1$.

For $-1 < r < 1$, $r^n \to 0$ as $n \to \infty$ and so the formula for the sum of a geometric series,

$$S_n = \frac{a(1 - r^n)}{1 - r}$$

may be rewritten for an infinite series as

$$S = \frac{a}{1 - r} \quad \text{or} \quad S_\infty = \frac{a}{1 - r}.$$

For an infinite series to have a finite sum, the value of r must be such that $-1 < r < 1$.

Discussion point

➔ What happens to an infinite series if $r = +1$ or $r = -1$?

Example 3.13

You saw this example in Chapter 1, on page 2.

Find the sum of the infinite series $2 + 1 + \frac{1}{2} + \frac{1}{4} + \frac{1}{8} + \ldots$.

Solution

This series is geometric with $a = 2$ and $r = \frac{1}{2}$ so the sum is given by

$$S = \frac{a}{1 - r} \text{ where } a = 2 \text{ and } r = \frac{1}{2}.$$

$$S = \frac{2}{1 - \frac{1}{2}}$$

$$= \frac{2}{\frac{1}{2}}$$

$$= 4.$$

Discussion point
You saw a diagram like
Figure 3.4 on page 3.

→ How does it illustrate
Example 3.13?

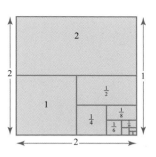

Figure 3.4

Example 3.14

The first term of a geometric progression is 20 and the sum to infinity is $13\frac{1}{3}$.
Find the common ratio.

Solution

$$S = \frac{a}{1-r}$$

$$\Rightarrow \quad 13\frac{1}{3} = \frac{20}{1-r}$$

$$\Rightarrow \quad \frac{40}{3}(1-r) = 20$$

$$\Rightarrow \quad 1 - r = 20 \times \frac{3}{40}$$

$$= \frac{3}{2}$$

$$\Rightarrow \quad r = -\frac{1}{2}$$

Example 3.15

The first three terms of an infinite geometric sequence are 2, 1.6 and 1.28.

(i)   State the common ratio of the sequence.

(ii)   Which is the first term of the sequence with a value less than 0.5?

(iii)   After how many terms will the sum be greater than 9?

**TECHNOLOGY**

Use a spreadsheet to
generate this sequence
and the sum of the
series. How quickly does
it converge? Try some
other geometric series
for which $-1 < r < 1$.

Solution

(i) $a = 2$ and $ar = 1.6$

$$\Rightarrow \quad r = \frac{1.6}{2}$$

$$\Rightarrow \quad r = 0.8$$

(ii)   The nth term $= 2 \times 0.8^{n-1}$.

$$2 \times 0.8^{n-1} < 0.5$$

$$\Rightarrow \quad 0.8^{n-1} < 0.25$$

$$\Rightarrow \quad \log 0.8^{n-1} < \log 0.25$$

$$\Rightarrow \quad (n-1)\log 0.8 < \log 0.25$$

The log of any number
less than 1 is negative,
so you need to reverse
the inequality since you
are dividing by a negative
amount.

$$\Rightarrow \quad (n-1) > \frac{\log 0.25}{\log 0.8}$$

$$\Rightarrow \quad (n-1) > 6.21\ldots$$

$$\Rightarrow \quad n > 7.21\ldots$$

n is a number of
terms so it must be
a whole number.

So it is the 8th term.

(iii) $S_n = \dfrac{2(1 - 0.8^n)}{(1 - 0.8)} = 10(1 - 0.8^n)$

$$10(1 - 0.8^n) > 9$$

$$\Rightarrow \quad 1 - 0.8^n > 0.9$$

$$\Rightarrow \quad 0.1 > 0.8^n$$

$$\Rightarrow \quad \log 0.1 > \log 0.8^n$$

$$\Rightarrow \quad \log 0.1 > n \log 0.8$$

$$\Rightarrow \quad \dfrac{\log 0.1}{\log 0.8} < n \quad \longleftarrow \boxed{\text{log } 0.8 \text{ is negative so reverse the inequality.}}$$

$$\Rightarrow \quad n > 10.31\ldots$$

The sum will be greater than 9 after 11 terms.

> **Note**
>
> Logarithms have been used in this solution but since the numbers are small, it would also have been reasonable to find the answer using a trial and improvement method.

Discussion point

You need to be very careful when dealing with infinite sequences.

Think about the following arguments:

(i) $\qquad S = 1 - 2 + 4 - 8 + 16 - 32 + 64 - \ldots$

$\Rightarrow \quad S = 1 - 2(1 - 2 + 4 - 8 + 16 - 32 \ldots)$

$\Rightarrow \quad S = 1 - 2S$

$\Rightarrow \quad 3S = 1$

$\qquad S = \frac{1}{3}$

(ii) $\qquad S = 1 + (-2 + 4) + (-8 + 16) + (-32 + 64) + \ldots$

$\Rightarrow \quad S = 1 + 2 + 8 + 16\ldots$

So S diverges towards $+\infty$.

(iii) $\qquad S = (1 - 2) + (4 - 8) + (16 - 32) + \ldots$

$\Rightarrow \quad S = -1 - 4 - 16 - \ldots$

So S diverges towards $-\infty$.

→ What is the sum of the series: is it $\frac{1}{3}, +\infty, -\infty$ or something else?

Modelling using sequences and series

Compound interest is an example of geometric growth: investing £100 at a rate of interest of 4% p.a. compounded annually will yield £100 × 1.04 after one year, £100 × 1.04² after two years and so on. In this example it is the geometric **sequence** that is important; the **series** has no practical relevance.

> **Discussion point**
>
> → What factors are likely to affect the annual percentage rate for population growth?

Although population growth may be thought to follow a similar pattern, the annual percentage rate is unlikely to be constant, since it is affected by many external factors.

Radioactive decay can also be modelled by a geometric sequence, as shown in the next example.

Example 3.16

A sample of radioactive material is decaying. Initially, the sample contains N nuclei. The probability that any nucleus will decay in the course of one year is p. What is the expected number of nuclei remaining after y years?

Solution

At the end of the first year $N_0 \times p$ nuclei will have decayed.

Number remaining, $N_1 = N_0 - N_0 p$
$$= N_0 (1 - p).$$

Similarly, after the second year, $N_2 = N_1(1 - p)$
$$= N_0(1 - p)^2.$$

Continuing this, the expected number of nuclei remaining after y years is $N_0(1 - p)^y$.

Exercise 3.3

① Are the following sequences geometric? If so, state the common ratio and the sixth term.

   (i) $5, 10, 20, 40, \ldots$

   (ii) $1, -1, 1, -1, 1, \ldots$

   (iii) $2, 4, 6, 8, \ldots$

   (iv) $1, 1.1, 1.11, 1.111, \ldots$

   (v) $6, 3, 1\frac{1}{2}, \frac{3}{4}, \ldots$

   (vi) $2, -4, 8, -16, \ldots$

② Write down the fifth term, the common ratio and the nth term of the following geometric progressions.

   (i) $3, -6, 12, \ldots$

   (ii) $2, 1, \frac{1}{2}, \ldots$

   (iii) $a, ab, ab^2, \ldots$

③ Which of the following geometric series converge? For each one that converges, find the sum to infinity.

   (i)   First term 8 and common ratio $-\frac{1}{2}$

   (ii)   First term 4 and second term 2

   (iii)   First term 0.1 and second term 1

   (iv)   First term 3 and sixth term $\frac{1}{81}$

   (v)   First term 4 and common ratio -1

④ Explain why the geometric series with $a = 1$ and $r = -1$ does not converge.

⑤ (i)   Find how many terms there are in the geometric sequence $2, 4, 8, \ldots 2048$.

   (ii)   Find the eighth term of this sequence.

   (iii)   Find the sum of all the terms of this sequence.

⑥ For each of the geometric sequences below,

   (a)   find the common ratio and the number of terms

   (b)   find the sum of the sequence.

   (i) $5, 10, 20, \ldots 10\,240$

   (ii) $2, 6, 18, \ldots 118\,098$

   (iii) $8, -4, 2, \ldots, -\dfrac{1}{4096}$

⑦ The value of a car when new is £10 000 and each year its value depreciates by 15%. Use logarithms to find how many complete years it will be before the car is worth less than £500.

⑧ The fifth term of a geometric sequence of positive numbers is 48 and the ninth term is 768.

   (i)   Find the common ratio.

   (ii)   Find the first term.

   (iii)   Find the sum of the first ten terms.

   (iv)   How many terms are needed if the sum is greater than a million?

⑨ The first three terms of an infinite geometric sequence are 9, 6 and 4.

   (i)   Find the common ratio.

   (ii)   Find the position of the first term in the sequence that is less than 1.

   (iii)   After how many terms is the sum greater than 25?

   (iv)   Find the sum to infinity of the terms of the sequence.

⑩ A geometric series has first term 20 and common ratio r. Find how many terms of the series are required for the sum to be within 1×10^{-6} of the sum to infinity in each of the following cases.

(i) $r = 0.8$

(ii) $r = -0.8$

⑪ Find the sum to n terms of each of the following series.

(i) $x + x^2 + x^3 + \dots$

(ii) $1 - y + y^2 - y^3 + \dots$

(iii) $1 - 2a + 4a^2 - 8a^3 + \dots$

⑫ The first three terms of an arithmetic sequence, a, $a + d$ and $a + 2d$, are the same as the first three terms a, ar, and ar^2 of a geometric sequences. ($a \neq 0$).
Show that this is only possible if $r = 1$ and $d = 0$.

⑬ A tank is filled with 20 litres of water. Half the water is removed and replaced with antifreeze and thoroughly mixed. Half this mixture is then removed and replaced with antifreeze. The process continues.

(i) Find the first five terms in the sequence of amounts of water in the tank at each stage.

(ii) After how many operations is there less than 5% of water remaining in the tank?

(iii) Find the first five terms of the sequence of amounts of antifreeze in the tank at each stage.

(iv) Is either of these sequences geometric? Explain your answer.

⑭ A pendulum is set swinging. Its first oscillation is through an angle of 30° and each succeeding oscillation is through 95% of the angle of the one before it.

(i) After how many swings is the angle through which it swings less than 1°?

(ii) What is the total angle it has swung through at the end of its tenth oscillation?

(iii) What is the total angle it has swung through before it stops?

⑮ The first two terms of a geometric sequence are $\sin\theta$ and $\tan\theta$ where $0 < \theta < \frac{\pi}{2}$.

(i) State the common ratio.

(ii) When $\theta = \frac{\pi}{3}$, find the sum of the first five terms, leaving your answer in surd form.

A second series has first term $\tan\theta$ and second term $\sin\theta$ where $0 < \theta < \frac{\pi}{2}$.

(iii) Find the sum to infinity of this series when $\theta = \frac{\pi}{3}$, leaving your answer in surd form.

A third series has first term $\sec\theta$ and second term $\cos\theta$ where $0 < \theta < \frac{\pi}{2}$.

(iv) Find the sum to infinity of this series.

(v) Is it possible for this sum to equal 1?

⑯ Figure 3.5 shows the steps involved in constructing a snowflake pattern.

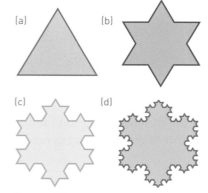

(a) (b)

(c) (d)

Figure 3.5

(a) This shows an equilateral triangle with each side of length 9 cm.

(b) Here each side is trisected and the centre section replaced with an equilateral triangle.

(c) The procedure is repeated for each of the six small triangles around (b).

(d) This construction is repeated until you have an infinite sequence.

(i) Calculate the length of the perimeter of the figure for the first five steps, starting with the original equilateral triangle.

(ii) What happens to the length of the perimeter as the number of steps increases?

(iii) Does the area of the figure increase without limit? Justify your answer.

KEY POINTS

1. A sequence is an ordered set of numbers, $a_1, a_2, a_3, \ldots, a_k, \ldots, a_n$, where a_k is the general term. It may be finite or infinite.

2. A series is the sum of the terms of a sequence:
$$a_1 + a_2 + a_3 + \ldots + a_n = \sum_{k=1}^{k=n} a_k$$

3. In an arithmetic sequence, $a_{k+1} = a_k + d$ where d is a fixed number called the common difference.

4. In a geometric sequence, $a_{k+1} = ra_k$ where r is a fixed number called the common ratio.

5. In a periodic sequence, $a_{k+p} = a_k$ for a fixed integer, p, called the period.

6. In an oscillating sequence the terms rise above and fall below a middle value.

7. For an arithmetic sequence with first term a, common difference d and n terms:
   - the kth term, $a_k = a + (k-1)d$
   - the last term, $l = a + (n-1)d$
   - the sum of the terms, $S_n = \frac{1}{2}n(a+l) = \frac{1}{2}n[2a + (n-1)d]$.

8. For a geometric sequence with first term a, common ratio r and n terms:
   - the kth term, $a_k = ar^{k-1}$
   - the last term, $a_n = ar^{n-1}$
   - the sum of the terms, $S_n = \dfrac{a(r^n - 1)}{(r-1)} = \dfrac{a(1 - r^n)}{(1-r)}$.

9. For an infinite geometric series to converge, $-1 < r < 1$.
   In this case, the sum of all the terms is given by $S_\infty = \dfrac{a}{(1-r)}$.

Review: Algebra 2

1 Equations and inequalities

Quadratic functions and equations

A quadratic equation is any equation which can be written in the form $ax^2 + bx + c = 0$ with $a \neq 0$. There are a number of ways of solving these:

TECHNOLOGY

Your calculator may have an equation solver which can be used to find the roots of a quadratic equation. You can use this to check answers.

- Factorising: $\quad 3x^2 - 8x + 4 = 0$

$$\Rightarrow \quad 3x^2 - 6x - 2x + 4 = 0$$

$$\Rightarrow \quad 3x(x - 2) - 2(x - 2) = 0$$

$$\Rightarrow \quad (3x - 2)(x - 2) = 0$$

$$\Rightarrow \quad x = \frac{2}{3} \quad \text{or} \quad x = 2$$

$3 \times 4 = 12$

Find two numbers that multiply to give 12 and add to give -8. These are -6 and -2.

- The quadratic formula: The roots of the general quadratic equation $ax^2 + bx + c = 0$ are given by $x = \dfrac{-b \pm \sqrt{b^2 - 4ac}}{2a}$.

This formula is derived using the method of completing the square.

  In the quadratic formula, the expression $b^2 - 4ac$ is referred to as the **discriminant**, since it discriminates between the types of solutions:

  ○ If $b^2 - 4ac > 0$ then the square root can be found and there are two distinct roots.

  ○ If $b^2 - 4ac = 0$ then both roots of the equation are equal.

  ○ If $b^2 - 4ac < 0$ there is no real value of the square root, so there are no real roots.

- Drawing a graph: The roots of the equation $x^2 + 2x - 5 = 0$ can be found by drawing the graph of $y = x^2 + 2x - 5$ and finding the x coordinates of the points where it crosses the x-axis.

This method will only give a very approximate solution.

- Using your calculator: Some calculators have the facility to solve quadratic equations. If you have such a calculator, make sure you know how to use it. It will be helpful for checking your answers. However, the three methods given above are all important and you must be able to carry them out without a calculator.

Completing the square

Example R.1 shows how the method of **completing the square** can be used to find the line of symmetry of a quadratic curve and the coordinates of its turning point.

Example R.1

Find the coordinates of the turning point and the equation of the line of symmetry for the curve $y = 2x^2 - 4x - 7$ and hence sketch the curve.

Solution

$$y = 2x^2 - 4x - 7 \Rightarrow y = 2[x^2 - 2x] - 7$$

$$\Rightarrow y = 2[x^2 - 2x + 1 - 1] - 7$$

$$\Rightarrow y = 2[(x-1)^2 - 1] - 7$$

$$\Rightarrow y = 2(x-1)^2 - 9$$

> The least value of this is when the squared term is zero, i.e. when $x = 1$ and $y = -9$.

> Add 1 to make $x^2 - 2x$ into a perfect square, and subtract 1 to compensate.

So there is a minimum turning point at $(1, -9)$.

The line of symmetry passes through the turning point, and so it is the line $x = 1$.

When $x = 0, y = -7$, so the graph passes through $(0, -7)$.

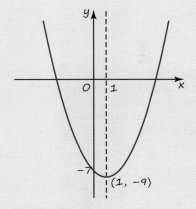

Figure R.1

Notes

1 When the coefficient of x^2 is negative, the procedure is the same, but in this case there is a maximum turning point.

2 This method can also be used to sketch the graph of a quadratic function which does not cross the x-axis.

💻 TECHNOLOGY

Your calculator may have an equation solver which can be used to solve simultaneous equations. You can use this to check answers.

Sometimes an equation needs to be re-written to form a quadratic equation before you can solve it.

Example R.2

Solve the equation $5^{2x} + 5^x - 6 = 0$

Solution

This is a quadratic equation in disguise.

Let $y = 5^x$

> $5^{2x} = (5^x)^2$
> $(5^x)^2 + 5^x - 6 = y^2 + y - 6$

So $y^2 + y - 6 = 0$

$\Rightarrow (y + 3)(y - 2) = 0$

$\Rightarrow \quad y = -3$ or $y = 2$

> $y = -3 \Rightarrow 5^x = -3$ which has no solutions. 5^x is always positive.

$y = 2 \Rightarrow 5^x = 2$

So $\log 5^x = \log 2$

$x \log 5 = \log 2$

$x = \dfrac{\log 2}{\log 5} = 0.431$ (3 s.f.)

Simultaneous equations

The **elimination method** is suitable for solving two linear simultaneous equations.

Example R.3

(i) Solve the simultaneous equations $3x - 2y = 17$

$7x - 3y = 28$

(ii) Check your answer.

Solution

You might prefer to write the equations in the opposite order before subtracting.

(i) $9x - 6y = 51$ ← Multiply the first equation by 3 and the second by 2 to give equations with the same coefficient of y.

$14x - 6y = 56$

$\underline{-5x \quad\quad = -5}$ ← Subtract.

$x \quad\quad = 1$

Substituting $x = 1$ into either of the original equations gives $y = -7$.
The solution is $x = 1$, $y = -7$.

(ii) Substitute $x = 1$, $y = -7$ into $7x - 3y = 28$.

$\text{LHS} = (7 \times 1) - (3 \times -7) = 7 - (-21)$

$= 28$

$= \text{RHS as required}$

Substitute your answers into the other one of the original equations you were given as a check that your values are correct.

The **substitution method** is used to solve simultaneous equations for which one is linear and the other is quadratic, although it may be used to solve two linear simultaneous equations.

To find the point(s) of intersection of two curves or lines you need to solve their equations simultaneously, using one of these two methods.

Example R.4

Find the points of intersection of the line $x - y = 1$ and the curve $x^2 + 2y^2 = 17$.

Solution

Make x the subject of the first equation: $x = y + 1$

Substitute into the second equation: $(y + 1)^2 + 2y^2 = 17$ ← Replace x with $y + 1$.

Expand and simplify: $y^2 + 2y + 1 + 2y^2 = 17$

$\Rightarrow \quad 3y^2 + 2y - 16 = 0$

$\Rightarrow \quad 3y^2 - 6y + 8y - 16 = 0$

$\Rightarrow \quad 3y(y - 2) + 8(y - 2) = 0$

$\Rightarrow \quad (3y + 8)(y - 2) = 0$

$\Rightarrow y = -\frac{8}{3}$ or $y = 2$

Substituting each value in turn into the original *linear* equation gives the points of intersection as $(3, 2)$ and $\left(-\frac{5}{3}, -\frac{8}{3}\right)$.

Inequalities

The rules for solving inequalities are similar to those for solving equations, *but* if you multiply both sides of an inequality by a negative number, then you need to change the direction of the inequality.

When dealing with quadratic inequalities, the easiest method is to sketch the associated graph.

Solve $x^2 + 6x + 2 \geqslant x + 8$.

Solution

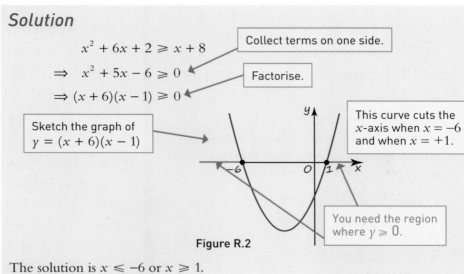

$$x^2 + 6x + 2 \geqslant x + 8$$

Collect terms on one side.

$$\Rightarrow \quad x^2 + 5x - 6 \geqslant 0$$

Factorise.

$$\Rightarrow \quad (x + 6)(x - 1) \geqslant 0$$

Sketch the graph of $y = (x + 6)(x - 1)$

This curve cuts the x-axis when $x = -6$ and when $x = +1$.

You need the region where $y \geqslant 0$.

Figure R.2

The solution is $x \leqslant -6$ or $x \geqslant 1$.

Note

This may also be expressed as $\{x:x \leqslant -6\} \cup \{x:x \geqslant 1\}$.

Notice that the points where $x = -1$ and $x = 6$ are marked with a solid circle in Figure R.2, to show that these values are included in the solution set. If you were asked to solve $x^2 + 6x + 2 > x + 8$ then the values -6 and $+1$ would not be included in the solution set and you would show those points using an open circle.

Inequalities may also be used to define **regions**, as in Example R.6.

Sketch the curve $y = x^2 + 5x - 6$ and indicate by shading the region where $y \geqslant x^2 + 5x - 6$.

Solution

$$y = x^2 + 5x - 6 = (x + 6)(x - 1)$$

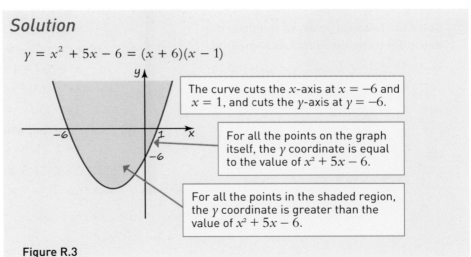

The curve cuts the x-axis at $x = -6$ and $x = 1$, and cuts the y-axis at $y = -6$.

For all the points on the graph itself, the y coordinate is equal to the value of $x^2 + 5x - 6$.

For all the points in the shaded region, the y coordinate is greater than the value of $x^2 + 5x - 6$.

Figure R.3

The region where $y \geqslant x^2 + 5x - 6$ includes the shaded region and the curve itself, so the curve is drawn with a solid line in Figure R.3. If the inequality had been $>$ and not $\geqslant$ then the curve would have been drawn with a broken line.

Review exercise R.1

① Solve the following equations by factorising.

   (i) $x^2 - 8x + 12 = 0$

   (ii) $a^2 + 11a + 30 = 0$

   (iii) $16 - x^2 = 0$

   (iv) $2p^2 + 5p + 2 = 0$

   (v) $4c^2 + 3c - 7 = 0$

   (vi) $3x^2 = 14x - 8$

② Solve the following equations by using the quadratic formula, giving your answers correct to 2 d.p.

   (i) $x^2 + x - 8 = 0$

   (ii) $2x^2 - 8x + 5 = 0$

   (iii) $x^2 + x = 10$

③ Write in the form $a(x + b)^2 + c$.

   (i) $3x^2 - 12x + 7$

   (ii) $2x^2 + 6x + 5$

   (iii) $5 + 8x - x^2$

④ Find the equation of the line of symmetry and the coordinates of the turning point for the following curves and sketch their graphs, including the line of symmetry.

   (i) $y = x^2 + 4x - 8$

   (ii) $y = 2x^2 + 8x - 3$

   (iii) $y = 4 + 2x - x^2$

⑤ Solve the following pairs of simultaneous equations using the elimination method.

   (i) $3x - 2y = 6$

        $5x + 6y = 38$

   (ii) $3x + 2y = 12$

        $4x + y = 11$

   (iii) $4x - 3y = 2$

        $5x - 7y = 9$

⑥ Solve the following pairs of simultaneous equations using the substitution method.

   (i) $x = 2y$

        $x^2 - y^2 + xy = 20$

   (ii) $x + y = 5$

        $x^2 + y^2 = 17$

   (iii) $2x - y + 3 = 0$

        $y^2 - 5x^2 = 20$

⑦ Express the following inequalities using set notation and illustrate them on number lines.

   (i) $-2 < x < 4$

   (ii) $-1 \geqslant x \geqslant -7$

   (iii) $x \geqslant 0$ or $x \leqslant -4$

⑧ In each case draw a sketch of the associated curve and identify the interval(s) on the x-axis where the inequality is true. Illustrate the interval(s) on number lines.

   (i) $x^2 - 2x - 8 > 0$

   (ii) $2 + x - x^2 \geqslant 0$

   (iii) $6x^2 - 13x + 6 < 0$

⑨ (i)   Sketch the curve $y = 2x^2 + x - 1$ and the line $y = 2x + 9$ on the same axes and find their points of intersection.

   (ii) On your graph, shade the region for which $y > 2x^2 + x - 1$ and $y < 2x + 9$.

⑩ Solve the following equations.

   (i) $x^4 - 5x^2 + 4 = 0$

   (ii) $4x^4 - 13x^2 + 3 = 0$

   (iii) $2x - 13\sqrt{x} + 15 = 0$

⑪ Solve the following equations.

   (i) $3^{2x} - 4 \times 3^x + 3 = 0$

   (ii) $3 \times 2^{2x} + 5 \times 2^x - 2 = 0$

   (iii) $6 \times 7^{2x} + 7 \times 7^x - 3 = 0$

2 Polynomials

Adding and subtracting polynomials

$$(3x^3 + 2x^2 - x + 4) + (x^3 + 5x - 6) = 3x^3 + x^3 + 2x^2 - x + 5x + 4 - 6$$

Collect like terms.

$$= 4x^3 + 2x^2 + 4x - 2$$

$$(3x^3 + 2x^2 - x + 4) - (x^3 + 5x - 6) = 3x^3 - x^3 + 2x^2 - x - 5x + 4 + 6$$

$$= 2x^3 + 2x^2 - 6x + 10$$

Multiplying polynomials

Multiply each term in the first polynomial by each term in the second.

$$(x + 2)(x^2 + 3x - 4) = x(x^2 + 3x - 4) + 2(x^2 + 3x - 4)$$

$$= x^3 + 3x^2 - 4x + 2x^2 + 6x - 8$$

$$= x^3 + 5x^2 + 2x - 8$$

Dividing polynomials

You can divide one polynomial by another to give a lower order polynomial.

Example R.7

Divide $(2x^3 - 5x^2 - 7x + 12)$ by $(x - 3)$.

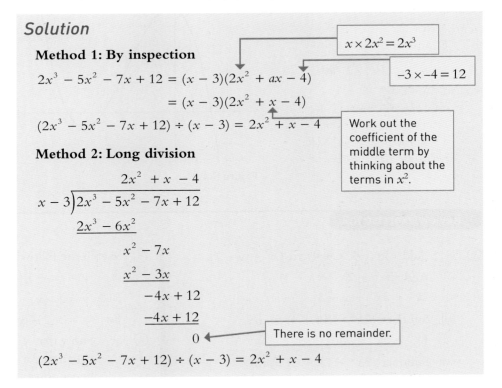

Solution

Method 1: By inspection

$x \times 2x^2 = 2x^3$

$-3 \times -4 = 12$

$$2x^3 - 5x^2 - 7x + 12 = (x - 3)(2x^2 + ax - 4)$$

$$= (x - 3)(2x^2 + x - 4)$$

$$(2x^3 - 5x^2 - 7x + 12) \div (x - 3) = 2x^2 + x - 4$$

Work out the coefficient of the middle term by thinking about the terms in x^2.

Method 2: Long division

$$
\begin{array}{r}
2x^2 + x - 4 \\
x - 3 \overline{\smash{\big)}\ 2x^3 - 5x^2 - 7x + 12} \\
\underline{2x^3 - 6x^2} \\
x^2 - 7x \\
\underline{x^2 - 3x} \\
-4x + 12 \\
\underline{-4x + 12} \\
0
\end{array}
$$

There is no remainder.

$$(2x^3 - 5x^2 - 7x + 12) \div (x - 3) = 2x^2 + x - 4$$

The factor theorem

The factor theorem is given by:

$$f(a) = 0 \Leftrightarrow (x - a) \text{ is a factor of } f(x).$$

The factor theorem is useful in solving polynomial equations.

Example R.8

(i)   Show that $x = -1$ is a root of the equation $x^3 - 4x^2 + x + 6 = 0$ and hence solve the equation.

(ii) Sketch the graph of $y = x^3 - 4x^2 + x + 6$.

Solution

(i)
$$f(x) = x^3 - 4x^2 + x + 6$$
$$f(-1) = (-1)^3 - 4(-1)^2 + (-1) + 6$$
$$= -1 - 4 - 1 + 6$$
$$= 0$$

So $x = -1$ is a root of $f(x) = 0$, and hence $(x + 1)$ is a factor of $f(x)$.

$$x^3 - 4x^2 + x + 6 = 0$$
$$\Rightarrow (x + 1)(x^2 - 5x + 6) = 0$$
$$\Rightarrow (x + 1)(x - 2)(x - 3) = 0$$

> The factorising is done by inspection here, but it could be done by long division instead.

The roots of the equation are $x = -1, x = 2, x = 3$.

(ii) The graph crosses the x-axis at $x = -1$, $x = 2$, $x = 3$.
The graph crosses the y-axis at $y = 6$.

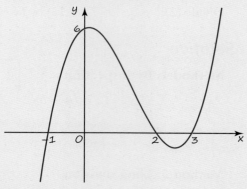

Figure R.4

Review exercise R.2

① (i)   Add $(3x^2 - 2x + 4)$ to $(x^3 + x - 3)$.

(ii) Add $(5x^4 + 2x^2 - x - 1)$ to $(x^3 + 2x^2 - x - 3)$.

(iii) Subtract $(2x^2 - 3x + 4)$ from $(x^3 + 3x^2 - 2)$.

(iv) Subtract $(3x^3 - 7x + 2)$ from $(6x^3 + 7x^2 - 10x + 3)$.

② (i)   Multiply $(2x^2 - 3x + 4)$ by $(x + 3)$.

(ii) Multiply $(x^2 + 2x + 1)$ by $(x^2 - 2x + 1)$.

(iii) Divide $(x^3 + 2x^2 - x - 2)$ by $(x - 1)$.

(iv) Divide $(2x^3 - 5x^2 - 11x - 4)$ by $(2x + 1)$.

③ Sketch the following polynomial curves.

(i) $y = (x - 1)(x + 2)(x - 3)$

(ii) $y = x^2(x - 3)$

(iii) $y = x(2 - x)(x + 3)$

④ (i)   Show that $x = -2$ is a root of the equation $2x^3 + 7x^2 + 4x - 4 = 0$ and hence solve the equation.

(ii) Sketch the graph of $y = 2x^3 + 7x^2 + 4x - 4$.

⑤ Find the equation of the curve shown in Figure R.5. Give your answer in the form $y = ax^3 + bx^2 + cx + d$.

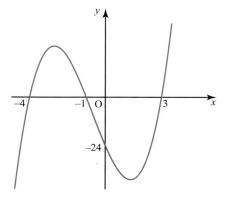

Figure R.5

⑥ Solve the following cubic equations.
   (i) $4x^3 - 8x^2 + x + 3 = 0$
   (ii) $3x^3 - 2x^2 - 19x - 6 = 0$

⑦ Leanne is dividing $x^4 - 16$ by $x + 2$.
She draws this table.

	x^3			
x	x^4			
$+2$				-16

Copy and complete the table and hence write down the result of dividing $x^4 - 16$ by $x + 2$.

KEY POINTS

1 For the quadratic equation $ax^2 + bx + c = 0$, the discriminant is given by $b^2 - 4ac$.
  ■ If the discriminant is positive, the equation has two real roots.
     If the discriminant is a perfect square, these roots are rational and the equation can be factorised.
  ■ If the discriminant is zero, the equation has a repeated real root.
  ■ If the discriminant is negative, the equation has no real roots.

2 Quadratic equations can be solved by
  ■ factorising, in cases where the discriminant is a perfect square
  ■ using the quadratic formula $x = \dfrac{-b \pm \sqrt{b^2 - 4ac}}{2a}$.

3 The vertex and the line of symmetry of a quadratic graph can be found by completing the square.

4 Simultaneous equations may be solved by
  ■ substitution
  ■ elimination
  ■ drawing graphs.

5 Linear inequalities are dealt with like equations *but* if you multiply or divide by a negative number, you must reverse the inequality sign: < reverses to > and ⩽ to ⩾.

6 When solving a quadratic inequality, it is helpful to start by sketching a graph.

7 The order, or degree, of a polynomial in x is the highest power of x which appears in the polynomial.

8 The factor theorem states that if $(x - a)$ is a factor of a polynomial $f(x)$ then $f(a) = 0$ and a is a root of the equation $f(x) = 0$.

Functions

Sometimes auto-tuning is used to improve a vocal track by correcting notes which are off-key.

The pitch of a musical note depends on the frequency of the sound wave.

→ How would the sound wave need to be transformed to change the pitch of a note?

Review: Graphs and transformations

Graphs of polynomial functions

The **order** of any polynomial function is the highest power of the variable, so $f(x) = 3x^2 - 4$ is of order 2 and $f(x) = 5x - x^3$ is of order 3, etc.

The graph of a polynomial function of order n can have up to $n - 1$ turning points, although often not all of these materialise. The general shape of a polynomial curve is as in Figure 4.1.

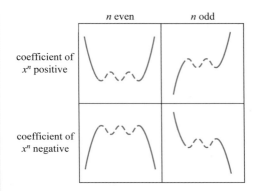

Figure 4.1

If a polynomial function $f(x)$ can be factorised, then each factor will give a root of the associated polynomial equation $f(x) = 0$. If there is a repeated factor then this will correspond to a repeated root.

Example 4.1

(i) (a) Solve $x(x - 2)(x + 4) = 0$.

　　(b) Sketch $y = x(x - 2)(x + 4)$. ◄

(ii) (a) Solve $(x + 1)^2(5 - x) = 0$.

　　(b) Sketch $y = (x + 1)^2(5 - x)$.

> Remember that in a sketch, the graph does not need to be drawn to scale, but it should show the main features, such as the points where the curve crosses the coordinate axes.

Solution

(i) (a) $x(x - 2)(x + 4) = 0 \Rightarrow x = 0, 2$ or -4.

(b)

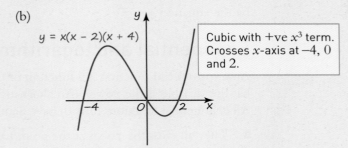

$y = x(x - 2)(x + 4)$

> Cubic with +ve x^3 term. Crosses x-axis at -4, 0 and 2.

Figure 4.2

(ii) (a) $(x + 1)^2(5 - x) = 0 \Rightarrow x = -1$ (repeated) or $x = 5$.

(b)

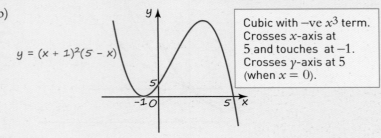

$y = (x + 1)^2(5 - x)$

> Cubic with −ve x^3 term. Crosses x-axis at 5 and touches at -1. Crosses y-axis at 5 (when $x = 0$).

Figure 4.3

Graphs of reciprocal functions

Reciprocal functions have the form $f(x) = \dfrac{a}{x}, f(x) = \dfrac{a}{x^2}$, etc.

> For both these functions, the lines $x = 0$ and $y = 0$ are **asymptotes** for the curve. The curve approaches an asymptote ever more closely, but never actually reaches it.

> Asymptotes are usually shown on a sketch using a dotted line but in this case the asymptotes are already there as the coordinate axes.

Discussion point

→ Figure 4.4 shows curves of the form

$$y = \frac{1}{x^n} \text{ for}$$

the cases $n = 1$ and $n = 2$. Describe the shapes of the curves for other positive integer values of n.

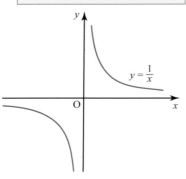 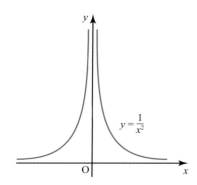

Figure 4.4

Proportional relationships and their graphs

> In both cases k is the **constant of proportionality**.

Two variables x and y are said to be (**directly**) **proportional** if $y = kx$ where k is a constant. They are **inversely proportional** if $y = \dfrac{k}{x}$.

The relationship $y = kx^n$ can be described by saying that y is **proportional** to x^n. In the same way, the statement 'y is **inversely proportional** to x^n', is represented by the equation $y = \dfrac{k}{x^n}$.

Exponential and logarithmic functions

An **exponential** function is a function of the form $y = a^x$, where $a > 0$, which has the variable as the power. (An alternative name for 'power' is 'exponent'.) All exponential functions ($a \neq 0$) have graphs that have a similar shape.

■ They all cross the y-axis at the point $(0,1)$.

■ They all have the x-axis as an asymptote.

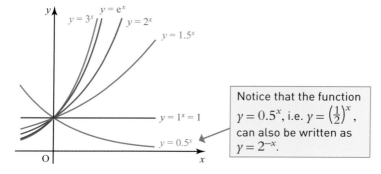

> Notice that the function $y = 0.5^x$, i.e. $y = \left(\frac{1}{2}\right)^x$, can also be written as $y = 2^{-x}$.

Figure 4.5

Although there are many exponential functions, the function $y = e^x$, where e is an irrational number approximately equal to 2.718, is referred to as *the* exponential function; the curve $y = e^x$ has the special property that for any value of x the value of y is the same as the gradient of the curve.

A relationship of the form $y = e^{kx}$, where $k > 0$, is described as **exponential growth**. If $k < 0$ the relationship is described as **exponential decay**.

Just as exponential graphs for a positive index are all very similar, so are the graphs of the **logarithm** function for different bases.

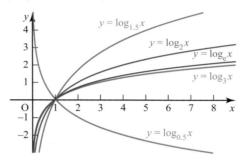

Figure 4.6

The logarithm function is the inverse of the exponential function.

Notation

In general the inverse of $y = a^x$ is written as $y = \log_a x$ but there are two special cases:

- The inverse of $y = e^x$ is $y = \log_e x$ which is more often written as $y = \ln x$.

- The inverse of $y = 10^x$ is $y = \log_{10} x$ which is often written just as $y = \log x$.

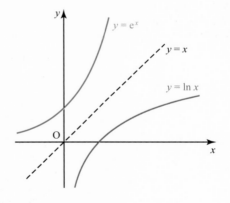

Figure 4.7

Transformations of graphs

A transformation is a relationship that is used in curve sketching. You can use stretches, translations and reflections to transform the graph of any function.

Table 4.1

TECHNOLOGY

Use graphing software to explore these transformations.

Function	Transformation
$f(x) \rightarrow f(x - t) + s$	Translation $\begin{pmatrix} t \\ s \end{pmatrix}$
$f(x) \rightarrow af(x)$	One-way stretch, parallel to y-axis, scale factor a
$f(x) \rightarrow f(ax)$	One-way stretch, parallel to x-axis, scale factor $\dfrac{1}{a}$
$f(x) \rightarrow -f(x)$	Reflection in x-axis
$f(x) \rightarrow f(-x)$	Reflection in y-axis

Example 4.2

Starting with the curve $y = \cos x$, show how transformations can be used to sketch the curves

(i) $y = \cos 3x$ (ii) $y = \cos\left(\dfrac{x}{2}\right)$ (iii) $y = \cos\left(x - \dfrac{\pi}{3}\right)$ (iv) $y = \cos x - 2$.

Solution

(i) The curve with equation $y = \cos 3x$ is obtained from the curve with equation $y = \cos x$ by a stretch of scale factor $\frac{1}{3}$ parallel to the x-axis.

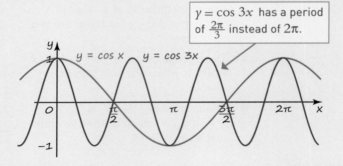

$y = \cos 3x$ has a period of $\frac{2\pi}{3}$ instead of 2π.

Figure 4.8

(ii) The curve of $y = \cos\left(\dfrac{x}{2}\right)$ is obtained from that of $y = \cos x$ by a stretch of scale factor 2 in the x direction.

$y = \cos\left(\frac{x}{2}\right)$ has a period of 4π instead of 2π.

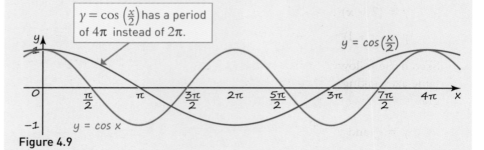

Figure 4.9

(iii)   The curve of $y = \cos\left(x - \dfrac{\pi}{3}\right)$ is obtained from that of $y = \cos x$

by a translation of $\begin{pmatrix} \dfrac{\pi}{3} \\ 0 \end{pmatrix}$.

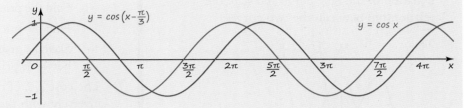

Figure 4.10

(iv)   The curve of $y = \cos x - 2$ is obtained from that of $y = \cos x$ by a

translation of $\begin{pmatrix} 0 \\ -2 \end{pmatrix}$.

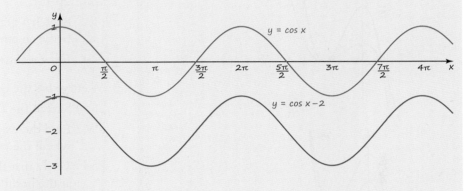

Figure 4.11

Review exercise

① Sketch the following curves.

(i) $y = (x + 1)(2x - 1)(x - 3)$

(ii) $y = (2 - x)(2 + x)(3 + x)$

(iii) $y = (x + 1)^2(x - 3)^2$

② Sketch the following pairs of curves on the same axes, taking care to clearly label each graph of the pair.

(i) $y = \dfrac{1}{x}$ and $y = \dfrac{3}{x}$

(ii) $y = \dfrac{1}{x^2}$ and $y = \dfrac{1}{x^4}$

③ Sketch, using the same scale on both axes, the following sets of curves.

(i) $y = x$, $y = e^x$ and $y = \ln x$

(ii) $y = 1^x$, $y = 2^x$ and $y = 3^x$

④ Sketch, on the same axes, the following sets of curves.

(i) $y = x^2$, $y = x^2 + 3$ and $y = (x + 3)^2$

(ii) $y = \sin x$, $y = \sin x - 2$ and

$y = \sin\left(x - \dfrac{\pi}{2}\right)$

⑤ Each part of Figure 4.12 shows the curve $y = (x - 2)^2(x + 1)$ (in red) and one other curve which has been obtained from $y = (x - 2)^2(x + 1)$ by one of the following transformations: translation, stretch or reflection. In each case write down the equation of the blue curve and state the relationship between the two curves.

(i)

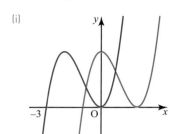

(ii)

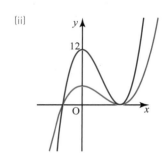

(iii)

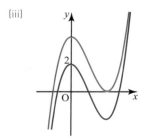

(iv)

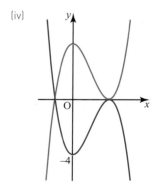

Figure 4.12

⑥ A new moisturiser is being marketed in cubical jars which have different sizes: 30 ml, 50 ml and 100 ml. The price £C is directly proportional to the *height* of the jar and the price of the 50 ml jar is £36.

(i) Find the height of the 30 ml jar.

(ii) Find the price of the 30 ml jar.

(iii) Find the price of the 100 ml jar.

(iv) Sketch a graph to show the relationship between price and volume.

⑦ The cost £C of my electricity bill varies inversely with the average temperature $\theta°$C over the period of the bill. One winter, when the average temperature was 8°C, my bill was £365 for the 3-month period.

(i) How much would I expect to pay in the summer, when the average temperature is 20°C?

(ii) Sketch a graph to show the relationship between temperature and electricity cost.

(iii) How realistic is this model?

⑧ A curve has the equation $f(x) = 1 + \ln x$.

(i) Sketch the curve.

(ii) Find the exact coordinates of the point where the curve crosses the x-axis.

(iii) Verify that the curve meets the line $y = x$ at the point $(1, 1)$.

⑨ (i) On a single set of axes, sketch the curves $y = e^x - 1$ and $y = 2e^{-x}$, indicating any asymptotes.

(ii) Find the exact coordinates of the point of intersection of these two curves.

⑩ Starting with the curve $y = \sin x$, show how transformations can be used to sketch the following curves.

(i) $y = \sin 2x$

(ii) $y = \sin\left(x + \dfrac{\pi}{2}\right)$

1 The language of functions

A **mapping** is any rule which associates
two sets of items, which are referred to
as the **object** and **image** or the **input**
and **output**.

For a mapping to make sense or to have
any practical application, the inputs and
outputs must each form a natural collection or set.

Figure 4.13

The set of possible inputs is called the **domain** of the mapping and the set of
possible outputs is called the **range** of the mapping.

Mappings

In mathematics, many (but not all) mappings can be expressed using algebra. A
mapping can be **one-to-one, one-to-many**, **many-to-one** or **many-to-many**.

Here are some examples of mathematical mappings.

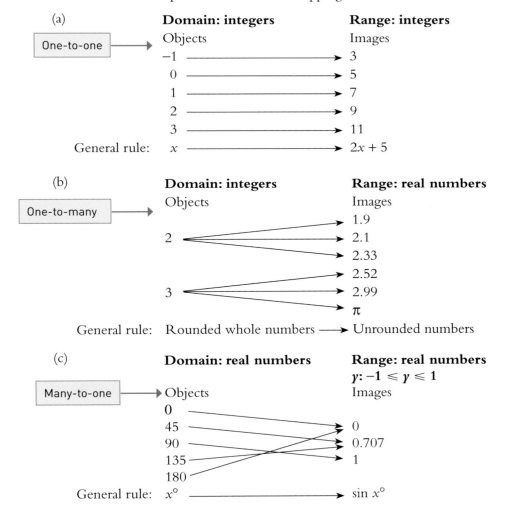

(d)

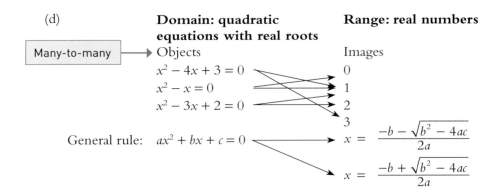

Domain: quadratic equations with real roots

Range: real numbers

Many-to-many

Objects

Images

$x^2 - 4x + 3 = 0$
$x^2 - x = 0$
$x^2 - 3x + 2 = 0$

0
1
2
3

General rule: $ax^2 + bx + c = 0$

$x = \dfrac{-b - \sqrt{b^2 - 4ac}}{2a}$

$x = \dfrac{-b + \sqrt{b^2 - 4ac}}{2a}$

Functions

Mappings which are one-to-one or many-to-one are of particular importance, since in these cases there is only one possible image for any object. Mappings of these types are called **functions**. For example, $x \to x^2$ and $x \to \cos x$ are both functions, because in each case for any value of x there is only one possible output.

The mapping of rounded whole numbers on to unrounded numbers is not a function since, for example, the rounded number 5 could map to any unrounded number between 4.5 and 5.5.

Notation

There are several different but equivalent ways of writing a function. For example, the function which maps x on to x^2 can be written in any of the following ways:

■ $f(x) = x^2$

■ $f: x \to x^2$ ← Read this as 'f maps x on to x^2'.

It can also be written as $y = x^2$ but this is not using function notation.

A function $y = f(x)$ may be defined for all values of x or only for a restricted set of values.

Strictly, when a function is written it should always be accompanied by its domain. When you write this, it is often helpful to use one of the following symbols to denote different types of numbers:

Note

In some texts you will see 0 excluded from $\mathbb{N}$.

■ $\mathbb{Z}^+$ is the set of counting numbers: 1, 2, 3, … .

■ $\mathbb{N}$ is the set of natural numbers: 0, 1, 2, 3, … .

■ $\mathbb{Z}$ is the set of integers: … −3, −2, −1, 0, 1, 2, 3, … .

■ $\mathbb{Q}$ is the set of rational numbers, i.e. fractions such as $-\frac{3}{4}$ and $\frac{23}{5}$, together with everything in $\mathbb{Z}$.

■ $\mathbb{R}$ is the set of real numbers, i.e. irrational numbers such as π and $\sqrt{2}$, together with everything in $\mathbb{Q}$.

If a function is written without its domain, the domain is usually taken to be the real numbers.

It is often helpful to represent a function graphically, as in the following example, which illustrates the importance of knowing the domain.

Example 4.3

Sketch the graph of $y = 3x + 2$ when the domain of x is given by

(i) $x \in \mathbb{R}$ ← This means 'x is in the set of real numbers'.

(ii) $x \in \mathbb{R}^+$ ← This means 'x is in the set of positive real numbers'.

(iii) $x \in \mathbb{N}$.

Solution

(i)   When $x \in \mathbb{R}$, all values of y are possible. The range is therefore $\mathbb{R}$, also.

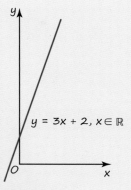

Figure 4.14

(ii) When $x \in \mathbb{R}^+$, so that x is restricted to positive values, all the values of y are greater than 2, so the range is $y > 2$.

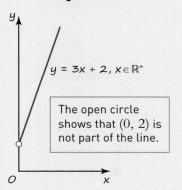

The open circle shows that $(0, 2)$ is not part of the line.

Figure 4.15

(iii) When $x \in \mathbb{N}$, the range is the set of points $\{2, 5, 8, \ldots\}$. These are all of the form $3x + 2$ where x is a natural number $(0, 1, 2, \ldots)$. This set can be written neatly as $\{3x + 2 : x \in \mathbb{N}\}$.

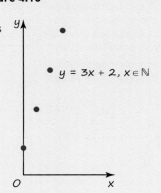

Figure 4.16

When you draw the graph of a mapping

■   the x coordinate of each point is an input value

■   the y coordinate is the corresponding output value.

Table 4.2 shows this for the mapping $x \rightarrow x^2$, or $y = x^2$, and Figure 4.17 shows the resulting points on a graph.

Table 4.2

Input (x)	Output (y)	Point plotted
−2	4	(−2, 4)
−1	1	(−1, 1)
0	0	(0, 0)
1	1	(1, 1)
2	4	(2, 4)

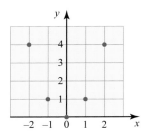

Figure 4.17

If the mapping is a function, there is a unique value of y for every value of x in the domain. Consequently the graph of a function is a simple curve or line going from left to right, with no doubling back.

Figure 4.18 illustrates some different types of mapping.

These two mappings are functions.

These two mappings are not functions as in each case there are two possible values of y corresponding to some or all values of x.

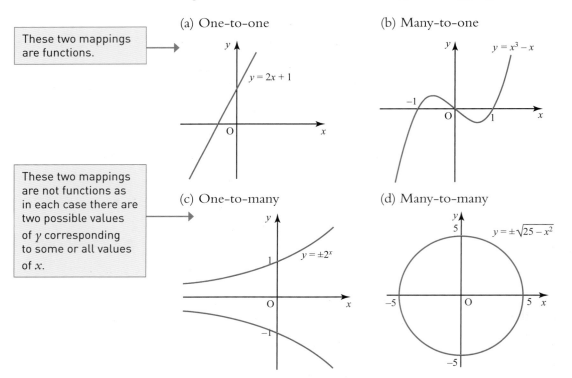

(a) One-to-one

$y = 2x + 1$

(b) Many-to-one

$y = x^3 - x$

(c) One-to-many

$y = \pm 2^x$

(d) Many-to-many

$y = \pm\sqrt{25 - x^2}$

Figure 4.18

Using transformations to sketch the curves of functions

In the review section you used translations and one-way stretches to relate the equation of a function to that of a standard function of the same form. This then allowed you to sketch the curve of your function.

It is possible to combine translations and stretches, but you must be careful over the order in which these are applied, as you will see in Activity 4.1.

ACTIVITY 4.1

Copy the triangle in Figure 4.19 and, for each of parts (i) to (v), perform the transformations in both of the orders given. In each case comment on whether the end results are the same or different.

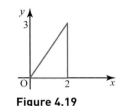

Figure 4.19

(i) (a) Translate the triangle by $\begin{pmatrix} 3 \\ 0 \end{pmatrix}$ and then stretch the image with a scale factor of 2 parallel to the x-axis.

   (b) Stretch the triangle with a scale factor of 2 parallel to the x-axis and then translate the image by $\begin{pmatrix} 3 \\ 0 \end{pmatrix}$.

(ii) (a) Translate the triangle by $\begin{pmatrix} 3 \\ 0 \end{pmatrix}$ and then stretch the image with a scale factor of 2 parallel to the y-axis.

   (b) Stretch the triangle with a scale factor of 2 parallel to the y-axis and then translate the image by $\begin{pmatrix} 3 \\ 0 \end{pmatrix}$.

(iii)(a) Translate the triangle by $\begin{pmatrix} 0 \\ 3 \end{pmatrix}$ and then stretch the image with a scale factor of 2 parallel to the x-axis.

   (b) Stretch the triangle with a scale factor of 2 parallel to the x-axis and then translate the image by $\begin{pmatrix} 0 \\ 3 \end{pmatrix}$.

(iv) (a) Translate the triangle by $\begin{pmatrix} 0 \\ 3 \end{pmatrix}$ and then stretch the image with a scale factor of 2 parallel to the y-axis.

   (b) Stretch the triangle with a scale factor of 2 parallel to the y-axis and then translate the image by $\begin{pmatrix} 0 \\ 3 \end{pmatrix}$.

(v) (a) Stretch the triangle with a scale factor of 2 parallel to the x-axis and then stretch the image with a scale factor of 3 parallel to the y-axis.

   (b) Stretch the triangle with a scale factor of 3 parallel to the y-axis and then stretch the image with a scale factor of 2 parallel to the x-axis.

! Activity 4.1 should have emphasised to you the importance of performing the transformations in the correct order. It is a good idea to check your results using a graphical calculator or graphing software whenever possible. **T**

Example 4.4

Starting with the curve $y = \cos x$, show how transformations can be used to sketch the following curves.

(i) $y = 2\cos 3x$ (ii) $y = \cos\left(2x - \dfrac{\pi}{3}\right)$

Solution

(i)

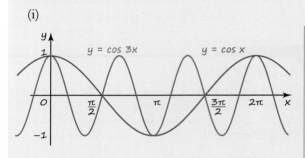

The curve with equation $y = \cos 3x$ is obtained from the curve with equation $y = \cos x$ by a stretch of scale factor $\frac{1}{3}$ parallel to the x-axis. There will therefore be one complete oscillation of the curve in $\frac{2\pi}{3}$ (instead of 2π).

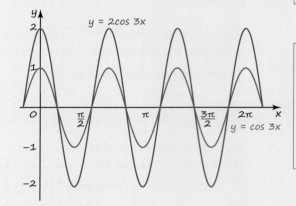

The curve of $y = 2\cos 3x$ is obtained from that of $y = \cos 3x$ by a stretch of scale factor 2 parallel to the y-axis. The curve therefore oscillates between $y = 2$ and $y = -2$ (instead of between $y = 1$ and $y = -1$).

Figure 4.20

(ii)

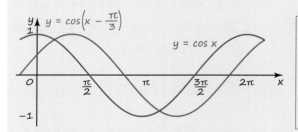

The curve with equation $y = \cos\left(x - \dfrac{\pi}{3}\right)$ is obtained from the curve with equation $y = \cos x$ by a transation of $\begin{pmatrix} \frac{\pi}{3} \\ 0 \end{pmatrix}$.

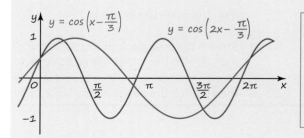

The curve of $y = \cos\left(2x - \dfrac{\pi}{3}\right)$ is obtained from that of $y = \cos\left(x - \dfrac{\pi}{3}\right)$ by a stretch of scale factor $\frac{1}{2}$ parallel to the x-axis.

Figure 4.21

Example 4.5

(i)   Write the equation $y = 1 + 4x - x^2$ in the form $y = a[(x + p)^2 + q]$.

(ii) Show how the graph of $y = 1 + 4x - x^2$ can be obtained from the graph of $y = x^2$ by a succession of transformations, and list the transformations in the order in which they are applied.

(iii) Sketch the graph.

Solution

(i)   If $1 + 4x - x^2 \equiv a[(x + p)^2 + q]$

then $-x^2 + 4x + 1 \equiv ax^2 + 2apx + a(p^2 + q)$.

Comparing coefficients of x^2: $a = -1$.

Comparing coefficients of x: $2ap = 4$, giving $p = -2$.

Comparing constant terms : $a(p^2 + q) = 1$, giving $q = -5$.

The equation is $y = -[(x - 2)^2 - 5]$.

(ii) The curve $y = x^2$ becomes the curve $y = (x - 2)^2 - 5$ by applying the translation

$\begin{pmatrix} 2 \\ -5 \end{pmatrix}$ as shown in Figure 4.22.

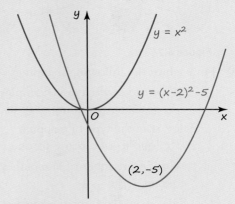

Figure 4.22

The curve $y = (x - 2)^2 - 5$ becomes the curve $y = -[(x - 2)^2 - 5]$ by applying a reflection in the x-axis (see Figure 4.23).

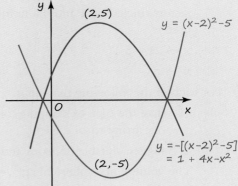

Figure 4.23

(iii) The blue curve in Figure 4.23 shows the graph.

Discussion point

➜ In Example 4.5, you could write the equation $y = 1 + 4x - x^2$ in the form $y = 5 - (2 - x)^2$. Describe a different set of transformations, suggested by this form, which would map $y = x^2$ to $y = 1 + 4x - x^2$.

Exercise 4.1

① Describe each of the mappings in Figure 4.24 as either one-to-one, many-to-one, one-to-many or many-to-many, and say whether it represents a function.

(i)   (ii)

(iii)   (iv)

(v)   (vi)

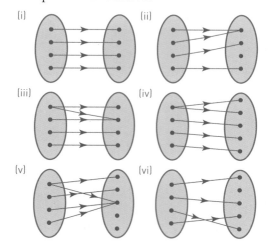

Figure 4.24

② For each of the following mappings:
- (a) write down a few examples of inputs and corresponding outputs
- (b) state the type of mapping (one-to-one, many-to-one, etc.).

(i) Words → number of letters they contain

(ii) Side of a square in cm → its perimeter in cm

(iii) Natural numbers → the number of factors (including 1 and the number itself)

(iv) $x \rightarrow 2x - 5$

(v) $x \rightarrow \sqrt{x}$

③ For each of the following mappings:
- (a) state the type of mapping (one-to-one, many-to-one, etc.)
- (b) suggest suitable domains and ranges.

(i) The volume of a sphere in cm³ → its radius in cm

(ii) The volume of a cylinder in cm³ → its height in cm

(iii) The length of a side of a regular hexagon in cm → its area in cm²

(iv) $x \rightarrow x^2$

④ (i) A function is defined by $f(x) = 2x - 5$. Write down the values of
- (a) f(0)   (b) f(7)   (c) f(−3).

(ii) A function is defined by g:(polygons) → (number of sides). What are the following:
- (a) g(triangle)   (b) g(pentagon)
- (c) g(decagon)

(iii) The function t maps Celsius temperatures on to Fahrenheit temperatures. It is defined by t: $C \rightarrow \dfrac{9C}{5} + 32$. Find
- (a) t(0)   (b) t(28)   (c) t(−10)
- (d) the value of C when $t(C) = C$.

⑤ Find the range of each of the following functions.

(You may find it helpful to draw the graph first.)

(i) $f(x) = 2 - 3x$ $x \geqslant 0$

(ii) $y = x^2 + 2$ $x \in \{0, 1, 2, 3, 4\}$

(iii) $f : x \rightarrow x^3 - 4$ $x \in \mathbb{R}$

(iv) $y = \tan \theta$ $0° < \theta < 90°$

(v) $y = \cos x$ $-\dfrac{\pi}{4} < x \leqslant \dfrac{\pi}{4}$

⑥ Find the range of each of the following functions.
(You may find it helpful to draw the graph first.)

(i) $f : x \rightarrow 2^x$ $x \in \{-1, 0, 1, 2\}$

(ii) $f(x) = \dfrac{1}{1 + x^2}$ $x \in \mathbb{R}$

(iii) $f(x) = \sqrt{x - 3} + 3$ $x \geqslant 3$

⑦ The mapping f is defined by

$$f(x) = \begin{cases} x^2 & 0 \leqslant x \leqslant 3 \\ 3x & 3 \leqslant x \leqslant 10. \end{cases}$$

The mapping g is defined by

$$g(x) = \begin{cases} x^2 & 0 \leqslant x \leqslant 2 \\ 3x & 2 \leqslant x \leqslant 10. \end{cases}$$

Explain why f is a function and g is not.

⑧ Starting with the graph of $y = x^2$, state the transformations which can be used to sketch each of the following curves. Specify the transformations in the order in which they are used and, where there is more than one stage in the sketching of the curve, state each stage.

In each case state the equation of the line of symmetry.

(i) $y = (x - 2)^2$ (ii) $y = 3(x - 2)^2$

(iii) $y = 3x^2 - 6x - 2$

⑨ Figure 4.25 shows a sketch of the graph of $y = f(x)$, where $f(x) = x^2 + 4x$.

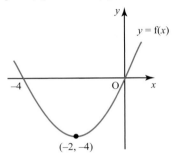

Figure 4.25

Draw separate sketches of the following functions.

In each case describe the transformations from $y = f(x)$.

(i) $y = f(x) - 2$ (ii) $y = f(x - 2)$

(iii) $y = 2f(x) + 3$ (iv) $y = f(2x) + 3$

(v) $y = 3f(x - 2)$

⑩ Figure 4.26 shows a sketch of the graph of $y = f(x)$, where $f(x) = 6x - x^2$.

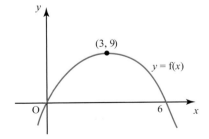

Figure 4.26

Use this graph to sketch the following curves on separate diagrams.

In each case indicate clearly where the graph crosses the x-axis and the coordinates of its highest point.

(i) $y = f(x - 2)$ (ii) $y = 2f(x - 2)$

(iii) $y = \frac{1}{2}f(x)$ (iv) $y = f\left(\dfrac{x}{2}\right)$

⑪ The circle with equation $x^2 + y^2 = 1$ is stretched with scale factor 3 parallel to the x-axis and with scale factor 2 parallel to the y-axis.

Sketch both curves on the same graph, and write down the equation of the new curve. (It is an ellipse.)

⑫ Figure 4.27 shows the graph of $y = f(x)$.

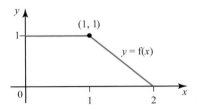

Figure 4.27

Sketch the graph of each of these functions.

(i) $y = f(2x)$ (ii) $y = f(x - 1)$

(iii) $y = 2f(x - 1)$ (iv) $y = 3f(x)$

(v) $y = f(3x)$ (vi) $y = f(3x - 1)$

⑬ Starting with the curve $y = \cos x$, state how transformations can be used to sketch these curves.

(i) $y = 3 \cos x$ (ii) $y = \cos 3x - 1$

(iii) $y = \cos(3x + 30°)$

⑭ For each of the following curves:

   (a)   sketch the curve

   (b)   identify the curve as being the same as one of the following:
      $y = \pm\sin x$, $y = \pm\cos x$ or $y = \pm\tan x$.

(i) $y = \cos(-x)$ (ii) $y = \tan(-x)$

(iii) $y = \sin(\pi - x)$ (iv) $y = \tan(\pi - x)$

(v) $y = \sin(-x)$

⑮ (i)   Write the expression $x^2 - 6x + 14$ in the form $(x - a)^2 + b$ where a and b are numbers which you are to find.

(ii)   Sketch the curves $y = x^2$ and $y = x^2 - 6x + 14$ and state the transformation which maps $y = x^2$ on to $y = x^2 - 6x + 14$.

(iii)   The curve $y = x^2 - 6x + 14$ is reflected in the x-axis. Write down the equation of the image.

⑯ Starting with the graph of $y = x^2$, state the transformations which can be used to sketch $y = 2x - 1 - x^2$.

Specify the transformations in the order in which they are used.

State the equation of the line of symmetry.

⑰ In Figure 4.28, $y = f(x)$ is a translation of the curve $y = x^2(x - 2)$ and curve $y = g(x)$ is the reflection of $y = f(x)$ in the x-axis.

Write down the equations of the two curves.

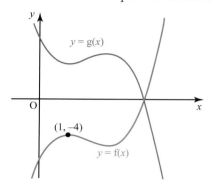

Figure 4.28

⑱ In Figure 4.29, $y = f(x)$ is the curve $y = \ln(2x + 1)$ and $y = g(x)$ is its reflection in the line $x = 2$.

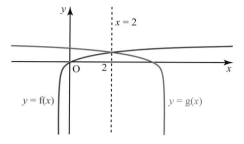

Figure 4.29

Write down the equation of $y = g(x)$.

2 Composite functions

It is possible to combine functions in several different ways, and you have already met some of these. For example, if $f(x) = x^2$ and $g(x) = 2x$, then you could write

$$f(x) + g(x) = x^2 + 2x.$$ ◄— | Here, two functions are added. |

Similarly if $f(x) = x$ and $g(x) = \sin x$, then

$$f(x) \cdot g(x) = x \sin x.$$ ◄— | Here, two functions are multiplied. |

Sometimes you need to apply one function and then apply another, so that the output of the first function is the input for the second one.

You are then creating a **composite function** or a **function of a function**.

| **Example 4.6** |

New parents are bathing their baby for the first time. They take the temperature of the bath water with a thermometer which reads in Celsius, but then have to convert the temperature to degrees Fahrenheit to apply the rule for correct bath water temperature taught to them by their own parents.

> At one o five
> He'll cook alive
> But ninety four
> Is rather raw.

Write down the two functions that are involved, and apply them to readings of

(i) 30°C   (ii) 38°C   (iii) 45°C.

Solution

The first function converts the Celsius temperature C into a Fahrenheit temperature, F.

$$F = \frac{9C}{5} + 32$$

The second function maps Fahrenheit temperatures on to the state of the bath.

$F \leqslant 94$	too cold
$94 < F < 105$	all right
$F \geqslant 105$	too hot

This gives

(i) $30°C \rightarrow 86°F \rightarrow$ too cold

(ii) $38°C \rightarrow 100.4°F \rightarrow$ all right

(iii) $45°C \rightarrow 113°C \rightarrow$ too hot.

In this case the composite function would be (to the nearest degree)

$C \leqslant 34°C$	too cold
$35°C \leqslant C \leqslant 40°C$	all right
$C \geqslant 41°C$	too hot.

In algebraic terms, a composite function is constructed as

input x $\xrightarrow{\ f\ }$ output f(x)

input f(x) $\xrightarrow{\ g\ }$ output g[f(x)]   (or gf(x))

> Read this as 'g of f of x.'

Thus the composite function gf(x) should be performed from right to left: start with x then apply f and then g.

▌ Notation

To indicate that f is being applied twice in succession, you could write ff(x) but you would usually use $f^2(x)$ instead. Similarly $g^3(x)$ means three applications of g.

In order to apply a function repeatedly its range must be completely contained within its domain.

Order of functions

It is often the case that the order of the individual functions in a composite function matters.

For example, if f(x) = x^2 and g(x) = $x + 1$, then

f(3) = 9	g(9) = 10	so gf(3) = 10
g(3) = 4	f(4) = 16	so fg(3) = 16.

More generally,

$$x \xrightarrow[\text{square}]{f} x^2 \xrightarrow[\text{add 1}]{g} x^2 + 1.$$

So $gf(x) = x^2 + 1$.

To find an expression for $fg(x)$ you must apply g first. In the example above, this would give:

$$x \xrightarrow[\text{add 1}]{g} (x + 1) \xrightarrow[\text{square}]{f} (x + 1)^2$$

and so $fg(x) = (x + 1)^2$.

Clearly this is *not* the same result.

Figure 4.30 illustrates the relationship between the domains and ranges of the functions f and g, and the range of the composite function gf.

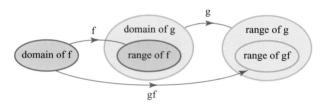

Figure 4.30

Notice that the range of f must be completely contained within the domain of g.

Given that $f(x) = 2x$, $g(x) = x^2$, and $h(x) = \dfrac{1}{x}$, find the following.

(i) $fg(3)$ (ii) $gh(2)$ (iii) $gf(x)$ (iv) $hfg(x)$

Solution

(i) $fg(3) = f[g(3)]$ (ii) $gh(2) = g[h(2)]$
$$= f(9)$$
$$= g\left(\tfrac{1}{2}\right)$$
$$= 18$$
$$= \tfrac{1}{4}$$

(iii) $gf(x) = g[f(x)]$ (iv) $fg(x) = f[g(x)]$
$$= g(2x)$$
$$= f(x^2)$$
$$= (2x)^2$$
$$= 2x^2$$
$$= 4x^2$$

So $hfg(x) = h(2x^2)$
$$= \frac{1}{2x^2}$$

Inverse functions

Look at the mapping $x \rightarrow x + 2$ with domain and range the set of integers.

Domain	Range
...	...
...	...
-1	-1
0	0
1	1
2	2
...	3
...	4
x	$x + 2$

The mapping is clearly a function, since for every input there is one and only one output, the number that is two greater than that input.

This mapping can also be seen in reverse. In that case, each number maps on to the number two less than itself: $x \rightarrow x - 2$. The reverse mapping is also a function because for any input there is one and only one output. The reverse mapping is called the **inverse function, f^{-1}**.

Function: $f : x \rightarrow x + 2$ $x \in \mathbb{Z}$

Inverse function: $f^{-1} : x \rightarrow x - 2$ $x \in \mathbb{Z}$.

For a mapping to be a function that also has an inverse function, every object in the domain must have one and only one image in the range, and vice versa. This can only be the case if the mapping is one-to-one.

So the condition for a function f to have an inverse function is that, over the given domain and range, f represents a one-to-one mapping. This is a common situation, and many inverse functions are self-evident as in the following examples, for all of which the domain and range are the real numbers.

$f : x \rightarrow x - 1;$ $f^{-1} : x \rightarrow x + 1$

$g : x \rightarrow 2x;$ $g^{-1} : x \rightarrow \frac{1}{2}x$

$h : x \rightarrow x^3;$ $h^{-1} : x \rightarrow \sqrt[3]{x}$

Discussion points

Some of these mappings (below) are functions that have inverse functions, and others are not.

→ Decide which mappings fall into each category, and for those that do not have inverse functions, explain why.

→ For those that have inverse functions, how can the functions and their inverses be written down algebraically?

(i)   Temperature measured in Celsius → temperature measured in Fahrenheit.

(ii) Marks in an examination → grade awarded.

(iii) Distance measured in light years → distance measured in metres.

(iv) Number of stops travelled on the London Underground → fare.

You can decide whether an algebraic mapping is a function, and whether it has an inverse function, by looking at its graph. The curve or line representing a one-to-one mapping does not double back on itself, has no turning points and covers the full domain. Figure 4.31 illustrates the functions f, g and h given on the previous page.

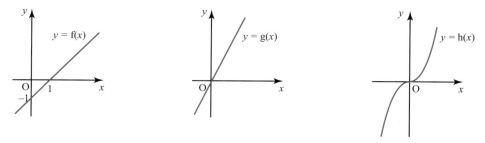

Figure 4.31

Now look at $f(x) = x^2$ for $x \in \mathbb{R}$ (Figure 4.32). You can see that there are two distinct input values giving the same output: for example $f(2) = f(-2) = 4$. When you want to reverse the effect of the function, you have a mapping which for a single input of 4 gives two outputs, -2 and $+2$. Such a mapping is not a function.

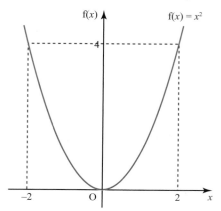

Figure 4.32

If the domain of $f(x) = x^2$ is restricted to $\mathbb{R}^+$ (the set of positive real numbers), you have the situation shown in Figure 4.33. This shows that the function which is now defined is one-to-one. The inverse function is given by $f^{-1}(x) = \sqrt{x}$, since the sign means 'the positive square root of'.

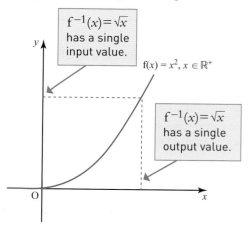

Figure 4.33

It is often helpful to restrict the domain of a function so that its inverse is also a function. When you use the inverse sin (i.e. $\sin^{-1}$ or arcsin) key on your calculator the answer is restricted to the range $-90°$ to $90°$, and is described as the **principal value**. Although there are infinitely many roots of the equation $\sin x = 0.5$ ($\dots, -330°, -210°, 30°, 150°, \dots$), only one of these, $30°$, lies in the restricted range of $-90°$ to $90°$ and this is the value your calculator will give you.

The graph of a function and its inverse

TECHNOLOGY

You can use graphing software or a graphical calculator.

ACTIVITY 4.2

For each of the following functions, work out the inverse function, and draw the graphs of both the original and the inverse on the same axes, using the same scale on both axes.

(i) $f(x) = x^2$, $x \in \mathbb{R}^+$ (ii) $f(x) = 2x$ (iii) $f(x) = x + 2$ (iv) $f(x) = x^3 + 2$

Look at your graphs. What pattern can you see emerging?
Try out a few more functions of your own to check your ideas.
Make a conjecture about the relationship between the graph of a function and that of its inverse.

You have probably realised by now that the graph of the inverse function is the same shape as that of the function, but reflected in the line $y = x$. To see why this is so, think of a function $f(x)$ mapping a on to b; (a, b) is clearly a point on the graph of $f(x)$. The inverse function, $f^{-1}(x)$, maps b on to a, and so (b, a) is a point on the graph of $f^{-1}(x)$.

The point (b, a) is the reflection of the point (a, b) in the line $y = x$.
This is shown for a number of points in Figure 4.34.

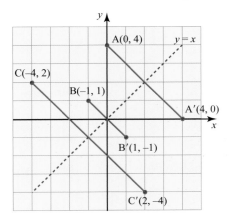

Figure 4.34

This result can be used to obtain a sketch of the inverse function without having to find its equation, provided that the sketch of the original function uses the same scale on both axes.

Finding the algebraic form of the inverse function

To find the algebraic form of the inverse of a function $f(x)$, you should start by changing notation and writing it in the form $y = \dots$.

Since the graph of the inverse function is the reflection of the graph of the original function in the line $y = x$, it follows that you may find its equation by interchanging y and x in the equation of the original function. You will then need to make y the subject of your new equation. This procedure is illustrated in Example 4.8.

Example 4.8

Find $f^{-1}(x)$ when $f(x) = 2x + 1$.

Solution

The function $f(x)$ is given by $\qquad\qquad\qquad y = 2x + 1$

Interchanging x and y gives $\qquad\qquad\qquad x = 2y + 1$

Rearranging to make y the subject: $\qquad\qquad y = \dfrac{x - 1}{2}$

So $\qquad\qquad\qquad\qquad\qquad\qquad\qquad f^{-1}(x) = \dfrac{x - 1}{2}$.

Sometimes the domain of the function f will not include the whole of $\mathbb{R}$. When any real numbers are excluded from the domain of f, it follows that they will be excluded from the range of f^{-1}, and vice versa.

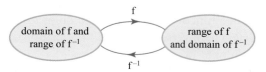

Figure 4.35

Example 4.9

Find $f^{-1}(x)$ when $f(x) = 2x - 3$ and the domain of f is $x \geqslant 4$.

Solution

	Domain	Range
Function: $y = 2x - 3$	$x \geqslant 4$	$y \geqslant 5$
Inverse function: $x = 2y - 3$	$x \geqslant 5$	$y \geqslant 4$

Rearranging the inverse function to make y the subject:

$$y = \frac{x + 3}{2}$$

The full definition of the inverse function is therefore

$$f^{-1}(x) = \frac{x + 3}{2} \quad \text{for } x \geqslant 5.$$

You can see in Figure 4.36 that the inverse function is the reflection of a restricted part of the line $y = 2x - 3$.

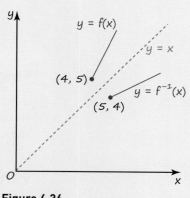

Figure 4.36

Example 4.10

(i)   Sketch the graph of $y = f(x)$ where $f(x) = x^2 + 2$, $x \geqslant 0$.
Sketch the graph of the inverse function on the same axes.

(ii)   Find $f^{-1}(x)$ algebraically.

(iii)   Find $f(7)$ and $f^{-1}f(7)$. What do you notice?

Solution

(i)

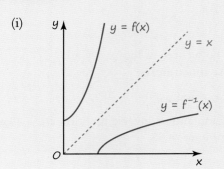

Figure 4.37

(ii)

	Domain	Range
Function: $y = x^2 + 2$	$x \geqslant 0$	$y \geqslant 2$
Inverse function: $x = y^2 + 2$	$x \geqslant 2$	$y \geqslant 0$

Rearranging the inverse function to make y its subject:

$$y^2 = x - 2.$$

This gives $y = \pm\sqrt{x - 2}$, but since you know the range of the inverse function to be $y \geqslant 0$ you can write:

$$y = +\sqrt{x - 2} \text{ or just } y = \sqrt{x - 2}.$$

The full definition of the inverse function is therefore:

$$f^{-1}(x) = \sqrt{x - 2} \text{ for } x \geqslant 2.$$

(iii) $f(7) = 7^2 + 2 = 51$

$f^{-1}f(7) = f^{-1}(51) = \sqrt{51 - 2} = 7$

Applying the function followed by its inverse brings you back to the original input value.

Note

Part (iii) of Example 4.10 illustrates an important general result. For any function $f(x)$ with an inverse $f^{-1}(x)$, $f^{-1}f(x) = x$. Similarly $ff^{-1}(x) = x$. The effects of a function and its inverse can be thought of as cancelling each other out.

Example 4.11

Find the inverse of the function $f(x) = 10^x$, and sketch $f(x)$ and $f^{-1}(x)$ on the same diagram.

Solution

The function $f(x)$ is given by $y = 10^x$.

Interchanging x and y, the inverse function is given by

$$x = 10^y.$$

This can be written as $\log_{10} x = y$, so the inverse function is

$$f^{-1}(x) = \log_{10} x.$$

The function and its inverse function are shown in Figure 4.38.

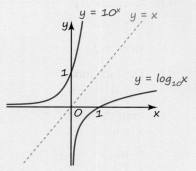

Figure 4.38

Note

The arcsin function is the inverse of the sin function. It is written as $\sin^{-1}$. Similarly arccos and arctan are the inverse functions of cos and tan respectively. These are covered in more detail in Chapter 6.

Discussion points

Many calculators have a function and its inverse on the same key, for example log and 10^x, $\sqrt{}$ and x^2, sin and arcsin, ln and e^x.

(i) With some calculators you can enter a number, apply x^2 and then $\sqrt{}$, and come out with a slightly different number. How is this possible?

(ii) Explain what happens if you find sin 199° and then find arcsin of the answer.

Exercise 4.2

① Simplify $\ln e^{x^2}$

② The functions f and g are defined by $f(x) = x^2$ and $g(x) = 2x + 1$.

Find each of the following.

(i) $f(2)$ (ii) $g(2)$ (iii) $gf(2)$ (iv) $fg(2)$

③ The functions f, g and h are defined by $f(x) = x^3$, $g(x) = 2x$ and $h(x) = x + 2$.

Find each of the following, in terms of x.

(i) fg (ii) gf (iii) fh (iv) hf
(v) g^2 (vi) h^2

④ Find the inverses of the following functions.

(i) $f(x) = 2x + 7$ (ii) $f(x) = 4 - x$

⑤ The function f is defined by $f(x) = (x - 2)^2 + 3$ for $x \geqslant 2$.

(i) Sketch the graph of $f(x)$.

(ii) On the same axes, sketch the graph of $f^{-1}(x)$ without finding its equation.

⑥ Express the following in terms of the functions $f: x \to \sqrt{x}$ and $g: x \to x + 4$.

(i) $x \to \sqrt{x + 4}$ (ii) $x \to x + 8$

(iii) $x \to \sqrt{x + 8}$ (iv) $x \to \sqrt{x} + 4$

⑦ The functions f and g are defined by $f(x) = \sin x$ and $g(x) = \cos x$ where x is measured in radians.

(i) Find the smallest positive value of θ so that $f(x) = g(x + \theta)$.

(ii) Find the smallest positive value of ϕ so that $g(x) = f(x + \phi)$.

⑧ The functions f, g and h are defined by $f(x) = x^3$, $g(x) = 2x$ and $h(x) = x + 2$. Find each of the following, in terms of x.

(i) fgh (ii) ghf (iii) $(fh)^2$

⑨ Find the inverses of the following functions.

(i) $f(x) = \dfrac{4}{2 - x}$

(ii) $f(x) = x^2 - 3$ for $x \geqslant 0$

⑩ The functions f, g and h are defined by

$$f(x) = \dfrac{3}{x - 4} \qquad g(x) = x^2$$
$$h(x) = \sqrt{2 - x}.$$

(i) For each function, state any real values of x for which it is not defined.

(ii) Find the inverse functions f^{-1} and h^{-1}.

(iii) Explain why g^{-1} does not exist when the domain of g is $\mathbb{R}$.

(iv) Suggest a suitable domain for g so that g^{-1} does exist.

(v) Is the domain for the composite function fg the same as for the composite function gf?
Give reasons for your answer.

⑪ A function f is defined by:

$$f: x \to \dfrac{1}{x} \qquad x \in \mathbb{R}, x \neq 0.$$

Find

(i) $f^2(x)$

(ii) $f^3(x)$

(iii) $f^{-1}(x)$

(iv) $f^{999}(x)$.

⑫ Two functions are defined as $f(x) = x^2$ and $g(x) = x^2 + 4x - 1$.

(i) Find a and b so that $g(x) = f(x + a) + b$.

(ii) Show how the graph of $y = g(x)$ is related to the graph of $y = f(x)$ and sketch the graph of $y = g(x)$.

(iii) State the range of the function $g(x)$.

(iv) State the least value of c so that $g(x)$ is one-to-one for $x \geqslant c$.

(v) With this restriction, sketch $y = g(x)$ and $y = g^{-1}(x)$ on the same axes.

⑬ (i) Write $2x^2 - 4x + 1$ in the form $a(x - 1)^2 + b$ where a and b are to be determined.

(ii) Sketch the graph of $y = 2x^2 - 4x + 1$, giving the equation of its line of symmetry and the coordinates of its vertex.

The function f is defined by $f : x \to 2x^2 - 4x + 1$ and has as its domain the set of real numbers.

(iii) Find the range of f.

(iv) Explain, with reference to your sketch, why f has no inverse with its given domain and suggest the largest domain of positive values of x for which $f(x)$ has an inverse.

(v) For this domain, sketch $f(x)$ and $f^{-1}(x)$ on the same axes together with their line of symmetry.

(vi) Find the exact value of the coordinates of the point where $f(x) = f^{-1}(x)$.

⑭ The functions $f(x)$ and $g(x)$ are defined by $f(x) = x^2$ and $g(x) = 2x - 1$ for all real values of x.

(i) State the ranges of $f(x)$ and $g(x)$ and explain why $f(x)$ has no inverse.

(ii) Find an expression for $g^{-1}(x)$ in terms of x and sketch the graphs of $y = g(x)$ and $y = g^{-1}(x)$ on the same axes.

(iii) Find expressions for $gf(x)$ and $fg(x)$ and solve the equation $gf(x) = fg(x)$.
Sketch the graphs of $y = gf(x)$ and $y = fg(x)$ on the same axes to illustrate your answer.

(iv) Find the range of values of a such that the equation $f(x + a) = g^2(x)$ has no solution.

3 The modulus function

Look at the graph of $y = f(x)$, where $f(x) = x$ (Figure 4.39).

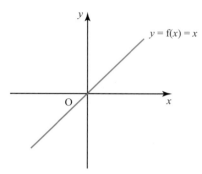

Figure 4.39

The function $f(x)$ is positive when x is positive and negative when x is negative.

Now look at Figure 4.40, the graph of $y = g(x)$, where $g(x) = |x|$.

Discussion points

➡ What is the value of $g(3)$ and $g(-3)$?

➡ What is the value of $|3 + 3|$, $|3 - 3|$, $|3| + |3|$ and $|3| + |-3|$?

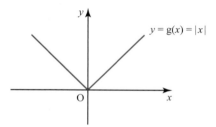

Figure 4.40

The function $g(x)$ is called the **modulus** of x. $g(x)$ always takes the positive numerical value of x. For example, when $x = -2$, $g(x) = 2$, so $g(x)$ is always positive. The modulus is also called the **magnitude** of the quantity.

Another way of writing the modulus function $g(x)$ is

$$g(x) = x \qquad \text{for } x \geqslant 0$$
$$g(x) = -x \qquad \text{for } x < 0.$$

The graph of $y = g(x)$ can be obtained from the graph of $y = f(x)$ by replacing values where $f(x)$ is negative by the equivalent positive values. This is the equivalent of reflecting that part of the line in the x-axis.

| Example 4.12 | Sketch the graphs of the following on separate axes. |

(i) $y = 1 - x$

(ii) $y = |1 - x|$

(iii) $y = 2 + |1 - x|$

Solution

(i) $y = 1 - x$ is the straight line through $(0,1)$ and $(1,0)$.

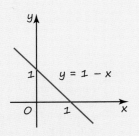

Figure 4.41

(ii) $y = |1 - x|$ is obtained by reflecting the part of the line for $x > 1$ in the x-axis, because that part is below the x-axis and so is negative.

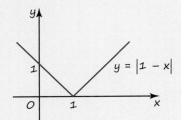

Figure 4.42

(iii) $y = 2 + |1 - x|$ is obtained from Figure 4.41 by applying the translation $\begin{pmatrix} 0 \\ 2 \end{pmatrix}$.

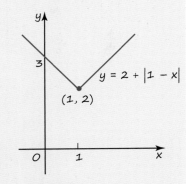

Figure 4.43

Discussion point

Look back at the graph of $y = |x|$ in Figure 4.40.

→ How does this show that $|x| < 2$ is equivalent to $-2 < x < 2$?

Inequalities involving the modulus sign

You will often meet inequalities involving the modulus sign.

| Example 4.13 | Solve the following. |

(i) $|x + 3| \leqslant 4$

(ii) $|2x - 1| > 9$

(iii) $5 - |x - 2| > 1$

Solution

(i) $|x + 3| \leqslant 4 \iff -4 \leqslant x + 3 \leqslant 4$

   $\iff -7 \leqslant x \leqslant 1$

> It is helpful to sketch the graph to check that your solution looks right.

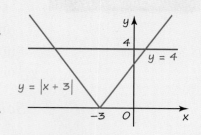

Figure 4.44

(ii) $|2x - 1| > 9 \iff 2x - 1 < -9 \quad \text{or} \quad 2x - 1 > 9$

> The solution represents two separate intervals, so it cannot be written as a single inequality.

   $\iff 2x < -8 \quad\quad \text{or} \quad 2x > 10$

   $\iff x < -4 \text{ or } x > 5$

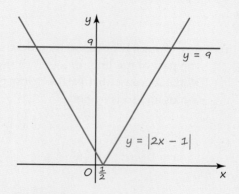

Figure 4.45

(iii) $5 - |x - 2| > 1 \iff 4 > |x - 2|$

   $\iff |x - 2| < 4$

   $\iff -4 < x - 2 < 4$

   $\iff -2 < x < 6$

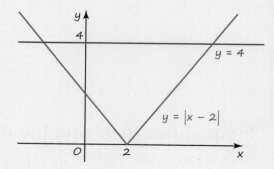

Figure 4.46

Example 4.14

Express the inequality $-2 < x < 6$ in the form $|x - a| < b$, where a and b are to be found.

Solution

$$|x - a| < b \quad \Leftrightarrow \quad -b < x - a < b$$
$$\Leftrightarrow \quad a - b < x < a + b$$

Comparing this with $-2 < x < 6$ gives

$$a - b = -2$$
$$a + b = 6.$$

Solving these simultaneously gives $a = 2, b = 4$, so $|x - 2| < 4$.

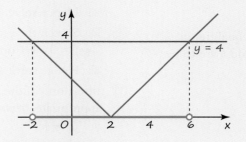

> Sketch the graph to check your solution.

Figure 4.47

Example 4.15

(i)   Sketch the graphs of $y = |x + 2|$ and $y = 2x + 1$ on the same axes.

(ii)   Hence solve the inequality $|x + 2| < 2x + 1$.

Solution

(i)

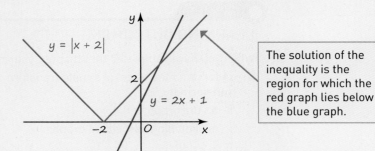

> The solution of the inequality is the region for which the red graph lies below the blue graph.

Figure 4.48

(ii)   From the graph, the critical point is where the line $y = 2x + 1$ meets the line $y = x + 2$. Solving these equations simultaneously:

$$2x + 1 = x + 2$$
$$\Rightarrow \quad x = 1$$

From the graph, the region required is given by $x > 1$.

Exercise 4.3

① Express $-3 < x < 3$ using modulus notation.

② Sketch each of the following graphs on a separate set of axes.

(i) $y = |x + 2|$

(ii) $y = |2x - 3|$

(iii) $y = |x| + 1$

③ Solve the following inequalities.

(i) $|x + 3| < 5$

(ii) $|x - 2| \leqslant 2$

(iii) $|x - 5| > 6$

(iv) $|x + 1| \geqslant 2$

④ Express each of the following inequalities in the form $|x - a| < b$, where a and b are to be found.

(i) $-1 < x < 3$

(ii) $2 < x < 8$

(iii) $-2 < x < 4$

⑤ Solve the following inequalities.

(i) $|2x - 3| < 7$

(ii) $|3x - 2| \leqslant 4$

(iii) $|2x + 3| > 5$

(iv) $|3x + 2| \geqslant 8$

⑥ Express each of the following inequalities in the form $|x - a| < b$, where a and b are to be found.

(i) $-1 < x < 6$

(ii) $9.9 < x < 10.1$

(iii) $0.5 < x < 7.5$

⑦ Sketch each of the following graphs on a separate set of axes.

(i) $y = |x + 2| - 2$

(ii) $y = |2x + 5| - 4$

(iii) $y = 3 + |x - 2|$

⑧ (i) Sketch the graphs of $y = |x + 3|$ and $y = 3x - 1$ on the same axes.

(ii) Hence solve $|x + 3| < 3x - 1$.

⑨ (i) Sketch the graphs of $y = |x - 3|$ and $y = 3x + 1$ on the same axes.

(ii) Hence solve $|x - 3| < 3x + 1$.

⑩ Insert one of the symbols $\Rightarrow$, $\Leftarrow$, or $\Leftrightarrow$, if appropriate, between these pairs of statements.

(i) $a^2 = b^2$ $|a| = |b|$

(ii) $|x - 3| > 4$ $(x - 3)^2 > 16$

(iii) $2x < |x - 1|$ $(2x)^2 < (x - 1)^2$

⑪ Solve $|x| > 2x - 1$.

⑫ Solve $|2x + 1| > |x - 2|$ by first sketching appropriate graphs.

⑬ Solve $|3x - 1| \leqslant |2x + 3|$.

○ **LEARNING OUTCOMES**

When you have completed this chapter, you should be able to:

➤ understand the effect of simple transformations, and combinations of transformations,

○ on the graph of f(x)

○ in sketching associated graphs

– $y = af(x)$

– $y = f(x) + a$

– $y = f(x + a)$

– $y = f(ax)$

➤ understand and use:

○ composite functions

○ inverse functions and their graphs

➤ sketch curves defined by simple equations involving the modulus function

➤ use functions in modelling:

○ consideration of limitations of the model

○ consideration of refinements of the model.

KEY POINTS

1 **Mappings and functions**
   - A mapping is any rule connecting input values (objects) and output values (images).
     It can be many-to-one, one-to-many, one-to-one or many-to-many.
   - A many-to-one or one-to-one mapping is called a function. It is a mapping for which each input value gives exactly one output value.
   - The domain of a mapping or function is the set of possible input values (values of x).
   - The range of a mapping or function is the set of output values.

2 **Transformations of the graphs of the function $y = f(x)$**

Table 4.3

Function	Transformation
$f(x) \rightarrow f(x - t) + s$	Translation $\begin{pmatrix} t \\ s \end{pmatrix}$
$f(x) \rightarrow af(x)$	One-way stretch, parallel to y-axis, scale factor a
$f(x) \rightarrow f(ax)$	One-way stretch, parallel to x-axis, scale factor $\dfrac{1}{a}$
$f(x) \rightarrow -f(x)$	Reflection in x-axis
$f(x) \rightarrow f(-x)$	Reflection in y-axis

When two transformations are combined, the order in which they are carried out matters if they are both in the same direction.

3 **Composite functions**
   A composite function is obtained when one function (say g) is applied to the output from another function (say f). The notation used is g[f(x)] or gf(x).

4 **Inverse functions**
   - For any one-to-one function f(x), there is an inverse function $f^{-1}(x)$.
   - The curves of a function and its inverse are reflections of each other in the line $y = x$.
     To illustrate this it is essential that the same scale is used on both axes.

5 **The modulus function**
   - The modulus of x, written $|x|$, means the positive value of x.
   - The modulus function is:
     $|x| = x,$ for $x \geqslant 0$
     $|x| = -x,$ for $x < 0$.
   - To sketch the modulus function, any part of the graph of the function that is below the x-axis is reflected in the x-axis.

FUTURE USES

You will use the language and concepts of functions and transformations throughout this book, particularly when you meet some new trigonometric functions in Chapter 6.

5

Differentiation

In the hourglass in the photo, the volume of sand in the bottom bulb is V, and the height of sand in the bottom bulb is h.

The rate of change of the volume of sand in the bottom bulb is given by $\dfrac{\mathrm{d}V}{\mathrm{d}t}$.

➔ What information is given by $\dfrac{\mathrm{d}h}{\mathrm{d}t}$ and by $\dfrac{\mathrm{d}V}{\mathrm{d}h}$?

➔ What information is given by $\dfrac{\mathrm{d}V}{\mathrm{d}h} \times \dfrac{\mathrm{d}h}{\mathrm{d}t}$?

Review: Differentiation

- The gradient of a curve at a point is given by the gradient of the tangent at the point.

- The gradient of the tangent at a point A on a curve is given by the limit of the gradient of chord AP as P approaches A along the curve.

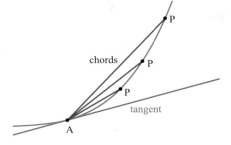

- The gradient function of a curve is the rate of change of y with respect to x and can be written as $\frac{dy}{dx}$.

Figure 5.1

- This is also called the **derived function** or **derivative**.

- In function notation, the derivative of $f(x)$ is written as $f'(x)$.

$$y = kx^n \qquad \Rightarrow \frac{dy}{dx} = nkx^{n-1}$$
$$y = f(x) + g(x) \Rightarrow \frac{dy}{dx} = f'(x) + g'(x)$$

For example, $\quad y = x^4 \Rightarrow \frac{dy}{dx} = 4x^3$

$$y = x^{-\frac{1}{2}} \Rightarrow \frac{dy}{dx} = -\tfrac{1}{2}x^{-\frac{3}{2}}.$$

Finding tangents and normals

You can use differentiation to find the equation of a tangent or a normal to a curve at any point.

Example 5.1

A curve has equation $y = x^4 - 20x + 1$.

(i) Find the gradient of the curve at the point where $x = 2$.

(ii) Find the equation of the tangent to the curve at this point.

(iii) Find the equation of the normal to the curve at this point.

Solution

(i) $\qquad y = x^4 - 20x + 1$

$\Rightarrow \dfrac{dy}{dx} = 4x^3 - 20$ ←— First differentiate to find the gradient function.

At $x = 2$, $\dfrac{dy}{dx} = 4(2)^3 - 20$ ←— Find the gradient when $x = 2$.

$\qquad\qquad = 12$

Therefore the gradient of the curve at the point where $x = 2$ is 12.

→

(ii)   When $x = 2, y = (2)^4 - 20(2) + 1$ ← Find the y coordinate when $x = 2$.

$\qquad = -23$

Equation of tangent is

$\qquad y - (-23) = 12(x - 2)$ ← Using $y - y_1 = m_1(x - x_1)$.

$\qquad 12x - y - 47 = 0.$

(iii)   Gradient of normal is $-\frac{1}{12}$. ← Using $m_2 = -\dfrac{1}{m_1}$.

Equation of normal is

$\qquad y - (-23) = -\frac{1}{12}(x - 2)$

$\qquad \Rightarrow 12y + x + 274 = 0.$

Increasing and decreasing functions, and stationary points

So when a function is increasing it has positive gradient, and when it is decreasing it has negative gradient.

A function is **increasing** for $a < x < b$ if $f'(x) > 0$ for $a < x < b$, and it is **decreasing** for $a < x < b$ if $f'(x) < 0$ for $a < x < b$.

At a stationary point, $\dfrac{dy}{dx} = 0$.

One method for determining the nature of a stationary point is to look at the sign of the gradient on either side of it.

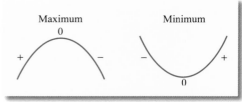

Another method for determining the nature of the stationary point is to consider the sign of the **second derivative**, $\dfrac{d^2 y}{dx^2}$, which is the rate of change of $\dfrac{dy}{dx}$ with respect to x.

- If $\dfrac{d^2 y}{dx^2} > 0$, the point is a **minimum**. If the second derivative is positive at the turning point then it means that the gradient function is increasing – it goes from being negative to zero and then positive.

- If $\dfrac{d^2 y}{dx^2} < 0$, the point is a **maximum**. If the second derivative is negative at the turning point then it means that the gradient function is decreasing – it goes from being positive to zero and then negative.

- If $\dfrac{d^2 y}{dx^2} = 0$, it is not possible to use this method to determine the nature of the stationary point.

Example 5.2

A function $f(x)$ is given by $f(x) = x^3 - 3x^2 - 9x + 20$.

(i)   Find the turning points of the curve $y = f(x)$, and determine their nature.

(ii) Sketch the curve.

(iii) State the range of values of x for which the function is increasing.

Solution

First differentiate.

(i) $\dfrac{dy}{dx} = 3x^2 - 6x - 9$

Turning points are where $\dfrac{dy}{dx} = 0$.

$3x^2 - 6x - 9 = 0$

$\Rightarrow 3(x - 3)(x + 1) = 0$

Substitute x values into the equation of the curve to find y values.

$\Rightarrow x = 3, x = -1$

When $x = 3$, $y = 3^3 - 3 \times 3^2 - 9 \times 3 + 20 = -7$.
When $x = -1$, $y = (-1)^3 - 3(-1)^2 - 9(-1) + 20 = 25$.

The turning points are $(3, -7)$ and $(-1, 25)$.

$\dfrac{d^2y}{dx^2} = 6x - 6$

Differentiate again to find the second derivative.

At $x = 3$, $\dfrac{d^2y}{dx^2} = 12 > 0$

Substitute x values into $\dfrac{d^2y}{dx^2}$ to see if turning points are maximum points or minimum points.

Therefore $(3, -7)$ is a minimum.

At $x = -1$, $\dfrac{d^2y}{dx^2} = -12 < 0$

Therefore $(-1, 25)$ is a maximum.

(ii)

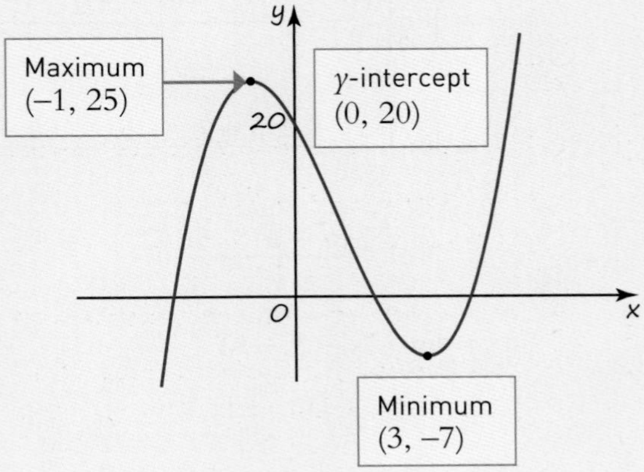

Maximum $(-1, 25)$

y-intercept $(0, 20)$

Minimum $(3, -7)$

Figure 5.2

(iii) From the sketch, $f(x)$ is increasing for $x < -1$, $x > 3$.

Differentiation from first principles

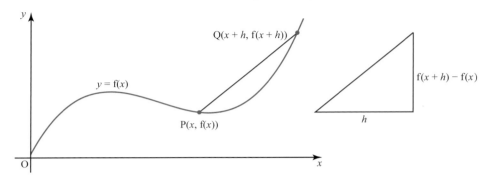

Figure 5.3

The gradient of the chord PQ in Figure 5.3 is given by $\dfrac{f(x + h) - f(x)}{h}$.

The derivative of $y = f(x)$ is defined as the limit of the gradient of the chord as h tends to 0.

This can be written formally as $f'(x) = \lim\limits_{h \to 0}\left(\dfrac{f(x + h) - f(x)}{h} \right)$.

This can also be expressed as $\dfrac{dy}{dx} = \lim\limits_{\delta x \to 0}\left(\dfrac{\delta y}{\delta x} \right)$, where δy is a small change in y and δx is a small change in x.

Example 5.3

Differentiate $y = x^2 - 5x$ from first principles.

Solution

$$\frac{dy}{dx} = \lim_{h \to 0}\left(\frac{\left((x + h)^2 - 5(x + h)\right) - \left(x^2 - 5x\right)}{h} \right)$$

$$= \lim_{h \to 0}\left(\frac{x^2 + 2xh + h^2 - 5x - 5h - x^2 + 5x}{h} \right)$$

$$= \lim_{h \to 0}\left(\frac{2xh + h^2 - 5h}{h} \right)$$

$$= \lim_{h \to 0}\left(2x - 5 + h \right)$$

$$= 2x - 5$$

① Find the gradient of each of the following curves at the point given.

(i) $y = x^2 - 2x + 1$, $(3,4)$

(ii) $y = 5x^3 - 6x + 7$, $(1,6)$

(iii) $y = 2\sqrt{x} - \dfrac{1}{x}$, $(1,1)$

② After t seconds, the number of thousands of bacteria, N, is modelled by the equation $N = 20t^3 - 8t^2 + 50t + 3$.
Find the rate of change of the number of bacteria with respect to time after 10 seconds.

③ Find the equations of the tangent and normal to the curve $y = x(2 - x)$ at the point on the curve where the gradient of the curve is -4.

④ Find $\dfrac{dy}{dx}$ and $\dfrac{d^2y}{dx^2}$ in the following cases.

(i) $y = 3x^4 - \dfrac{x^2}{2}$

(ii) $y = 2x^4 - \dfrac{x}{3} + 7$

(iii) $y = 2\sqrt{x} - x^2 + \dfrac{1}{x}$

⑤ Differentiate $y = x^2 - 3x + 2$ from first principles.

⑥ Sketch the gradient function of the curve in Figure 5.4.

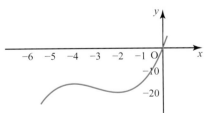

Figure 5.4

⑦ In each case, find the set of values of x for which y is increasing.

(i) $y = 2 - 4x - x^2$

(ii) $y = x^3 - 48x$

(iii) $y = x^2 - x^3$

⑧ Find $\dfrac{dy}{dx}$ and $\dfrac{d^2y}{dx^2}$ in the following cases.

(i) $y = \sqrt{x}\left(2x^2 - 4\right)$

(ii) $y = \dfrac{10x^5 - 3x^3}{2x^2}$

⑨ Given that $y = 3x - x^3$

(i) find $\dfrac{dy}{dx}$ and $\dfrac{d^2y}{dx^2}$

(ii) find the coordinates of any stationary points and identify their nature

(iii) sketch the curve.

⑩ Find the equation of the normal to the curve $y = \dfrac{2}{\sqrt{x}} + 4\sqrt{x}$ at the point where $x = 4$.

⑪ Find the coordinates of the turning point of the curve $y = 2x\sqrt{x} - 9x + 6$, $x \geqslant 0$, and determine its nature.

⑫ Given that $y = ax^3 + bx^2 + 15x - 2$ and that when $x = 1$, $y = 17$ and $\dfrac{dy}{dx} = 24$

(i) find a and b

(ii) show that the curve has no stationary points.

⑬ Given that $y = 3x^4 - 2x^6$

(i) find $\dfrac{dy}{dx}$ and $\dfrac{d^2y}{dx^2}$

(ii) find the coordinates of any stationary points and identify their nature

(iii) sketch the curve.

PS ⑭ Find the values of c for which the line $y = 5x + c$ is a tangent to the curve

$$y = \dfrac{x^3}{3} - \dfrac{x^2}{2} - x + 4.$$

PS ⑮ A rectangular sheet of cardboard measures 24 cm by 15 cm. Equal squares of side x cm are removed from each corner of the rectangle. The edges are then turned up to make an open box of volume V cm³.

Find the maximum possible volume of the box and the corresponding value of x.

1 The shape of curves

Stationary points of inflection

Figure 5.5 shows the curve $y = x^3 - 3x^2 + 3x + 1$.

Discussion point

→ What can you say about the curve at the point $(1, 2)$?

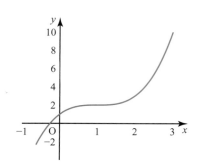

Figure 5.5

To understand the behaviour of the curve better, look at the first and second derivatives.

$$y = x^3 - 3x^2 + 3x + 1 \Rightarrow \frac{dy}{dx} = 3x^2 - 6x + 3$$

At the point $(1, 2)$, $\frac{dy}{dx} = 3 - 6 + 3$

$$= 0$$

So the gradient is zero at this point. ← You can see this from the graph.

$$\frac{d^2y}{dx^2} = 6x - 6$$

At the point $(1, 2)$, $\frac{d^2y}{dx^2} = 6 - 6$

$$= 0$$

So the second derivative is also zero at this point.

You can see from the graph that the gradient is positive on both sides of $(1, 2)$. You can verify this by working out the gradient of the curve at a point just to the left of $(1, 2)$ and a point just to the right of $(1, 2)$.

At $x = 0.9$, $\frac{dy}{dx} = 3(0.9)^3 - 6(0.9) + 3 > 0$.

At $x = 1.1$, $\frac{dy}{dx} = 3(1.1)^2 - 6(1.1) + 3 > 0$.

The point $(1, 2)$ is neither a maximum point nor a minimum point. It is a **stationary point of inflection**.

Having located stationary points, you can use the method of finding the gradient of the curve close to, and on either side of, the stationary point to identify whether a stationary point is a point of inflection.

If a stationary point is a point of inflection then the gradient will be either

- positive just before and positive just after the stationary point, as in Figure 5.6

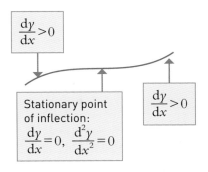

Figure 5.6

- or negative just before and negative just after the stationary point, as in Figure 5.7.

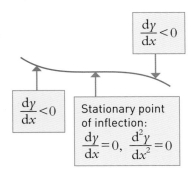

Figure 5.7

Example 5.4

You are given that $y = 3x^5 - 5x^3 - 2$.

(i)   Find the stationary points on the curve.

(ii) Identify the nature of the stationary points.

(iii) Sketch the curve.

Solution

(i) $y = 3x^5 - 5x^3 - 2 \implies \dfrac{dy}{dx} = 15x^4 - 15x^2$

> Stationary points are where $\dfrac{dy}{dx} = 0$.

At stationary points, $15x^4 - 15x^2 = 0$

$\implies \quad 15x^2(x^2 - 1) = 0$

$\implies 15x^2(x + 1)(x - 1) = 0$ ← Solve the equation.

$\implies x = 0, 1, -1$.

> Substitute x values into the original equation to find y coordinates of stationary points.

When $x = 0, \quad y = 3(0)^5 - 5(0)^3 - 2 = -2$.

When $x = 1, \quad y = 3(1)^5 - 5(1)^3 - 2 = -4$.

When $x = -1, \quad y = 3(-1)^5 - 5(-1)^3 - 2 = 0$.

Stationary points are $(0, -2), (1, -4)$ and $(-1, 0)$.

➡

(ii)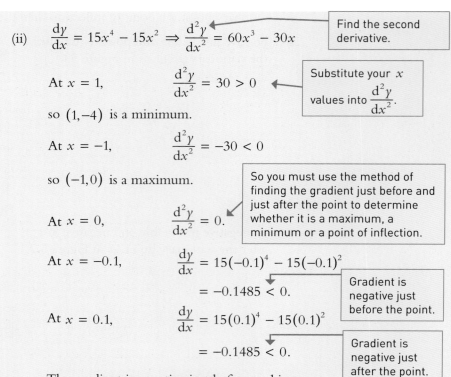

$$\frac{dy}{dx} = 15x^4 - 15x^2 \Rightarrow \frac{d^2y}{dx^2} = 60x^3 - 30x$$

Find the second derivative.

At $x = 1$, $\quad \frac{d^2y}{dx^2} = 30 > 0$

Substitute your x values into $\frac{d^2y}{dx^2}$.

so $(1, -4)$ is a minimum.

At $x = -1$, $\quad \frac{d^2y}{dx^2} = -30 < 0$

so $(-1, 0)$ is a maximum.

At $x = 0$, $\quad \frac{d^2y}{dx^2} = 0$.

So you must use the method of finding the gradient just before and just after the point to determine whether it is a maximum, a minimum or a point of inflection.

At $x = -0.1$, $\quad \frac{dy}{dx} = 15(-0.1)^4 - 15(-0.1)^2$

$$= -0.1485 < 0.$$

Gradient is negative just before the point.

At $x = 0.1$, $\quad \frac{dy}{dx} = 15(0.1)^4 - 15(0.1)^2$

$$= -0.1485 < 0.$$

Gradient is negative just after the point.

The gradient is negative just before and just after the point, so $(0, -2)$ is a point of inflection.

(iii)

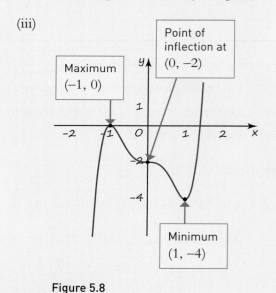

Maximum $(-1, 0)$

Point of inflection at $(0, -2)$

Minimum $(1, -4)$

Figure 5.8

Concave and convex curves

ACTIVITY 5.1

The section of curve shown in Figure 5.9 has a gradient that is initially positive. The gradient is increasing as x increases.

Figure 5.9

(i) Sketch sections of curves with the following properties:
  (a) gradient is initially negative and gradient is increasing (i.e. becoming less negative or going from negative to positive)
  (b) gradient is initially positive and gradient is decreasing
  (c) gradient is initially negative and gradient is decreasing (i.e. becoming more negative).

(ii) Describe the shape of:
  (a) the two curves (including Figure 5.9) with increasing gradient
  (b) the two curves with decreasing gradient.

When the gradient is increasing, the rate of change of the gradient is positive:

$$\frac{\mathrm{d}\left(\frac{\mathrm{d}y}{\mathrm{d}x}\right)}{\mathrm{d}x} = \frac{\mathrm{d}^2y}{\mathrm{d}x^2} > 0.$$

When the gradient is decreasing, the rate of change of the gradient is negative:

$$\frac{\mathrm{d}\left(\frac{\mathrm{d}y}{\mathrm{d}x}\right)}{\mathrm{d}x} = \frac{\mathrm{d}^2y}{\mathrm{d}x^2} < 0.$$

A section of curve with $\frac{\mathrm{d}^2y}{\mathrm{d}x^2} > 0$ is **concave upwards** (or convex downwards). If you join the points at the two ends of the curve with a straight line then the curve is under the line, as in Figure 5.10.

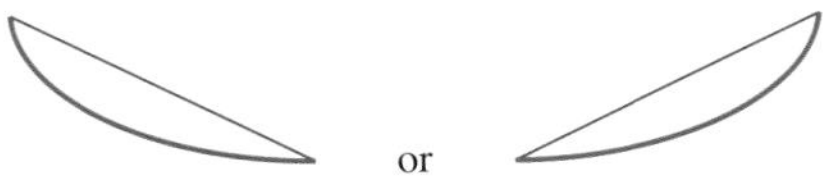

or

Figure 5.10

Discussion point

→ A section of curve has $\frac{\mathrm{d}^2y}{\mathrm{d}x^2} = 0$ at all points. How would you describe it? Justify your answer.

A section of curve with $\frac{\mathrm{d}^2y}{\mathrm{d}x^2} < 0$ is **concave downwards** (or convex upwards). If you join the points at the two ends of the curve with a straight line then the curve is above the line, as in Figure 5.11.

or

Figure 5.11

Example 5.5

Find the set of values of x for which the curve $y = x^3 - 6x^2 + 6x - 5$ is concave upwards.

Solution

$$y = x^3 - 6x^2 + 6x - 5$$

$$\Rightarrow \frac{dy}{dx} = 3x^2 - 12x + 6$$

Differentiate twice to find $\frac{d^2y}{dx^2}$.

$$\Rightarrow \frac{d^2y}{dx^2} = 6x - 12$$

The curve is concave upwards when $\frac{d^2y}{dx^2} > 0$

Substitute your expression for $\frac{d^2y}{dx^2}$.

$$6x - 12 > 0$$

$$6x > 12$$

Solve the inequality.

$$x > 2$$

The curve is concave upwards when $x > 2$.

You have already met stationary points of inflection. However, a point of inflection does not have to be a stationary point. More generally, a **point of inflection** is a point at which a curve goes from being concave upwards to concave downwards, or vice versa. So a point of inflection is a point at which $\frac{d^2y}{dx^2}$ goes from being positive to being negative, or from being negative to being positive. At the point of inflection itself, $\frac{d^2y}{dx^2} = 0$.

Discussion point

The statement 'Point A is a point of inflection $\Rightarrow \frac{d^2y}{dx^2} = 0$ at point A' is true.

→ Is the converse statement '$\frac{d^2y}{dx^2} = 0$ at point A $\Rightarrow$ point A is a point of inflection' also true?

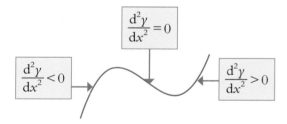

Figure 5.12

In Figure 5.12, on the left the gradient is decreasing; the curve is concave downwards. On the right the gradient is increasing; the curve is concave upwards. $\frac{d^2y}{dx^2}$ goes from negative to positive, and in between there is a point where it is zero. This is a point of inflection. However, it is clear from the diagram that the gradient of the curve is not zero at this point, and therefore it is a non-stationary point of inflection.

Example 5.6

You are given that $y = x^3 - 6x^2 + 15x - 10$.

(i) Find the coordinates of the point at which the second derivative is zero, and show that it is a point of inflection.

(ii) Find the gradient of the curve at the point of inflection.

(iii) Show that the curve has no turning points.

(iv) Find the coordinates of the point where the curve crosses the x-axis.

(v) Sketch the curve.

Solution

(i)
$$y = x^3 - 6x^2 + 15x - 10$$

$$\Rightarrow \frac{dy}{dx} = 3x^2 - 12x + 15$$

> Differentiate twice to find $\frac{d^2y}{dx^2}$.

$$\Rightarrow \frac{d^2y}{dx^2} = 6x - 12$$

At the point of inflection,

$$\frac{d^2y}{dx^2} = 0$$

$$6x - 12 = 0$$

$$x = 2$$

> Substitute the x value into the original equation to find the y coordinate of point of inflection.

When $x = 2$, $y = (2)^3 - 6(2)^2 + 15(2) - 10 = 4$.

The point where $\frac{d^2y}{dx^2} = 0$ is $(2, 4)$.

At $x = 1.9$, $\frac{d^2y}{dx^2} = 6 \times 1.9 - 12 < 0$.

At $x = 2.1$, $\frac{d^2y}{dx^2} = 6 \times 2.1 - 12 > 0$.

> To show that this is a point of inflection, evaluate $\frac{d^2y}{dx^2}$ at a point just before and a point just after $(2, 4)$.

Therefore the curve goes from being concave downwards to concave upwards at $(2, 4)$, and so it is a point of inflection.

(ii) When $x = 2$, $\frac{dy}{dx} = 3(2)^2 - 12(2) + 15 = 3$.

The gradient of the curve is 3 at the point of inflection.

(iii)
$$\frac{dy}{dx} = 3x^2 - 12x + 15$$

$$3x^2 - 12x + 15 = 0$$

> At a turning point, $\frac{dy}{dx} = 0$.

$$\Rightarrow x^2 - 4x + 5 = 0$$

$$b^2 - 4ac = (-4)^2 - 4 \times 1 \times 5$$

$$= -4$$

Since the discriminant of the quadratic is negative, it has no real solutions. So the curve has no turning points.

$\rightarrow$

TECHNOLOGY

Use graphing software to draw this curve. Add a point to the curve and construct a tangent through this point. Move the point along the curve. What happens to the tangent as it passes through the point of inflection?

(iv) $x^3 - 6x^2 + 15x - 10 = 0$ ← The curve crosses the x-axis when $y = 0$.

$f(x) = x^3 - 6x^2 + 15x - 10$ ← Using the factor theorem to solve the cubic equation.

$f(0) = -10$ ← $x = 0$ is not a root.

$f(1) = 0$ ← $x = 1$ is a root.

Because the curve has no turning points it only crosses the x-axis in one place. So the curve crosses the axis at $(1,0)$.

(v) You know that the curve has a non-stationary point of inflection at $(2, 4)$ and that it has no turning points.

You also know that the curve crosses the y-axis at $(0,-10)$, and the x-axis at $(1,0)$.

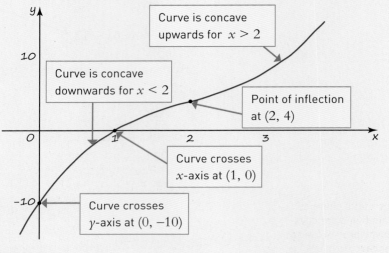

Curve is concave upwards for $x > 2$

Curve is concave downwards for $x < 2$

Point of inflection at $(2, 4)$

Curve crosses x-axis at $(1, 0)$

Curve crosses y-axis at $(0, -10)$

Figure 5.13

Example 5.7

You are given that $y = x^4 - 2x^3$.

(i) Find the coordinates of the stationary points of the curve and determine their nature.

(ii) Find the coordinates of the point at which the second derivative is zero, and show that it is a point of inflection.

(iii) Sketch the curve.

Solution

(i) $y = x^4 - 2x^3 \Rightarrow \dfrac{dy}{dx} = 4x^3 - 6x^2$

$4x^3 - 6x^2 = 0$ ← Stationary points are where $\dfrac{dy}{dx} = 0$.

$2x^2(2x - 3) = 0$

$x = 0, \ x = \frac{3}{2}$

When $x = 0$, $\quad y = 0$.

When $x = \frac{3}{2}$, $\quad y = -\frac{27}{16}$.

> Substituting x values into the original equation.

The curve has stationary points at $(0,0)$ and $\left(\frac{3}{2}, -\frac{27}{16}\right)$.

$$\frac{dy}{dx} = 4x^3 - 6x^2 \Rightarrow \frac{d^2y}{dx^2} = 12x^2 - 12x$$

> To determine the nature of the stationary points find the second derivative.

When $x = \frac{3}{2}$, $\qquad \frac{d^2y}{dx^2} = 9 > 0$

therefore $\left(\frac{3}{2}, -\frac{27}{16}\right)$ is a minimum point.

When $x = 0$, $\qquad \frac{d^2y}{dx^2} = 0$.

When $x = -0.1$, $\qquad \frac{dy}{dx} = -\frac{8}{125} < 0$.

> Since $\frac{d^2y}{dx^2} = 0$, you must use the method of looking at the gradient at a point just before and a point just after the stationary point to determine its nature.

When $x = 0.1$, $\qquad \frac{dy}{dx} = -\frac{7}{125} < 0$.

The gradient is negative before and after the stationary point, so $(0, 0)$ is a stationary point of inflection.

(ii)
$$\frac{d^2y}{dx^2} = 12x^2 - 12x$$

$$12x^2 - 12x = 0$$

$$12x(x - 1) = 0$$

$$\Rightarrow x = 0, x = 1$$

> You already know that there is a stationary point of inflection at $x = 0$.

When $x = 0.9$, $\quad \frac{d^2y}{dx^2} = -\frac{27}{25} < 0$.

When $x = 1.1$, $\quad \frac{d^2y}{dx^2} = \frac{33}{25} > 0$.

> To check that there is a point of inflection at $x = 1$, you must look at the sign of $\frac{d^2y}{dx^2}$ at a point just before and a point just after $x = 1$.

So $x = 1$ is a non-stationary point of inflection.

When $x = 1$, $\quad y = -1$

so the coordinates of the non-stationary point of inflection are $(1, -1)$.

> The curve changes from being concave downwards to being concave upwards at $x = 1$.

Note

It is important to remember that not all points of inflection are stationary points.

To show that a non-stationary point is a point of inflection, it is not enough to show that $\frac{d^2y}{dx^2} = 0$ at the point: you must show that the sign of $\frac{d^2y}{dx^2}$ changes from positive to negative, or vice versa, at the point.

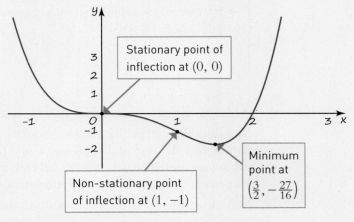

Stationary point of inflection at $(0, 0)$

Non-stationary point of inflection at $(1, -1)$

Minimum point at $\left(\frac{3}{2}, -\frac{27}{16}\right)$

Figure 5.14

Exercise 5.1

① Explain the difference between a stationary point of inflection and a non-stationary point of inflection.

② Find the set of values of x for which the following curves are concave downwards.

(i) $y = x^3 - 9x^2 + 6x - 1$

(ii) $y = x^3 + 21x^2 - 24x + 9$

(iii) $y = -2x^3 + 3x^2 - x - 2$

③ Find the set of values of x for which the following curves are concave upwards.

(i) $y = x^3 + x^2 - 19x - 41$

(ii) $y = x^3 - 12x^2 - 11x + 5$

(iii) $y = -x^3 + 2x^2 + 7x - 17$

④ For each of the following, find the coordinates and nature of any stationary points and sketch the curve.

(i) $y = -x^3 + 3x^2 - 3x - 3$

(ii) $y = 8x^3 + 12x^2 + 6x + 2$

(iii) $y = -x^3 - 6x^2 - 12x$

⑤ Find the set of values of x for which the following curves are concave upwards.

(i) $y = x^4 - 4x^3 - 18x^2 + 11x + 1$

(ii) $y = -x^4 + 2x^3 + 12x^2 + 5x - 7$

⑥ Given that $y = x^3 - x^2$

(i) find $\dfrac{dy}{dx}$ and $\dfrac{d^2y}{dx^2}$

(ii) find the coordinates of the stationary points and determine their nature

(iii) show that there is a point of inflection and find its coordinates

(iv) find the gradient of the curve at the point of inflection

(v) sketch the curve.

⑦ Given that $y = 4x^3 - x^4$

(i) find the stationary points on the curve and determine their nature

(ii) sketch the curve

(iii) find the coordinates of the non-stationary point of inflection.

⑧ Given that $y = 3x^4 - 8x^3 + 6x^2 - 3$

(i) find the stationary points on the curve and determine their nature

(ii) sketch the curve

(iii) find the coordinates of the non-stationary point of inflection.

PS ⑨ Find the coordinates of all the turning points and points of inflection of the curve $y = 3x^5 - 25x^3 + 60x$, and sketch the curve.

2 The chain rule

> **Discussion point**
>
> → How would you differentiate these expressions?
>
> (i) $y = \left(x^2 + 1\right)^2$
>
> (ii) $y = \left(x^2 + 1\right)^{12}$
>
> (iii) $y = \sqrt{x^2 + 1}$
>
> (iv) $y = \left(x^2 + 1\right)^{-1}$

You can differentiate the first two expressions by expanding the brackets, although in the case of part (ii) this would be very tedious! However, you cannot use this approach in parts (iii) and (iv). Instead you need to think of the expression as a **composite function**, that is, a 'function of a function'.

You can separate out the two functions, and use the notation u for the first function to be applied.

In the case of part (iii), $u = x^2 + 1$ and $y = u^{\frac{1}{2}}$.

ACTIVITY 5.2

Copy and complete Table 5.1.

Table 5.1

$y = f(x)$	u	$y = f(u)$
$y = \tan\left(\dfrac{1}{x^3}\right)$	$u = \dfrac{1}{x^3}$	$y = \tan u$
(i) $y = \left(x^3 + 3\right)^4$		
(ii) $y = \sin\left(x^2 - 4\right)$		
(iii) $y = e^{2x+5}$		
(iv) $y = \ln\left(x^2 - 2x\right)$		

This row shows an example.

You will learn to differentiate functions involving exponentials, logarithms and trigonometric functions in Chapter 9.

You can differentiate composite functions such as these using the **chain rule**.

In the review section you saw that

$$\frac{dy}{dx} = \lim_{\delta x \to 0}\left(\frac{\delta y}{\delta x}\right),$$

It is important to remember that while $\dfrac{\delta y}{\delta x}$ means 'δy divided by δx', $\dfrac{dy}{dx}$ is not a fraction but a notation for the rate of change of y with respect to x.

where δy is a small change in y and δx is a small change in x.

Think about the hourglass shown at the beginning of the chapter. A small change in time, δt, leads to a small change in the height of sand in the bottom bulb, δh, and to the volume of sand in the bottom bulb, δV.

You can see that

$$\frac{\delta V}{\delta t} = \frac{\delta V}{\delta h} \times \frac{\delta h}{\delta t}$$

Just simple algebra, as these terms are 'real' fractions.

Taking the limit as δt tends to zero gives us the relationship

$$\frac{dV}{dt} = \frac{dV}{dh} \times \frac{dh}{dt}$$

This is usually written using the variables x, y and u:

This result is known as the **chain rule**.

$$\frac{dy}{dx} = \frac{dy}{du} \times \frac{du}{dx}$$

Although at first glance this may look obvious, you must remember that these terms are not actually fractions.

Example 5.8

(i)   Differentiate $y = \left(x^3 - 3\right)^2$

    (a)   by expanding the brackets

    (b)   by using the chain rule.

(ii)   Show that the two methods give the same result.

➜

Solution

(i) (a) $y = (x^3 - 3)^2$

$= (x^3 - 3)(x^3 - 3)$ ← First expand the brackets.

$= x^6 - 6x^3 + 9$

$\Rightarrow \dfrac{dy}{dx} = 6x^5 - 18x^2$ ← Now differentiate.

(b) Let $u = x^3 - 3$, then $y = u^2$. ← Expressing $y = (x^3 - 3)^2$ as a function of a function.

$\dfrac{dy}{du} = 2u$

$= 2(x^3 - 3)$ ← Differentiating y with respect to u.

$\dfrac{du}{dx} = 3x^2$ ← Differentiating u with respect to x.

By the chain rule

$\dfrac{dy}{dx} = \dfrac{dy}{du} \times \dfrac{du}{dx}$

$\dfrac{dy}{dx} = 2(x^3 - 3) \times 3x^2$

$= 6x^2(x^3 - 3)$

(ii) $6x^2(x^3 - 3) = 6x^5 - 18x^2$ ← Expanding the answer to (i) (b).

Therefore the two methods do give the same result.

Example 5.9

Differentiate $y = (x^2 + 1)^{\frac{1}{2}}$.

Solution

Let $u = x^2 + 1$, then $y = u^{\frac{1}{2}}$. ← Expressing $y = (x^2 + 1)^{\frac{1}{2}}$ as a function of a function.

$\dfrac{dy}{du} = \dfrac{1}{2}u^{-\frac{1}{2}}$ ← Differentiating y with respect to u.

$= \dfrac{1}{2\sqrt{u}}$

$= \dfrac{1}{2\sqrt{x^2 + 1}}$ ← Since the question was in terms of y and x not u, your answer must also be given in terms of x.

$\dfrac{du}{dx} = 2x$ ← Differentiating u with respect to x.

By the chain rule

$\dfrac{dy}{dx} = \dfrac{dy}{du} \times \dfrac{du}{dx}$

$\dfrac{dy}{dx} = \dfrac{1}{2\sqrt{x^2 + 1}} \times 2x$

$= \dfrac{x}{\sqrt{x^2 + 1}}$

In the following example you need to find the gradient at a point, so you must first find the gradient function. You could do this by expanding $\left(x^3 - 2\right)^4$, but it is much quicker to use the chain rule.

Example 5.10

Find the gradient of the curve $y = \left(x^3 - 2\right)^4$ at the point where $x = 1$.

Solution

Let $u = x^3 - 2$, then $y = u^4$. ← First express this composite function as a function of a function.

$$\frac{dy}{du} = 4u^3$$

$$= 4\left(x^3 - 2\right)^3$$ ← Remember to rewrite in terms of x.

$$\frac{du}{dx} = 3x^2$$

By the chain rule

$$\frac{dy}{dx} = \frac{dy}{du} \times \frac{du}{dx}$$

This is the gradient function of the curve. You must now find the gradient at $x = 1$.

$$\Rightarrow \quad \frac{dy}{dx} = 4\left(x^3 - 2\right)^3 \times 3x^2$$

$$= 12x^2 \left(x^3 - 2\right)^3$$

Substituting $x = 1$ into the expression for $\frac{dy}{dx}$.

$$x = 1 \quad \Rightarrow \quad \frac{dy}{dx} = 12(1)^2 \left((1)^3 - 2\right)^3 = 12 \times -1 = -12$$

Therefore the gradient of the curve $y = \left(x^3 - 2\right)^4$ at the point where $x = 1$ is -12.

With practice, you may find that you can do the necessary substitution mentally, and just write down the answer, leading to the result below. If you have any doubt, however, you should write down the full method.

Discussion point

The general rule for differentiating a composite function can be written as

$$y = g\left(f\left(x\right)\right) \Rightarrow \frac{dy}{dx} = g'\left(f\left(x\right)\right) \times f'\left(x\right).$$

→ Put this rule into words.

→ Do you find this rule helpful compared with using $\dfrac{dy}{dx} = \dfrac{dy}{du} \times \dfrac{du}{dx}$?

Exercise 5.2

① Which of these two functions would you differentiate using the chain rule? Explain your answer.

$$f(x) = 2x(1 + x)^9 \qquad f(x) = 2(1 + x)^9$$

② To differentiate each of the following functions, decide whether you

A   must use the chain rule

B   could use the chain rule

C   would not use the chain rule.

If you choose B or C, describe an alternative method for differentiating the function.

(i) $y = (2x - 5)^4$

(ii) $y = x^3 - 3x + 2$

(iii) $y = \dfrac{1}{\sqrt{x^2 + 3}}$

③ Use the chain rule to differentiate the following functions.

(i) $y = (3x + 2)^{-1}$

(ii) $y = (2x^2 + 6)^7$

(iii) $y = \sqrt{6x - 2}$

④ $y = (x^2 - 2)^3$

(i)   By first expanding the brackets, find $\dfrac{dy}{dx}$.

(ii) Now use the chain rule to find $\dfrac{dy}{dx}$.

(iii) Show that your answers to (i) and (ii) are equal.

⑤ (i)   Figure 5.15 shows the curve $y = (4x^2 - 3)^6$.

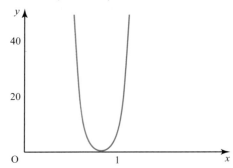

Figure 5.15

Find the gradient of the curve at the point where $x = 1$.

(ii) Figure 5.16 shows the curve $y = \sqrt{6x - 8}$.

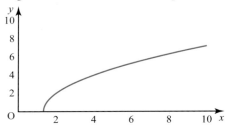

Figure 5.16

Find the gradient of the curve at the point where $x = 2$.

(iii) Figure 5.17 shows the curve $y = \dfrac{1}{x^3 + 2}$.

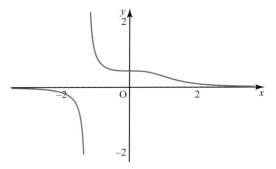

Figure 5.17

Find the gradient of the curve at the point where $x = -1$.

⑥ Figure 5.18 shows the curve $y = (3x - 5)^3$.

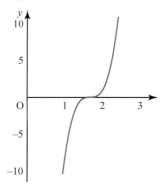

Figure 5.18

(i)   Use the chain rule to find $\dfrac{dy}{dx}$.

(ii) Find the equation of the tangent to the curve at the point where $x = 2$.

(iii) Show that the equation of the normal to the curve at the point where $x = 1$ can be written in the form $36y + x + 287 = 0$.

⑦ (i) Given that $x = \left(\sqrt{t} - 1\right)^4$, find $\dfrac{dx}{dt}$.

(ii) Given that $z = \sqrt{\dfrac{1}{y} + y}$, find $\dfrac{dz}{dy}$.

(iii) Given that $p = \dfrac{1}{2r\sqrt{r} - 6}$, find $\dfrac{dp}{dr}$.

⑧ Given that $y = (2x - 1)^4$

(i) find $\dfrac{dy}{dx}$

(ii) find the coordinates of any turning points and determine their nature

(iii) sketch the curve.

⑨ Given that $y = \left(x^2 - x - 2\right)^4$

(i) find $\dfrac{dy}{dx}$

(ii) find the coordinates of any turning points and determine their nature

(iii) sketch the curve.

PS ⑩ The graph of $y = \left(x^3 - x^2 + 2\right)^3$ is shown in Figure 5.19.

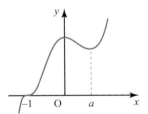

Figure 5.19

(i) Find the gradient function $\dfrac{dy}{dx}$.

(ii) Verify, showing your working clearly, that when $x = -1$ the curve has a stationary point of inflection and that when $x = 0$ the curve has a maximum.

(iii) The curve has a minimum when $x = a$. Find a and verify that this corresponds to a minimum.

(iv) Find the gradient at $(1, 8)$ and the equation of the tangent to the curve at this point.

PS ⑪ Some students on an expedition reach the corner of a large rectangular area of heathland, which is 4 km long and 2 km wide. They need to reach the opposite corner as quickly as possible as they are behind schedule.

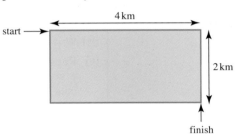

Figure 5.20

They estimate that they could walk along the edges of the heath at $5\,\text{km}\,\text{h}^{-1}$ and across the heath at $4\,\text{km}\,\text{h}^{-1}$.

Giving your answers to the nearest minute, how long will it take the students to cross the heath if

(i) they walk along the edges of the heath

(ii) they cross the heath diagonally from start to finish?

The students decide to walk some of the way along the longer edge, and then cross diagonally to the finish point, in such a way as to make their time as small as possible.

(iii) To the nearest metre, how far along the edge should they walk and, to the nearest minute, how long does it take them to reach the finish point?

3 Connected rates of change

The relationship between $\frac{dy}{dx}$ and $\frac{dx}{dy}$

ACTIVITY 5.3

(i) Differentiate $y = x^3$ to find $\frac{dy}{dx}$.

(ii) Rearrange to make x the subject, and find $\frac{dx}{dy}$ in terms of y.

(iii) Rearrange $y = x^3$ to make x the subject.

(iv) Write down the relationship between $\frac{dy}{dx}$ and $\frac{dx}{dy}$ for $y = x^3$.

(v) Repeat this process with other functions such as $y = x^2$, $y = x^4$ and $y = 2x$. Does the relationship you wrote down for (iv) seem to be a general rule?

The general rule connecting $\frac{dy}{dx}$ and $\frac{dx}{dy}$ is

$$\frac{dx}{dy} = \frac{1}{\frac{dy}{dx}}$$

Once again, although this may look obvious, you must remember that these terms are not actually fractions.

Discussion point

$\frac{dy}{dx}$ is the rate of change of y with respect to x, and $\frac{dx}{dy}$ is the rate of change of x with respect to y. In Figure 5.21 the line $y = 2x$ is drawn in two different ways, first with the axes the normal way around, and then with the axes interchanged.

→ How does this demonstrate the result $\frac{dx}{dy} = \frac{1}{\frac{dy}{dx}}$?

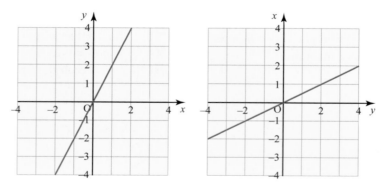

Figure 5.21

Finding connected rates of change

The chain rule makes it possible to differentiate with respect to a variable that does not feature in the original expression. For example, the volume V of a spherical balloon of radius r is given by $V = \frac{4}{3}\pi r^3$. Differentiating this with respect to r gives the rate of change of volume with radius, $\frac{dV}{dr} = 4\pi r^2$.

However, you might be more interested in finding the rate of change of volume with time, t.

To find this, you would use the chain rule:

$$\frac{dV}{dt} = \frac{dV}{dr} \times \frac{dr}{dt}$$

$$= 4\pi r^2 \times \frac{dr}{dt}$$

You have now differentiated V with respect to t.

Example 5.11

The radius $r\,\mathrm{cm}$ of a circular ripple made by dropping a stone into a pond is increasing at a rate of $8\,\mathrm{cm\,s^{-1}}$.

At what rate is the area $A\,\mathrm{cm^2}$ enclosed by the ripple increasing when the radius is $25\,\mathrm{cm}$?

Figure 5.22

Solution

When $r = 25$, $\dfrac{dr}{dt} = 8$. ← Information given in the question.

$A = \pi r^2$ ← The ripple is a circle.

$\dfrac{dA}{dr} = 2\pi r$

You want to know $\dfrac{dA}{dt}$, the rate of change of area with respect to time.

You know that $\qquad \dfrac{dA}{dt} = \dfrac{dA}{dr} \times \dfrac{dr}{dt}$ ← Use the chain rule to link the information you have to what you are trying to find out.

so $\qquad\qquad \dfrac{dA}{dt} = 2\pi r \times \dfrac{dr}{dt}$.

When $r = 25$ and $\dfrac{dr}{dt} = 8$, $\quad \dfrac{dA}{dt} = 2\pi \times 25 \times 8 = 1260$.

The area is increasing at a rate of $1260\,\mathrm{cm^2\,s^{-1}}$ (3 s.f.).

Exercise 5.3

① Given that $\dfrac{dy}{dx} = \dfrac{2}{3}$, write down $\dfrac{dx}{dy}$.

② Given that $\dfrac{dy}{dx} = 6$ and $\dfrac{dx}{dt} = -2$, find the following.

(i) $\dfrac{dx}{dy}$ (ii) $\dfrac{dt}{dx}$ (iii) $\dfrac{dy}{dt}$

③ Given that $\dfrac{dy}{dx} = \dfrac{1}{3}$, $\dfrac{dz}{dx} = -1$ and $\dfrac{dy}{dt} = -\dfrac{1}{2}$, find the following.

(i) $\dfrac{dy}{dz}$ (ii) $\dfrac{dx}{dt}$ (iii) $\dfrac{dz}{dt}$

④ Given that $A = 5x^2$ and $\dfrac{dx}{dt} = \dfrac{1}{2}$, find the following.

(i) $\dfrac{dA}{dx}$

(ii) An expression for $\dfrac{dA}{dt}$ in terms of x.

(iii) The value of $\dfrac{dA}{dt}$ when $x = 6$.

PS ⑤ The lengths of the sides of a square are increasing at a rate of $2\,\text{cm}\,\text{s}^{-1}$.

Find the rate at which the area of the square is increasing when the square has sides of length $4\,\text{cm}$.

⑥ The radius of a circular fungus is increasing at a uniform rate of $0.5\,\text{cm}$ per day. At what rate is the area increasing when the radius is $1\,\text{m}$?

PS ⑦ The force, F newtons, between two magnetic poles is given by the formula $F = \dfrac{1}{500r^2}$, where r m is their distance apart.

Find the rate of change of the force when the poles are $0.2\,\text{m}$ apart and the distance between them is increasing at a rate of $0.03\,\text{m}\,\text{s}^{-1}$.

PS ⑧ The area of a circular oil slick is increasing at a constant rate of $4\,\text{m}^2\,\text{s}^{-1}$.

Find the rate of increase of the radius when the area is $100\pi\,\text{m}^2$.

PS ⑨ The volume, $V\,\text{m}^3$, of water in a container is given by the expression $V = 8h^2$, where $h\,\text{m}$ is the depth of water in the container. The volume of water in the container is increasing at a rate of $2\,\text{m}^3$ per hour.

Find the rate of increase of the depth of water in the container when $h = 6$.

PS ⑩ The lengths of the sides of a cube are increasing at the rate of $0.1\,\text{cm}\,\text{s}^{-1}$.

Find the rate of increase of the volume of the cube when the outer surface area is $24\,\text{cm}^2$.

PS ⑪ The volume of a sphere is decreasing at a rate of $60\,\text{mm}^3\,\text{s}^{-1}$. Find the rate of decrease of the surface area when the volume is $800\,\text{mm}^3$.

4 The product and quotient rules

The product rule

Figure 5.23 shows the curve $y = 20x(x-1)^6$.

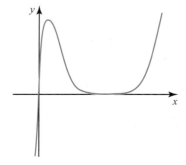

Figure 5.23

If you wanted to find the gradient function, $\dfrac{dy}{dx}$, for the curve, you could expand the right-hand side then differentiate it term by term, but this is long and tedious.

There are other functions like this, made up of the product of two or more simpler functions, which you cannot expand. One such function is

$$y = (x + 1)^{\frac{1}{2}}(2x - 3)^4.$$

Clearly you need a technique for differentiating functions that are products of simpler ones, and a suitable notation with which to express it.

When differentiating products, the most commonly used notation involves writing $y = uv$, where u and v are functions of x.

In the example above, $u = (x + 1)^{\frac{1}{2}}$ and $v = (2x - 3)^4$.

ACTIVITY 5.4

Jack says that if $y = uv$, $\dfrac{dy}{dx} = \dfrac{du}{dx} \times \dfrac{dv}{dx}$.

Use the function $y = x^3 \times x^6$ to show that Jack is wrong.

(i)   Simplify the function and differentiate to find $\dfrac{dy}{dx}$.

(ii) Identify u and v, and find $\dfrac{du}{dx}$ and $\dfrac{dv}{dx}$.

(iii) Confirm that, in this case, $\dfrac{dy}{dx} \neq \dfrac{du}{dx} \times \dfrac{dv}{dx}$.

To find the rule to differentiate products, you can look at the change in area of a rectangle with increasing sides (see Figure 5.24).

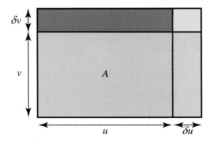

Figure 5.24

The original (blue) rectangle has an area of $y = uv$.

When you make a small increase to u and to v, the increase in the area of the rectangle is given by the sum of the areas of the red, green and yellow rectangles:

$$\delta A = u\delta v + v\delta u + \delta u \delta v.$$

Because both δu and δv are small, the yellow rectangle has a very small area. So

Divide through by δx. $\longrightarrow$ $\dfrac{\delta A}{\delta x} = u\dfrac{\delta v}{\delta x} + v\dfrac{\delta u}{\delta x} + \delta u\dfrac{\delta v}{\delta x}.$

Taking the limit as δx, δu and δv tend to zero leads to

$$\frac{\mathrm{d}A}{\mathrm{d}x} = u\frac{\mathrm{d}v}{\mathrm{d}x} + v\frac{\mathrm{d}u}{\mathrm{d}x}.$$

This result is called the **product rule**.

$$y = uv \implies \frac{\mathrm{d}y}{\mathrm{d}x} = u\frac{\mathrm{d}v}{\mathrm{d}x} + v\frac{\mathrm{d}u}{\mathrm{d}x}$$

Example 5.12

Differentiate $y = 20x(x-1)^6$.

Solution

First identify u and v.

Differentiate u and v, using the chain rule if necessary.

$u = 20x$	$v = (x-1)^6$
$\dfrac{\mathrm{d}u}{\mathrm{d}x} = 20$	$\dfrac{\mathrm{d}v}{\mathrm{d}x} = 6(x-1)^5$

Using the product rule

$$\frac{\mathrm{d}y}{\mathrm{d}x} = u\frac{\mathrm{d}v}{\mathrm{d}x} + v\frac{\mathrm{d}u}{\mathrm{d}x}$$

$$\frac{\mathrm{d}y}{\mathrm{d}x} = 20x \times 6(x-1)^5 + (x-1)^6 \times 20$$

To simplify the answer, look for common factors. In this case both terms contain 20 and $(x-1)^5$.

$$\implies \frac{\mathrm{d}y}{\mathrm{d}x} = 20(x-1)^5\left[6x + (x-1)\right]$$

Factorise by taking these common factors out.

$$\implies \frac{\mathrm{d}y}{\mathrm{d}x} = 20(x-1)^5(7x-1)$$

The quotient rule

ACTIVITY 5.5

If $y = \dfrac{u}{v}$, does

$$\frac{\mathrm{d}y}{\mathrm{d}x} = \frac{\dfrac{\mathrm{d}u}{\mathrm{d}x}}{\dfrac{\mathrm{d}v}{\mathrm{d}x}}?$$

Use the function $y = \dfrac{x^7}{x^3}$ to investigate.

Suppose you want to differentiate a function that is the quotient of two simpler functions, for example $y = \dfrac{3x+1}{\left(x^2-2\right)^2}$. You can write this in the form $y = \dfrac{u}{v}$, where u and v are functions of x. In this example, $u = 3x+1$ and

$$v = \left(x^2 - 2\right)^2.$$

To find the rule to differentiate quotients, you can rewrite $y = \dfrac{u}{v}$ as $y = uv^{-1}$, and use the product rule.

Because v is a function of x, to differentiate v^{-1}, you must use the chain rule.

$$\frac{\mathrm{d}\left(v^{-1}\right)}{\mathrm{d}x} = -v^{-2} \times \frac{\mathrm{d}v}{\mathrm{d}x}$$

You might find it easier to see this if you put brackets around the v: $v^{-1} = (v)^{-1}$.

u	v^{-1}
$\dfrac{\mathrm{d}u}{\mathrm{d}x}$	$-v^{-2}\,\dfrac{\mathrm{d}v}{\mathrm{d}x}$

So $\qquad y = \dfrac{u}{v} \Rightarrow \dfrac{\mathrm{d}y}{\mathrm{d}x} = v^{-1}\dfrac{\mathrm{d}u}{\mathrm{d}x} + \left(-v^{-2}\dfrac{\mathrm{d}v}{\mathrm{d}x}\right)u$ ← Applying the product rule.

$$= \frac{1}{v}\frac{\mathrm{d}u}{\mathrm{d}x} - \frac{u}{v^2}\frac{\mathrm{d}v}{\mathrm{d}x}$$

$$= \frac{v\dfrac{\mathrm{d}u}{\mathrm{d}x} - u\dfrac{\mathrm{d}v}{\mathrm{d}x}}{v^2}$$ ← Putting fractions over a common denominator.

$$\frac{\mathrm{d}y}{\mathrm{d}x} = \frac{v\dfrac{\mathrm{d}u}{\mathrm{d}x} - u\dfrac{\mathrm{d}v}{\mathrm{d}x}}{v^2}$$

$$y = \frac{u}{v} \Rightarrow \frac{\mathrm{d}y}{\mathrm{d}x} = \frac{v\dfrac{\mathrm{d}u}{\mathrm{d}x} - u\dfrac{\mathrm{d}v}{\mathrm{d}x}}{v^2}$$ ← This result is called the **quotient rule.**

Discussion point

You have already seen that when using the product rule, it does not matter which function you call u and which v.

➔ Does it matter which function you call u and which v when using the quotient rule?

Example 5.13

Given that $y = \dfrac{x^2 + 1}{3x - 1}$, find $\dfrac{\mathrm{d}y}{\mathrm{d}x}$ using the quotient rule.

Solution

$u = x^2 + 1$	$v = 3x - 1$
$\dfrac{\mathrm{d}u}{\mathrm{d}x} = 2x$	$\dfrac{\mathrm{d}v}{\mathrm{d}x} = 3$

← First identify u and v.

← Differentiate u and v.

Using the quotient rule

$$\frac{\mathrm{d}y}{\mathrm{d}x} = \frac{v\dfrac{\mathrm{d}u}{\mathrm{d}x} - u\dfrac{\mathrm{d}v}{\mathrm{d}x}}{v^2}$$

$$\frac{\mathrm{d}y}{\mathrm{d}x} = \frac{(3x - 1) \times 2x - (x^2 + 1) \times 3}{(3x - 1)^2}$$

$$\Rightarrow \quad \frac{\mathrm{d}y}{\mathrm{d}x} = \frac{6x^2 - 2x - 3x^2 - 3}{(3x - 1)^2}$$ ← Expand the brackets in the numerator.

$$\Rightarrow \quad \frac{\mathrm{d}y}{\mathrm{d}x} = \frac{3x^2 - 2x - 3}{(3x - 1)^2}$$ ← Simplify to get the final answer.

1

ACTIVITY 5.6

You could have differentiated $y = \dfrac{x^2 + 1}{3x - 1}$ by

rewriting it as $y = \left(x^2 + 1\right)(3x - 1)^{-1}$
and using the product rule.

Verify that this method gives you the same answer as Example 5.13.

Exercise 5.4

① Copy and complete Table 5.2 to show how you would differentiate the following functions.

Table 5.2

This row shows an example.

Function	Product rule, quotient rule or neither?	u	v
$y = \dfrac{(x + 1)^6}{\sqrt[3]{x - 1}}$	Quotient	$u = (x + 1)^6$	$v = \sqrt[3]{x - 1}$
(i) $\quad y = \sqrt{\left(x^2 - 1\right)}\left(x^3 + 3\right)^2$			
(ii) $\quad y = \left(2x^3 - 3\right)^5$			
(iii) $\quad y = \dfrac{\sqrt{x + 1}}{x^2}$			
(iv) $\quad y = x^2\sqrt{x - 2}$			

② Use the product rule to differentiate the following functions.

  (i) $\quad y = x^2\left(2x - 1\right)^4$

  (ii) $\quad y = (x + 1)\sqrt{x - 1}$

  (iii) $\quad y = \sqrt{x}\left(x - 3\right)^3$

③ Use the quotient rule to differentiate the following functions.

  (i) $\quad y = \dfrac{x^2}{3x - 1}$

  (ii) $\quad y = \dfrac{(1 - 2x)^3}{x^3}$

  (iii) $\quad y = \dfrac{x}{\sqrt{x} + 1}$

④ (i) Figure 5.25 shows the curve

$$y = \frac{x + 1}{x + 2}.$$

Find the gradient at the point $\left(0, \frac{1}{2}\right)$.

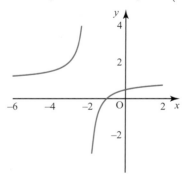

Figure 5.25

(ii) Figure 5.26 shows the curve

$$y = (x + 1)^3 (\sqrt{x} - 2), \; x \geqslant 0.$$

Find the gradient at the point $(1, -8)$.

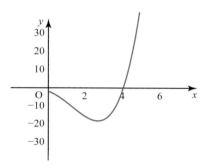

Figure 5.26

(iii) Figure 5.27 shows the curve

$$y = \frac{\sqrt{x + 1}}{x^2}, \; x \geqslant -1.$$

Find the gradient at the point $\left(3, \frac{2}{9}\right)$.

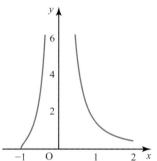

Figure 5.27

⑤ Given that $y = (x + 1)(x - 2)^2$

(i) find $\dfrac{dy}{dx}$

(ii) find the coordinates of the turning points of the curve $y = (x + 1)(x - 2)^2$ and determine their nature

(iii) sketch the curve.

⑥ Figure 5.28 shows the curve $y = \dfrac{3x}{2x - 3}$.

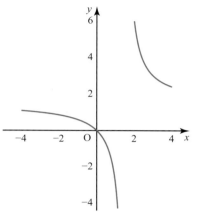

Figure 5.28

(i) Find $\dfrac{dy}{dx}$.

(ii) Find the equation of the tangent to the curve at the point where $x = 1$.

(iii) Find the equation of the normal to the curve at the origin.

(iv) Find the coordinates of the point where the normal to the curve at the origin cuts the curve again.

⑦ Figure 5.29 shows the curve

$$y = \frac{x^2 - 2x - 5}{2x + 3}.$$

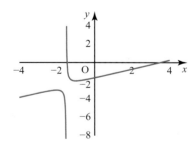

Figure 5.29

(i) Find $\dfrac{dy}{dx}$.　　(ii) Find $\dfrac{d^2 y}{dx^2}$.

(iii) Find the coordinates of the turning points of the curve

$$y = \frac{x^2 - 2x - 5}{2x + 3}.$$

PS ⑧ Find the coordinates and nature of the turning points of the curve $y = x^2(x - 3)^4$, and sketch the curve.

⑨ The graph of the function $y = \dfrac{2x}{\sqrt{x} - 1}$ is undefined for $x < 0$ and $x = 1$. P is a stationary point.

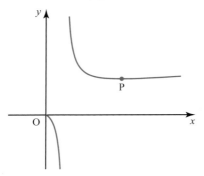

Figure 5.30

(i) Find $\dfrac{dy}{dx}$.

(ii) Find the gradient of the curve at the point A with coordinates $(9, 9)$, and show that the equation of the normal at A is $y = -4x + 45$.

(iii) Find the coordinates of P and verify that it is a minimum point.

PS (iv) The point Q is the intersection of the normal at A and the tangent at P. The point R is the intersection of the normal at A and the normal at P.

Find the area of triangle PQR.

⑩ A function is given by

$$f(x) = x\sqrt{9 - 2x^2}, \quad 0 \le x \le \frac{3\sqrt{2}}{2}.$$

(i) Find the coordinates of the maximum point on the curve $y = f(x)$.

(ii) What is the gradient of the curve at the origin?

(iii) What is the value of y when $x = \frac{3}{2}\sqrt{2}$? What is the gradient of the curve at this point?

(iv) Sketch the curve.

LEARNING OUTCOMES

When you have completed this chapter, you should be able to:

➤ understand and use the second derivative:
- as the rate of change of gradient
- and its connection to convex and concave sections of the curve
- for finding points of inflection

➤ differentiate using:
- the product rule
- the quotient rule
- the chain rule

➤ differentiate to solve problems using:
- connected rates of change
- inverse functions.

KEY POINTS

1 When $\dfrac{d^2y}{dx^2} > 0$, a curve is concave upwards.

When $\dfrac{d^2y}{dx^2} < 0$, a curve is concave downwards.

Figure 5.31

2 At a point of inflection, a curve changes between from being concave upwards to being concave downwards, or vice versa.

At a point of inflection, $\dfrac{d^2y}{dx^2} = 0$.

Some points of inflection are also stationary points, where $\dfrac{dy}{dx} = 0$.

3 Chain rule: $\dfrac{dy}{dx} = \dfrac{dy}{du} \times \dfrac{du}{dx}$

4 Product rule: $y = uv \implies \dfrac{dy}{dx} = u\dfrac{dv}{dx} + v\dfrac{du}{dx}$

5 Quotient rule: $y = \dfrac{u}{v} \implies \dfrac{dy}{dx} = \dfrac{v\dfrac{du}{dx} - u\dfrac{dv}{dx}}{v^2}$

6 $\dfrac{dx}{dy} = \dfrac{1}{\dfrac{dy}{dx}}$

FUTURE USES

You will apply these techniques for differentiation to a wider range of functions in Chapter 9.

MULTIPLE CHOICE QUESTIONS: CHAPTERS 1 TO 5

① Which, of the following, is not implied by the statement 'n is an integer':

A $2n + 1$ is odd

B $10n$ is even

C $4n + 6$ is a multiple of 4

D $6n$ is a multiple of 3

② Which of the following is not true for an arithmetic sequence?

A It increases/decreases in equal steps

B Its sum converges to a limit

C Its sum tends to infinity (positive or negative)

D It is a linear sequence

③ Which of the following is equivalent to $-1 \leqslant x \leqslant 3$?

A $|x - 1| < 2$

B $|x + 1| \leqslant 4$

C $|x + 1| \leqslant 2$

D $|x - 1| \leqslant 2$

④ Which of the following would be used when differentiating $y = (1 - \frac{1}{x})^6$ using the chain rule?

A $y = u^6, u = \frac{1}{x}$

B $y = 1 - u^6, u = \frac{1}{x}$

C $y = u^6, u = 1 - \frac{1}{x}$

D $y = \frac{1}{u}, u = x^6$

① Figure 1 shows a sector of a circle with radius r cm. The angle subtended at the centre of the circle is θ radians.

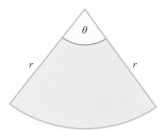

Figure 1

The perimeter of the sector is 3 cm.

(i) Show that the area of the sector is $\frac{3}{2}r - r^2$ cm². [3 marks]

(ii) Find the value of θ when the area of the sector is a maximum, justifying your answer. [4 marks]

② (i) Find the value of the y coordinate on each of the curves

$y = \cos x$ and $y = \cos\left(x - \frac{\pi}{6}\right) + 0.5$ when $x = -\frac{\pi}{3}$ and

when $x = -\frac{\pi}{2}$. [2 marks]

(ii) Give the coordinates of **one** point with positive x coordinate

where the curves $y = \cos x$ and $y = \cos\left(x - \frac{\pi}{6}\right) + 0.5$ cross. [2 marks]

(iii) Describe a transformation that maps the curve $y = \cos x$ on

to the curve $y = \cos\left(x - \frac{\pi}{6}\right) + 0.5$. [2 marks]

③ You are given that $y = \left(x^2 - 1\right)^4$.

(i) Show that $\frac{d^2y}{dx^2} = 8(x^2 - 1)^2(7x^2 - 1)$. [4 marks]

(ii) State the set of values of x for which $y = \left(x^2 - 1\right)^4$ is concave downwards. [2 marks]

④ You are given that $f(x) = \frac{x + 3}{x - 1}$.

(i) Find $ff(x)$. Hence write down $f^{-1}(x)$. [4 marks]

(ii) Deduce the equation of a line of symmetry of the

graph $y = f(x)$. [1 mark]

(iii) Write $f(x)$ in the form $a + \frac{b}{x - c}$ where a, b and c are integers. [2 marks]

(iv) Starting with the curve $y = \frac{1}{x}$, write down a sequence of

transformations that would result in the curve $y = f(x)$. [3 marks]

T ⑤ Figure 2 shows a spreadsheet with the information about day length on different dates in a town in northern England. The day length is the time between sunrise and sunset. It is given in hours and minutes in column C and in hours in decimal form in column D. Some of the rows in the spreadsheet are missing and some of the cells have not been completed. The graph in Figure 3 was drawn by the spreadsheet; it shows the number of hours of daylight plotted against day number for the whole year.

	A	B	C	D
1	Day number (x)	Date	Actual day length (hours:mins)	Actual day length (y hours)
2	0	31 Dec	07:40	7.666667
3	1	1 Jan	07:41	7.683333
4	2	2 Jan	07:42	7.7
5	3	3 Jan	07:43	7.716667
154	152	1 June	16:24	16.4
155	153	2 June	16:26	16.43333
		21 June	16:45	
		21 Dec	07:35	

Figure 2

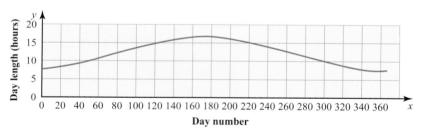

Figure 3

(i) A model of the form $y = a - b \cos x°$ is suggested for the length y hours of day x.

■ The shortest day is 21st December.

■ The longest day is 21st June.

Find values of a and b consistent with this information.   [5 marks]

(ii) Compare the length of day which the model predicts for 21st June with the actual day length.   [3 marks]

(iii) Explain why each of the two following transformations of the graph of $y = a - b \cos x°$ would make it a better model for day length.

(a) Stretch parallel to x-axis.

(b) Translation parallel to x-axis.   [2 marks]

(P) ⑥ A three term arithmetic sequence has a non-zero common difference. Prove by contradiction that the three terms, in the same order as for the arithmetic sequence, cannot form a geometric sequence. [5 marks]

(PS) ⑦ A geometric sequence has first term 1. The product of the first 9 terms is 262 144.

Find the possible values for the sum of the first 9 terms. [6 marks]

(PS) (P) ⑧ Figure 4 shows the curve $y = x^4 - 1.5x^2 + 3$.

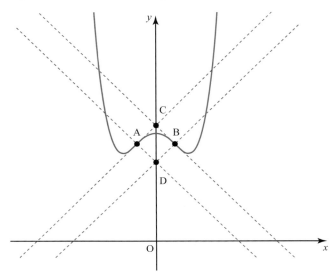

Figure 4

The curve has points of inflection at A and B. The tangents and normals to the curve at points A and B are shown. The tangents intersect at C. The normals intersect at D.

(i) Find the coordinates of A and B. [5 marks]

(ii) Show that ACBD is a square. [5 marks]

R

Review: The sine and cosine rules

When people thought the Earth was flat, they were wrong. When people thought the Earth was spherical, they were wrong. But if you think that thinking the Earth is spherical is just as wrong as thinking the Earth is flat, then your view is wronger than both of them put together.

Isaac Asimov (1920–1992)

1 Working with triangles

In a *right-angled* triangle, for $0° < \theta < 90°$, the trigonometric functions are

$$\sin\theta = \frac{\text{opposite}}{\text{hypotenuse}}$$

$$\cos\theta = \frac{\text{adjacent}}{\text{hypotenuse}}$$

$$\tan\theta = \frac{\text{opposite}}{\text{adjacent}}$$

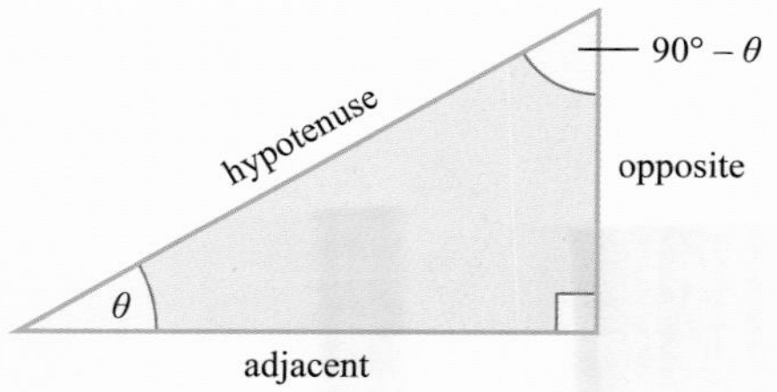

Figure R.1

Also $\sin\theta = \cos(90° - \theta)$

and $\cos\theta = \sin(90° - \theta)$.

Here is a reminder of the general formula for the area of a triangle, and the sine and cosine rules, which can be used for *any* triangle ABC.

Figure R.2

Area

Area of a triangle $= \frac{1}{2}ab\sin C$

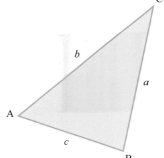

The cosine rule

For an unknown side: $a^2 = b^2 + c^2 - 2bc\cos A$

For an unknown angle: $\cos A = \dfrac{b^2 + c^2 - a^2}{2bc}$

The sine rule

For an unknown side: $\dfrac{a}{\sin A} = \dfrac{b}{\sin B} = \dfrac{c}{\sin C}$

For an unknown angle: $\dfrac{\sin A}{a} = \dfrac{\sin B}{b} = \dfrac{\sin C}{c}$

Always check for the **ambiguous case** when you are given two sides and a non-included angle of a triangle and are using the sine rule to find one of the other angles. In such cases θ and $180° - \theta$ can both be correct answers.

Example R.1

Two hikers set off from point A.

Some time later Ben has walked 8 km on a bearing of 065° to point B.
Carla has walked 6 km on a bearing of 135° to point C.

The points A, B and C are on level ground.

Find

(i) the distance between Ben and Carla

(ii) the bearing of Carla from Ben at that time.

Solution

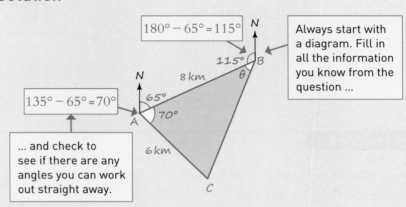

Figure R.3

(i) You know two sides and the angle between them and you want to find the third side, so use the cosine rule.

$$BC^2 = AB^2 + AC^2 - 2 \times AB \times BC \times \cos A$$
$$= 8^2 + 6^2 - 2 \times 8 \times 6 \times \cos 70°$$
$$= 67.16...$$

$$\Rightarrow \quad BC = 8.195... \quad \longleftarrow \boxed{\text{Keep the answer stored in your calculator.}}$$

$$= 8.20 \, \text{km (to 3 s.f.)}$$

Ben and Carla are 8.20 km (to 3 s.f.) apart.

(ii) You need to find θ first before you can work out the bearing.

You know three sides and need to find an angle – so you could use the cosine rule again. Or you could use the sine rule.

$$\frac{\sin \theta}{6} = \frac{\sin 70°}{8.195...} \quad \longleftarrow \boxed{\begin{array}{l}\text{Use the exact value} \\ \text{calculated in part (i).}\end{array}}$$

$$\Rightarrow \quad \sin \theta = \frac{6 \sin 70°}{8.195...} = 0.687...$$

$$\Rightarrow \quad \theta = 43.5° \quad \text{(to 3 s.f.)}$$

So the bearing is $360° - 115° - 43.5° = 201.5°$.

Discussion point

➜ Check that using the cosine rule would give you the same answer.

Example R.2

Find the area of triangle ABC.

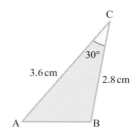

Figure R.4

Solution

Area of a triangle $= \frac{1}{2} ab \sin C$ ← You know two sides and the angle between them, so you can use the general formula for the area of a triangle.

$= \frac{1}{2} \times 3.6 \times 2.8 \times \sin 30°$

$= 2.52 \, \text{cm}^2$

Review exercise R.1

① For each triangle in Figure R.5 find
   (a) the angle θ (b) the area.

   (i)

   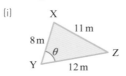

   (ii)

   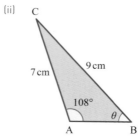

   Figure R.5

② Find the length x in each triangle in Figure R.6.

   (i)

   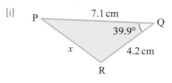

   (ii)

   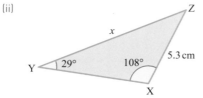

   (iii)

   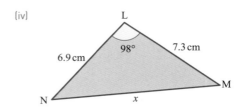

   (iv)

   Figure R.6

③ A tower 60 m high stands on the top of a hill. From a point on the ground at sea level, the angles of elevation of the top and bottom of the tower are 49° and 37° respectively.

   Find the height of the hill.

④ Find the angle θ in Figure R.7.

   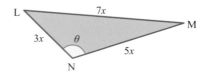

   Figure R.7

⑤ In triangle ABC, AB = 8 cm, BC = 7 cm and angle CAB = 56°.

Find the possible sizes for the angle BCA.

⑥ Three points A, B and C lie in a straight line on level ground with B between A and C. A vertical mast BD stands at B and is supported by wires, two of which are along the lines AD and CD.

Given that ∠DAB = 55°, ∠DCB = 42° and AC = 85 m, find the lengths of the wires AD and CD and the height of the mast.

⑦ A yacht sets off from A and sails 3 km on a bearing of 045° to a point B. It then sails 1 km on a bearing of 322° to a point C.

Find the distance AC.

⑧ The lengths of the hands of a clock are 7 cm and 10 cm.

Find the distance between the tips of the hands at 8 p.m.

⑨ X, Y, and Z are three points on level ground. Point Y is 2 km from X on a bearing of 117°, and Z is 5 km from X on a bearing of 204°.

Find

(i) ∠YXZ

(ii) the distance YZ.

⑩ A triangle has sides of length 3 cm, 8 cm and 7 cm.

Find the exact area of the triangle.

⑪ The area of triangle ABC is $12\sqrt{3}$ cm². AB = 6 cm and AC = 8 cm. Angle BAC is acute.

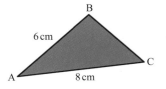

Figure R.8

Find the exact perimeter of the triangle.

Triples

(T)

Look at the triangle in Figure 1.
You can use Pythagoras' theorem to
show that it is right angled.

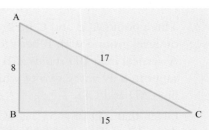

Figure 1

$$8^2 + 15^2 = 17^2$$

Alternatively you can use the cosine rule
on angle B.

$$\cos B = \frac{8^2 + 15^2 - 17^2}{2 \times 8 \times 15} = 0 \quad \Rightarrow \quad B = 90°.$$

The sides of the triangle are all integers and so the set $\{8, 15, 17\}$ is called a
Pythagorean triple. There are formulae for generating them, for example
$\{2n, n^2 - 1, n^2 + 1\}$ where n is an integer greater than 1.

Your first question is:

How can you be certain that this
generator works?

Now look at the triangle in Figure 2.

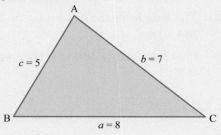

Figure 2

In this case the angle B is given by

$$\cos B = \frac{c^2 + a^2 - b^2}{2ca} = \frac{25 + 64 - 49}{2 \times 5 \times 8} = \frac{40}{80} = \frac{1}{2} \text{ and so } B = 60°.$$

The set $\{8, 7, 5\}$ forms a different sort of triple, giving an angle of $60°$ rather
than a right angle. This set of integers is a **60° triple**.

Equilateral triangles obviously give rise to these triples, for example
$\{5, 5, 5\}$ but there are others too.

Your second question is:
How can you find these other $60°$ triples?

1 **Problem specification and analysis**

   At the moment the problem is somewhat open-ended. Some preliminary
   ideas will help to tie it down.

   If $\{a, b, c\}$ is a $60°$ triple with angle $B = 60°$, the cosine rule gives
   $$\cos B = \frac{c^2 + a^2 - b^2}{2ca} = \frac{1}{2}.$$
   After a few lines of algebra this can be rearranged as the equation
   $$(b + c - a)(b - c + a) = ac.$$
   So multiplying $(b + c - a)$ by $(b - c + a)$ gives ac.

   There are three options for analysing this. The first two are

   a Set $(b + c - a) = a$ and so $(b - c + a) = c$, or vice versa.

   b Set $(b + c - a) = na$ and so $(b - c + a) = \dfrac{c}{n}$ where n is an integer greater
          than 1.

   These two will provide the basis for your investigation and information
   collection.

2 **Information collection**

The first question Start by setting up a spreadsheet for the first 100 Pythagorean triples generated by $\{2n, n^2 - 1, n^2 + 1\}$ for $n \geqslant 2$. Use an extra column to check that the generator works for all these values of n.

You should find that it does work for these 100 Pythagorean triples but to be certain you should also prove it algebraically.

The second question Show that option **a** for the analysis of 60° triples always gives you equilateral triangles.

Now prove that eliminating c between the equations in option **b** leads to the equation $b = a\left(\dfrac{n^2 + n + 1}{2n + 1}\right)$.

Use this to find the 60° triple generator with $a = 2n + 1$.

Set this up on a spreadsheet.

3 **Processing and representation**

For the first question, the essential processing and representation have been done on the spreadsheet. You may choose to copy the listing of Pythagorean triples from the spreadsheet into another document. However, this task is not complete without an algebraic proof that the generator works.

For the second question, the first task is essentially algebra.

The next task begins with algebra and then goes into a spreadsheet. Again you may wish to copy the spreadsheet listing into another document. This list is of 60° triples and you will want to check if you have listed them all.

A way of checking is to notice that, apart from the equilateral triangles, each triple has a partner triangle.

You can see this in Figure 3 for the blue $\{8, 7, 5\}$ triangle.

Fitting it into the equilateral triangle of side 8 reveals its red partner triangle with a triple of $\{8, 7, 3\}$.

So you now have a possible check of whether your generator has caught all the 60° triples.

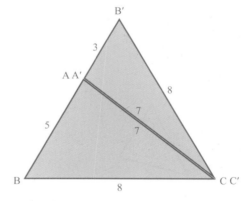

Figure 3

4 **Interpretation**

At this stage you will want to tie up any loose ends, particularly if you have found that your generator missed some of the 60° triples.

In the Problem specification and analysis section, you were told there are three options but were only given two of them to work on, **a** and **b**. Now it is time to investigate option **c**.

c Set $(b + c - a) = \dfrac{n}{m}a$ and so $(b - c + a) = \dfrac{m}{n}c$ where m and n are both integers greater than 1.

KEY POINTS

1 For any triangle ABC

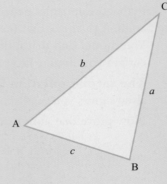

- $\dfrac{a}{\sin A} = \dfrac{b}{\sin B} = \dfrac{c}{\sin C}$ (sine rule)

- $a^2 = b^2 + c^2 - 2bc\cos A$ (cosine rule)

- area $= \dfrac{1}{2}bc\sin A$.

2   **Table R.1** Summary of the use of the sine and cosine rules

Triangle	You know	You want	Use
	3 sides	Any angle $\cos A = \dfrac{b^2 + c^2 - a^2}{2bc}$	Cosine rule
	2 sides + **included** angle	3rd side $a^2 = b^2 + c^2 - 2bc\cos A$	
	2 angles + 1 side	Any side $\dfrac{a}{\sin A} = \dfrac{b}{\sin B}$	Sine rule
	2 sides + 1 angle	Any angle $\dfrac{\sin A}{a} = \dfrac{\sin B}{b}$	

Don't forget that once you know two angles in a triangle, you can use 'angles sum to 180°' to find the third angle!

Check for the **ambiguous case**: sometimes $180° - \theta$ is also a solution.

6 Trigonometric functions

→ How does the photo here of a time lapse of the sun's position in the sky show how the sine and cosine functions can model real life situations?

→ Where else do you encounter sine and cosine waves in real life?

1 Reciprocal trigonometric functions

As well as the three main trigonometric functions, $\sin\theta$, $\cos\theta$ and $\tan\theta$, there are three more which are commonly used. These are their reciprocals – cosecant (cosec), secant (sec) and cotangent (cot), defined by

$$\operatorname{cosec}\theta = \frac{1}{\sin\theta}; \quad \sec\theta = \frac{1}{\cos\theta}; \quad \cot\theta = \frac{1}{\tan\theta}\left(=\frac{\cos\theta}{\sin\theta}\right).$$

> Use the 3rd letter of each reciprocal function to remind you which trigonometric function it is paired with.

Each of these is undefined for certain values of θ. For example, $\operatorname{cosec}\theta$ is undefined for $\theta = 0°, 180°, 360°, \dots$ since $\sin\theta$ is zero for these values of θ.

Figure 6.1 overleaf shows the graphs of these functions. Notice how all three of the functions have asymptotes at intervals of $180°$.

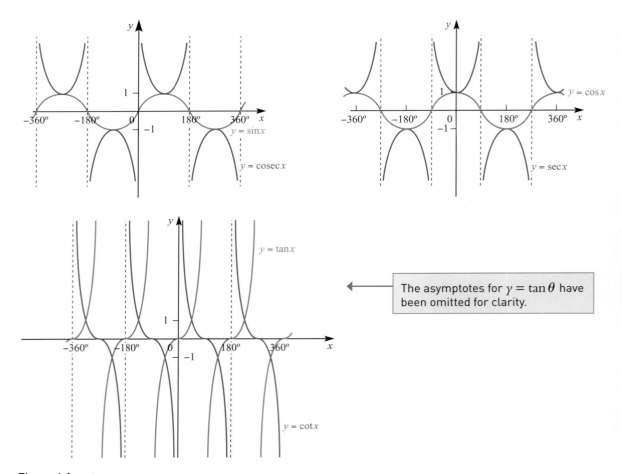

Figure 6.1

The asymptotes for $y = \tan\theta$ have been omitted for clarity.

Discussion point

→ Use the diagram of the unit circle in Figure 6.2 to explain why these identities are true.

For any angle (in degrees or radians):

$$\sin\theta = y, \cos\theta = x, \tan\theta = \frac{y}{x}$$

You have already met the identities

■ $\tan\theta \equiv \dfrac{\sin\theta}{\cos\theta}$, for $\cos\theta \neq 0$,

■ $\sin^2\theta + \cos^2\theta \equiv 1$.

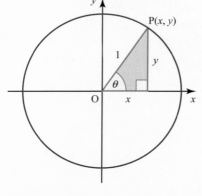

Figure 6.2

Using the definitions of the reciprocal functions two alternative trigonometric forms of Pythagoras' theorem can be obtained.

> You will use this identity in mechanics, when you use the equation of the trajectory of a projectile to find the angle of projection.

(i) Start with $\qquad\qquad\qquad\qquad\sin^2\theta + \cos^2\theta \equiv 1.$

Divide both sides by $\cos^2\theta$: $\dfrac{\sin^2\theta}{\cos^2\theta} + \dfrac{\cos^2\theta}{\cos^2\theta} \equiv \dfrac{1}{\cos^2\theta}$

$\Rightarrow \ \tan^2\theta + 1 \equiv \sec^2\theta.$

(ii) Start with $\qquad\qquad\qquad\qquad\sin^2\theta + \cos^2\theta \equiv 1.$

Divide both sides by $\sin^2\theta$: $\dfrac{\sin^2\theta}{\sin^2\theta} + \dfrac{\cos^2\theta}{\sin^2\theta} \equiv \dfrac{1}{\sin^2\theta}$

$\Rightarrow \ 1 + \cot^2\theta \equiv \operatorname{cosec}^2\theta.$

> $\cot\theta = \dfrac{1}{\tan\theta} = \dfrac{\cos\theta}{\sin\theta}$

These three trigonometric identities are called the Pythagorean identities, as they are all derived from Pythagoras' theorem.

The Pythagorean identities

$\sin^2\theta + \cos^2\theta \equiv 1; \qquad \tan^2\theta + 1 \equiv \sec^2\theta; \qquad 1 + \cot^2\theta \equiv \operatorname{cosec}^2\theta.$

> If you can't remember the second and third identities, you can easily work them out from the first one, as shown above.

Questions involving reciprocal functions are usually most easily solved by considering the related function, as in the following example.

Example 6.1

Find $\operatorname{cosec} 120°$ leaving your answer in surd form.

Solution

$\operatorname{cosec} 120° = \dfrac{1}{\sin 120°}$ $\qquad\longleftarrow$ $\boxed{\sin 120° = \sin(180° - 120°) = \sin 60°}$

$= 1 \div \dfrac{\sqrt{3}}{2}$ $\qquad\longleftarrow$ $\boxed{\text{It is easier to write it like this rather than dealing with a 'triple-decker' fraction.}}$

$= \dfrac{2}{\sqrt{3}} = \dfrac{2\sqrt{3}}{3}$

Example 6.2

Given that $\sec\theta = -\dfrac{5}{3}$ and θ is reflex, find the exact value of

(i) $\sin\theta$ $\qquad\qquad$ (ii) $\cot\theta$.

$\boxed{\text{A \textbf{reflex} angle is between } 180° \text{ and } 360°.}$

Solution

(i) $\sec\theta = -\dfrac{5}{3} \Rightarrow \cos\theta = -\dfrac{3}{5}$ $\qquad\longleftarrow$ $\boxed{\sec\theta = \dfrac{1}{\cos\theta}}$

1

Use the identity $\sin^2\theta + \cos^2\theta \equiv 1$.

$\Rightarrow \quad \sin^2\theta \equiv 1 - \cos^2\theta$

$\Rightarrow \quad \sin^2\theta = 1 - \left(\frac{3}{5}\right)^2$

So $\quad \sin^2\theta = 1 - \frac{9}{25} = \frac{16}{25}$

So $\quad \sin\theta = \pm\sqrt{\frac{16}{25}} = \pm\frac{4}{5}$

Use the CAST diagram to work out the correct sign for $\sin\theta$.

$\cos\theta$ is negative so θ must lie in either the 2nd or the 3rd quadrant.

But θ is reflex, so θ is in the 3rd quadrant.

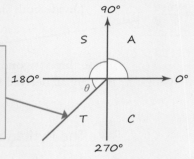

Figure 6.3

Hence $\sin\theta = -\frac{4}{5}$.

$$\cot\theta \equiv \frac{1}{\tan\theta} \quad \text{and} \quad \tan\theta \equiv \frac{\sin\theta}{\cos\theta}.$$

(ii)   Use the identity $\cot\theta \equiv \dfrac{\cos\theta}{\sin\theta}$.

$\Rightarrow \quad \cot\theta = \dfrac{-\frac{4}{5}}{-\frac{3}{5}} = \dfrac{4}{3}$

$\boxed{-\frac{4}{5} \div -\frac{3}{5} = -\frac{4}{5} \times -\frac{5}{3} = \frac{4}{3}}$

Exercise 6.1

①

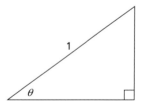

Write down, in terms of θ, the lengths of the two shorter sides of this triangle.

②

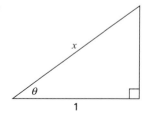

Find the length of the hypotenuse of this triangle in terms of θ.

③   Without using a calculator, write the following in exact form.

(i)   (a) $\sin 135°$ (b) $\cos 135°$

(c) $\cot 135°$ (d) $\sec 135°$

(ii)   (a) $\sin(-150)°$ (b) $\sec 150°$

(c) $\tan 150°$ (d) $\csc 150°$

(iii)   (a) $\tan 420°$ (b) $\cos(-420)°$

(c) $\sin 420°$ (d) $\cot(-420°)$

(iv)   (a) $\cos^2 210°$ (b) $\sin^2 210°$

(c) $\cot^2(-210°)$ (d) $\sec^2 210°$

④   In triangle ABC, angle $A = 90°$ and $\sec B = 2$.

(i)   Find the angles B and C.

(ii)   Find $\tan B$.

(iii)   Show that $1 + \tan^2 B = \sec^2 B$.

⑤ In triangle LMN, angle $M = 90°$ and $\cot N = 1$.

(i) Find the angles L and N.

(ii) Find $\sec L$, $\operatorname{cosec} L$ and $\tan L$.

(iii) Show that $1 + \tan^2 L = \sec^2 L$.

⑥ Malini is 1.5 m tall.

At 8 p.m. one evening her shadow is 6 m long.

Figure 6.4

Given that the angle of elevation of the sun at that moment is α

(i) show that $\cot \alpha = 4$

(ii) find α.

⑦ Find the domain and range of

(i) $f(x) = \sec x$

(ii) $f(x) = \operatorname{cosec} x$

(iii) $f(x) = \cot x$.

⑧ Given that $\cos \theta = \dfrac{\sqrt{3}}{2}$ and θ is acute, find the exact value of

(i) $\sin \theta$ (ii) $\tan \theta$.

⑨ Given that $\cot \theta = \dfrac{\sqrt{7}}{3}$ and θ is reflex, find the exact value of

(i) $\cos \theta$

(ii) $\operatorname{cosec} \theta$.

⑩ Given that $\operatorname{cosec} \theta = k$ and θ is acute, show that

(i) $\cos \theta = \dfrac{\sqrt{k^2 - 1}}{k}$

(ii) $\tan^2 \theta = \dfrac{1}{k^2 - 1}$.

⑪ For each of the following functions

(a) find the range, expressing each answer as an inequality

(b) state the period

(c) sketch the graph for $0° \leqslant x \leqslant 360°$.

(i) $f(x) = \sec 3x$

(ii) $f(x) = 3\cot x$

(iii) $f(x) = \operatorname{cosec} x - 3$

(iv) $f(x) = 3 + \sec \dfrac{x}{3}$

⑫ Starting with the graph of $y = \sec x$, state the transformations which can be used to sketch each of the following curves.

(i) $y = -2\sec x$

(ii) $y = -\sec 2x$

(iii) $y = 2 + \sec(-x)$

(iv) $y = 2\sec \dfrac{x}{2}$

(v) $y = -\sec(x + 30)$

(vi) $y = 2 - \sec x$

⑬ Sketch the graphs of the following functions for $0 \leqslant x \leqslant 360°$.

(i) $f(x) = 2 + \sec x$

(ii) $f(x) = -\cot x$

(iii) $f(x) = \operatorname{cosec} 2x$

(iv) $f(x) = \sec(x + 30°)$

⑭ Figure 6.5 shows $f(x) = \sin x$ and $g(x) = a + b\operatorname{cosec} cx$ for $0° \leqslant x \leqslant 360°$.

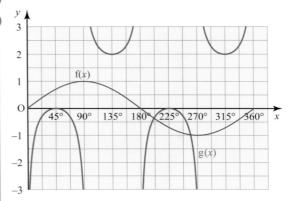

Figure 6.5

(i) Find the values of a, b and c.

(ii) Use the graph to solve $f(x) = g(x)$.

(iii) How many roots does the equation $f(2x) = g(x)$ have?
Explain your reasoning.

2 Working with trigonometric equations and identities

Inverse trigonometric functions

The functions sine, cosine and tangent are all many-to-one mappings, so their inverse mappings are one-to-many. So the problem 'find $\sin 30°$' has only one solution, 0.5, but 'solve $\sin\theta = 0.5$' has infinitely many roots. You can see this from the graph of $y = \sin\theta$ (Figure 6.6).

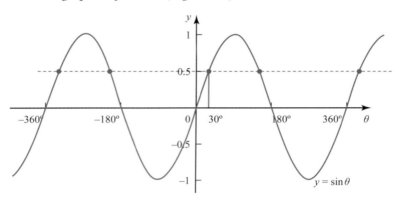

Figure 6.6

In order to define inverse functions for sine, cosine and tangent, a restriction has to be placed on the domain of each so that it becomes a one-to-one mapping. The restricted domains are listed in Table 6.1.

Table 6.1

Function	Restricted domain (degrees)	Restricted domain(radians)
$y = \sin\theta$	$-90° \leqslant \theta \leqslant 90°$	$-\dfrac{\pi}{2} \leqslant \theta \leqslant \dfrac{\pi}{2}$
$y = \cos\theta$	$0° \leqslant \theta \leqslant 180°$	$0 \leqslant \theta \leqslant \pi$
$y = \tan\theta$	$-90° < \theta < 90°$	$-\dfrac{\pi}{2} < \theta < \dfrac{\pi}{2}$

Remember that $\tan\pm 90°$ is undefined.

> **Notation**
>
> **Table 6.2**
>
> | Function | Inverse |
> |---|---|
> | $\sin x$ | $\arcsin x$ or $\sin^{-1} x$ or $\text{invsin}\, x$ |
> | $\cos x$ | $\arccos x$ or $\cos^{-1} x$ or $\text{invcos}\, x$ |
> | $\tan x$ | $\arctan x$ or $\tan^{-1} x$ or $\text{invtan}\, x$ |

When you use your calculator to find $\arcsin 0.5$ it will return just one answer – this value is called the **principal value** and it lies in the restricted domain. To solve a trigonometric equation, you need to find all the roots in a given range.

Figure 6.7 shows the graph of each trigonometric function over its restricted domain, and that of its corresponding inverse function.

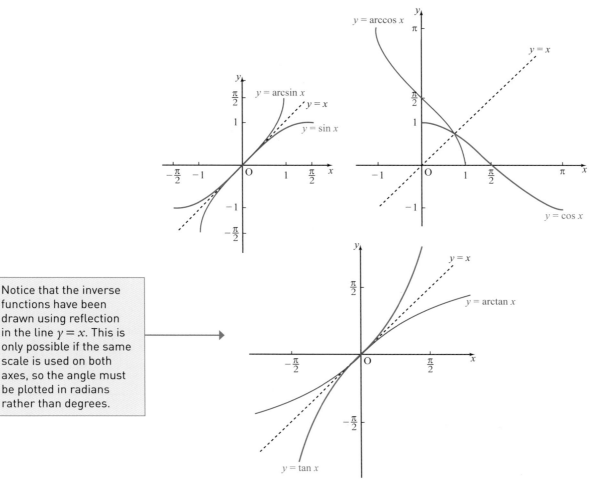

Notice that the inverse functions have been drawn using reflection in the line $y = x$. This is only possible if the same scale is used on both axes, so the angle must be plotted in radians rather than degrees.

Figure 6.7

Solving trigonometric equations

Example 6.3

Solve the following.

(i) $\sin x = 0.5$ for $0° \leqslant x \leqslant 360°$

(ii) $\sin(\theta + 40°) = 0.5$ for $0° \leqslant \theta \leqslant 360°$

Solution

(i) $\sin x = 0.5$

From your calculator: $x = 30°$. ◄—— This is the principal value.

From the graph of $y = \sin x$ then a second root is
$x = 180° - 30° = 150°$.

So $x = 30°$ or $x = 150°$.

➜

(ii) $\sin(\theta + 40°) = 0.5$

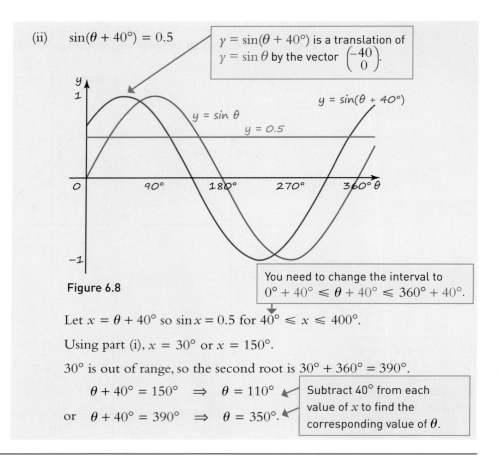

$y = \sin(\theta + 40°)$ is a translation of $y = \sin \theta$ by the vector $\begin{pmatrix} -40 \\ 0 \end{pmatrix}$.

Figure 6.8

You need to change the interval to $0° + 40° \leqslant \theta + 40° \leqslant 360° + 40°$.

Let $x = \theta + 40°$ so $\sin x = 0.5$ for $40° \leqslant x \leqslant 400°$.

Using part (i), $x = 30°$ or $x = 150°$.

$30°$ is out of range, so the second root is $30° + 360° = 390°$.

$$\theta + 40° = 150° \quad \Rightarrow \quad \theta = 110°$$

or $\theta + 40° = 390° \quad \Rightarrow \quad \theta = 350°.$

Subtract 40° from each value of x to find the corresponding value of θ.

Example 6.4

Solve $\cos 2\theta = \dfrac{\sqrt{2}}{2}$ for $0° \leqslant \theta \leqslant 360°$.

Solution

First sketch the graph.

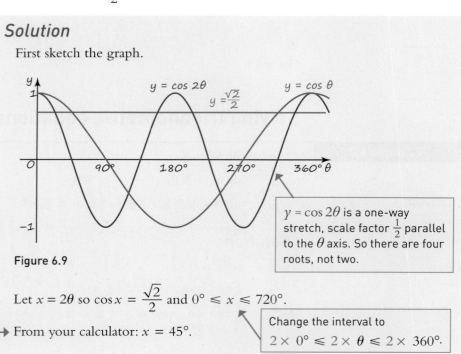

Figure 6.9

$y = \cos 2\theta$ is a one-way stretch, scale factor $\frac{1}{2}$ parallel to the θ axis. So there are four roots, not two.

Let $x = 2\theta$ so $\cos x = \dfrac{\sqrt{2}}{2}$ and $0° \leqslant x \leqslant 720°$.

Find the principal value from your calculator.

From your calculator: $x = 45°$.

Change the interval to $2 \times 0° \leqslant 2 \times \theta \leqslant 2 \times 360°$.

From the graph of $y = \cos x$ a second root is $x = 360° - 45° = 315°$.

So $x = 45°, 315°, 405°, 675°$ ◄──

> Add $360°$ to these two x values until you have found all the roots in the interval $0° \leqslant x \leqslant 720°$.

and $\theta = 22.5°, 157.5°, 202.5°, 337.5°$.

> Since $x = 2\theta$, halve each x value to find the roots for θ.

Sometimes you will need to use one or more trigonometric identities to help you solve an equation.

| **Example 6.5** | Find values of θ in the interval $0 \leqslant \theta \leqslant 360°$ for which $\sec^2 \theta = 4 + 2\tan \theta$. |

Solution

First you need to obtain an equation containing only one trigonometric function.

> Replace $\sec^2 \theta$ with $\tan^2 \theta + 1$.

$$\sec^2 \theta = 4 + 2\tan \theta$$

> Now you have a quadratic equation in $\tan \theta$, so rearrange so it equals 0.

$$\Rightarrow \quad \tan^2 \theta + 1 = 4 + 2\tan \theta$$

$$\Rightarrow \quad \tan^2 \theta - 2\tan \theta - 3 = 0$$

> You can factorise this in the same way as $x^2 - 2x - 3 = (x - 3)(x + 1)$.

$$\Rightarrow \quad (\tan \theta - 3)(\tan \theta + 1) = 0$$

$$\Rightarrow \quad \tan \theta = 3 \text{ or } \tan \theta = -1.$$

> From your calculator.

$$\tan \theta = 3 \quad \Rightarrow \quad \theta = 71.6°$$

> See Figure 6.10.

$$\text{or} \quad \theta = 71.6° + 180° = 251.6°.$$

$$\tan \theta = -1 \quad \Rightarrow \quad \theta = -45°$$

> Add $180°$ to find other roots.

$$\text{or} \quad \theta = -45° + 180° = 135°$$

$$\text{or} \quad \theta = 135° + 180° = 315°$$

The values of θ are $71.6°, 135°, 251.6°, 315°$.

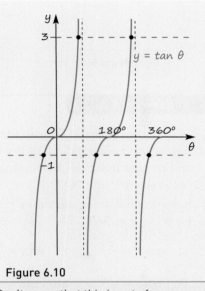

Figure 6.10

> Don't worry that this is out of range – you still use it to find the other roots.

Using trigonometric identities

You can use the identities

$$\sin^2 \theta + \cos^2 \theta \equiv 1, \qquad \tan^2 \theta + 1 \equiv \sec^2 \theta, \qquad 1 + \cot^2 \theta \equiv \operatorname{cosec}^2 \theta$$

$$\tan \theta \equiv \frac{\sin \theta}{\cos \theta}, \qquad \cot \theta \equiv \frac{\cos \theta}{\sin \theta}$$

to prove that other identities are true.

Example 6.6

Prove the identity $\cot\theta + \tan\theta \equiv \sec\theta\,\mathrm{cosec}\,\theta$.

Solution

It is sensible to work with one side of the identity at a time.

Working with the LHS: ◄────── Left-**H**and **S**ide

$$\cot\theta + \tan\theta = \frac{\cos\theta}{\sin\theta} + \frac{\sin\theta}{\cos\theta}$$

Rewrite using sin and cos.
Use $\cot\theta \equiv \dfrac{\cos\theta}{\sin\theta}$ and $\tan\theta \equiv \dfrac{\sin\theta}{\cos\theta}$.

$$= \frac{\cos^2\theta}{\cos\theta\sin\theta} + \frac{\sin^2\theta}{\sin\theta\cos\theta}$$

$$= \frac{\cos^2\theta + \sin^2\theta}{\cos\theta\sin\theta}$$

Use $\sin^2\theta + \cos^2\theta \equiv 1$.

$$= \frac{1}{\cos\theta\sin\theta}$$

$\dfrac{1}{\cos\theta\sin\theta} = \sec\theta \times \mathrm{cosec}\,\theta$

$$= \sec\theta\,\mathrm{cosec}\,\theta$$

Right-**H**and **S**ide ──────► $= \text{RHS}$ as required.

Exercise 6.2

① Why is the restricted domain for the cosine function between 0° and 180° while the sine and tangent functions have restricted domains between −90° and 90°?

② Solve the following equations for $0° \leqslant x \leqslant 360°$.

(i) $\cos x = \dfrac{\sqrt{3}}{2}$ (ii) $\sin x = \dfrac{\sqrt{3}}{2}$

(iii) $\tan x = \sqrt{3}$ (iv) $\cos x = -\dfrac{\sqrt{3}}{2}$

(v) $\sin x = -\dfrac{\sqrt{3}}{2}$ (vi) $\tan x = -\sqrt{3}$

③ Write down the number of roots for each of the following equations.

(i) $\cos x = 0.4$
     for $0° \leqslant x \leqslant 360°$

(ii) $\sin x = -0.4$
     for $0° \leqslant x \leqslant 360°$

(iii) $\tan x = -0.4$
     for $0° \leqslant x \leqslant 180°$

(iv) $\cos 2x = -0.4$
     for $0° \leqslant x \leqslant 360°$

(v) $2\sin x = -0.4$
     for $-360° \leqslant x \leqslant 360°$

(vi) $\tan(x - 30°) = -0.4$
     for $-360° \leqslant x \leqslant 360°$

④ Write down the domain and range of the following functions, where x is in radians.

(i) $f(x) = \arccos x$
(ii) $f(x) = \arcsin x$
(iii) $f(x) = \arctan x$

⑤ Solve the following equations for $0° \leqslant x \leqslant 360°$.

(i) $\mathrm{cosec}\,x = 1$ (ii) $\sec x = 2$

(iii) $\cot x = 4$ (iv) $\sec x = -3$

(v) $\cot x = -1$ (vi) $\mathrm{cosec}\,x = -2$

⑥ Prove each of the following identities.

(i) $\cot\theta\sin\theta \equiv \cos\theta$

(ii) $\dfrac{\tan x}{\sec x} \equiv \sin\theta$

(iii) $\cos^2\theta - \sin^2\theta \equiv 2\cos^2\theta - 1$

(iv) $\sec^2\theta - \mathrm{cosec}^2\theta \equiv \tan^2\theta - \cot^2\theta$

⑦ Solve the following equations for $0° \leqslant x \leqslant 360°$.

(i) $\cos x = \sec x$

(ii) $\mathrm{cosec}\,x = \sec x$

(iii) $2\sin x = 3\cot x$

(iv) $\mathrm{cosec}^2 x + \cot^2 x = 2$

(v) $3\sec^2 x - 10\tan x = 0$

(vi) $1 + \cot^2 x = 2\tan^2 x$

⑧ Prove each of the following identities.

(i) $\sec^2\theta + \mathrm{cosec}^2\theta \equiv \sec^2\theta\,\mathrm{cosec}^2\theta$

(ii) $\dfrac{\tan^2\theta - 1}{\tan^2\theta + 1} \equiv 2\sin^2\theta - 1$

(iii) $\dfrac{\cos\theta}{1 - \sin\theta} \equiv \sec\theta + \tan\theta$

(iv) $\dfrac{1}{1 - \sin\theta} + \dfrac{1}{1 + \sin\theta} \equiv 2\sec^2\theta$

(v) $\sec^4\theta - \tan^4\theta \equiv \sec^2\theta + \tan^2\theta$

3 Solving equations involving radians

! Make sure you put your calculator into radians mode.

You can solve trigonometric equations in radians in a similar way to solving them in degrees. Sometimes the roots can be given in exact form in terms of π, as in the next example.

Example 6.7

Solve $\sec 2\theta = 2$ for $0 < \theta < 2\pi$ giving your answers as multiples of π.

Solution

So you need to find all the roots in the interval $0 < x < 4\pi$.

→ Let $x = 2\theta$, then $\sec x = 2$. ← Remember that $\sec x = \dfrac{1}{\cos x}$.

$$\Rightarrow \frac{1}{\cos x} = 2$$

You might find it helpful to think in degrees.

$\cos x = \frac{1}{2} \Rightarrow x = 60°$

or $360° - 60° = 300°$.

$$\Rightarrow \cos x = \frac{1}{2}$$

$$\Rightarrow \qquad x = \frac{\pi}{3} \text{ or } 2\pi - \frac{\pi}{3} = \frac{5\pi}{3}$$

So $\qquad x = \dfrac{\pi}{3} + 2\pi = \dfrac{7\pi}{3}$

Keep adding 2π until you have found all the roots for x in the interval $0 < x < 4\pi$.

and $\dfrac{5\pi}{3} + 2\pi = \dfrac{11\pi}{3}$

So $\quad \theta = \dfrac{\pi}{6}, \dfrac{5\pi}{6}, \dfrac{7\pi}{6}, \dfrac{11\pi}{6}.$

Halve all of the values for x that are in range to find θ.

Exercise 6.3

① Find the exact values of

(i) $\sec \dfrac{\pi}{3}$

(ii) $\operatorname{cosec} \dfrac{\pi}{3}$

(iii) $\cot \dfrac{\pi}{3}$

(iv) $\sec \dfrac{\pi}{6}$

(v) $\operatorname{cosec} \dfrac{5\pi}{6}$

(vi) $\cot \dfrac{5\pi}{3}$

(vii) $\sec \dfrac{5\pi}{4}$

(viii) $\operatorname{cosec} \dfrac{7\pi}{4}$

(ix) $\cot \dfrac{7\pi}{4}$

② Solve the following equations for $0 \leqslant \theta \leqslant 2\pi$, giving your answers as multiples of π.

(i) $\cos \theta = \dfrac{\sqrt{3}}{2}$

(ii) $\tan \theta = 1$

(iii) $\sin \theta = \dfrac{1}{\sqrt{2}}$

(iv) $\operatorname{cosec} \theta = -2$

(v) $\sec \theta = -\sqrt{2}$

(iv) $\cot \theta = \dfrac{1}{\sqrt{3}}$

③ Solve the following equations for $-\pi \leqslant \theta \leqslant \pi$.

(i) $\sin \theta = 0.2$

(ii) $\cos \theta = 0.74$

(iii) $\tan \theta = 3$

(iv) $4 \sin \theta = -1$

(v) $3 + \cos \theta = 2.6$

(vi) $2 \tan \theta + 1 = 0$

④ Solve the following equations for $0 \leqslant \theta \leqslant 2\pi$.

(i) $\sin 2\theta = 0.6$

(ii) $3 \cos (\theta - 1) = 0.9$

(iii) $2 \cot 2\theta = 5$

(iv) $\sin \left(2\theta - \dfrac{\pi}{2} \right) = \dfrac{\sqrt{2}}{2}$

(v) $\sqrt{3} \tan \left(\dfrac{\theta}{2} \right) - 1 = 0$

(vi) $\sqrt{3} \operatorname{cosec}(\pi - \theta) = 2$

⑤ Solve the following equations for $0 \leqslant \theta \leqslant \pi$.

(i) $\sin^2 \theta = \dfrac{1}{2}$

(ii) $\cos^2 \theta = \dfrac{1}{2}$

(iii) $\tan^2\theta = 1$

(iv) $(\tan\theta - \sqrt{3})(3\tan\theta + \sqrt{3}) = 0$

(v) $4\cos^2\theta - 1 = 0$

(vi) $2\cos^2\theta + \cos\theta - 1 = 0$

(vii) $3\cos^2\theta + 2\sin\theta = 3$

(viii) $\tan^2\theta - 5\sec\theta + 7 = 0$

⑥   Solve $(1 - \cos x)(1 + \sec x) = \sin x$ for $0 \leqslant x \leqslant 2\pi$.

⑦   Solve $\cos\theta = 1 + \sec\theta$ for $0 \leqslant \theta \leqslant 2\pi$.

⑧   (i)   Show that $\dfrac{\csc^2\theta - 1}{\csc^2\theta} = \cos^2\theta$.

(ii)   Hence solve $\dfrac{4\csc^2\theta - 4}{\csc^2\theta} = 3$ for $0 \leqslant \theta \leqslant 2\pi$.

LEARNING OUTCOMES

When you have completed this chapter, you should be able to:

➤ understand and use the definitions of:
  ○ secant   ○ arcsin
  ○ cosecant   ○ arccos
  ○ cotangent   ○ arctan

➤ know the relationships of secant, cosecant, cotangent, arcsin, arccos, arctan to sine, cosine and tangent

➤ understand the graphs of secant, cosecant, cotangent, arcsin, arccos, arctan

➤ understand the domains and ranges of secant, cosecant, cotangent, arcsin, arccos, arctan

➤ understand and use:
  ○ $\sin^2\theta + \cos^2\theta = 1$
  ○ $1 + \tan^2\theta = \sec^2\theta$
  ○ $1 + \cot^2\theta = \csc^2\theta$

➤ solve simple trigonometric equations in a given interval, expressed in radians, including:
  ○ quadratic equations in sin, cos, tan
  ○ equations involving multiples of the unknown angle

➤ construct proofs involving trigonometric functions and identities.

KEY POINTS

1   The reciprocal trigonometric functions are

$$\csc\theta = \frac{1}{\sin\theta}; \qquad \sec\theta = \frac{1}{\cos\theta}; \qquad \cot\theta = \frac{1}{\tan\theta}\left(= \frac{\cos\theta}{\sin\theta}\right).$$

2   $\sin^2\theta + \cos^2\theta \equiv 1; \qquad \tan^2\theta + 1 \equiv \sec^2\theta;$

$1 + \cot^2\theta \equiv \csc^2\theta \qquad \tan\theta \equiv \dfrac{\sin\theta}{\cos\theta}$

3   Table 6.3

Function	Restricted domain (degrees)	Restricted domain (radians)	Inverse function
$y = \sin\theta$	$-90° \leqslant \theta \leqslant 90°$	$-\dfrac{\pi}{2} \leqslant \theta \leqslant \dfrac{\pi}{2}$	$\arccos\theta$
$y = \cos\theta$	$0° \leqslant \theta \leqslant 180°$	$0 \leqslant \theta \leqslant \pi$	$\arcsin\theta$
$y = \tan\theta$	$-90° < \theta < 90°$	$-\dfrac{\pi}{2} < \theta < \dfrac{\pi}{2}$	$\arctan\theta$

Table 6.4 Steps in solving a trigonometric equation

	Use your calculator to find the 1st root (principal value)	Find a 2nd root for sin or cos	Find all other roots in range
sin	θ_1 Principal value is $-90° \leqslant \theta \leqslant 90°$ or $\quad -\dfrac{\pi}{2} \leqslant \theta \leqslant \dfrac{\pi}{2}$	The curve is symmetrical about $x = 90°$ or $x = \dfrac{\pi}{2}$ radians, so $\theta_2 = 180° - \theta_1$ or $\theta_2 = \pi - \theta_1$	$\theta_1 \pm 360°$ and $\theta_2 \pm 360°$ or $\theta_1 \pm 2\pi$ and $\theta_2 \pm 2\pi$
cos	θ_1 Principal value is $0° \leqslant \theta \leqslant 180°$ or $\quad 0 \leqslant \theta \leqslant \pi$	The curve is symmetrical about the y-axis, so $\theta_2 = -\theta_1$	$\theta_1 \pm 360°$ and $\theta_2 \pm 360°$ or $\theta_1 \pm 2\pi$ and $\theta_2 \pm 2\pi$
tan	θ_1 Principal value is $-90° < \theta < 90°$ or $\quad -\dfrac{\pi}{2} < \theta < \dfrac{\pi}{2}$		$\theta_1 \pm 180°$ or $\theta_1 \pm \pi$

Further algebra

→ How would you find $\sqrt{101}$ correct to 3 decimal places without using a calculator?

One possibility, using a pen and paper, would be to use trial and improvement; another is that there is a structured method, which is rather like long division.

In the days before calculators and computers, some people needed to develop a very high degree of skill in mental arithmetic, particularly those whose work often called for quick reckoning. Some, like bookmakers, still do. There are also those who have quite exceptional innate skills. Shakuntala Devi (pictured above), born in 1929 in India, was known as 'the human calculator' because of her astonishing ability to perform complex calculations mentally. In 1977 she gave the 23rd root of a 201-digit number in 50 seconds.

While most mathematicians do not have Shakuntala Devi's high level of talent with numbers, they do acquire a sense of when something looks right or wrong. This often involves finding approximate values of numbers, such as $\sqrt{101}$, using methods that are based on series expansions, and these are the subject of the first part of this chapter.

Review: Pascal's triangle and the binomial expansion

You have already met the binomial expansion, initially using Pascal's triangle to expand expressions of the form $(x + y)^n$.

The coefficients in this type of expansion are called the binomial coefficients. They form **Pascal's triangle**.

```
                    (1)
                1        1                      Each number
            1       2        1                  is obtained by
        1       3        3        1              adding the two
    1       4        6        4        1         numbers above it
1       5       10       10        5        1
```

Example 7.1

(i) Expand $(x + 1)^4$ using the coefficients from Pascal's triangle.

(ii) Extend the result in (i) to write down the expansion of $(x + y)^4$.

Solution

> You use the row starting 1, 4, ... since the power of x in $(x + 1)^4$ is 4.

(i) From Pascal's triangle, the coefficients are 1, 4, 6, 4, 1.

$$(x + 1)^4 = 1x^4 + 4x^3 + 6x^2 + 4x + 1$$

> The first term, taking the x from all four brackets, is x^4 and thereafter the powers of x decrease

(ii) The expansion of $(x + y)^4$ uses the same row of coefficients as in (i), but as the powers of x decrease, powers of y are introduced.

$$(x + y)^4 = x^4 + 4x^3y^1 + 6x^2y^2 + 4x^1y^3 + y^4$$

> Notice that the sum of the powers in each term is 4.

Notation

$n!$ is 'n factorial', where
$$1! = 1, \quad 2! = 1 \times 2,$$
$$3! = 1 \times 2 \times 3,$$
$$n! = 1 \times 2 \times 3 \times \ldots \times n.$$

Special case: $0! = 1$

$$= \binom{n}{r} = \frac{n!}{r!(n-r)!};$$

$$= \binom{n}{0} = \binom{n}{n} = 1$$

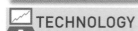
TECHNOLOGY

Most calculators have a button for $_nC_r$.

When expanding an expression of the form $(1 + x)^n$, Pascal's triangle becomes progressively larger and more tedious row by row, as the value of n increases.

The coefficients for the expansion of $(x + y)^n$ can be found using the formula

$$\binom{n}{r} = \frac{n!}{r!(n-r)!}$$

$\binom{n}{r}$ is also sometimes written as $_nC_r$ or ^nC_r.

Using this notation, $(x + y)^n$ can be written as

$$(x+y)^n = \binom{n}{0}x^n + \binom{n}{1}x^{n-1}y + \binom{n}{2}x^{n-2}y^2 + \binom{n}{3}x^{n-3}y^3 + \ldots + \binom{n}{n-1}x^1y^{n-1} + \binom{n}{n}y^n,$$

$_nC_r$ gives the number of ways of choosing r objects from n.

This is covered in more detail in Chapter 16 on probability distributions.

Example 7.2

Find the term in x^4 in the expansion of $(2x - 3)^7$.

Solution

The binomial coefficient of x^4 is given by $\binom{7}{4} = \frac{7!}{4!3!} = \frac{7 \times 6 \times 5}{3 \times 2 \times 1} = 35.$

$$\text{Term in } x^4 = \binom{7}{3} = (2x)^4(-3)^3 \qquad \text{Be careful with signs.}$$

$$= 35 \times 16x^4 \times (-27)$$

$$= -15120x^4$$

Review exercise

① (i) Write out Pascal's triangle as far as the row starting $1, 6, \ldots$.

   (ii) Add up the numbers in each row. What do you notice?

   (iii) What would be the sum of the numbers in the row starting $1, 12, \ldots$?

② Expand each of the following as a series of ascending powers of x.

   (i) $(1 + x)^3$

   (ii) $(1 + 2x)^3$

   (iii) $(1 - 2x)^3$

③ (i) Expand each of the following as a series of ascending powers of x.

   (a) $(1 + 2x)^4$ (b) $(1 - x)^4$

   (ii) Hence find the first three terms in the expansion of $[(1 + 2x)(1 - x)]^4$.

④ Write down the term indicated in the binomial expansion of each of the following functions.

   (i) $(1 - 2x)^3$, 3rd term

   (ii) $(2 - x)^4$, 3rd term

   (iii) $(3 - 2x)^5$, 4th term

⑤ Find the binomial expansion of $\left(1 - \dfrac{x}{2}\right)^7$ up to and including the term in x^4.

1 The general binomial expansion

ACTIVITY 7.2

Show that this form of the binomial expansion (right) gives the correct result for $(1 + x)^4$.

Explain why it results in an expansion which terminates after the term in x^4.

In general, the binomial coefficient $\dbinom{n}{r} = \dfrac{n!}{r!(n-r)!}$ can be written as

$$\frac{n \times (n-1) \times \ldots \times (n-r+1)}{r!}.$$

So the binomial expansion of $(1 + x)^n$ can be written as

$$(1 + x)^n = 1 + nx + \frac{n(n-1)}{2!}x^2 + \frac{n(n-1)(n-2)}{3!}x^3 + \ldots$$

This form of the binomial expansion can also be used to find the expansion of $(1 + x)^n$ if n is a negative number or a fraction. However, in such cases there will be an infinite number of terms.

You will have seen in Activity 7.3 that for $x = 0.1$, taking the first three terms of the binomial expansion gives a good approximation, but for $x = 10$ it does not.

Discussion point

➜ Explain why the binomial coefficient

$$\binom{n}{2}$$

can be written

as $\dfrac{n(n-1)}{1 \times 2}$ and the binomial coefficient

$$\binom{n}{3}$$

can be written

as

$$\frac{n(n-1)(n-2)}{1 \times 2 \times 3}.$$

ACTIVITY 7.3

(i) Use the formula above to find the first three terms of

    (a) $(1 + x)^{\frac{1}{2}}$ (b) $(1 + x)^{-2}$.

(ii) Substitute $x = 0.1$ into each of your expansions in (i) to find approximate values for

    (a) $\sqrt{1.1}$ (b) $\dfrac{1}{1.1^2}$.

(iii) Use a calculator to check that these answers are approximately correct.

(iv) What happens if you substitute $x = 10$ into the expansions to try to find approximate values for $\sqrt{11}$ and $\dfrac{1}{11^2}$?

When x is a small number, the terms in x^n get smaller and smaller as n gets larger, and so taking just the first few terms means that you are only disregarding small numbers. This is not the case when x is a larger number.

In fact, the binomial expansion when n is negative or fractional is valid only for $|x| < 1$.

This gives the general binomial theorem:

$$(1 + x)^n = 1 + nx + \frac{n(n-1)}{2!}x^2 + \frac{n(n-1)(n-2)}{3!}x^3 + \ldots$$

$$+ \frac{n(n-1)(n-2)\ldots(n-r+1)}{r!}x^r + \ldots$$

This is valid **when n is any real number,** provided that $|x| < 1$.

Example 7.3

Expand $(1 - x)^{-2}$ as a series of ascending powers of x up to and including the term in x^3, stating the set of values of x for which the expansion is valid.

Solution

$$(1 + x)^n = 1 + nx + \frac{n(n - 1)}{2!}x^2 + \frac{n(n - 1)(n - 2)}{3!}x^3 + \dots$$

Replacing n by -2 and x by $(-x)$ gives

$$(1 + (-x))^{-2} = 1 + (-2)(-x) + \frac{(-2)(-3)}{2!}(-x)^2 + \frac{(-2)(-3)(-4)}{3!}(-x)^3 + \dots$$

when $|-x| < 1$

which leads to

> It is important to put brackets round the term $-x$ since, for example, $(-x)^2$ is not the same as $-x^2$.

$$(1 - x)^{-2} \approx 1 + 2x + 3x^2 + 4x^3 \quad \text{when } |x| < 1.$$

> In this case the infinite series can be written in sigma notation as $\sum_{r=1}^{\infty} rx^{r-1}$.

Sometimes you need to rewrite an expression so that it is in the form $(1 + x)^n$ before using the binomial expansion. This is shown in the next example.

Example 7.4

Find the first three terms of a series expansion for $\dfrac{1}{2 + y}$ and state the values of y for which the expansion is valid.

Solution

$$\frac{1}{2 + y} = (2 + y)^{-1}$$

$$= \left(2\left[1 + \frac{y}{2}\right]\right)^{-1}$$

> Take out a factor of 2, but remember that the index still applies to this 2.

$$= 2^{-1}\left(1 + \frac{y}{2}\right)^{-1}$$

$$= \tfrac{1}{2}\left(1 + \frac{y}{2}\right)^{-1}$$

Using the binomial expansion,

> You only need to go as far as the x^2 term since the question asks for the first three terms.

$$(1 + x)^n = 1 + nx + \frac{n(n - 1)}{2!}x^2 + \dots,$$

replacing n by -1 and x by $\dfrac{y}{2}$ gives

> Remember to put brackets round the term $\frac{y}{2}$ since $\left(\frac{y}{2}\right)^2 = \frac{y^2}{4}$.

$$(2 + y)^{-1} = \tfrac{1}{2}\left[1 + (-1)\left(\frac{y}{2}\right) + \frac{(-1)(-2)}{2!}\left(\frac{y}{2}\right)^2 + \dots\right]$$

$$\approx \tfrac{1}{2}\left(1 - \frac{y}{2} + \frac{y^2}{4}\right) \quad \text{valid when } \left|\frac{y}{2}\right| < 1$$

$$\Rightarrow \quad \frac{1}{2 + y} \approx \frac{1}{2} - \frac{y}{4} + \frac{y^2}{8} \quad \text{when } |y| < 2.$$

Example 7.5

For each of the functions below

(a) write down the first three terms in the binomial expansion as a series of ascending powers of x,

(b) state the values of x for which your expansion is valid,

(c) substitute $x = 0.1$ in both the function and its expansion and calculate the percentage error, where

$$\text{percentage error} = \frac{\text{absolute error}}{\text{true value}} \times 100\%.$$

(i) $(1 + x)^{-3}$ (ii) $\sqrt{(1 + x^2)}$

Solution

(i) (a) $(1 + x)^{-3} = 1 + (-3)x + \dfrac{(-3)(-4)}{2!}x^2 + \ldots$

$$\approx 1 - 3x + 6x^2 \qquad \text{(first three terms)}$$

(b) The expansion is valid for $|x| < 1$.

(c) Substituting $x = 0.1$, the true value is $1.1^{-3} = \dfrac{1000}{1331}$.

The approximate value is $1 - 3(0.1) + 6(0.1)^2 = 0.76$

$$\Rightarrow \text{absolute error} = 0.76 - \frac{1000}{1331}$$

$$= 0.008\,685$$

Relative error $= \dfrac{0.008\,685}{\frac{1000}{1331}} \times 100\% = 1.16\% \qquad \text{(to 3 s.f.)}.$

(ii) (a) Writing $\sqrt{(1 + x^2)}$ as $(1 + x^2)^{\frac{1}{2}}$

$$(1 + x^2)^{\frac{1}{2}} = 1 + \tfrac{1}{2}x^2 + \frac{\left(\frac{1}{2}\right)\left(-\frac{1}{2}\right)}{2!}(x^2)^2 + \ldots$$

$$\approx 1 + \frac{x^2}{2} - \frac{x^4}{8} \qquad \text{(first three terms)}$$

(b) The expansion is valid for $|x^2| < 1 \Rightarrow |x| < 1$.

(c) Substituting $x = 0.1$, the true value is $\sqrt{1.01}$.

The approximate value is $1 + \dfrac{0.01}{2} - \dfrac{0.0001}{8} = 1.004\,987\,5$

$$\Rightarrow \text{absolute error} = \sqrt{1.01} - 1.004\,987\,5$$

$$= 0.000\,000\,062$$

Relative error $= \dfrac{0.000\,000\,062}{\sqrt{1.01}} \times 100\% = 6.17 \times 10^{-6}\%$

> **Note**
>
> This example shows how quickly this approximation may converge to the true value.

> **Discussion point**
>
> ➜ What helps to make the convergence so rapid?

Example 7.6

(i) Write $\sqrt{4 + 3x}$ in the form $a^n\left(1 + \dfrac{bx}{a}\right)^n$.

(ii) Hence find the first three terms in the expansion of $\sqrt{4 + 3x}$, stating the values of x for which the expansion is valid.

Solution

(i) $\sqrt{4 + 3x} = \sqrt{4\left(1 + \dfrac{3x}{4}\right)}$

$= 4^{\frac{1}{2}}\left(1 + \dfrac{3x}{4}\right)^{\frac{1}{2}}$

$= 2\left(1 + \dfrac{3x}{4}\right)^{\frac{1}{2}}$

(ii) $2\left(1 + \dfrac{3x}{2}\right)^{\frac{1}{2}} = 2\left[1 + \dfrac{1}{2}\left(\dfrac{3x}{4}\right) + \dfrac{\left(\frac{1}{2}\right)\left(-\frac{1}{2}\right)}{2!}\left(\dfrac{3x}{4}\right)^2 + \ldots\right]$ ← Only the first three terms are required.

$= 2\left[1 + \dfrac{3x}{8} + \left(-\dfrac{1}{8}\right)\left(\dfrac{9x^2}{16}\right) + \ldots\right]$

$\approx 2 + \dfrac{3x}{4} - \dfrac{9x^2}{64}$ (first three terms)

Expansion valid when $\left|\dfrac{3x}{2}\right| < 1$

$\Rightarrow |x| < \dfrac{2}{3}$.

Example 7.7

Find a, b and c such that

$$\frac{1}{(1 - 2x)(1 + 3x)} \approx a + bx + cx^2$$

and state the values of x for which the expansion is valid.

Solution

$$\frac{1}{(1 - 2x)(1 + 3x)} = (1 - 2x)^{-1}(1 + 3x)^{-1}$$

Using the binomial expansion:

$(1 - 2x)^{-1} = 1 + (-1)(-2x) + \dfrac{(-1)(-2)}{2!}(-2x)^2 + \ldots$

for $|-2x| < 1$, i.e. $|x| < \dfrac{1}{2}$

and

$(1 + 3x)^{-1} = 1 + (-1)(3x) + \dfrac{(-1)(-2)}{2!}(3x)^2 + \ldots$

for $|3x| < 1$, i.e. $|x| < \dfrac{1}{3}$

Each bracket involves an infinite series, but when you multiply out you can ignore any terms which are higher than quadratic.

$\Rightarrow (1 - 2x)^{-1}(1 + 3x)^{-1} = (1 + 2x + 4x^2 + \ldots)(1 - 3x + 9x^2 + \ldots)$

$= 1 + 2x + 4x^2 - 3x - 6x^2 + 9x^2 + \ldots$

$\approx 1 - x + 7x^2$ (ignoring higher powers of x),

giving $a = 1$, $b = -1$ and $c = 7$.

Both the restrictions are satisfied if $|x| < \dfrac{1}{3}$, which is the stricter restriction.

The binomial expansion may also be used when the first term is the variable. For example

$$(x + 2)^{-1} \text{ may be written as } (2 + x)^{-1} = 2^{-1}\left(1 + \frac{x}{2}\right)^{-1}$$

$$\text{and} \quad (2x - 1)^{-3} = [(-1)(1 - 2x)]^{-3}$$
$$= (-1)^{-3}(1 - 2x)^{-3}$$
$$= -(1 - 2x)^{-3}.$$

Exercise 7.1

① Expand each of the following as a series of ascending powers of x up to and including the term in x^3, stating the set of values of x for which the expansion is valid.

(i) $(1 + x)^{-3}$ (ii) $(1 + 2x)^{-3}$

(iii) $(1 - 2x)^{-3}$

② Expand each of the following as a series of ascending powers of x up to and including the term in x^2, stating the set of values of x for which the expansion is valid.

(i) $(1 + x)^{\frac{1}{2}}$ (ii) $(1 + x)^{-\frac{1}{2}}$

(iii) $(1 + x)^{\frac{1}{4}}$ (iv) $(1 + x)^{-\frac{1}{4}}$

③ Expand each of the following as a series of ascending powers of x up to and including the term in x^3, stating the set of values of x for which the expansion is valid.

(i) $\left(1 + \frac{x}{3}\right)^{-2}$ (ii) $\left(1 + \frac{2x}{3}\right)^{-2}$

(iii) $\left(1 - \frac{2x}{3}\right)^{-2}$

④ For each of the expressions below

(a) write down the first three non-zero terms in their expansions as a series of ascending powers of x

(b) state the values of x for which the expansions is valid

(c) substitute $x = 0.1$ in both the function and its expansion and calculate the percentage error, where

$$\text{percentage error} = \frac{\text{absolute error}}{\text{true value}} \times 100\%.$$

(i) $(1 + x)^{-2}$ (ii) $(1 + 2x)^{-1}$

(iii) $\sqrt{1 - x^2}$

⑤ (i) Write down the expansion of $(1 - x)^3$.

(ii) Find the first three terms in the expansion of $(1 + x)^{-4}$ in ascending powers of x. For what values of x is this expansion valid?

(iii) When the expansion is valid, $\dfrac{(1 - x)^3}{(1 + x)^4}$ can be written as

$1 + ax + bx^2 +$ higher powers of x.

Find the values of a and b.

⑥ (i) Show that $\dfrac{1}{\sqrt{4 + x}} = \frac{1}{2}\left(1 + \frac{x}{4}\right)^{-\frac{1}{2}}$

(ii) Write down the first three terms in the binomial expansion of

$\left(1 + \frac{x}{4}\right)^{-\frac{1}{2}}$ in ascending powers of x, stating the range of values of x for which this expansion is valid.

(iii) Find the first three terms in the expansion of $\dfrac{2(1 - x)}{\sqrt{4 + x}}$ in ascending powers of x, for small values of x.

⑦ (i) Write down the first three terms in the binomial expansion of $\dfrac{1}{(1 + 2x)(1 + x)}$ in ascending powers of x.

(ii) For what values of x is this expansion valid?

⑧ Find a quadratic approximation for $\dfrac{(3 - x)(1 + x)}{(4 - x)}$ and state the values of x for which this is a valid approximation.

In the next section you will learn an alternative way of approaching this question.

⑨ The expansion of $(a + bx)^{-3}$ may be approximated by $\frac{1}{8} + \frac{3}{16}x + cx^2$.

  (i)   Find the values of the constants a, b and c.

  (ii) For what range of values of x is the expansion valid?

⑩ Find a cubic approximation for $\frac{2}{(x + 1)(x^2 + 1)}$, stating the range of values of x for which the expansion is valid.

⑪ (i)   Find a quadratic function that approximates to $\frac{1}{\sqrt[3]{(1 - 3x)^2}}$ for values of x close to zero.

  (ii) For what values of x is the approximation valid?

⑫ (i)   Find the first three terms in the binomial expansion of $\frac{4 + 2x}{(2x - 1)(x^2 + 1)}$

  (ii) For what range of values is the expansion valid?

⑬ (i)   Write $\sqrt{7}$ in the form $a\sqrt{b}$ where a is an integer and $b < 1$.

  (ii) Use this to find an approximate value for $\sqrt{7}$.

  (iii) Comment on how good the approximation is.

⑭ (i)   Rearrange $\sqrt{x - 1}$, where $x > 1$, in a form where the binomial expansion can be used.

  (ii) Using the first four terms of the expansion, find an approximation for $\sqrt{15}$ to 3 decimal places.

2 Simplifying algebraic expressions

If f(x) and g(x) are polynomials, the expression $\frac{\text{f}(x)}{\text{g}(x)}$ is an **algebraic fraction** and so it is a **rational expression**. There are many occasions in mathematics when a problem reduces to the manipulation of algebraic fractions, and the rules are exactly the same as those for numerical fractions.

Simplifying fractions

To simplify a fraction, you look for a factor common to both the numerator (top line) and the denominator (bottom line) and cancel by it. Remember that when you cancel in this way, you are dividing the top and bottom of the fraction by the same quantity.

For example, in arithmetic

$$\frac{15}{20} = \frac{5 \times 3}{5 \times 4} = \frac{3}{4}$$ ← Dividing top and bottom by 5.

and in algebra

$$\frac{6a}{9a^2} = \frac{2 \times 3 \times a}{3 \times 3 \times a \times a} = \frac{2}{3a}$$ ← Dividing top and bottom by $3a$.

Notice how you must **factorise** both the numerator and denominator before cancelling, since it is only possible to cancel by a **common factor**. In some cases this involves putting brackets in.

$$\frac{2a + 4}{a^2 - 4} = \frac{2(a + 2)}{(a + 2)(a - 2)} = \frac{2}{(a - 2)}$$

Multiplying and dividing fractions

Multiplying fractions is shown here.

$$\frac{10a}{3b^2} \times \frac{9ab}{25} = \frac{10a \times 9ab}{3b^2 \times 25}$$

Multiply the numerators and multiply the denominators.

$$= \frac{2 \times 5 \times a \times 3 \times 3 \times a \times b}{3 \times b \times b \times 5 \times 5}$$

Look for common factors to cancel.

$$= \frac{2 \times a \times 3 \times a}{b \times 5}$$

Divide top and bottom by $3 \times 5 \times b$.

$$= \frac{6a^2}{5b}$$

As with simplifying, it is often necessary to factorise any algebraic expression first.

$$\frac{a^2 + 3a + 2}{9} \times \frac{12}{a + 1} = \frac{(a + 1)(a + 2)}{3 \times 3} \times \frac{3 \times 4}{(a + 1)}$$

$$= \frac{4(a + 2)}{3}$$

Dividing by a fraction is the same as multiplying by the reciprocal of the fraction, just as dividing by 2 is the same as multiplying by $\frac{1}{2}$.

For example:

$$\frac{12}{x^2 - 1} \div \frac{4}{x + 1} = \frac{12}{(x + 1)(x - 1)} \times \frac{(x + 1)}{4}$$

Dividing by $\frac{4}{x + 1}$ is equivalent to multiplying by $\frac{x + 1}{4}$.

$$= \frac{12(x + 1)}{4(x + 1)(x - 1)}$$

Divide top and bottom by $4(x + 1)$.

$$= \frac{3}{(x - 1)}$$

Addition and subtraction of fractions

To add or subtract two fractions they must be replaced by equivalent fractions, both of which have the same denominator.

For example, in arithmetic

$$\frac{2}{3} + \frac{1}{4} = \frac{8}{12} + \frac{3}{12} = \frac{11}{12}$$

Similarly, in algebra

$$\frac{2x}{3} + \frac{x}{4} = \frac{8x}{12} + \frac{3x}{12} = \frac{11x}{12}$$

and

$$\frac{2}{3x} + \frac{1}{4x} = \frac{8}{12x} + \frac{3}{12x} = \frac{11}{12x}$$

Notice how you only need $12x$ here, not $12x^2$.

You must take particular care when the subtraction of fractions introduces a sign change. For example:

$$\frac{4x - 3}{6} - \frac{2x + 1}{4} = \frac{2(4x - 3) - 3(2x + 1)}{12}$$

$$= \frac{8x - 6 - 6x - 3}{12}$$

$$= \frac{2x - 9}{12}$$

Notice how, in addition and subtraction, the new denominator is the **lowest common multiple** of the original denominators. When two denominators have **no common factor**, their product gives the new denominator. For example:

$$\frac{2}{y+3} + \frac{3}{y-2} = \frac{2(y-2) + 3(y+3)}{(y+3)(y-2)}$$

$$= \frac{2y - 4 + 3y + 9}{(y+3)(y-2)}$$

$$= \frac{5y + 5}{(y+3)(y-2)}$$

$$= \frac{5(y+1)}{(y+3)(y-2)}$$

It may be necessary to factorise denominators in order to identify common factors, as shown here.

$$\frac{2b}{a^2 - b^2} - \frac{3}{a+b} = \frac{2b}{(a+b)(a-b)} - \frac{3}{(a+b)}$$

$(a+b)$ is a common factor.

$$= \frac{2b - 3(a-b)}{(a+b)(a-b)}$$

$$= \frac{5b - 3a}{(a+b)(a-b)}$$

Algebraic division

Example 7.8

Divide $2x^2 - 2x - 12$ by $x + 2$.

Solution

The division can be written as $\dfrac{2x^2 - 2x - 12}{x+2}$.

$$\frac{2x^2 - 2x - 12}{x+2} = \frac{2(x^2 - x - 6)}{x+2}$$

There is a numerical factor of 2 in the numerator.

$$= \frac{2(x-3)(x+2)}{(x+2)}$$

You can now factorise the numerator.

$$= 2(x-3)$$

Now you can divide top and bottom by $(x+2)$.

Prior knowledge

You need to know the factor theorem and be confident in algebraic division. This is covered in Review: Algebra (2) (page 61).

Sometimes factorising the numerator is not as straightforward, as in the following example.

Example 7.9

Simplify $\dfrac{2x^3 - 3x^2 + x - 6}{x - 2}$.

Solution

> The **factor theorem** says that when $(x - a)$ is a factor of $f(x)$ then $f(a) = 0$.

Start by using the factor theorem to check if $(x - 2)$ is a factor of $2x^3 - 3x^2 + x - 6$.

Substituting $x = 2$ gives $2(2)^3 - 3(2)^2 + 2 - 6 = 0$
so $(x - 2)$ is a factor.

Using algebraic division or inspection gives
$$2x^3 - 3x^2 + x - 6 = (x - 2)(2x^2 + x + 3)$$

So $\dfrac{2x^3 - 3x^2 + x - 6}{x - 2} = \dfrac{(x - 2)(2x^2 + x + 3)}{x - 2}$

$$= 2x^2 + x + 3$$

> Divide top and bottom by $(x - 2)$.

> **Note**
>
> A **proper** algebraic fraction is a fraction where the order (highest power) of the numerator (top line) is strictly less than that of the denominator (bottom line). The following, for example, are proper fractions:
> $$\dfrac{2}{1 + x}, \quad \dfrac{5x - 1}{x^2 - 3}, \quad \dfrac{7x}{(x + 1)(x - 2)}.$$
> Examples of **improper** fractions are
> $$\dfrac{2x}{x + 1} \text{ (which can be written as } 2 - \dfrac{2}{x + 1}\text{)}$$
> and
> $$\dfrac{x^2}{x - 2} \text{ (which can be written as } x + 2 + \dfrac{4}{x - 2}\text{)}.$$

Example 7.10

(i) Write $\dfrac{x + 1}{x + 2}$ in the form $a + \dfrac{b}{x + 2}$.

(ii) Hence show how the graph of $y = \dfrac{x + 1}{x + 2}$ can be sketched by starting with the graph of $y = \dfrac{1}{x}$.

Solution

(i) $\dfrac{x + 1}{x + 2} = \dfrac{(x + 2) - 1}{x + 2}$

$$= \dfrac{x + 2}{x + 2} - \dfrac{1}{x + 2}$$

$$= 1 - \dfrac{1}{x + 2}$$

> **Note**
>
> Alternatively this could be done using long division.

→

(ii) Starting with the graph of $y = \frac{1}{x}$, the graph of $y = \frac{1}{x+2}$ is obtained

using a translation of $\begin{pmatrix} -2 \\ 0 \end{pmatrix}$.

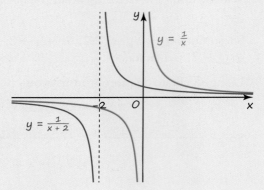

Figure 7.1

Next reflect this in the x-axis to give $y = -\dfrac{1}{x+2}$.

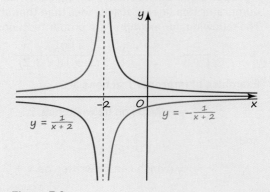

Figure 7.2

Finally, a translation of $\begin{pmatrix} 0 \\ 1 \end{pmatrix}$ will give the graph of $y = 1 - \dfrac{1}{x+2}$.

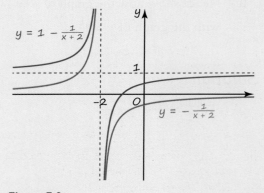

Figure 7.3

① Describe the transformation that transforms $y = \dfrac{1}{x}$ into $y = \dfrac{1}{x-3}$.

② Simplify the following algebraic fractions.

(i) $\dfrac{(x+2)(x-1)}{(x-1)}$ (ii) $\dfrac{(x-1)}{(x+2)(x-1)}$

(iii) $\dfrac{(x-1)^2}{(x+2)(x-1)}$ (iv) $\dfrac{(x+2)^2}{(x+2)(x-1)}$

③ Simplify the following algebraic fractions.

(i) $\dfrac{x^2-9}{x^2-9x+18}$

(ii) $\dfrac{4x^2-25}{4x^2+20x+25}$

(iii) $\dfrac{a^2-b^2}{2a^2+ab-b^2}$

④ Write each of the expressions as a single fraction in its simplest form.

(i) $\dfrac{1}{4x}+\dfrac{1}{5x}$ (ii) $\dfrac{1}{a+1}+\dfrac{1}{a-1}$

(iii) $\dfrac{x}{3}+\dfrac{x+1}{4}$

⑤ Write each of the following expressions as a single fraction in its simplest form.

(i) $\dfrac{x}{x^2-4}-\dfrac{1}{x-2}$

(ii) $\dfrac{2}{b^2+2b+1}-\dfrac{3}{b+1}$

(iii) $\dfrac{2}{a+2}-\dfrac{a-2}{2a^2+a-6}$

⑥ Simplify the following algebraic fractions.

(i) $\dfrac{4x^2+6x+2}{x+1}$ (ii) $\dfrac{2x^3+3x^2+x}{x+1}$

(iii) $\dfrac{2ax^2+3ax+2a}{x+1}$

⑦ Simplify the following algebraic fractions.

(i) $\dfrac{1}{x-2}+\dfrac{1}{x}+\dfrac{1}{x+2}$

(ii) $\dfrac{1}{3-a}+\dfrac{1}{a}+\dfrac{1}{3+a}$

(iii) $\dfrac{1}{b}+\dfrac{1}{(b-1)}+\dfrac{1}{(b-1)^2}$

⑧ Write each of the following in the form $y = a + \dfrac{b}{f(x)}$ and hence sketch the graph in each case.

(i) $y = \dfrac{5x-3}{x+1}$ (ii) $y = \dfrac{4x+2}{x-2}$

(iii) $y = \dfrac{6x-3}{2x+1}$

⑨ Write each of the following in the form $ax + b + \dfrac{c}{f(x)}$.

(i) $\dfrac{2x^2+3x+5}{x+1}$ (ii) $\dfrac{3x^2+x-6}{x+2}$

(iii) $\dfrac{x^2-6x+7}{x-1}$

⑩ Write $\dfrac{3x^3+2x^2-5x-4}{x+2}$ in the form $ax^2+bx+c+\dfrac{d}{x+2}$.

⑪ Use transformations to sketch the graph of $y = \dfrac{2-3x}{x}$, starting with the graph of $y = \dfrac{1}{x}$.

3 Partial fractions

ACTIVITY 7.4

(i) Write $\dfrac{2}{3x+1}+\dfrac{3}{2x-3}$ as a single fraction.

(ii) Find the first three terms in the binomial expansions of both the original expression and the single fraction.

(iii) Which did you find easier?

You probably discovered in Activity 7.4 that, when using a binomial expansion, it is easier to work with two or more separate fractions than with a single more complicated one.

There are other situations where it is useful to be able to split a fraction into two or more simpler ones. For example, in Chapter 10 you will learn to integrate an expression such as $\dfrac{1}{(1 + 2x)(1 + x)}$ by first writing it as $\dfrac{2}{(1 + 2x)} - \dfrac{1}{(1 + x)}$.

The process of taking an expression such as $\dfrac{1}{(1 + 2x)(1 + x)}$ and writing it in the form $\dfrac{2}{(1 + 2x)} - \dfrac{1}{(1 + x)}$ is called expressing the algebraic fraction in **partial fractions**, and you will now look at how this can be done.

It can be shown that, when a proper algebraic fraction is decomposed into its partial fractions, each of the partial fractions will be a proper fraction.

When finding partial fractions you must always assume the most general numerator possible, and the method for doing this is illustrated in the following examples.

Type 1: Denominators of the form $(ax + b)(cx + d)$

Example 7.11

Express $\dfrac{4 + x}{(1 + x)(2 - x)}$ as a sum of partial fractions.

Discussion point

➜ How could you have worked out these values for A and B?

Solution

Assume $\dfrac{4 + x}{(1 + x)(2 - x)} \equiv \dfrac{A}{1 + x} + \dfrac{B}{2 - x}$

> The numerators must be constants, so that these are proper fractions.

Multiplying both sides by $(1 + x)(2 - x)$ gives

$4 + x \equiv A(2 - x) + B(1 + x)$

> This is an identity: it is true for all values of x.

$A = 1$ and $B = 2$

> These values for A and B make the identity true for all values of x.

So $\dfrac{4 + x}{(1 + x)(2 - x)} = \dfrac{1}{1 + x} + \dfrac{2}{2 - x}$

In general you would need to use a structured method to find the constants A and B rather than hope that you could guess them correctly. There are two possible methods, and the following example will use each of these in turn.

Example 7.12

Express $\dfrac{x}{(x - 2)(x - 3)}$ as a sum of partial fractions.

Solution

Let $\dfrac{x}{(x - 2)(x - 3)} \equiv \dfrac{A}{(x - 2)} + \dfrac{B}{(x - 3)}$

> Multiply through by $(x - 2)(x - 3)$.

$\Rightarrow \qquad x \equiv A(x - 3) + B(x - 2)$

Method 1: Substitution

Substituting $x = 2 \quad \Rightarrow \quad 2 = A(-1) + B(0)$

$\Rightarrow \quad A = -2$

Substituting $x = 3 \quad \Rightarrow \quad 3 = A(0) + B(1)$

$\Rightarrow \quad B = 3$

> You can substitute any two values of x, but the easiest to use in this case are $x = 2$ and $x = 3$, since each makes the value of one bracket in the identity equal to zero.

So $\dfrac{x}{(x - 2)(x - 3)} = \dfrac{3}{x - 3} - \dfrac{2}{x - 2}$

$$x \equiv A(x - 3) + B(x - 2)$$
$$\Rightarrow \quad x \equiv Ax - 3A + Bx - 2B$$
$$\Rightarrow \quad x \equiv (A + B)x - 3A - 2B$$

> Write the right-hand side as a polynomial in x, and then compare coefficients.

Equating the constant terms: $0 = -3A - 2B$

Equating the coefficients of x: $1 = A + B$

> These are simultaneous equations in A and B.

Solving these simultaneous equations gives $A = -2$ and $B = 3$, as before.

Discussion point

In each of these methods the identity ($\equiv$) was later replaced by an equality ($=$).

→ Why was this done?

In some cases it is necessary to factorise the denominator before finding the partial fractions.

Example 7.13

Express $\dfrac{2}{4 - x^2}$ as a sum of partial fractions.

Solution

Let $\quad \dfrac{2}{4 - x^2} \equiv \dfrac{2}{(2 + x)(2 - x)}$

> Factorising the denominator.

$$\equiv \dfrac{A}{2 + x} + \dfrac{B}{2 - x}$$

Multiplying both sides by $(2 + x)(2 - x)$ gives

$$2 \equiv A(2 - x) + B(2 + x).$$

Substituting $x = 2$ gives $2 = 4B$, so $B = \frac{1}{2}$.

Substituting $x = -2$ gives $2 = 4A$, so $A = \frac{1}{2}$.

Using these values:

$$\dfrac{2}{(2 + x)(2 - x)} \equiv \dfrac{\frac{1}{2}}{(2 + x)} + \dfrac{\frac{1}{2}}{(2 - x)}$$

$$\equiv \dfrac{1}{2(2 + x)} + \dfrac{1}{2(2 - x)}$$

In the next example there are three factors in the denominator.

Example 7.14

Express $\dfrac{11x + 1}{(x + 1)(x - 1)(2x + 1)}$ as a sum of partial fractions.

Solution

Let $\quad \dfrac{11x + 1}{(x + 1)(x - 1)(2x + 1)} \equiv \dfrac{A}{x + 1} + \dfrac{B}{x - 1} + \dfrac{C}{2x + 1}$

Multiplying both sides by $(x + 1)(x - 1)(2x + 1)$ gives

$$11x + 1 \equiv A(x - 1)(2x + 1) + B(x + 1)(2x + 1) + C(x + 1)(x - 1)$$

Substituting $x = 1$ gives $12 = 6B$, so $B = 2$.

Substituting $x = -1$ gives $-10 = 2A$, so $A = -5$.

Substituting $x = -\frac{1}{2}$ gives $-4.5 = \dfrac{-3C}{4}$, so $C = 6$.

Using these values:

$$\frac{11x + 1}{(x + 1)(x - 1)(2x + 1)} \equiv \frac{2}{x - 1} - \frac{5}{x + 1} + \frac{6}{2x + 1}$$

Type 2: Denominators of the form $(ax + b)(cx + d)^2$

The factor $(cx + d)^2$ is of order 2, so it would have an numerator of order 1 in the partial fractions. However, in the case of a repeated factor, there is a simpler form.

Consider
$$\frac{4x + 5}{(2x + 1)^2}$$

This can be written as
$$\frac{2(2x + 1) + 3}{(2x + 1)^2}$$

Note

In this form, both the numerators are constant.

$$\equiv \frac{2(2x + 1)}{(2x + 1)^2} + \frac{3}{(2x + 1)^2}$$

$$\equiv \frac{2}{(2x + 1)} + \frac{3}{(2x + 1)^2}$$

In the same way, any fraction of the form $\dfrac{px + q}{(cx + d)^2}$ can be written as

$$\frac{A}{(cx + d)} + \frac{B}{(cx + d)^2}$$

When expressing an algebraic fraction in partial fractions, you are aiming to find the simplest partial fractions possible, so you would want the form where the numerators are constant.

Example 7.15

Express $\dfrac{x + 1}{(x - 1)(x - 2)^2}$ as a sum of partial fractions.

Note

Substituting another value of x, or equating the coefficients of a different term would give a third equation but it would involve all three of A, B and C.

Solution

Let $\dfrac{x + 1}{(x - 1)(x - 2)^2} \equiv \dfrac{A}{(x - 1)} + \dfrac{B}{(x - 2)} + \dfrac{C}{(x - 2)^2}$

Multiplying both sides by $(x - 1)(x - 2)^2$ gives

$$x + 1 \equiv A(x - 2)^2 + B(x - 1)(x - 2) + C(x - 1)$$

Substituting $x = 1 \implies 2 = A(-1)^2 \implies A = 2$

Substituting $x = 2 \implies 3 = C$

Equating coefficients of $x^2 \implies 0 = A + B \implies B = -2$

This gives

$$\frac{x + 1}{(x - 1)(x - 2)^2} \equiv \frac{2}{x - 1} - \frac{2}{x - 2} + \frac{3}{(x - 2)^2}$$

① Simplify $\dfrac{3}{(x-1)} + \dfrac{2}{x+5}$

② (i) Write $\dfrac{1}{x(x-1)}$ in the form

$\dfrac{A}{x} + \dfrac{B}{x-1}$

(ii) Write $\dfrac{1}{x(x+1)}$ in the form $\dfrac{A}{x} + \dfrac{B}{x+1}$

(iii) Write $\dfrac{2}{(x+1)(x-1)}$ in the form

$\dfrac{A}{x+1} + \dfrac{B}{x-1}$

③ Express each of the following as a sum of partial fractions.

(i) $\dfrac{3}{(x-1)(x+2)}$ (ii) $\dfrac{5}{(x-2)(x+3)}$

(iii) $\dfrac{7}{(x-3)(x+4)}$

④ Express each of the following as a sum of partial fractions.

(i) $\dfrac{3}{(x-2)(2x-1)}$

(ii) $\dfrac{8}{(x-3)(3x-1)}$

(iii) $\dfrac{15}{(x-4)(4x-1)}$

⑤ Express each of the following as a sum of partial fractions.

(i) $\dfrac{2}{x^2-2x}$ (ii) $\dfrac{2}{x^2-4x+3}$

(iii) $\dfrac{2}{x^2-6x+8}$

⑥ Express each of the following as a sum of partial fractions.

(i) $\dfrac{5-2x}{(x-1)^2(x+2)}$

(ii) $\dfrac{4}{(1-3x)(1-x)^2}$

⑦ Express each of the following as a sum of partial fractions.

(i) $\dfrac{2x-2}{(1+x)(1+2x)(1+3x)}$

(ii) $\dfrac{2x-2}{(1+x)(2+x)(3+x)}$

⑧ Express each of the following as a sum of partial fractions.

(i) $\dfrac{7x+3}{x(x-1)(x+1)}$

(ii) $\dfrac{32-4x}{x(2-x)(2+x)}$

⑨ (i) Express $\dfrac{4}{(1-3x)(1-x)^2}$ as a sum of partial fractions.

(ii) Hence find the first three terms in the binomial expansion of $\dfrac{4}{(1-3x)(1-x)^2}$

(iii) For what values of x is the expansion valid?

⑩ (i) Write $\dfrac{6x-8}{(x^2+1)(x+1)}$ in the form

$\dfrac{Ax+B}{x^2+1} + \dfrac{C}{x+1}$

(ii) Hence find the first three terms in the binomial expansion of

$\dfrac{6x-8}{(x^2+1)(x+1)}$, stating the values of

x for which the expansion is valid.

⑪ (i) Write $\dfrac{2x^2-3x+4}{(x-1)^3}$ in the form

$\dfrac{A}{(x-1)} + \dfrac{B}{(x-1)^2} + \dfrac{C}{(x-1)^3}$

(ii) Hence find the first three terms in the binomial expansion of

$\dfrac{2x^2-3x+4}{(x-1)^3}$, stating the values of

x for which the expansion is valid.

Partial fractions

LEARNING OUTCOMES

When you have completed this chapter, you should be able to:

➤ extend the binomial expansion to any rational n

➤ use the binomial expansion with any rational n for approximation

➤ be aware that the expansion is valid for $\left|\dfrac{bx}{a}\right| < 1$

➤ simplify rational expressions by:

  ○ factorising and cancelling

  ○ algebraic division (linear expressions only)

➤ decompose rational functions into partial fractions (denominators not more complicated than squared linear terms and with no more than 3 terms, numerators constant or linear).

KEY POINTS

1 The general binomial expansion for $n \in \mathbb{R}$ is
$$(1 + x)^n = 1 + nx + \frac{n(n - 1)}{2!}x^2 + \frac{n(n - 1)(n - 2)}{3!}x^3 + \dots$$
In the special case when $n \in \mathbb{N}$, the series expansion is finite and valid for all n. When $n \notin \mathbb{N}$, the series expansion is non-terminating (infinite) and valid only if $|x| < 1$.

2 When $n \notin \mathbb{N}$, $(a + x)^n$ should be written as $a^n\left(1 + \dfrac{x}{a}\right)^n$ before obtaining the binomial expansion.

3 When multiplying algebraic fractions, you can only cancel when the same factor occurs in both the numerator and the denominator.

4 When adding or subtracting algebraic fractions, you first need to find a common denominator.

5 A proper algebraic fraction with a denominator which factorises can be decomposed into a sum of proper partial fractions.

6 The following forms of partial fractions should be used:
$$\frac{px + q}{(ax + b)(cx + d)} \equiv \frac{A}{ax + b} + \frac{B}{cx + d}$$
$$\frac{px + q}{(ax + b)(cx + d)^2} \equiv \frac{A}{ax + b} + \frac{B}{cx + d} + \frac{C}{(cx + d)^2}.$$

FUTURE USES

Partial fractions are often useful in integration. You will use them in Chapter 10.

Trigonometric identities

8

Many waves can be modelled as a sine curve. Estimate the wavelength and the amplitude in metres of the wave in the picture above (see Figure 8.1).

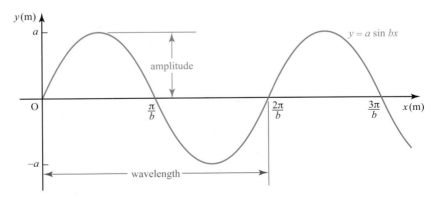

Figure 8.1

→ Use your estimates to suggest values of a and b which would make $y = a \sin bx$ a suitable model for the curve.

→ Do you think a sine curve is a good model for the wave?

You need to be able to use the general formula for the area of a triangle $\left(\frac{1}{2}ab\sin C\right)$ and use exact values of sin, cos and tan for common angles such as $60°$ or $\frac{\pi}{3}$ radians. You also need to be able to solve a trigonometric equation and use small-angle approximations – see Review: The sine and cosine rules, and Chapter 2 and Chapter 6.

The photograph on the previous page shows just one of the countless examples of waves and oscillations that are part of the world around us. Because such phenomena are often modelled by trigonometric (and especially sine and cosine) functions, trigonometry has an importance in mathematics far beyond its origins in right-angled triangles.

1 Compound angle formulae

ACTIVITY 8.1

Find an acute angle θ such that $\sin(\theta + 60°) = \cos(\theta - 60°)$.

Hint: Try drawing graphs and searching for a numerical solution.

You should be able to find the solution to Activity 8.1 using the suggested method, but replacing $60°$ by, for example, $35°$ would make it more difficult to find an accurate value for θ. In this chapter you will meet some formulae which help you to solve such equations more efficiently.

To find an expression for $\sin(\theta + 60°)$, you would use the **compound angle formula**

$$\sin(\theta + \phi) = \sin\theta\cos\phi + \cos\theta\sin\phi.$$

ϕ is the Greek letter phi.

This is proved below in the case when θ and ϕ are acute angles. It is, however, true for all values of the angles. It is an **identity**.

> It is tempting to think that $\sin(\theta + 60°)$ should equal $\sin\theta + \sin 60°$, but this is not the case. For example, putting $\theta = 30°$ gives $\sin(\theta + 60°) = 1$, but $\sin\theta + \sin 60° \approx 1.366$.

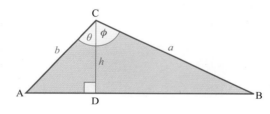

Figure 8.2

For the triangle in Figure 8.2

Area of a triangle $= \frac{1}{2}ab\sin C.$

$$\text{area of } \Delta ABC = \text{area of } \Delta ADC + \text{area of } \Delta DBC$$

$h = a\cos\phi$ from ΔBDC.

$$\frac{1}{2}ab\sin(\theta + \phi) = \frac{1}{2}bh\sin\theta + \frac{1}{2}ah\sin\phi$$

$h = b\cos\theta$ from ΔADC.

$$\Rightarrow \frac{1}{2}ab\sin(\theta + \phi) = \frac{1}{2}b(a\cos\phi)\sin\theta + \frac{1}{2}a(b\cos\theta)\sin\phi$$

$$\Rightarrow \frac{1}{2}ab\sin(\theta + \phi) = \frac{1}{2}ab\sin\theta\cos\phi + \frac{1}{2}ab\cos\theta\sin\phi$$

Dividing through by $\frac{1}{2}ab$ gives

$$\sin(\theta + \phi) = \sin\theta\cos\phi + \cos\theta\sin\phi \qquad ①$$

This is the first of the compound angle formulae (or expansions), and it can be used to prove several more. These are true for all values of θ and ϕ.

Replacing ϕ by $-\phi$ in ① gives

The graph of $y = \cos\phi$ is symmetrical about the y-axis so $\cos(-\phi) = \cos\phi$.

$$\sin(\theta - \phi) = \sin\theta\cos(-\phi) + \cos\theta\sin(-\phi)$$

The graph of $y = \sin\phi$ has rotational symmetry about the origin so $\sin(-\phi) = -\sin\phi$.

$$\Rightarrow \sin(\theta - \phi) = \sin\theta\cos\phi - \cos\theta\sin\phi$$

Hint:
$\sin(90° − A) = \cos A$
and
$\cos(90° − A) = \sin A.$

Hint: Divide the top and bottom lines by $\cos\theta\cos\phi$ to give an expansion in terms of $\tan\theta$ and $\tan\phi$.

ACTIVITY 8.2

1 There are four more compound angle formulae. Work through this activity in order to derive them.

(i) To find an expansion for $\cos(\theta − \phi)$ replace θ by $(90° − \theta)$ in the expansion of $\sin(\theta + \phi)$.

(ii) To find an expansion for $\cos(\theta + \phi)$ replace ϕ by $(−\phi)$ in the expansion of $\cos(\theta − \phi)$.

(iii) To find an expansion for $\tan(\theta + \phi)$, write $\tan(\theta + \phi)$ as

$\dfrac{\sin(\theta + \phi)}{\cos(\theta + \phi)}$ and use the expansions of $\sin(\theta + \phi)$ and $\cos(\theta + \phi)$.

(iv) To find an expansion for $\tan(\theta − \phi)$ in terms of $\tan\theta$ and $\tan\phi$, replace ϕ by $(−\phi)$ in the expansion of $\tan(\theta + \phi)$.

2 Are your results valid for all values of θ and ϕ? Test your results with $\theta = 60°$, $\phi = 30°$.

Check that your results work for angles in radians.

The four results obtained in Activity 8.2, together with the two previous results, form the set of compound angle formulae.

Compound angle formulae

$$\sin(\theta + \phi) = \sin\theta\cos\phi + \cos\theta\sin\phi$$
$$\sin(\theta − \phi) = \sin\theta\cos\phi − \cos\theta\sin\phi$$
$$\cos(\theta + \phi) = \cos\theta\cos\phi − \sin\theta\sin\phi$$
$$\cos(\theta − \phi) = \cos\theta\cos\phi + \sin\theta\sin\phi$$
$$\tan(\theta + \phi) = \frac{\tan\theta + \tan\phi}{1 − \tan\theta\tan\phi} \qquad (\theta + \phi) \neq 90°, 270°, \dots$$
$$\tan(\theta − \phi) = \frac{\tan\theta − \tan\phi}{1 + \tan\theta\tan\phi} \qquad (\theta − \phi) \neq 90°, 270°, \dots$$

You can now solve Activity 8.1 more easily. To find an acute angle θ such that

$$\sin(\theta + 60°) = \cos(\theta − 60°),$$

expand each side using the compound angle formulae.

Expand the left-hand side:

$$\sin(\theta + 60°) = \sin\theta\cos 60° + \cos\theta\sin 60°$$
$$= \tfrac{1}{2}\sin\theta + \frac{\sqrt{3}}{2}\cos\theta \qquad ①$$

Expand the right-hand side:

$$\cos(\theta − 60°) = \cos\theta\cos 60° + \sin\theta\sin 60°$$
$$= \tfrac{1}{2}\cos\theta + \frac{\sqrt{3}}{2}\sin\theta \qquad ②$$

Compound angle formulae

> Be careful when you divide by a trigonometric function — you must always check that the function is not equal to zero, otherwise you will miss some of the roots to the equation. In this case you know that $\cos\theta$ is not equal to zero (since $\cos\theta$ is not equal to $\sin\theta$ when $\cos\theta = 0$) so you are safe to divide by it. Example 8.3 on page 175 is another illustration of this.

Equating ① and ② gives

$$\tfrac{1}{2}\sin\theta + \tfrac{\sqrt{3}}{2}\cos\theta = \tfrac{1}{2}\cos\theta + \tfrac{\sqrt{3}}{2}\sin\theta$$

So $\quad \sin\theta + \sqrt{3}\cos\theta = \cos\theta + \sqrt{3}\sin\theta \quad$ ← Multiply each term by 2.

$\Rightarrow \quad \sqrt{3}\cos\theta - \cos\theta = \sqrt{3}\sin\theta - \sin\theta$

$\Rightarrow \quad (\sqrt{3}-1)\cos\theta = (\sqrt{3}-1)\sin\theta \quad$ ← Collect like terms.

$\Rightarrow \quad \cos\theta = \sin\theta \quad$ ← Divide each side by $(\sqrt{3}-1)$.

You need to rewrite the equation so it is terms of just one trigonometric ratio so you can solve it.

$$\cos\theta = \sin\theta$$

$\Rightarrow \quad 1 = \dfrac{\sin\theta}{\cos\theta} \quad$ ← Divide each side by $\cos\theta$.

$\Rightarrow \quad 1 = \tan\theta \quad$ ← Use $\tan\theta \equiv \dfrac{\sin\theta}{\cos\theta}$

$\Rightarrow \quad \theta = 45°$

Since an acute angle ($\theta < 90°$) was required, this is the only root.

Example 8.1

Simplify $\cos\theta\cos 3\theta - \sin\theta\sin 3\theta$.

Solution

The formula which has the same pattern of $\cos\cos - \sin\sin$ is

$$\cos(\theta + \phi) = \cos\theta\cos\phi - \sin\theta\sin\phi$$

Using this, and replacing ϕ by 3θ, gives

$$\cos\theta\cos 3\theta - \sin\theta\sin 3\theta = \cos(\theta + 3\theta)$$
$$= \cos 4\theta$$

Example 8.2

Find an approximate expression for $\cos\left(\dfrac{\pi}{3} - \theta\right)$ for small values of θ.

Solution

Expanding gives

$$\cos\left(\tfrac{\pi}{3} - \theta\right) = \cos\tfrac{\pi}{3}\cos\theta + \sin\tfrac{\pi}{3}\sin\theta$$
$$= \tfrac{1}{2}\cos\theta + \tfrac{\sqrt{3}}{2}\sin\theta \quad$$ ← Use exact values for $\cos\tfrac{\pi}{3}$ and $\sin\tfrac{\pi}{3}$.

So when θ is small

$$\cos\theta \approx 1 - \tfrac{\theta^2}{2} \quad \text{and} \quad \sin\theta \approx \theta \quad$$ ← For small angles in radians

$$\cos\left(\tfrac{\pi}{3} - \theta\right) \approx \tfrac{1}{2}\left(1 - \tfrac{\theta^2}{2}\right) + \tfrac{\sqrt{3}}{2}\theta$$
$$\approx \tfrac{1}{2} + \tfrac{\sqrt{3}}{2}\theta - \tfrac{1}{4}\theta^2$$

172

① Use the compound angle formulae to expand each of the following expressions.

(i) $\sin(\theta + 45°)$ (ii) $\sin(45° - \theta)$

(iii) $\cos(\theta - 45°)$ (iv) $\cos(\theta + 45°)$

(v) $\tan(\theta + 45°)$ (vi) $\tan(45° - \theta)$

② Match together the equivalent expressions.

$\cos\theta\cos 3\theta - \sin\theta\sin 3\theta$

$\sin 2\theta\cos\theta - \cos 2\theta\sin\theta$

$\cos\theta\cos\theta - \sin\theta\sin\theta$

$\cos\theta\cos\theta + \sin\theta\sin\theta$

$\sin 2\theta\cos\theta + \cos 2\theta\sin\theta$

$\sin\theta$

1

$\sin 3\theta$

$\cos 4\theta\cos 2\theta + \sin 4\theta\sin 2\theta$

$\cos\theta\cos\theta - \sin\theta\sin\theta$

$\cos 4\theta$

$\cos 2\theta$

③ Use the compound angle formulae to find the exact value of each of the following.

(i) $\sin 120°\cos 60° + \cos 120°\sin 60°$

(ii) $\sin 120°\cos 60° - \cos 120°\sin 60°$

(iii) $\cos 120°\cos 60° + \sin 120°\sin 60°$

(iv) $\cos 120°\cos 60° - \sin 120°\sin 60°$

④ (i) By writing 75° as 45° + 30° find the exact values of

(a) $\sin 75°$ (b) $\cos 75°$ (c) $\tan 75°$

(ii) By writing 15° as 45° − 30° find the exact values of

(a) $\sin 15°$ (b) $\cos 15°$ (c) $\tan 15°$

⑤ Use the compound angle formulae to write each of the following as surds.

(i) $\sin 165°$ (ii) $\cos 105°$

(iii) $\tan 285°$

⑥ Solve the following equations for values of θ in the range $0° \le \theta \le 180°$.

(i) $\cos(60° + \theta) = \sin\theta$

(ii) $\sin(45° - \theta) = \cos\theta$

(iii) $\tan(45° + \theta) = \tan(45° - \theta)$

(iv) $2\sin\theta = 3\cos(\theta - 60°)$

(v) $\sin\theta = \cos(\theta + 120°)$

⑦ Solve the following equations for values of θ in the range $0 \le \theta \le \pi$.
Give your answers as multiples of π where appropriate.

(i) $\sin\left(\theta + \dfrac{\pi}{4}\right) = \cos\theta$

(ii) $\cos\left(\theta - \dfrac{\pi}{4}\right) = \sin\left(\theta - \dfrac{\pi}{4}\right)$

(iii) $2\cos\left(\theta - \dfrac{\pi}{3}\right) = \cos\left(\theta + \dfrac{\pi}{2}\right)$

⑧ (i) Prove that
$\sin(A + B) + \sin(A - B) \equiv 2\sin A\cos B$.

(ii) Hence solve
$\sin\left(\theta + \dfrac{\pi}{6}\right) + \sin\left(\theta - \dfrac{\pi}{6}\right) = \sqrt{3}$
for $0 \le \theta \le 2\pi$.

⑨ Prove the following identities.

(i) $\cos(A + B) + \cos(A - B) \equiv 2\cos A\cos B$

(ii) $\tan A + \tan B \equiv \dfrac{\sin(A + B)}{\cos A\cos B}$

⑩ When θ is small enough for θ^3 to be ignored, find approximate expressions for the following.

(i) $2\cos\left(\dfrac{\pi}{3} + \theta\right)$

(ii) $\sin(\alpha + \theta)\sin\theta$ where α is not small

(iii) $\tan\left(\dfrac{\pi}{4} - \theta\right)$

⑪ (i) When θ is small enough for θ^3 to be ignored, find approximate expressions for the following.

(a) $\sin\theta\left[\sin\left(\dfrac{\pi}{6} + \theta\right) - \sin\dfrac{\pi}{6}\right]$

(b) $1 - \cos 2\theta$

(ii) Hence find

$\displaystyle\lim_{\theta\to 0}\dfrac{\sin\theta\left[\sin\left(\dfrac{\pi}{6} + \theta\right) - \sin\dfrac{\pi}{6}\right]}{1 - \cos 2\theta}$

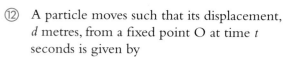

⑫ A particle moves such that its displacement, d metres, from a fixed point O at time t seconds is given by

$$d = \frac{\sqrt{3}}{2}\sin\frac{t}{12} - \frac{1}{2}\cos\frac{t}{12} \text{ for } 0 \leqslant t \leqslant 60.$$

(i) Find the displacement at $t = 0$.

(ii) Show that $d = \sin\left(\frac{t}{12} - k\right)$ and find the smallest positive value of k.

(iii) Find the exact times that the particle is at O.

(iv) State the maximum displacement of the particle from O. Find the exact time that the particle is furthest from O.

(v) Is the particle ever -1 metres from O? Explain your reasoning clearly.

(vi) Sketch the curve

$$d = \frac{\sqrt{3}}{2}\sin\frac{t}{12} - \frac{1}{2}\cos\frac{t}{12} \text{ for } 0 \leqslant t \leqslant 60.$$

⑬ (i) Use a compound angle formula to write down an expression for $\sin(x + h)$.

(ii) Rewrite your answer to part (i) using small-angle approximations for $\sin h$ and $\cos h$ where h is small.

(iii) Use your answer to part (ii) to write down an expression for $\dfrac{\sin(x + h) - \sin x}{h}$.

(iv) State $\displaystyle\lim_{h \to 0} \frac{\sin(x + h) - \sin x}{h}$.

(v) Explain the significance of your answer to part (iv).

⑭ (i) Simplify $\tan\left(\frac{\pi}{4} + \theta\right)$ when θ is small.

(ii) Use the binomial expansion for $(1 + \theta)^{-1}$ to find a quadratic approximation for $\tan\left(\frac{\pi}{4} + \theta\right)$ when θ is small.

2 Double angle formulae

Discussion points

→ As you work through these proofs, think about how you can check the results.

→ Is a check the same as a proof?

Substituting $\phi = \theta$ in the relevant compound angle formulae leads immediately to expressions for $\sin 2\theta$, $\cos 2\theta$ and $\tan 2\theta$, as follows.

(i) Starting with $\sin(\theta + \phi) = \sin\theta\cos\phi + \cos\theta\sin\phi$,

when $\phi = \theta$, this becomes

$$\sin(\theta + \theta) = \sin\theta\cos\theta + \cos\theta\sin\theta$$

giving $\sin 2\theta = 2\sin\theta\cos\theta$.

(ii) Starting with $\cos(\theta + \phi) = \cos\theta\cos\phi - \sin\theta\sin\phi$,

when $\phi = \theta$, this becomes

$$\cos(\theta + \theta) = \cos\theta\cos\theta - \sin\theta\sin\theta$$

giving $\cos 2\theta = \cos^2\theta - \sin^2\theta$.

Using the Pythagorean identity $\cos^2\theta + \sin^2\theta = 1$, two other forms for $\cos 2\theta$ can be obtained.

$$\cos 2\theta = (1 - \sin^2\theta) - \sin^2\theta \quad \Rightarrow \quad \cos 2\theta = 1 - 2\sin^2\theta$$
$$\cos 2\theta = \cos^2\theta - (1 - \cos^2\theta) \quad \Rightarrow \quad \cos 2\theta = 2\cos^2\theta - 1$$

These alternative forms are often more useful since they contain only one trigonometric function.

(iii) Starting with $\tan(\theta + \phi) = \dfrac{\tan\theta + \tan\phi}{1 - \tan\theta\tan\phi}$ $(\theta + \phi) \neq 90°, 270°, \dots$,

when $\phi = \theta$ this becomes

$$\tan(\theta + \theta) = \frac{\tan\theta + \tan\theta}{1 - \tan\theta\tan\theta}$$

giving $\tan 2\theta = \dfrac{2\tan\theta}{1 - \tan^2\theta}$ $\theta \neq 45°, 135°, \dots$

Double angle formulae

$$\sin 2\theta = 2\sin\theta\cos\theta$$
$$\cos 2\theta = \cos^2\theta - \sin^2\theta$$
$$\cos 2\theta = 1 - 2\sin^2\theta$$
$$\cos 2\theta = 2\cos^2\theta - 1$$
$$\tan 2\theta = \frac{2\tan\theta}{1 - \tan^2\theta} \qquad \theta \neq 45°, 135°, \dots$$

> You will need to decide which of the three formulae for $\cos 2\theta$ is best for each situation.

Compound angle and double angle formulae are used

- in solving trigonometric equations and proving identities
- to help you rewrite trigonometric functions into a form that you can integrate or into a more convenient form for differentiation
- in modelling, for example finding the standard formula for the range of a projectile in mechanics. See Chapter 21.

Example 8.3

Solve the equation $\sin 2\theta = \sin\theta$ for $0° \leqslant \theta \leqslant 360°$.

Solution

$$\sin 2\theta = \sin\theta \qquad \text{Use } \sin 2\theta = 2\sin\theta\cos\theta.$$
$$\Rightarrow \qquad 2\sin\theta\cos\theta = \sin\theta \qquad \text{Gather terms on one side.}$$
$$\Rightarrow \quad 2\sin\theta\cos\theta - \sin\theta = 0 \qquad \text{Factorise.}$$
$$\Rightarrow \qquad \sin\theta(2\cos\theta - 1) = 0$$
$$\Rightarrow \qquad \sin\theta = 0 \text{ or } \cos\theta = \tfrac{1}{2}$$

> ! Do not divide by $\sin\theta$, otherwise you will lose the roots to $\sin\theta = 0$.

$\sin\theta = 0 \quad \Rightarrow \quad \theta = 0°$ (principal value) or $180°$ or $360°$ (see Figure 8.3).

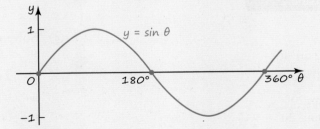

Figure 8.3

$\cos\theta = \tfrac{1}{2} \quad \Rightarrow \quad \theta = 60°$ (principal value) or $300°$ (see Figure 8.4).

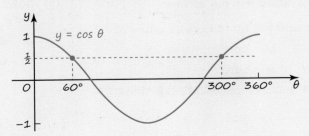

Figure 8.4

The full set of roots for $0° \leqslant \theta \leqslant 360°$ is $\theta = 0°, 60°, 180°, 300°, 360°$.

When an equation contains $\cos 2\theta$, you will save time if you take care to choose the most suitable expansion.

Example 8.4

Solve $2 + \cos 2\theta = \sin \theta$ for $0 \le \theta \le 2\pi$.

> **!** Since you are asked for roots in the range $0 \le \theta \le 2\pi$, you need to give the roots in radians.

Solution

Using $\cos 2\theta = 1 - 2\sin^2 \theta$ gives

> Using this form gives an equation in $\sin \theta$ only.

$$2 + (1 - 2\sin^2\theta) = \sin \theta$$

$$\Rightarrow \quad 2\sin^2\theta + \sin\theta - 3 = 0$$

$$\Rightarrow \quad (2\sin\theta + 3)(\sin\theta - 1) = 0$$

$$\Rightarrow \quad \sin\theta = -\tfrac{3}{2} \quad \text{(not valid since } -1 \le \sin\theta \le 1)$$

$$\text{or} \quad \sin\theta = 1.$$

Figure 8.5 shows that the principal value $\theta = \frac{\pi}{2}$ is the only root for $0 \le \theta \le 2\pi$.

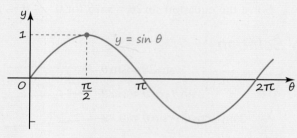

Figure 8.5

Exercise 8.2

① Simplify

(i) $1 - 2\sin^2 40°$

(ii) $\sin 40° \cos 40°$

(iii) $\dfrac{2\tan 20°}{1 - \tan^2 20°}$

② Simplify $\dfrac{1 + \cos 2\theta}{\sin 2\theta}$

③ *Do not use a calculator in this question.*

Given that $\sin \theta = \frac{3}{5}$ find the exact values of

(a) $\cos\theta$ (b) $\tan\theta$ (c) $\sin 2\theta$

(d) $\cos 2\theta$ (e) $\tan 2\theta$

when θ is (i) an acute angle and (ii) an obtuse angle.

④ Solve the following equations for $0° \le \theta \le 360°$.

(i) $2\sin 2\theta = \cos\theta$

(ii) $\tan 2\theta = 4\tan\theta$

(iii) $\cos 2\theta + \sin\theta = 0$

(iv) $\tan\theta \tan 2\theta = 1$

(v) $2\cos 2\theta = 1 + \cos\theta$

⑤ Solve the following equations for $-\pi \le \theta \le \pi$.

(i) $\sin 2\theta = 2\sin\theta$

(ii) $\tan 2\theta = 2\tan\theta$

(iii) $\cos 2\theta - \cos\theta = 0$

(iv) $1 + \cos 2\theta = \sin 2\theta$

(v) $\sin 4\theta = \cos 2\theta$

> Hint: Write the expression as an equation in 2θ.

⑥ By first writing $\sin 3\theta$ as $\sin (2\theta + \theta)$, express $\sin 3\theta$ in terms of $\sin\theta$.

Hence solve the equation $\sin 3\theta = \sin\theta$ for $0 \leqslant \theta \leqslant 2\pi$.

⑦ Solve $\cos 3\theta = 1 - 3\cos\theta$ for $0° \leqslant \theta \leqslant 360°$.

⑧ Prove the following identities.

(i) $\dfrac{1 - \tan^2\theta}{1 + \tan^2\theta} \equiv \cos 2\theta$

(ii) $\cos^4\theta - \sin^4\theta \equiv \cos 2\theta$

(iii) $\dfrac{\tan\theta(3 - \tan^2\theta)}{1 - 3\tan^2\theta} \equiv \tan 3\theta$

(iv) $\sin\frac{1}{2}\theta \equiv \pm\sqrt{\dfrac{1 - \cos\theta}{2}}$

(v) $\left(\sin\frac{1}{2}\theta + \cos\frac{1}{2}\theta\right)^2 \equiv 1 + \sin\theta$

⑨ (i) Show that $\tan\left(\dfrac{\pi}{4} + \theta\right)\tan\left(\dfrac{\pi}{4} - \theta\right) = 1$.

(ii) Given that $\tan 26.6° = 0.5$, solve $\tan\theta = 2$ without using your calculator. Give θ to 1 decimal place, where $0° < \theta < 90°$.

Prior knowledge

You need to be able to use transformations to sketch trigonometric functions – see Chapter 6.

3 The forms $r\cos(\theta \pm \alpha)$, $r\sin(\theta \pm \alpha)$

ACTIVITY 8.3

Use graphing software to plot $y = a\sin\theta + b\cos\theta$ for three different sets of values of a and b. What shape is the resulting curve?

For each of your three curves, find values of r and α such that

(i) $y = r\sin(\theta + \alpha)$ (ii) $y = r\cos(\theta + \alpha)$

give the same curve.

You probably noticed that expressions of the form $a\sin\theta + b\cos\theta$ are the same shape as the sine and cosine graphs, but they have been translated and stretched. This suggests that expressions of this form can be written in the form $r\cos(\theta + \alpha)$ or $r\sin(\theta + \alpha)$ where r and α are constants.

For example, to find a single expression for $4\sin\theta + 3\cos\theta$, you can match it to the expression

$$r\sin(\theta + \alpha) = r(\sin\theta\cos\alpha + \cos\theta\sin\alpha).$$

> This is because the expansion of $r\sin(\theta + \alpha)$ has $\sin\theta$ in the first term, $\cos\theta$ in the second term and a plus sign in between them.

It is then possible to find the values of r and α.

$$4\sin\theta + 3\cos\theta \equiv r(\sin\theta\cos\alpha + \cos\theta\sin\alpha)$$

Equating coefficients of $\sin\theta$: $4 = r\cos\alpha$
Equating coefficients of $\cos\theta$: $3 = r\sin\alpha$.

You can now draw a right-angled triangle (Figure 8.6) to help you find the values for r and α.

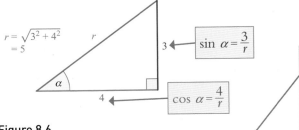

$r = \sqrt{3^2 + 4^2} = 5$ r $3 \leftarrow$ $\sin\alpha = \dfrac{3}{r}$ α $4 \leftarrow$ $\cos\alpha = \dfrac{4}{r}$

> Note that the symbol $\sqrt{}$ means 'the positive square root of'. This is consistent with r being the hypotenuse of the triangle and so positive.

Figure 8.6

In this triangle, the hypotenuse, r, is $\sqrt{4^2 + 3^2} = 5$.

The angle α is given by

$$\sin\alpha = \frac{3}{5} \quad \text{and} \quad \cos\alpha = \frac{4}{5} \quad \text{and} \quad \tan\alpha = \frac{3}{4} \implies \alpha = 36.9°.$$

So the expression becomes

$$4\sin\theta + 3\cos\theta = 5\sin(\theta + 36.9°).$$

Discussion point

Describe the transformations which map $y = \sin x$ on to $y = 5\sin(\theta + 36.9°)$.

The steps involved in this procedure can be generalised to write

$$a\sin\theta + b\cos\theta = r\sin(\theta + \alpha)$$

where

$$r = \sqrt{a^2 + b^2} \quad \text{and}$$

$$\sin\alpha = \frac{b}{\sqrt{a^2 + b^2}} \qquad \cos\alpha = \frac{a}{\sqrt{a^2 + b^2}} \qquad \tan\alpha = \frac{b}{a}$$

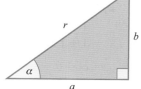

Figure 8.7

ACTIVITY 8.4

(i) Write $y = 4\sin\theta + 3\cos\theta$ as a cosine function.

Start by rewriting $4\sin\theta + 3\cos\theta$ as $3\cos\theta + 4\sin\theta$ and use the expansion of $\cos(\theta - \beta)$.

(ii) Explain using transformations why the equation you found in part (i) and $y = 5\sin(\theta + 36.9°)$ give the same graph.

Notice that the expansion of $\cos(\theta - \beta)$

- starts with $\cos\theta$ ...
- and has a plus sign in the middle ...
- ... just like the expression $3\cos\theta + 4\sin\theta$.

Note

The value of r will always be positive, but $\cos\alpha$ and $\sin\alpha$ may be positive or negative, depending on the values of a and b. In all cases, it is possible to find an angle α for which $-180° < \alpha < 180°$.

The method used in Activity 8.4 can be generalised to give the result

$$a\cos\theta + b\sin\theta = r\cos(\theta - \alpha)$$

where

$$r = \sqrt{a^2 + b^2} \quad \text{and}$$

$$\sin\alpha = \frac{b}{r} \qquad \cos\alpha = \frac{a}{r} \qquad \tan\alpha = \frac{b}{a}$$

It is a good idea to sketch a right-angled triangle to help you work out r and α.

You can derive alternative expressions of this type based on other compound angle formulae if you wish α to be an acute angle, as is done in the next example.

Example 8.5

Notice that the range of α is given in radians. This is telling you to work the whole question in radians.

(i) Express $\sqrt{3}\sin\theta - \cos\theta$ in the form $r\sin(\theta - \alpha)$, where $r > 0$ and $0 < \alpha < \frac{\pi}{2}$.

(ii) State the maximum and minimum values of $\sqrt{3}\sin\theta - \cos\theta$.

(iii) Sketch the graph of $y = \sqrt{3}\sin\theta - \cos\theta$ for $0 \le \theta \le 2\pi$.

(iv) Solve the equation $\sqrt{3}\sin\theta - \cos\theta = 1$ for $0 \le \theta \le 2\pi$.

Solution

(i)　　$r\sin(\theta - \alpha) = r(\sin\theta\cos\alpha - \cos\theta\sin\alpha)$　　⟵ Expand $r\sin(\theta - \alpha)$.

　　　　$= (r\cos\alpha)\sin\theta - (r\sin\alpha)\cos\theta$

Compare the expansion with $\sqrt{3}\sin\theta - \cos\theta$. ⟶

$\Rightarrow (r\cos\alpha)\sin\theta - (r\sin\alpha)\cos\theta = \sqrt{3}\sin\theta - \cos\theta$

$r\cos\alpha = \sqrt{3}$　and　$r\sin\alpha = 1$

Draw a right-angled triangle to help you find r and α.

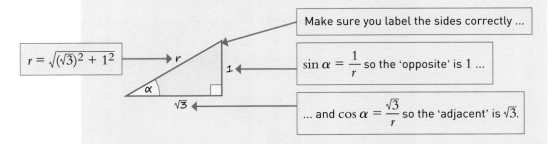

$r = \sqrt{(\sqrt{3})^2 + 1^2}$

Make sure you label the sides correctly ...

$\sin\alpha = \dfrac{1}{r}$ so the 'opposite' is 1 ...

... and $\cos\alpha = \dfrac{\sqrt{3}}{r}$ so the 'adjacent' is $\sqrt{3}$.

Figure 8.8

From the triangle in Figure 8.8

$$r = \sqrt{3 + 1} = 2 \quad \text{and} \quad \tan\alpha = \frac{1}{\sqrt{3}} \quad \Rightarrow \quad \alpha = \frac{\pi}{6}$$

so　$\sqrt{3}\sin\theta - \cos\theta = 2\sin\left(\theta - \dfrac{\pi}{6}\right).$　⟵ Substitute r and α into $r\sin(\theta - \alpha)$.

(ii)　The sine function oscillates between 1 and −1, so $2\sin\left(\theta - \dfrac{\pi}{6}\right)$ oscillates between 2 and −2.

Maximum value = 2.
Minimum value = −2.

(iii)　The graph of $y = 2\sin\left(\theta - \dfrac{\pi}{6}\right)$ is obtained from the graph of

$y = \sin\theta$ by a translation of $\begin{pmatrix} \dfrac{\pi}{6} \\ 0 \end{pmatrix}$ and a stretch of scale factor 2

parallel to the y-axis, as shown in Figure 8.9.

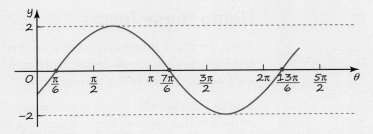

Figure 8.9

(iv)   The equation $\sqrt{3}\sin\theta - \cos\theta = 1$ is equivalent to

$$2\sin\left(\theta - \frac{\pi}{6}\right) = 1$$

$$\Rightarrow \quad \sin\left(\theta - \frac{\pi}{6}\right) = \tfrac{1}{2}$$

Let $x = \left(\theta - \dfrac{\pi}{6}\right)$ and solve $\sin x = \tfrac{1}{2}$.

> Find all the values of x in the interval $0 - \dfrac{\pi}{6} \leqslant \theta \leqslant 2\pi - \dfrac{\pi}{6}$.

Solving $\sin x = \tfrac{1}{2}$ gives $x = \dfrac{\pi}{6}$ (principal value)

or $x = \pi - \dfrac{\pi}{6} = \dfrac{5\pi}{6}$ (from the graph in Figure 8.10).

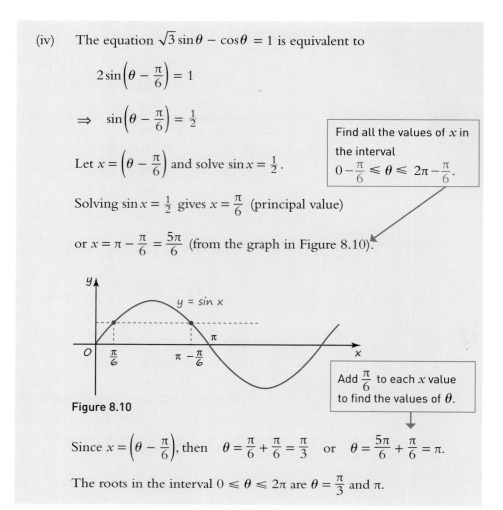

Figure 8.10

> Add $\dfrac{\pi}{6}$ to each x value to find the values of θ.

Since $x = \left(\theta - \dfrac{\pi}{6}\right)$, then $\quad \theta = \dfrac{\pi}{6} + \dfrac{\pi}{6} = \dfrac{\pi}{3} \quad$ or $\quad \theta = \dfrac{5\pi}{6} + \dfrac{\pi}{6} = \pi.$

The roots in the interval $0 \leqslant \theta \leqslant 2\pi$ are $\theta = \dfrac{\pi}{3}$ and π.

! Make sure you don't miss out any roots.

- To solve $\sin(\theta - \alpha) = c$ for $0 \leqslant \theta \leqslant 2\pi$ by first solving $\sin x = c$, you need to find all possible values of x in the interval $0 + \alpha \leqslant x \leqslant 2\pi + \alpha$.

- Find **all possible values** of x in the new interval before working out the corresponding values of θ.

Using these forms

There are many situations that produce expressions that can be tidied up using these forms. They are also particularly useful for solving equations involving both the sine and cosine of the same angle.

The fact that $a\cos\theta + b\sin\theta$ can be written as $r\cos(\theta - \alpha)$ is an illustration of the fact that any two waves of the same frequency, whatever their amplitudes, can be added together to give a single combined wave, also of the same frequency.

① For each triangle in Figure 8.11, find the exact value of r and the angle α, where α is in degrees.

(i)   (ii)

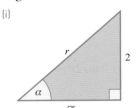

Figure 8.11

② Expand each of the following.

(i) $\sqrt{3}\cos(\theta - 30°)$

(ii) $\sqrt{3}\cos(\theta + 30°)$

(iii) $\sqrt{3}\sin(\theta - 30°)$

(iv) $\sqrt{3}\sin(\theta + 30°)$

③ For each of the following functions find

(a)   the maximum and minimum values

(b)   the smallest positive value of x for which the function is a maximum.

(i) $f(x) = 2\cos\left(x - \dfrac{\pi}{3}\right)$

(ii) $f(x) = \sqrt{2}\cos\left(x + \dfrac{\pi}{3}\right)$

(iii) $\dfrac{1}{\sqrt{2}}\sin\left(\theta + \dfrac{\pi}{3}\right)$

(iv) $\dfrac{1}{2 + \sqrt{2}\sin\left(\theta + \dfrac{\pi}{3}\right)}$

④ (i)   Express each of the following in the form $r\cos(\theta - \alpha)$, where $r > 0$ and $0° < \alpha < 90°$.

     (a) $\cos\theta + \sin\theta$

     (b) $3\cos\theta + 4\sin\theta$

(ii)   Express each of the following in the form $r\sin(\theta - \alpha)$, where $r > 0$ and $0° < \alpha < 90°$.

     (a) $\sin\theta - \cos\theta$

     (b) $3\sin\theta - 4\cos\theta$

⑤ Express each of the following in the form

(a) $r\cos(\theta + \alpha)$, where $r > 0$ and
$$0 < \alpha < \frac{\pi}{2}$$

(b) $r\sin(\theta - \alpha)$, where $r > 0$ and
$$0 < \alpha < \frac{\pi}{2}.$$

(i) $\cos\theta - \sin\theta$

(ii) $\sqrt{3}\cos\theta - \sin\theta$

⑥ Express each of the following in the form $r\cos(\theta - \alpha)$, where $r > 0$ and $-180° < \alpha < 180°$.

(i) $\cos\theta - \sqrt{3}\sin\theta$

(ii) $2\sqrt{2}\cos\theta - 2\sqrt{2}\sin\theta$

(iii) $\sin\theta + \sqrt{3}\cos\theta$

(iv) $5\sin\theta + 12\cos\theta$

(v) $\sin\theta - \sqrt{3}\cos\theta$

(vi) $\sqrt{2}\sin\theta - \sqrt{2}\cos\theta$

⑦ (i)   Express $5\cos\theta - 12\sin\theta$ in the form $r\cos(\theta + \alpha)$, where $r > 0$ and $0° < \alpha < 90°$.

(ii)   State the maximum and minimum values of $5\cos\theta - 12\sin\theta$.

(iii) Sketch the graph of $y = 5\cos\theta - 12\sin\theta$ for $0° \leqslant \theta \leqslant 360°$.

(iv) Solve the equation $5\cos\theta - 12\sin\theta = 4$ for $0° \leqslant \theta \leqslant 360°$.

⑧ (i)   Express $3\sin\theta - \sqrt{3}\cos\theta$ in the form $r\sin(\theta - \alpha)$, where $r > 0$ and $0 < \alpha < \dfrac{\pi}{2}$.

(ii)   State the maximum and minimum values of $3\sin\theta - \sqrt{3}\cos\theta$ and the smallest positive values of θ for which they occur.

(iii) Sketch the graph of $y = 3\sin\theta - \sqrt{3}\cos\theta$ for $0 \leqslant \theta \leqslant 2\pi$.

(iv) Solve the equation $3\sin\theta - \sqrt{3}\cos\theta = \sqrt{3}$ for $0 \leqslant \theta \leqslant 2\pi$.

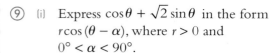

⑨ (i) Express $\cos\theta + \sqrt{2}\sin\theta$ in the form $r\cos(\theta - \alpha)$, where $r > 0$ and $0° < \alpha < 90°$.

(ii) State the maximum and minimum values of $\cos\theta + \sqrt{2}\sin\theta$ and the smallest positive values of θ for which they occur.

(iii) Sketch the graph of $y = \cos\theta + \sqrt{2}\sin\theta$ for $0° \leqslant \theta \leqslant 360°$.

(iv) State the maximum and minimum values of

$$\frac{1}{3 + \cos\theta + \sqrt{2}\sin\theta}$$

and the smallest positive values of θ for which they occur.

⑩ (i) Express $2\sin 2\theta + 3\cos 2\theta$ in the form $r\sin(2\theta + \alpha)$, where $r > 0$ and $0° < \alpha < 90°$.

(ii) State the maximum and minimum values of $2\sin 2\theta + 3\cos 2\theta$ and the smallest positive values of θ for which they occur.

(iii) Sketch the graph of $y = 2\sin 2\theta + 3\cos 2\theta$ for $0° \leqslant \theta \leqslant 360°$.

(iv) Solve the equation $2\sin 2\theta + 3\cos 2\theta = 1$ for $0° \leqslant \theta \leqslant 360°$.

Summary exercise

The previous exercises in this chapter and in Chapter 6 have each concentrated on just one technique at a time. However, you will often have to choose the correct technique from the many that you have met.

In this exercise you will need to select appropriate formulae and techniques.

① Simplify each of the following.

(i) $2\sin 3\theta \cos 3\theta$

(ii) $\cos^2 3\theta - \sin^2 3\theta$

(iii) $\cos^2 3\theta + \sin^2 3\theta$

(iv) $1 - 2\sin^2 \dfrac{\theta}{2}$

(v) $\sin(\theta - \alpha)\cos\alpha + \cos(\theta - \alpha)\sin\alpha$

(vi) $3\sin\theta \cos\theta$

(vii) $\dfrac{\sin 2\theta}{2\sin\theta}$

(viii) $\cos 2\theta - 2\cos^2\theta$

② Express

(i) $(\cos x - \sin x)^2$ in terms of $\sin 2x$

(ii) $2\cos^2 x - 3\sin^2 x$ in terms of $\cos 2x$.

③ Prove that

(i) $\dfrac{1 - \cos 2\theta}{1 + \cos 2\theta} \equiv \tan^2\theta$

(ii) $\operatorname{cosec} 2\theta + \cot 2\theta \equiv \cot\theta$

(iii) $2\sin^2(\theta + 45) - \sin 2\theta \equiv 1$

(iv) $\tan 4\theta \equiv \dfrac{4t(1 - t^2)}{1 - 6t^2 + t^4}$ where $t = \tan\theta$.

④ Solve the following equations.

(i) $\sin(\theta + 40°) = 0.7$
for $0° \leqslant \theta \leqslant 360°$

(ii) $3\cos^2\theta + 5\sin\theta - 1 = 0$
for $0° \leqslant \theta \leqslant 360°$

(iii) $2\cos\left(\theta - \dfrac{\pi}{6}\right) = 1$
for $-\pi \leqslant \theta \leqslant \pi$

(iv) $\cos(45° - \theta) = 2\sin(30° + \theta)$
for $-180° \leqslant \theta \leqslant 180°$

(v) $\cos 2\theta + 3\sin\theta = 2$
for $0 \leqslant \theta \leqslant 2\pi$

(vi) $\cos\theta + 3\sin\theta = 2$
for $0° \leqslant \theta \leqslant 360°$

(vii) $\sec^2\theta - 3\tan\theta - 5 = 0$
for $0° \leqslant \theta \leqslant 180°$

LEARNING OUTCOMES

When you have completed this chapter, you should be able to:

➤ use formulae for sin $(A \pm B)$, cos $(A \pm B)$ and tan $(A \pm B)$
➤ understand geometrical proofs of these formulae
➤ understand and use double angle formulae
➤ understand and use expressions for $a\cos\theta + b\sin\theta$ in the equivalent forms of $r\cos(\theta \pm \alpha)$ or $r\sin(\theta \pm \alpha)$
➤ solve quadratic equations in sin, cos and tan of an unknown angle
➤ construct proofs involving trigonometric functions and identities.

KEY POINTS

1 Compound angle formulae

$$\sin(\theta + \phi) = \sin\theta\cos\phi + \cos\theta\sin\phi$$

$$\sin(\theta - \phi) = \sin\theta\cos\phi - \cos\theta\sin\phi$$

$$\cos(\theta + \phi) = \cos\theta\cos\phi - \sin\theta\sin\phi$$

$$\cos(\theta - \phi) = \cos\theta\cos\phi + \sin\theta\sin\phi$$

$$\tan(\theta + \phi) = \frac{\tan\theta + \tan\phi}{1 - \tan\theta\tan\phi} \qquad (\theta + \phi) \neq 90°, 270°, \ldots$$

$$\tan(\theta - \phi) = \frac{\tan\theta - \tan\phi}{1 + \tan\theta\tan\phi} \qquad (\theta - \phi) \neq 90°, 270°, \ldots$$

2 Double angle formulae

$$\sin 2\theta = 2\sin\theta\cos\theta$$
$$\cos 2\theta = \cos^2\theta - \sin^2\theta$$
$$\cos 2\theta = 1 - 2\sin^2\theta$$
$$\cos 2\theta = 2\cos^2\theta - 1$$

$$\tan 2\theta = \frac{2\tan\theta}{1 - \tan^2\theta} \qquad \theta \neq 45°, 135°, \ldots$$

3 The r, α form

$$\left.\begin{array}{l} a\sin\theta + b\cos\theta = r\sin(\theta + \alpha) \\ a\sin\theta - b\cos\theta = r\sin(\theta - \alpha) \\ a\cos\theta + b\sin\theta = r\cos(\theta - \alpha) \\ a\cos\theta - b\sin\theta = r\cos(\theta + \alpha) \end{array}\right\} \quad \text{where} \quad r = \sqrt{a^2 + b^2}$$

$$\cos\alpha = \frac{a}{r}$$

$$\sin\alpha = \frac{b}{r}$$

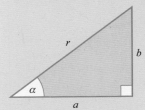

Figure 8.12

9 Further differentiation

Many physical systems, such as a simple pendulum or swing or a mass on an elastic spring, can be modelled as having displacement–time graphs which have a sine wave shape.

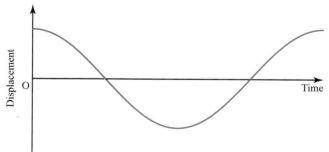

Figure 9.1

To be able to perform calculations involving velocity and acceleration for these systems, you need to be able to differentiate the sine function.

➔ Think of some other situations in which it would be useful to be able to differentiate functions other than polynomials.

1 Differentiating exponentials and logarithms

Differentiating exponential functions

The exponential function $y = e^x$ has the important property that its gradient at any point is equal to the y coordinate of that point.

TECHNOLOGY
Use graphing software to verify this result.

$$y = e^x \Rightarrow \frac{dy}{dx} = e^x$$

Note
You can prove this result using the chain rule.

This result can be extended to functions of the form $y = e^{kx}$.

$$y = e^{kx} \Rightarrow \frac{dy}{dx} = ke^{kx}$$

You can use the chain rule to differentiate more complicated exponential expressions.

Example 9.1

Differentiate $y = e^{x^2-1}$.

Solution

Since this is a composite function you use the chain rule to differentiate.

$$u = x^2 - 1, \quad y = e^u$$ ← Write down u as a function of x and y as a function of u.

$$\frac{du}{dx} = 2x, \quad \frac{dy}{du} = e^u$$ ← Differentiate both functions.

$$\frac{dy}{dx} = \frac{dy}{du} \times \frac{du}{dx}$$ ← The chain rule

$$= 2xe^u$$

$$= 2xe^{x^2-1}$$ ← Rewrite in terms of x.

Discussion point

→ What is the derivative of the general expression $e^{f(x)}$?

Differentiating logarithmic functions

To differentiate $y = \ln x$, you can use the fact that the inverse of $\ln x$ is e^x.

$$y = \ln x \Rightarrow x = e^y$$ ← Rearrange.

$$\Rightarrow \frac{dx}{dy} = e^y$$

$$\Rightarrow \frac{dy}{dx} = \frac{1}{e^y}$$ ← Using $\frac{dy}{dx} = \frac{1}{\frac{dx}{dy}}$

$$\Rightarrow \frac{dy}{dx} = \frac{1}{x}$$ ← Rewrite in terms of x.

To differentiate $y = \ln kx$, where k is a constant, you can use log rules.

$$y = \ln kx = \ln k + \ln x$$

$$\frac{\mathrm{d}y}{\mathrm{d}x} = \frac{\mathrm{d}(\ln k)}{\mathrm{d}x} + \frac{\mathrm{d}(\ln x)}{\mathrm{d}x}$$

$$\Rightarrow \quad \frac{\mathrm{d}y}{\mathrm{d}x} = 0 + \frac{1}{x}$$

> In k is a constant so differentiating it gives 0.

$$\Rightarrow \quad \frac{\mathrm{d}y}{\mathrm{d}x} = \frac{1}{x}$$

$$y = \ln x \;\Rightarrow\; \frac{\mathrm{d}y}{\mathrm{d}x} = \frac{1}{x}$$

$$y = \ln kx \;\Rightarrow\; \frac{\mathrm{d}y}{\mathrm{d}x} = \frac{1}{x}$$

> **Discussion point**
>
> The gradient function for $y = \ln kx$ is the same for all values of k.
>
> → What does this tell you about the graphs of these functions?

You can use the chain rule to differentiate more complicated natural logarithmic expressions.

Example 9.2

Differentiate $y = \ln(x^3 - 4x)$.

Solution

Since this is a composite function you use the chain rule to differentiate.

$$u = x^3 - 4x, \qquad y = \ln u$$

> Write down u as a function of x, and y as a function of u.

$$\frac{\mathrm{d}u}{\mathrm{d}x} = 3x^2 - 4, \qquad \frac{\mathrm{d}y}{\mathrm{d}u} = \frac{1}{u}$$

> Differentiate both functions.

$$\frac{\mathrm{d}y}{\mathrm{d}x} = \frac{\mathrm{d}y}{\mathrm{d}u} \times \frac{\mathrm{d}u}{\mathrm{d}x}$$

> The chain rule

$$= \frac{3x^2 - 4}{u}$$

$$= \frac{3x^2 - 4}{x^3 - 4x}$$

> Rewrite in terms of x.

> **Discussion point**
>
> → What is the derivative of $\ln(f(x))$?

Differentiating $y = a^x$ and $y = a^{kx}$

To differentiate $y = a^x$, where a is any positive constant, you can use logarithms to rearrange to make x the subject.

$$y = a^x \;\Rightarrow\; \ln y = \ln(a^x)$$

> Take natural logs of both sides.

$$\Rightarrow\; \ln y = x \ln a$$

> Use log rules.

$$\Rightarrow\; x = \frac{\ln y}{\ln a} = \frac{1}{\ln a} \times \ln y$$

$$\Rightarrow\; \frac{\mathrm{d}x}{\mathrm{d}y} = \frac{1}{\ln a} \times \frac{1}{y}$$

> Differentiate both sides, remembering that $\frac{1}{\ln a}$ is just a constant.

$$= \frac{1}{y \ln a}$$

So $\qquad \dfrac{\mathrm{d}y}{\mathrm{d}x} = y \ln a$

$\qquad\qquad\qquad = a^x \ln a \longleftarrow$ | Rewrite in terms of x. |

Now that you can differentiate $y = a^x$, you can differentiate $y = a^{kx}$ using the chain rule.

Let $\quad y = a^u, u = kx$.

By the chain rule,

$$\dfrac{\mathrm{d}y}{\mathrm{d}x} = \dfrac{\mathrm{d}y}{\mathrm{d}u} \times \dfrac{\mathrm{d}u}{\mathrm{d}x}$$

$$\Rightarrow \dfrac{\mathrm{d}y}{\mathrm{d}x} = a^u \ln a \times k$$

$$= ka^{kx} \ln a$$

$$y = a^{kx} \Rightarrow \dfrac{\mathrm{d}y}{\mathrm{d}x} = ka^{kx} \ln a$$

Example 9.3

Differentiate $y = \dfrac{\ln x}{2^x}$.

Solution

Using the quotient rule,

$$u = \ln x \Rightarrow \dfrac{\mathrm{d}u}{\mathrm{d}x} = \dfrac{1}{x} \longleftarrow$$ | Remember that with the quotient rule u must be the numerator ... |

$$v = 2^x \Rightarrow \dfrac{\mathrm{d}v}{\mathrm{d}x} = 2^x \ln 2 \longleftarrow$$ | ... and v must be the denominator. |

$$\dfrac{\mathrm{d}y}{\mathrm{d}x} = \dfrac{v\dfrac{\mathrm{d}u}{\mathrm{d}x} - u\dfrac{\mathrm{d}v}{\mathrm{d}x}}{v^2} \longleftarrow$$ | The quotient rule |

$$= \dfrac{2^x \times \dfrac{1}{x} - \ln x \times 2^x \ln 2}{\left(2^x\right)^2}$$

$$= \dfrac{\dfrac{1}{x} - \ln 2 \ln x}{2^x} \longleftarrow$$ | Dividing top and bottom by 2^x. |

$$= \dfrac{1 - x \ln 2 \ln x}{2^x x} \longleftarrow$$ | Multiplying top and bottom by x. |

Exercise 9.1

① Differentiate the following functions.

    (i) $\quad y = e^{3x+1}$

    (ii) $\quad y = e^{x^2+3x+1}$

② Differentiate the following functions.

    (i) $\quad y = 7^x$ (ii) $\quad y = 7^{5x}$

③ Differentiate the following functions.

    (i) $\quad y = \ln(8x)$

    (ii) $\quad y = \ln(x^2 + 1)$

    (iii) $\quad y = \ln(\sqrt{x})$

④ Differentiate the following functions.

(i) $y = xe^{4x}$ (ii) $y = \dfrac{\ln x}{x^2}$ (iii) $y = \dfrac{x}{e^x}$

⑤ For each of the following, find the value of the gradient of the curve at the given point.

(i) $y = 2x^3 e^{-x}$ at the point $\left(1, \dfrac{2}{e}\right)$

(ii) $y = \left(e^{2x} + 1\right)^3$ at the point $(0,8)$

(iii) $y = \ln \sqrt{x^2 - 1}$ at the point $\left(3, \dfrac{3\ln 2}{2}\right)$

⑥ The graph of $y = xe^x$ is shown in Figure 9.2.

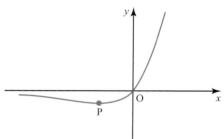

Figure 9.2

(i) Find $\dfrac{dy}{dx}$ and $\dfrac{d^2y}{dx^2}$.

(ii) Find the coordinates of the minimum point, P.

(iii) Find the coordinates of the point of inflection.

⑦ (i) Find the coordinates of the point of intersection of the curves $y = \ln(4x - 1)$ and $y = \ln(2x + 3)$.

(ii) Find an equation for the tangent to the curve $y = \ln(2x + 3)$ at this point.

(iii) Sketch the two curves on the same axes.

⑧ The graph of $f(x) = x\ln\left(x^2\right)$ is shown in Figure 9.3.

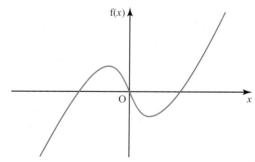

Figure 9.3

(i) Describe and justify any symmetries of the graph.

(ii) Find $f'(x)$ and $f''(x)$.

(iii) Find the coordinates of any turning points.

⑨ The curve $y = \ln(3x - 2)^3$ crosses the x-axis at the point $(a,0)$.

(i) Find the value of a.

(ii) Find an equation of the normal to the curve at this point.

⑩ Figure 9.4 shows the curve with equation $y = x^2 - 3x + \ln x + 2$, $x > 0$.

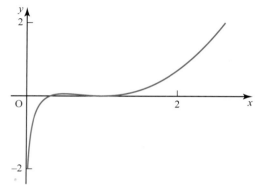

Figure 9.4

(i) Find $\dfrac{dy}{dx}$ and $\dfrac{d^2y}{dx^2}$.

(ii) Find the coordinates of the two turning points and use calculus to show their nature.

⑪ Sketch the curve $y = \dfrac{e^x}{x}$, giving the coordinates of the turning point and the equations of any asymptotes.

⑫ The equation for the Normal distribution is usually given in statistics as

$$\phi(z) = \dfrac{1}{\sqrt{2\pi}} e^{-\frac{1}{2}z^2}$$

The Normal curve is thus given by $y = \phi(z)$ with z on the horizontal axis and y on the vertical axis.

(i) Use calculus to prove that $y = \phi(z)$ has

(a) a maximum when $z = 0$

(b) no other turning points

(c) non-stationary points of inflection when $z = \pm 1$.

(ii) Mark the maximum point and the points of inflection on a sketch of the Normal curve.

2 Differentiating trigonometric functions

ACTIVITY 9.1

Differentiating $\sin x$ and $\cos x$

Figure 9.5 shows the graph of $y = \sin x$, with x measured in radians, together with the graph of $y = x$. You are going to sketch the graph of the gradient function for the graph of $y = \sin x$.

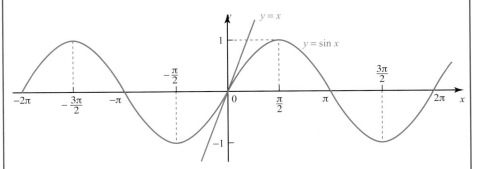

Figure 9.5

Draw a horizontal axis for x, marked from -2π to 2π, and a vertical axis for the gradient, marked from -1 to 1, as shown in Figure 9.6.

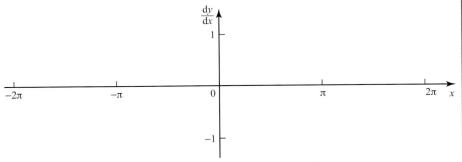

Figure 9.6

First, look for the values of x for which the gradient of $y = \sin x$ is zero. Mark zeros at these values of x on your gradient graph.

Decide which parts of $y = \sin x$ have a positive gradient and which have a negative gradient. This will tell you whether your gradient graph should be above or below the y-axis at any point.

Look at the part of the graph of $y = \sin x$ near $x = 0$ and compare it with the graph of $y = x$. What do you think the gradient of $y = \sin x$ is at this point? Mark this point on your gradient graph. Also mark on any other points with plus or minus the same gradient.

Now, by considering whether the gradient of $y = \sin x$ is increasing or decreasing at any particular point, sketch in the rest of the gradient graph.

The gradient graph that you have drawn should look like a familiar graph. What graph do you think it is?

Now sketch the graph of $y = \cos x$, with x measured in radians, and use it as above to obtain a sketch of the graph of the gradient function of $y = \cos x$.

Activity 9.1 showed you that the graph of the gradient function of $y = \sin x$ resembled the graph of $y = \cos x$. You will also have found that the graph of the gradient function of $y = \cos x$ looks like the graph of $y = \sin x$ reflected in the x-axis to become $y = -\sin x$.

Activity 9.1 suggests the following results:

$$y = \sin x \Rightarrow \frac{dy}{dx} = \cos x$$

$$y = \cos x \Rightarrow \frac{dy}{dx} = -\sin x$$

You can prove the result for $y = \sin x$ using differentiation from first principles.

The gradient of the chord PQ in Figure 9.7 is $\dfrac{\sin(x + h) - \sin x}{h}$.

Taking the limit as the points P and Q move closer together gives

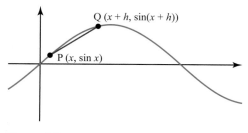

Q $(x + h, \sin(x + h))$

P $(x, \sin x)$

Figure 9.7

$$y = \sin x \Rightarrow \frac{dy}{dx} = \lim_{h \to 0} \left(\frac{\sin(x + h) - \sin x}{h} \right)$$

Using the compound angle formula,

$$\sin(x + h) = \sin x \cos h + \cos x \sin h$$

> $\sin(A + B) = \sin A \cos B + \cos A \sin B$

Therefore

$$\sin(x + h) \approx (\sin x)\left(1 - \tfrac{1}{2}h^2\right) + (\cos x)h$$

> As h is small, you can replace $\sin h$ and $\cos h$ by their small-angle approximations:
> $$\sin h = h, \quad \cos h = 1 - \tfrac{1}{2}h^2$$

$$\Rightarrow \sin(x + h) = \sin x - \tfrac{1}{2}h^2 \sin x + h \cos x$$

Substituting this into $\dfrac{dy}{dx} = \lim_{h \to 0} \left(\dfrac{\sin(x + h) - \sin x}{h} \right)$ gives

$$\frac{dy}{dx} = \lim_{h \to 0} \left(\frac{\sin x - \tfrac{1}{2}h^2 \sin x + h \cos x - \sin x}{h} \right)$$

$$\Rightarrow \frac{dy}{dx} = \lim_{h \to 0} \left(\frac{-\tfrac{1}{2}h^2 \sin x + h \cos x}{h} \right)$$

$$\Rightarrow \frac{dy}{dx} = \lim_{h \to 0} \left(-\tfrac{1}{2}h \sin x + \cos x \right) = \cos x$$

The result

$$y = \cos x \Rightarrow \frac{dy}{dx} = -\sin x$$

> You will do this in question 11 of Exercise 9.2.

can also be proved using differentiation from first principles in a similar way.

Activity 9.2 leads to the following result:

$$y = \tan x \Rightarrow \frac{dy}{dx} = \sec^2 x$$

Example 9.4

Differentiate $y = \sin 3x$.

Solution

Using the chain rule, ⟵ $\boxed{\sin 3x \text{ is a function of a function.}}$

$$u = 3x \implies \frac{\mathrm{d}u}{\mathrm{d}x} = 3$$

$$y = \sin u \implies \frac{\mathrm{d}y}{\mathrm{d}u} = \cos u$$

$$\frac{\mathrm{d}y}{\mathrm{d}x} = \frac{\mathrm{d}u}{\mathrm{d}x} \times \frac{\mathrm{d}y}{\mathrm{d}u}$$

$$\implies \frac{\mathrm{d}y}{\mathrm{d}x} = 3\cos u = 3\cos 3x$$

Using the chain rule as in Example 9.4 on the more general expression $y = \sin kx$ gives

$$\frac{\mathrm{d}y}{\mathrm{d}x} = k\cos kx.$$

ACTIVITY 9.3

Find the derivatives of the general expression $\sin(f(x))$ and the general expression $\cos(f(x))$.

Similarly, $y = \cos kx \implies \dfrac{\mathrm{d}y}{\mathrm{d}x} = -k\sin kx$

and $\quad y = \tan kx \implies \dfrac{\mathrm{d}y}{\mathrm{d}x} = k\sec^2 kx.$

Example 9.5

Differentiate $y = x^2 \cos x$.

Solution

Using the product rule,

$$u = x^2 \implies \frac{\mathrm{d}u}{\mathrm{d}x} = 2x$$

$$v = \cos x \implies \frac{\mathrm{d}v}{\mathrm{d}x} = -\sin x$$

$$\frac{\mathrm{d}y}{\mathrm{d}x} = u\frac{\mathrm{d}v}{\mathrm{d}x} + v\frac{\mathrm{d}u}{\mathrm{d}x}$$

$$= -x^2 \sin x + 2x\cos x$$

Example 9.6

Differentiate $y = \tan(x^3 - 1)$.

Solution

Using the chain rule,

$$u = x^3 - 1, \quad y = \tan u$$

$$\frac{\mathrm{d}y}{\mathrm{d}x} = \frac{\mathrm{d}y}{\mathrm{d}u} \times \frac{\mathrm{d}u}{\mathrm{d}x}$$

$$= \sec^2 u \times 3x^2$$

$$= 3x^2 \sec^2(x^3 - 1)$$

$\boxed{y = \tan u \implies \dfrac{\mathrm{d}y}{\mathrm{d}u} = \sec^2 u}$

$\boxed{u = x^3 - 1 \implies \dfrac{\mathrm{d}u}{\mathrm{d}x} = 3x^2}$

Differentiating trigonometric functions

Exercise 9.2

① Differentiate the following functions with respect to x.

   (i) $y = 2\cos x + \sin x$

   (ii) $y = 5\sin x - 3\cos x$

   (iii) $y = 3\tan x - 2x$

② Differentiate the following functions with respect to x.

   (i) $y = \tan 3x$

   (ii) $y = \sin\left(\dfrac{x}{2}\right)$

   (iii) $y = \cos(-6x)$

③ Use the chain rule to differentiate the following functions with respect to x.

   (i) $y = \sin\left(x^2\right)$

   (ii) $y = \mathrm{e}^{\tan x}$

   (iii) $y = \ln(\cos x)$

④ Use the product rule to differentiate the following functions with respect to x.

   (i) $y = x\tan x$

   (ii) $y = \mathrm{e}^x\cos x$

   (iii) $y = \sin x\cos x$

⑤ Use the quotient rule to differentiate the following functions with respect to x.

   (i) $y = \dfrac{\sin x}{x}$

   (ii) $y = \dfrac{\mathrm{e}^x}{\cos x}$

   (iii) $y = \dfrac{x + \cos x}{\sin x}$

⑥ Use an appropriate method to differentiate each of the following functions.

   (i) $y = \sqrt{\sin 3x}$

   (ii) $y = \mathrm{e}^{\cos 2x}$

   (iii) $y = \dfrac{\ln x}{\tan x}$

⑦ Verify the following results.

   (i) $y = \cot x \implies \dfrac{\mathrm{d}y}{\mathrm{d}x} = -\operatorname{cosec}^2 x$

   (ii) $y = \sec x \implies \dfrac{\mathrm{d}y}{\mathrm{d}x} = \sec x\tan x$

   (iii) $y = \operatorname{cosec} x \implies \dfrac{\mathrm{d}y}{\mathrm{d}x} = -\operatorname{cosec} x\cot x$

⑧ Figure 9.8 shows the curve $y = x\cos x$.

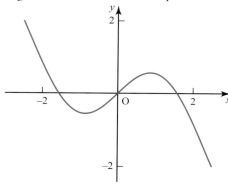

Figure 9.8

   (i)   Differentiate $y = x\cos x$.

   (ii)   Find the gradient of the curve $y = x\cos x$ at the point where $x = \pi$.

   (iii)   Find the equation of the tangent to the curve $y = x\cos x$ at the point where $x = \pi$.

   (iv)   Find the equation of the normal to the curve $y = x\cos x$ at the point where $x = \pi$.

⑨ Figure 9.9 shows the curve $y = \sin x + \cos 2x$.

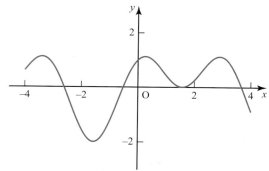

Figure 9.9

   (i)   Find $\dfrac{\mathrm{d}y}{\mathrm{d}x}$ and $\dfrac{\mathrm{d}^2 y}{\mathrm{d}x^2}$.

   (ii)   Find the coordinates of the three turning points in the range $0 \leqslant x \leqslant \pi$, giving your answers to 3 decimal places where appropriate.

   (iii)   Use calculus to show the nature of these turning points.

   (iv)   Find the coordinates of the two points of inflection in the range $0 \leqslant x \leqslant \pi$, giving your answers to 3 decimal places.

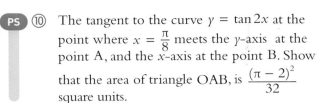

⑩ The tangent to the curve $y = \tan 2x$ at the point where $x = \frac{\pi}{8}$ meets the y-axis at the point A, and the x-axis at the point B. Show that the area of triangle OAB, is $\frac{(\pi - 2)^2}{32}$ square units.

⑪ By using differentiation from first principles, prove that

$$y = \cos x \implies \frac{dy}{dx} = -\sin x$$

using the following steps.

(i) For $y = \cos(x + h)$, write an expression for $\frac{dy}{dx}$ as a limit as h tends to zero.

(ii) Use the compound angle expression for $\cos(A + B)$ to expand $\cos(x + h)$.

(iii) Use your answer to (ii) and the small-angle approximations for $\sin h$ and $\cos h$ to rewrite your expression for $\frac{dy}{dx}$ from (i), and simplify your answer.

(iv) Find the limit as h tends to zero.

⑫ A curve has the equation $y = 3x - 2\cos x$, $0 \leqslant x \leqslant \frac{\pi}{2}$.

(i) Find the coordinates of the point on this curve where the normal is parallel to the line $4y + x = 0$.

(ii) The tangent to the curve at this point meets the y-axis at the point Y. Find the exact value of the distance OY.

3 Implicit differentiation

All the functions you have differentiated so far have been of the form $y = f(x)$. However, many functions, for example $x^3 + y^3 = xy$, cannot be written in this way at all. Others can look clumsy when you try to make y the subject.

An example of this is the circle $x^2 + y^2 = 4$, illustrated in Figure 9.10, which is much more easily recognised in this form than in the equivalent form $y = \pm\sqrt{4 - x^2}$.

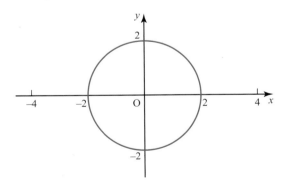

Figure 9.10

A function which is not given in the form $y = f(x)$ is called an **implicit function**, and to differentiate it you use **implicit differentiation**. Implicit differentiation uses the chain rule to differentiate term by term with respect to x, using the product and quotient rules where necessary.

An example of an implicit function that you would differentiate in this way is $y^3 + xy = 2$. As you go through the expression differentiating term by term, the first term that you need to differentiate with respect to x is y^3.

Using the chain rule,

$$\frac{d\left(y^3\right)}{dx} = \frac{d\left(y^3\right)}{dy} \times \frac{dy}{dx}$$

| Differentiate y^3 with respect to y. |

$$= 3y^2 \frac{dy}{dx}$$

So to differentiate a function of y with respect to x, you differentiate with respect to y and multiply by $\frac{dy}{dx}$. Of course, this technique can be extended to other variables.

ACTIVITY 9.4

Write down

(i)   the derivative of x^7 with respect to x
(ii) the derivative of $2y^5$ with respect to x
(iii) the derivative of $-z^2$ with respect to x
(iv) the derivative of $3z - 6$ with respect to y
(v)   the derivative of $2x^4$ with respect to z.

Example 9.7

The equation of a curve is given by $y^3 + xy = 2$.

(i)   Find an expression for $\frac{dy}{dx}$ in terms of x and y.

(ii) Hence find the gradient of the curve at $(1,1)$.

Solution

(i) Differentiate both sides of the equation term by term, remembering to use the product rule to differentiate xy.

$$\frac{d}{dx}(y^3) + \frac{d}{dx}(xy) = 0$$

$$3y^2 \frac{dy}{dx} + \left(x\frac{dy}{dx} + y \right) = 0$$

To differentiate xy by the product rule:

$u = x \;\Rightarrow\; \frac{du}{dx} = 1$

$v = y \;\Rightarrow\; \frac{dv}{dx} = \frac{dy}{dx}$

To make $\frac{dy}{dx}$ the subject, collect up the terms that have it as a factor on to one side of your equation.

$$\frac{dy}{dx}\left(3y^2 + x\right) = -y$$

$$\frac{dy}{dx} = \frac{-y}{3y^2 + x}$$

(ii) At $(1,1)$, $\frac{dy}{dx} = \frac{-1}{3 + 1} = -\frac{1}{4}$

Turning points

As you already know, turning points occur where $\frac{dy}{dx} = 0$.

When working with an implicit function, putting $\frac{dy}{dx} = 0$ will not usually give values of x directly, but it will give a relationship between x and y. You can then solve this simultaneously with the equation of the curve to find the coordinates of the turning points.

Example 9.8

Find the turning points on the curve $x^2 + 4xy + y^2 = -48$.

Solution

$$x^2 + 4xy + y^2 = -48$$

$$\Rightarrow 2x + \left(4y + 4x\frac{dy}{dx}\right) + 2y\frac{dy}{dx} = 0 \quad \longleftarrow \boxed{\text{Differentiate both sides of the equation.}}$$

At the turning points, $\dfrac{dy}{dx} = 0$ $\qquad \boxed{\text{It is not necessary to rearrange for } \dfrac{dy}{dx} \text{ at this point.}}$

$$\Rightarrow 2x + 4y = 0$$

$$\Rightarrow \qquad x = -2y$$

To find the coordinates of the turning points, solve $x^2 + 4xy + y^2 = -48$ and $x = -2y$ simultaneously.

$$\left(-2y\right)^2 + 4\left(-2y\right)y + y^2 = -48 \quad \longleftarrow \boxed{\text{Substituting for } x.}$$

$$\Rightarrow \quad -3y^2 = -48$$

$$\Rightarrow \quad y^2 = 16$$

$$\Rightarrow \quad y = \pm 4$$

$x = -2y$ so the coordinates of the turning points are $\left(-8, 4\right)$ and $\left(8, -4\right)$.

Exercise 9.3

① Differentiate the following with respect to x.

(i) $\sqrt{y}$

(ii) $\dfrac{4}{y^2}$

(iii) $z^3 + y^9$

② Differentiate the following with respect to x.

(i) e^{y^2}

(ii) $\sin\left(3y - 2\right)$

(iii) $\ln\left(y^2 + 2\right)$

③ Differentiate the following with respect to x.

(i) xe^y

(ii) $\sin x \cos y$

(iii) $\dfrac{x^3}{\sin y}$

④ For each of the following functions, find an expression for $\dfrac{dy}{dx}$ in terms of y and x.

(i) $xy + x + y = 9$

(ii) $e^{2y} + xy^2 = 2$

(iii) $\cos x + \cos y = 1$

⑤ Find the gradient of the curve $xy^3 = 5\ln y$ at the point $\left(0, 1\right)$.

⑥ (i) Find the gradient of the curve $x^2 + 3xy + y^2 = x + 3y$ at the point $\left(2, -1\right)$.

(ii) Hence find the equation of the tangent to the curve at this point.

⑦ Find the equations of the tangent and normal to the curve $y^2e^x - 2x^2 = 4$ at the point $\left(0, 2\right)$.

⑧ Given that $y = \sin^{-1}x$, show that $\dfrac{dy}{dx} = \dfrac{1}{\cos y}$ by

(i) writing $x = \sin y$ and finding $\dfrac{dx}{dy}$

(ii) writing $\sin y = x$ and using implicit differentiation.

(iii) Show further that $\dfrac{dy}{dx}$ can be written as $\dfrac{1}{\sqrt{\left(1 - x^2\right)}}$.

⑨ Use both methods of question 8 to show that, if $y = \tan^{-1} x$, then $\dfrac{dy}{dx} = \dfrac{1}{1 + x^2}$.

⑩ Find the coordinates of all the stationary points on the curve $x^2 + y^2 + xy = 3$.

⑪ A curve has the equation $(x - 6)(y + 4) = 2$.

(i) By using the product rule, find an expression for $\dfrac{dy}{dx}$ in terms of x and y.

(ii) Find the equation of the normal to the curve at the point $(7, -2)$.

(iii) Find the coordinates of the point where the normal meets the curve again.

⑫ A curve has equation $\sin x + \sin y = 1$ for $0 \leqslant x \leqslant \pi, 0 \leqslant y \leqslant \pi$.

(i) Differentiate the equation of the curve with respect to x and hence find the coordinates of any turning points.

(ii) Find the y coordinates of the points where $x = 0$ and $x = \pi$.

(iii) The curve $y = \sin x$ is symmetrical in the line $x = \dfrac{\pi}{2}$.

Use this to write down the equations of two lines of symmetry of the curve $\sin x + \sin y = 1$, $0 \leqslant x \leqslant \pi, 0 \leqslant y \leqslant \pi$.

(iv) Sketch the curve.

LEARNING OUTCOMES

When you have completed this chapter, you should be able to:

➤ understand and use differentiation from first principles for $\sin x$ and $\cos x$

➤ differentiate:
   ○ e^{kx} and a^{kx} and related sums, differences and constant multiples
   ○ $\sin kx$, $\cos kx$, $\tan kx$ and related sums, differences and constant multiples

➤ understand and use the derivative of $\ln x$

➤ differentiate simple functions defined implicitly for first derivative only.

KEY POINTS

1. $y = \ln x \Rightarrow \dfrac{dy}{dx} = \dfrac{1}{x}$

2. $y = e^x \Rightarrow \dfrac{dy}{dx} = e^x$

3. $y = a^x \Rightarrow \dfrac{dy}{dx} = a^x \ln a$

4. $y = \sin x \Rightarrow \dfrac{dy}{dx} = \cos x$

   $y = \cos x \Rightarrow \dfrac{dy}{dx} = -\sin x$

   $y = \tan x \Rightarrow \dfrac{dy}{dx} = \sec^2 x$

5. Implicit differentiation, for example: $\dfrac{d}{dx}\big[f(y)\big] = \dfrac{d}{dy}\big[f(y)\big] \times \dfrac{dy}{dx}$

10 Integration

The mathematical process has a reality and virtue in itself, and once discovered it constitutes a new and independent factor.

Winston Churchill (1876–1965)

→   How would you estimate the volume of water in a sea wave?

So far, all the functions you have integrated have been polynomial functions. However, you sometimes need to integrate other functions. For example, you might model a sea wave as a sine wave, so finding the volume of water in the wave might involve integrating a sine function.

Review: Integration

Indefinite and definite integration

Integration is the inverse of differentiation; that is

$$\frac{dy}{dx} = x^n \Rightarrow y = \frac{x^{n+1}}{n+1} + c, \qquad n \neq -1,$$

where c is an arbitrary constant.

Indefinite integration results in a function of x together with an arbitrary constant. The notation used is

$$\int x^n \, dx = \frac{x^{n+1}}{n+1} + c.$$

Definite integration involves limits on the integral sign which are substituted into the integrated function, giving a number as the answer:

$$\int_a^b x^n \, dx = \left[\frac{x^{n+1}}{n+1} \right]_a^b = \frac{b^{n+1} - a^{n+1}}{n+1}, \qquad n \neq -1.$$

Example 10.1

Find the following integrals.

(i) $\displaystyle\int x^2 \, dx$ (ii) $\displaystyle\int_0^3 x^2 \, dx$

Solution

(i) $\displaystyle\int x^2 \, dx = \frac{x^3}{3} + c$ | The outcome from indefinite integration is a function of x and an arbitrary constant.

(ii) $\displaystyle\int_0^3 x^2 \, dx = \left[\frac{x^3}{3} \right]_0^3$

$$= \frac{3^3}{3} - \frac{0^3}{3}$$

$$= 9$$ | The outcome from definite integration is a number.

Sometimes you may need to rewrite an expression before you can integrate it, as in the next example.

Example 10.2

1

Find $\displaystyle\int \frac{(x + 1)^2}{x^4}\, dx$.

Solution

$$\int \frac{(x + 1)^2}{x^4}\, dx = \int \frac{x^2 + 2x + 1}{x^4}\, dx$$

$$= \int \left(x^{-2} + 2x^{-3} + x^{-4}\right) dx \longleftarrow \boxed{\text{Write as a sum of powers of } x.}$$

$$= \frac{x^{-1}}{-1} + \frac{2x^{-2}}{-2} + \frac{x^{-3}}{-3} + c \longleftarrow \boxed{\text{Integrate term by term.}}$$

$$= -\frac{1}{x} - \frac{1}{x^2} - \frac{1}{3x^3} + c$$

The area under a curve

Definite integration can be used to find areas bounded by a curve and the x-axis.

Example 10.3

(i) Sketch the curve $y = -x^2 + 5x - 4$, and shade the region bounded by the curve, the x-axis, and the lines $x = 2$ and $x = 5$.

(ii) Find the area of the shaded region.

> ❗ When integrating to find an area, you must be careful if some of the area is below the x-axis.

Solution

(i) $y = -x^2 + 5x - 4$

$= -(x^2 - 5x + 4)$

$= -(x - 1)(x - 4)$

The graph crosses the coordinate axes at $(1, 0)$, $(4, 0)$ and $(0, -4)$.

Since the coefficient of x^2 is negative, the graph is $\cap$ shaped.

> You must find areas of regions above and below the x-axis separately.

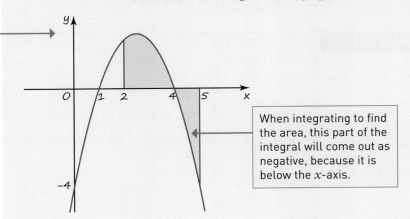

> When integrating to find the area, this part of the integral will come out as negative, because it is below the x-axis.

Figure 10.1

> ! Be particularly careful if you are using the Integration command on your calculator. Enter the area above and below the axes separately.

(ii)   Area above the x-axis

$$= \int_2^4 \left(-x^2 + 5x - 4\right) \, dx$$

> Substitute in upper and lower limits and subtract.

$$= \left[-\frac{x^3}{3} + \frac{5x^2}{2} - 4x\right]_2^4$$

$$= \left(-\frac{4^3}{3} + \frac{5 \times 4^2}{2} - 4 \times 4\right) - \left(-\frac{2^3}{3} + \frac{5 \times 2^2}{2} - 4 \times 2\right)$$

$$= \frac{10}{3}$$

> Area of region above the axis is $\frac{10}{3}$.

Area below the x-axis

$$= \int_4^5 \left(-x^2 + 5x - 4\right) \, dx$$

$$= \left[-\frac{x^3}{3} + \frac{5x^2}{2} - 4x\right]_4^5$$

$$= \left(-\frac{5^3}{3} + \frac{5 \times 5^2}{2} - 4 \times 5\right) - \left(-\frac{4^3}{3} + \frac{5 \times 4^2}{2} - 4 \times 4\right)$$

$$= -\frac{11}{6}$$

> This is negative because the region is below the x-axis. Area of region is $\frac{11}{6}$.

Total area of shaded region $= \frac{10}{3} + \frac{11}{6} = \frac{31}{6}$.

Finding the equation of a curve from its gradient function

The function $\frac{dy}{dx}$ is the gradient function of a curve, and integrating it gives the equation of a family of curves.

Example 10.4

A curve has gradient function $\frac{dy}{dx} = \frac{2}{\sqrt{x}}$, $x \geqslant 0$.

(i)   Integrate $\frac{dy}{dx}$ to find an expression for y.

(ii)   Sketch three members of the family of solution curves.

(iii)   Find the equation of the particular curve that passes through the point $(1, 3)$.

Solution

(i) $\dfrac{dy}{dx} = 2x^{-\frac{1}{2}}$ ← [Rewrite the equation in index form.]

$\Rightarrow y = 2 \times \dfrac{x^{\left(-\frac{1}{2}+1\right)}}{-\frac{1}{2}+1} + c$

$\Rightarrow y = 2 \times \dfrac{x^{\frac{1}{2}}}{\frac{1}{2}} + c$ ← [Using $\dfrac{dy}{dx} = x^n \Rightarrow y = \dfrac{x^{n+1}}{n+1} + c, \quad n \neq -1.$ In this case $n + 1 = -\frac{1}{2} + 1 = \frac{1}{2}.$]

$= 4x^{\frac{1}{2}} + c$ ← [This is the general equation of the family of solution curves.]

(ii)   Figure 10.2 shows three members of the family of solution curves.

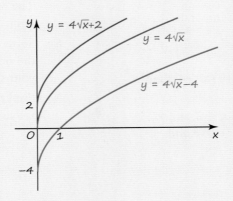

Figure 10.2

(iii)   When $x = 1, y = 3$, so ← [Substitute coordinates into the general equation.]

$3 = 4\sqrt{1} + c$ ← [Rearrange to find the value of c.]

$\Rightarrow c = 3 - 4 = -1$ ← [Substitute the value of c into the general equation to get the particular equation.]

Therefore the particular equation is $y = 4\sqrt{x} - 1.$

Review exercise

① Integrate the following functions with respect to x.

(i) $12x^7$ (ii) $4x^2 + 1$

(iii) $2x - x^4 + 3x^8$

② Evaluate the following definite integrals.

(i) $\displaystyle\int_1^2 2x^3 \, dx$ (ii) $\displaystyle\int_0^3 \left(x^3 + x - 2\right) dx$

(iii) $\displaystyle\int_{-2}^0 \left(3x^5 - 2x + 1\right) dx$

③ Integrate the following functions with respect to x.

(i) $2\sqrt{x}$ (ii) $\dfrac{3}{x^2}$ (iii) $3x^2\sqrt{x}$

④ Evaluate the following definite integrals.

(i) $\displaystyle\int_1^4 \left(\dfrac{1}{x^2} - 2\sqrt{x}\right) dx$

(ii) $\displaystyle\int_{-1}^1 \left(\dfrac{y^4 - 2}{3y^2}\right) dy$

⑤ A curve has gradient function $\frac{dy}{dx} = 2x - 1$.

(i) Find the general equation of the family of solution curves.

(ii) Sketch three members of the family of solution curves.

(iii) Find the equation of the particular curve that passes through the point $(-2, 7)$.

⑥ A curve has gradient function $\frac{dy}{dx} = \frac{4}{x^3} - \frac{1}{\sqrt{x}}$ and passes through the point with coordinates $(1, -2)$. Find the equation of the curve.

⑦ Find the following indefinite integrals.

(i) $\int \left(y + \frac{1}{y} \right)^2 dy$ (ii) $\int \left(\frac{2t^4 - \sqrt{t}}{t} \right) dt$

⑧ $f'(x) = 3x^2 - 4x + 1$ and $f(0) = -4$. Find $f(x)$.

⑨ Sketch the curve $y = -x^2 + x + 2$, and find the area bounded by the curve and the x-axis.

⑩ Sketch the curve $y = 2x^2 - 3x + 1$, and find the area bounded by the curve and the x-axis.

⑪ (i) Factorise $x^3 + 3x^2 - x - 3$.

(ii) Sketch the curve $y = x^3 + 3x^2 - x - 3$ and shade the two regions bounded by the curve and the x-axis.

(iii) Find the total area of the shaded regions.

⑫ A curve has equation $y = \sqrt{x}$, $x \geq 0$.

(i) Find the equation of the normal to the curve at the point A, where $x = 1$.

(ii) The normal in (i) meets the x-axis at point B. Find the coordinates of B.

(iii) Show on the same sketch the curve and the normal, and mark points A and B on your sketch.

(iv) Find the area bounded by the curve, the x-axis and the line segment AB.

1 Finding areas

Integration as the limit of a sum

You can estimate the area under a curve by dividing it up into rectangular strips. For example, the area under the curve $y = 16 - x^2$ between $x = 0$ and $x = 4$ is shown on Figure 10.3.

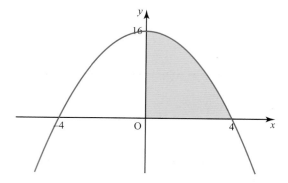

Figure 10.3

Its area is approximately equal to the total area of the four rectangles in Figure 10.4.

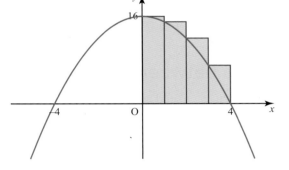

Discussion points

➜ How good is this approximation?
➜ How could you improve the approximation?

Figure 10.4

The height of each of the rectangles can be found by substituting the appropriate x value into the equation of the curve. For example, when $x = 0$, $y = 16 - 0^2 = 16$.

So the area under the curve is approximately equal to

$(1 \times 16) + (1 \times 15) + (1 \times 12) + (1 \times 7) = 50$ square units.

This method can be generalised, as illustrated in Figures 10.5 and 10.6. Each strip has height y and thickness δx and so its area is given by $\delta A = y \times \delta x$. The total area of the region is thus given by

$$\sum_{\text{All strips}} \delta A = \sum_{\text{All strips}} y \, \delta x.$$

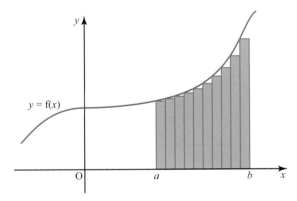

Figure 10.5

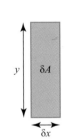

Figure 10.6

For any given region, the greater the number of strips the thinner they become and the more accurate will be the estimate. You can keep on increasing the number of strips, approaching the situation where there is an infinite number of infinitesimally thin strips and the area is exact.

In this limiting situation δx is written as $\mathrm{d}x$ and the Σ symbol is replaced by the integral sign, $\int$.

This can be written formally as

$$\text{Area} = \lim_{\delta x \to 0} \left(\sum_{x=a}^{x=b} y\,\delta x \right).$$

The definite integral $\int_a^b y\,\mathrm{d}x$ is defined as $\lim_{\delta x \to 0} \left(\sum_{x=a}^{x=b} y\,\delta x \right)$.

The area between a curve and the y-axis

So far you have calculated areas between curves and the x-axis. You can also use integration to calculate the area between a curve and the y-axis.

Figure 10.7 shows the area between the line $y = x^2$, the y-axis and the line $y = 4$.

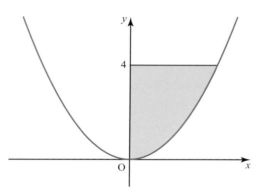

Figure 10.7

Instead of strips of width δx and height y, you now sum strips of width δy and length x (see Figure 10.8).

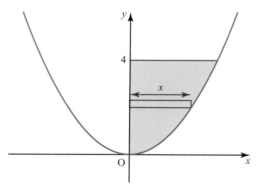

Figure 10.8

Note

- In such cases, the integral involves $\mathrm{d}y$ and not $\mathrm{d}x$. It is therefore necessary to write x in terms of y wherever it appears. The integration is then said to be carried out with respect to y instead of x.
- The limits of the integration are y values rather than x values.

The shaded area can be written as $\lim_{\delta y \to 0} \left(\sum_{y=0}^{y=4} x\,\delta y \right)$, which is equal to the integral $\int_0^4 x\,\mathrm{d}y.$

To carry out the integration, you need to express x in terms of y. In this case,

$$y = x^2 \implies x = y^{\frac{1}{2}}.$$

So the shaded area $= \int_0^4 x\,\mathrm{d}y$

$$= \int_0^4 y^{\frac{1}{2}}\,\mathrm{d}y$$

$$= \left[\frac{2}{3} y^{\frac{3}{2}} \right]_0^4$$

$$= \frac{2}{3} \times 4^{\frac{3}{2}} - 0$$

$$= \frac{16}{3} \text{ square units.}$$

Finding the area between two curves

In the review section at the beginning of this chapter, you were reminded how to use integration to find the areas of regions between curves and the x-axis, including regions below the x-axis, and of regions partly above and partly below the x-axis.

These ideas will now be extended to find the area between two curves.

Example 10.5

(i) Find the points of intersection of the line $y = x + 1$ and the curve $y = (x - 1)^2$.

(ii) Sketch the line and the curve on the same axes.

(iii) Find the area under the line $y = x + 1$ between the x values found in (i).

(iv) Find the area under the curve $y = (x - 1)^2$ between the x values found in (i).

(v) Find the difference between your answer to (iii) and your answer to (iv). What area does this represent? Shade the region on your sketch.

Solution

(i) $x + 1 = x^2 - 2x + 1$ ← Solve the equations simultaneously to find where the line and the curve intersect.

$\Rightarrow x^2 - 3x = 0$

$\Rightarrow x(x - 3) = 0$

$\Rightarrow x = 0, x = 3$

Substitute the x values into either equation to find the y values.

When $x = 0$, $y = 1$.
When $x = 3$, $y = 4$.

The line and the curve intersect at the points $(0, 1)$ and $(3, 4)$.

(ii)

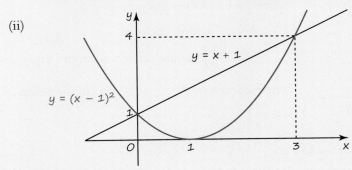

Figure 10.9

(iii) $\displaystyle\int_0^3 (x+1)\,dx = \left[\dfrac{x^2}{2} + x\right]_0^3$

> Integrate the equation of the line.

$\qquad\qquad\qquad = \left(\dfrac{9}{2} + 3\right) - 0$

> Substitute in the limits.

$\qquad\qquad\qquad = \dfrac{15}{2}$

The area under the line is $\dfrac{15}{2}$ square units.

> You don't actually need to use integration in this case but can find the answer as the area of a trapezium.

> Integrate the equation of the curve.

(iv) $\displaystyle\int_0^3 (x^2 - 2x + 1)\,dx = \left[\dfrac{x^3}{3} - x^2 + x\right]_0^3$

$\qquad\qquad\qquad = \left(\dfrac{27}{3} - 9 + 3\right) - 0$

> Substitute in the limits.

$\qquad\qquad\qquad = 3$

The area under the curve is 3 square units.

(v) The difference between these two areas is $\dfrac{9}{2}$ square units. This is the area of the region enclosed by the line and the curve, as shaded in Figure 10.10.

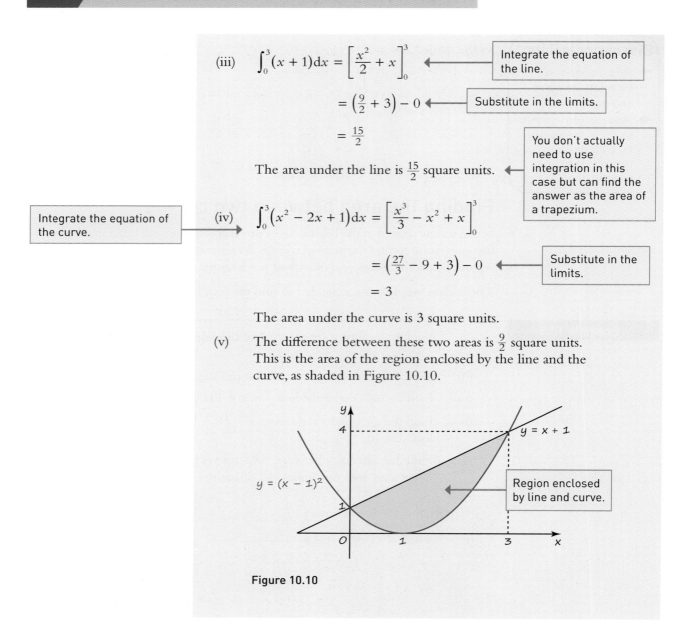

> Region enclosed by line and curve.

Figure 10.10

Instead of integrating each equation separately and then subtracting to find the area enclosed, it is usually more efficient to subtract the equations of the two curves first and then integrate.

Area of region enclosed

$$= \int_a^b \left(\text{equation of top curve} - \text{equation of bottom curve}\right) dx$$

where a and b are the x coordinates of the points of intersection of the two curves.

Example 10.6

10

Chapter 10 Integration

(i) Find the points of intersection of the curve $y = x^2 - 4x + 3$ and the curve $y = 3 + 4x - x^2$.

(ii) Sketch the curves on the same axes, shading the area enclosed by the two curves.

(iii) Find the area of the shaded region.

Solution

(i) $x^2 - 4x + 3 = 3 + 4x - x^2$ ← Solve the equations simultaneously to find where the two curves intersect.

$$2x^2 - 8x = 0$$

$$2x(x - 4) = 0$$

$$x = 0, x = 4$$

When $x = 0$, $y = 3$. ← Substitute x values into either equation to find y values.

When $x = 4$, $y = 3$.

The points of intersection are $(0, 3)$ and $(4, 3)$.

(ii)

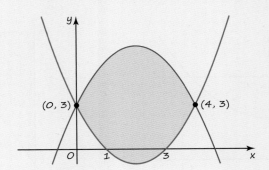

Figure 10.11

(iii)   Area of shaded region

$$= \int_0^4 \left(\text{equation of top curve} - \text{equation of bottom curve} \right) dx$$

$$= \int_0^4 \left((3 + 4x - x^2) - (x^2 - 4x + 3) \right) dx$$ ← First subtract the equations of the curves and simplify.

$$= \int_0^4 (8x - 2x^2) \, dx$$

$$= \left[4x^2 - \frac{2x^3}{3} \right]_0^4$$ ← Then integrate.

$$= \left(4 \times 4^2 - \frac{2 \times 4^3}{3} \right) - \left(4 \times 0^2 - \frac{2 \times 0^3}{3} \right)$$

$$= \frac{64}{3}$$

The area enclosed between the curves is $\frac{64}{3}$ square units.

In Example 10.6, notice that part of the shaded area is below the x-axis. However, the function that you are integrating (which is the distance between the blue curve and the red curve) is always positive, so you do not need to split the calculation up into different parts.

> ### Discussion points
>
> ➡ What would happen if you tried to calculate the shaded area by first finding the area under the blue curve, and then subtracting the area under the red curve?
>
> ➡ How could you find the shaded area by splitting it up?

Exercise 10.1

① Figure 10.12 shows the line $y = 2$ and the curve $y = x^2 + 1$, which intersect at the points $(-1, 2)$ and $(1, 2)$.

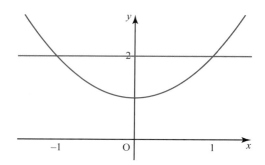

Figure 10.12

Find the area of the region enclosed by the line and the curve.

② Figure 10.13 shows the region between the line $y = 3x + 1$, the y-axis and the line $y = 7$.

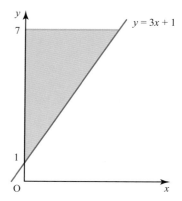

Figure 10.13

(i) Show that the area of the shaded region can be written as $\int_{1}^{7} \frac{1}{3}(y - 1)\, dy$.

(ii) Evaluate this integral to find the area.

(iii) Find the area of the shaded region using the formula for the area of a triangle, and check that this is the same answer as for part (ii).

③ (i) Find the points of intersection of the curve $y = x^2$ and the line $y = 9$.

(ii) Sketch the curve and the line on the same axes.

(iii) Find the area bounded by the curve and the line.

④ (i) Find the points of intersection of the curve $6x - x^2$ and the line $y = 5$.

(ii) Sketch the curve and the line on the same axes.

(iii) Find the area bounded by the curve and the line.

⑤ (i) Find the points of intersection of the curve $y = 4x - x^2$ and the line $y = x$.

(ii) Sketch the curve and the line on the same axes.

(iii) Find the area bounded by the curve and the line.

⑥ Figure 10.14 shows the curve $y = x^2 - 1$ and the line $y = 3$, which intersect at the points $(-2, 3)$ and $(2, 3)$.

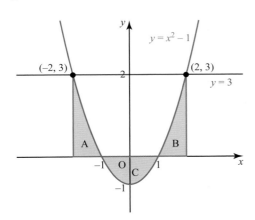

Figure 10.14

(i) Find the areas of the regions labelled A, B and C.

(ii) Deduce the area between the line and the curve.

(iii) Explain why the area between the line and the curve is given by $\int_{-2}^{2}(4 - x^2)\,dx$.

(iv) Evaluate this integral and verify that your answer is the same as that for (ii).

⑦ Find the area of the region bounded by the curve $y = \sqrt{x - 1}$, the y-axis, the x-axis and the line $y = 2$.

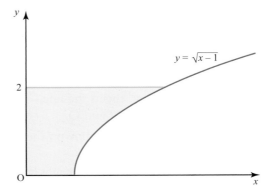

Figure 10.15

⑧ (i) Find the points of intersection of the curve $y = x^2 - 4x$ and the line $y = 3 - 2x$.

(ii) Sketch the curve and the line on the same axes.

(iii) Find the area bounded by the curve and the line.

⑨ Figure 10.16 shows the curves $y = x^2$ and $y = 8 - x^2$, which intersect at the points $(2, 4)$ and $(-2, 4)$.

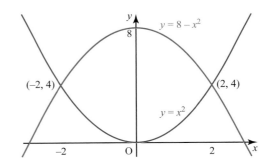

Figure 10.16

Find the area enclosed by the two curves.

⑩ (i) Find the points of intersection of the curves $y = x^2 + 3$ and $y = 5 - x^2$.

(ii) Sketch the two curves on the same axes.

(iii) Find the area bounded by the two curves.

PS ⑪ Find the area of the region enclosed by the curve $y = -x^2 - 1$ and the curve $y = -2x^2$.

PS ⑫ Find the area of the region enclosed by the curve $y = x^2 - 16$ and the curve $y = 4x - x^2$.

PS ⑬ Sketch the curve with equation $y = x^3 + 1$ and the line $y = 4x + 1$, and find the total area of the two regions enclosed by the line and the curve.

⑭ The curve $y = 3x - x^2$ meets the line $y = 6 - 2x$ at the points P and Q.

(i) Sketch the curve and the line on the same axes.

(ii) Find the exact length of the straight line PQ.

(iii) Find the area enclosed by the line and the curve.

⑮ Figure 10.17 shows the curve $y = 4 + 3x - x^2$. The area under the curve between $x = 0$ and $x = 4$ is to be estimated.

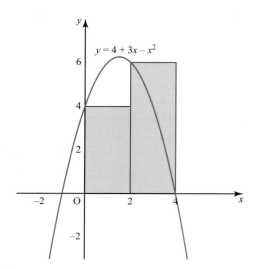

Figure 10.17

(i) An initial estimate is made using the two rectangular strips in the diagram. Find the total area of the two rectangles.

(ii) Looking at the diagram, can you say whether the area from (i) is smaller or larger than the true area? Or is it impossible to say?
Explain your answer.

(iii) Find an approximation for the area under the curve between $x = 0$ and $x = 4$ using four rectangular strips.

(iv) Find an approximation for the area using eight rectangular strips.

(v) Use integration to find an exact value for the area, and comment on your answer.

⑯ Figure 10.18 shows the curves $y = (x - 2)^2$ and $y = 4 + 4x - x^2$.

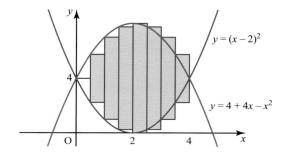

Figure 10.18

Samir wants to find the approximate area of the region enclosed by the two curves. Samir has divided the region into 8 rectangular strips as shown in Figure 10.19; the first one has zero height and so is a straight line.

(i) Find the total area of the 8 strips shown in the diagram.

(ii) Repeat the process using 16 strips.

(iii) Find the exact area of the region by integration.

(iv) Comment on your results.

⑰ The area enclosed by the curve $y = 10x - x^2$ and the curve $y = x^2 + 3x - 4$ lies partially above and partially below the x-axis.
Find the ratio of the area above the x-axis to the area below the x-axis.

2 Integration by substitution

Suppose you want to find $\int (5x-1)^3 dx$.

One approach is to start by expanding the brackets and then integrate term by term, but this is time consuming. A more efficient approach is to use the method of **integration by substitution**.

Because you already know how to integrate $\int u^3 du$, you can use the substitution $u = 5x - 1$ to transform the integral into one that you can do. This approach is shown in Example 10.7.

Example 10.7

(i)   Find $\int (5x-1)^3 dx$.

(ii) Check your answer by differentiation.

Solution

(i)   Use the substitution $u = 5x - 1$.

> You will change the integral involving the variable x into an integral involving the variable u.

First use this substitution to replace dx.

> You are now integrating with respect to a new variable, and you must replace every x, including the dx, with your new variable.

$$u = 5x - 1 \Rightarrow \frac{du}{dx} = 5$$

$$\frac{du}{dx} = 5 \quad \Rightarrow \quad dx = \tfrac{1}{5} du$$

Note that $\frac{du}{dx}$ is not really a fraction, but it is usual to treat it like one in this situation in order to find what to substitute for dx.

$$\int (5x-1)^3 dx = \int (5x-1)^3 \times \tfrac{1}{5} du$$

> Now replace the other terms involving x with terms involving u.

$$= \int \tfrac{1}{5} u^3 du$$

$$= \tfrac{1}{20} u^4 + c$$

> Integrate.

> Finally rewrite in terms of x.

$$= \tfrac{1}{20}(5x-1)^4 + c$$

(ii) To differentiate $\tfrac{1}{20}(5x-1)^4$ you use the chain rule, $\frac{dy}{dx} = \frac{dy}{du} \times \frac{du}{dx}$.

$$y = \tfrac{1}{20}(5x-1)^4$$

$$u = 5x - 1 \Rightarrow \frac{du}{dx} = 5$$

$$y = \tfrac{1}{20} u^4 \quad \Rightarrow \quad \frac{dy}{du} = \tfrac{1}{5} u^3$$

$$\frac{dy}{dx} = \tfrac{1}{5} u^3 \times 5$$

$$= (5x-1)^3$$

So the integral found in (i) is correct.

Integration of $(ax + b)^n$ by recognition

You have probably noticed that you can integrate any function of the type $(ax + b)^n$ by using the substitution $u = ax + b$. With experience and practice, you may feel no need to write down all the working, and you may be able to go straight to the answer. This process is sometimes called **recognition** or **inspection**. Remember that you can always check your answer by differentiation.

You will see shortly that recognition is possible for other types of integral. However, if you are in doubt you should use the substitution method.

ACTIVITY 10.1

(i) Integrate the following functions.

(a) $(2x + 3)^6$

(b) $(5 - 3x)^3$

(c) $\sqrt{1 + 2x}$

(ii) What is the general result for $\int (ax + b)^n \, \mathrm{d}x$?

Example 10.8

Evaluate the following definite integral, giving your answer to 2 decimal places.

$$\int_{-1}^{1} (2x - 1)^2 \, \mathrm{d}x$$

Solution

$$\int_{-1}^{1} (2x - 1)^2 \, \mathrm{d}x = \left[\frac{1}{2 \times 3} (2x - 1)^3 \right]_{-1}^{1}$$

First integrate the function.

$$= \left[\tfrac{1}{6} (2x - 1)^3 \right]_{-1}^{1}$$

Then simplify.

$$= \left(\tfrac{1}{6} (1)^3 \right) - \left(\tfrac{1}{6} (-3)^3 \right)$$

Now substitute in the limits.

$$= 4.67 \ (2 \text{ d.p.})$$

Integrating other functions by substitution

The method of integration by substitution can be used to integrate many other types of function as well as those of the form $(ax + b)^n$. With experience, you will be able to see how to choose a suitable substitution to use.

Example 10.9

(i) Use the substitution $u = x^2 + 3$ to find $\int x(x^2 + 3)^5 \, dx$.

(ii) Use differentiation to check your answer.

Solution

(i) First use the given substitution to replace dx.

$$u = x^2 + 5 \Rightarrow \frac{du}{dx} = 2x$$

$$\frac{du}{dx} = 2x \Rightarrow dx = \frac{1}{2x} \, du$$

$$\int x(x^2 + 3)^5 \, dx = \int x(x^2 + 3)^5 \times \frac{1}{2x} \, du \longleftarrow$$

> Substitute for dx first to see if anything cancels.

$$= \int \frac{1}{2}(x^2 + 3)^5 \, du \longleftarrow$$

> Cancel and simplify.

$$= \int \frac{1}{2} u^5 \, du \longleftarrow$$

> Now replace the other terms involving x with terms involving u.

> Integrate.

$$= \frac{1}{12} u^6 + c$$

> Finally rewrite in terms of x.

$$= \frac{1}{12}(x^2 + 3)^6 + c$$

(ii) To differentiate $\frac{1}{12}(x^2 + 3)^6$, use the chain rule, $\frac{dy}{dx} = \frac{dy}{du} \times \frac{du}{dx}$.

$$u = x^2 + 3 \Rightarrow \frac{du}{dx} = 2x$$

$$y = \frac{1}{12} u^6 \Rightarrow \frac{dy}{du} = \frac{1}{2} u^5$$

$$\frac{dy}{dx} = 2x \times \frac{1}{2} u^5$$

$$= x(x^2 + 3)^5$$

So the integral found in (i) is correct.

Integration of this 'reverse chain rule' type can always be done by using the appropriate substitution. Notice that in Example 10.9, $x(x^2 + 3)^5$ can be written as $\frac{1}{2} \times 2x \times (x^2 + 3)^5$, and that $2x$ is the derivative of $x^2 + 3$. Any function of the form $k \times f'(x) \times [f(x)]^n$, where k is a constant, can be integrated using the substitution $u = f(x)$.

Alternatively, as you become more familiar with this type of integration, you may find that you are able to do it by recognition, by seeing what function would differentiate to the given one by using the chain rule.

Definite integration by substitution

When doing definite integration by substitution, you must remember that the limits you are given are values of x. When you rewrite the integral in terms of the variable u, you must also replace the limits with values of u.

Example 10.10

Use an appropriate substitution to evaluate $\int_0^2 x^2 \sqrt{x^3 + 1} \, dx$.

Solution

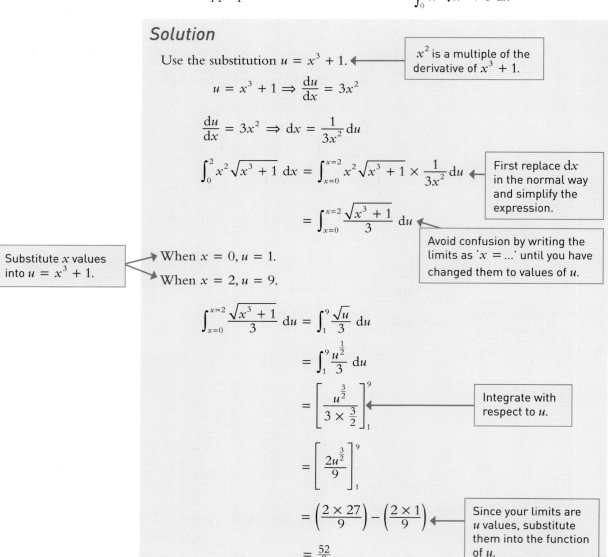

Use the substitution $u = x^3 + 1$. ← x^2 is a multiple of the derivative of $x^3 + 1$.

$$u = x^3 + 1 \Rightarrow \frac{du}{dx} = 3x^2$$

$$\frac{du}{dx} = 3x^2 \Rightarrow dx = \frac{1}{3x^2} \, du$$

$$\int_0^2 x^2 \sqrt{x^3 + 1} \, dx = \int_{x=0}^{x=2} x^2 \sqrt{x^3 + 1} \times \frac{1}{3x^2} \, du$$

First replace dx in the normal way and simplify the expression.

$$= \int_{x=0}^{x=2} \frac{\sqrt{x^3 + 1}}{3} \, du$$

Avoid confusion by writing the limits as '$x = ...$' until you have changed them to values of u.

Substitute x values into $u = x^3 + 1$.

When $x = 0, u = 1$.

When $x = 2, u = 9$.

$$\int_{x=0}^{x=2} \frac{\sqrt{x^3 + 1}}{3} \, du = \int_1^9 \frac{\sqrt{u}}{3} \, du$$

$$= \int_1^9 \frac{u^{\frac{1}{2}}}{3} \, du$$

$$= \left[\frac{u^{\frac{3}{2}}}{3 \times \frac{3}{2}} \right]_1^9$$

Integrate with respect to u.

$$= \left[\frac{2u^{\frac{3}{2}}}{9} \right]_1^9$$

$$= \left(\frac{2 \times 27}{9} \right) - \left(\frac{2 \times 1}{9} \right)$$

Since your limits are u values, substitute them into the function of u.

$$= \frac{52}{9}$$

An alternative method for definite integration by substitution involves leaving the limits as x values, and instead rewriting the integrated function in terms of x before substituting in the limits.

ACTIVITY 10.2

(i) Use the substitution shown in Example 10.10 to find the indefinite integral $\int x^2 \sqrt{x^3 + 1} \, dx$, expressing your answer in terms of x.

(ii) Use your answer from (i) to evaluate $\int_0^2 x^2 \sqrt{x^3 + 1} \, dx$, and check that your answer is the same as that given in Example 10.10.

Exercise 10.2

① (i) Use the substitution $u = x - 7$ to find $\int (x - 7)^4 \, dx$.

(ii) Use the substitution $u = 2x - 7$ to find $\int (2x - 7)^4 \, dx$.

(iii) Use the substitution $u = 5x - 7$ to find $\int (5x - 7)^4 \, dx$.

Check your answers by differentiation.

② Integrate the following functions.

(i) $(x - 2)^9$ (ii) $(2x + 3)^3$

(iii) $(1 - 6x)^5$

Check your answers by differentiation.

③ Evaluate the following definite integrals.

(i) $\int_0^1 (2x - 1)^4 \, dx$

(ii) $\int_1^3 (4 - x)^3 \, dx$

(iii) $\int_{-1}^1 (1 - 3x)^3 \, dx$

④ Figure 10.19 shows the curve $y = (3x - 1)^3$.

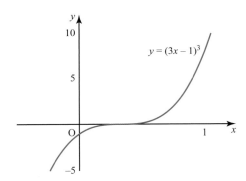

Figure 10.19

Find the area of the region bounded by the curve, the x-axis, and the lines $x = 0.6$ and $x = 1$.

⑤ Integrate the following functions.

(i) $\sqrt{x + 3}$ (ii) $\dfrac{1}{(3x + 1)^2}$

(iii) $\dfrac{1}{\sqrt{2x - 1}}$

Check your answers by differentiation.

PS ⑥ A curve has gradient function $\dfrac{dy}{dx} = \dfrac{1}{\sqrt{2x + 3}}$ and passes through the point $(0, 2\sqrt{3})$. Find the equation of the curve.

⑦ Find the area of the shaded region in Figure 10.20

(i) by treating it as an area between the line $y = 2$ and the curve $y = \sqrt{x + 1}$

(ii) by treating it as an area between the curve $y = \sqrt{x + 1}$ and the y-axis.

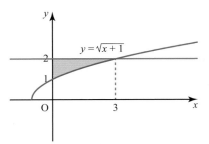

Figure 10.20

⑧ Evaluate $\int_1^2 x^2 (x^3 - 4)^3 \, dx$ using the substitution $u = x^3 - 4$.

⑨ Integrate the following functions using either recognition or a suitable substitution.

(i) $x(x^2 - 1)^6$ (ii) $x^2 \sqrt{x^3 + 1}$

(iii) $\dfrac{x^3}{(1 - x^4)^2}$

⑩ Using a suitable substitution, or otherwise, find the exact value of the following:

$$\int_0^1 \dfrac{x^2}{(2 - x^3)^2} \, dx$$

⑪ (i) Expand $(2x + 1)^4$ and hence find $\int (2x + 1)^4 \, dx$.

(ii) Find $\int (2x + 1)^4 \, dx$ using substitution or recognition.

(iii) Show that the coefficients of each power of x are the same in the answers to parts (i) and (ii).

(iv) Comment on the constant terms in the two answers.

⑫ Figure 10.21 shows the curve $y = \dfrac{x}{\sqrt{x^2 - 1}}$.

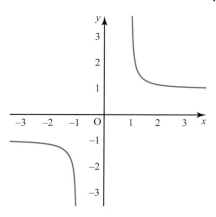

Figure 10.21

(i) Explain from the equation why y is undefined for $-1 \le x \le 1$.

(ii) Explain from the equation why the value of y cannot be between -1 and 1.

(iii) Find the exact value of the area of the region bounded by the curve, the x-axis and the lines $x = -3$ and $x = -2$.

Prior knowledge

You should be confident in working with radians (covered in Chapter 2), and you should know how to differentiate trigonometric functions (covered in Chapter 9).

3 Integrating other functions

Integrating trigonometric functions

ACTIVITY 10.3

In this activity, x is measured in radians.

1 Differentiate the following with respect to x.
   (i) $\sin x$ (ii) $\cos x$ (iii) $\tan x$
2 Use your answers to question 1 to write down the following indefinite integrals.
   (i) $\displaystyle\int \cos x \, dx$ (ii) $\displaystyle\int \sin x \, dx$ (iii) $\displaystyle\int \sec^2 x \, dx$
3 Now differentiate the following with respect to x.
   (i) $\sin 4x$ (ii) $\cos 5x$ (iii) $\tan 5x$
4 Use your answers to question 3 to write down the following indefinite integrals.
   (i) $\displaystyle\int \cos 4x \, dx$ (iii) $\displaystyle\int \sin 5x \, dx$ (iii) $\displaystyle\int \sec^2 5x \, dx$

The results from Activity 10.3 can be generalised to the following standard results, where k and c are constants.

ACTIVITY 10.4

Use integration by substitution, with $u = kx$, to confirm the results shown here.

$$\int \sin kx \, dx = -\frac{1}{k}\cos kx + c$$

$$\int \cos kx \, dx = \frac{1}{k}\sin kx + c$$

$$\int \sec^2 kx \, dx = \frac{1}{k}\tan kx + c$$

Example 10.11

Find the following indefinite integrals.

(i) $\displaystyle\int \sin 2x \, \mathrm{d}x$

(ii) $\displaystyle\int \sec^2 3x \, \mathrm{d}x$

Solution

(i) $\displaystyle\int \sin 2x \, \mathrm{d}x = -\frac{1}{2}\cos 2x + c$ ← Check your answer by differentiating.

(ii) $\displaystyle\int \sec^2 3x \, \mathrm{d}x = \frac{1}{3}\tan 3x + c$ ← Remember that
$$y = \tan x \Rightarrow \frac{\mathrm{d}y}{\mathrm{d}x} = \sec^2 x.$$

Definite integration with trigonometric functions

When doing definite integration with trigonometric functions, it is important to remember that the limits are always in radians, not degrees.

Example 10.12

Evaluate $\displaystyle\int_0^\pi \sin\left(\frac{x}{2}\right) \mathrm{d}x$.

Solution

$$\int_0^\pi \sin\left(\frac{x}{2}\right) \mathrm{d}x = \left[-\frac{1}{\frac{1}{2}}\cos\left(\frac{x}{2}\right)\right]_0^\pi$$ ← First integrate the function.

$$= \left[-2\cos\left(\frac{x}{2}\right)\right]_0^\pi$$ ← Then simplify.

$$= \left(-2\cos\frac{\pi}{2}\right) - (-2\cos 0)$$ ← Now substitute in the limits.

$$= (-2 \times 0) - (-2 \times 1)$$

$$= 2$$

Using trigonometric identities

Sometimes you can use a trigonometric identity to rewrite an integral in a form that makes it easy to integrate.

Example 10.13

Find $\int \tan^2 x \, dx$.

ACTIVITY 10.5

1 Differentiate the following with respect to x.
   (i) e^x (ii) e^{3x}

2 Use your answers to question 1 to write down the following indefinite integrals.
   (i) $\int e^x \, dx$
   (ii) $\int e^{3x} \, dx$

Solution

$$\int \tan^2 x \, dx = \int (\sec^2 x - 1) \, dx$$

$$= \tan x - x + c$$

> Using the identity $\sec^2 x \equiv 1 + \tan^2 x$.

> The integral of $\sec^2 x$ is $\tan x$.

Integrating exponential functions

The results from Activity 10.5 can be generalised to the following standard result, where k and c are constants.

$$\int e^{kx} \, dx = \frac{1}{k} e^{kx} + c$$

Example 10.14

Find $\int e^{-2x} \, dx$.

Solution

$$\int e^{-2x} \, dx = -\frac{1}{2} e^{-2x} + c$$

> Check your answer by differentiating.

Using substitution with exponential and trigonometric functions

Example 10.15

Use the substitution $u = \cos x$ to find $\int \sin x \cos^2 x \, dx$.

Solution

Use the substitution $u = \cos x$.

$$\frac{du}{dx} = -\sin x \Rightarrow dx = -\frac{1}{\sin x} \, du$$

$$\int \sin x \cos^2 x \, dx = \int \sin x \cos^2 x \times -\frac{1}{\sin x} \, du$$

> Substitute for dx first to see what cancels.

$$= \int -\cos^2 x \, du$$

$$= \int -u^2 \, du$$

$$= -\frac{u^3}{3} + c$$

$$= -\frac{\cos^3 x}{3} + c$$

> Remember to write your final answer in terms of x.

Notice that in Example 10.15, $\sin x$ is the derivative of $\cos x$, so you can think of $\sin x \cos^2 x$ as

a constant $\times$ the derivative of $\cos x \times$ a function of $\cos x$.

This shows you that the substitution $u = \cos x$ will be helpful.

This integral is of the 'reverse chain rule' type; differentiating the answer directly by the chain rule gives the original function. Integrals of this form can be found by using substitution as shown in Example 10.15. Alternatively, as you become more familiar with this type of integration, you may be able to do them by recognition.

In the next example, the expression to be integrated follows a similar pattern.

Example 10.16

Evaluate $\displaystyle\int_{\frac{\pi}{6}}^{\frac{\pi}{2}} \cos x\ \mathrm{e}^{\sin x}\ \mathrm{d}x$.

> $\cos x$ is the derivative of $\sin x$, and $\mathrm{e}^{\sin x}$ is a function of $\sin x$, so the substitution $u = \sin x$ will be helpful.

Solution

Use the substitution $u = \sin x$.

$$\frac{\mathrm{d}u}{\mathrm{d}x} = \cos x \Rightarrow \mathrm{d}x = \frac{1}{\cos x}\,\mathrm{d}u$$

$$\int_{\frac{\pi}{6}}^{\frac{\pi}{2}} \cos x\ \mathrm{e}^{\sin x}\ \mathrm{d}x = \int_{x=\frac{\pi}{6}}^{x=\frac{\pi}{2}} \cos x\ \mathrm{e}^{\sin x} \frac{1}{\cos x}\,\mathrm{d}u$$

When $x = \dfrac{\pi}{6}$, $u = \sin\left(\dfrac{\pi}{6}\right) = \dfrac{1}{2}$.

When $x = \dfrac{\pi}{2}$, $u = \sin\left(\dfrac{\pi}{2}\right) = 1$.

$$\int_{x=\frac{\pi}{6}}^{x=\frac{\pi}{2}} \cos x\ \mathrm{e}^{\sin x} \frac{1}{\cos x}\,\mathrm{d}u = \int_{\frac{1}{2}}^{1} \mathrm{e}^{\sin x}\ \mathrm{d}u$$

> Swap your limits from x values to u values.

$$= \int_{\frac{1}{2}}^{1} \mathrm{e}^{u}\ \mathrm{d}u$$

$$= \left[\mathrm{e}^{u}\right]_{\frac{1}{2}}^{1}$$

$$= \mathrm{e}^{1} - \mathrm{e}^{\frac{1}{2}}$$

$$= \mathrm{e} - \sqrt{\mathrm{e}}$$

ACTIVITY 10.6

Find general expressions for the following.

(i) $\displaystyle\int \mathrm{f}'(x)\mathrm{e}^{\mathrm{f}(x)}\ \mathrm{d}x$

(ii) $\displaystyle\int \mathrm{f}'(x)\big(\mathrm{f}(x)\big)^{n}\ \mathrm{d}x$

Exercise 10.3

① Find $\int \sin 2x \, dx$?

② Integrate the following functions with respect to x.

　(i) e^{2x}　　(ii) e^{3x+5}　　(iii) e^{3-2x}

③ Integrate the following functions with respect to x.

　(i) $\sin 6x$

　(ii) $\cos(2x+1)$

　(iii) $\sin(1-3x)$

④ Integrate the following functions with respect to x.

　(i) $\cos x - \sin x$

　(ii) $e^x + e^{-x}$

　(iii) $3\sin x - e^{2x}$

⑤ Evaluate the following definite integrals, giving your answers in exact form.

　(i) $\int_0^{\frac{\pi}{4}} \sec^2 x \, dx$

　(ii) $\int_0^{\frac{\pi}{2}} (1 - \cos x) \, dx$

　(iii) $\int_0^2 e^{\frac{1}{2}x} \, dx$

⑥ Find the following integrals using the given substitution.

　(i) $\int x e^{x^2+3} dx, \ u = x^2 + 3$

　(ii) $\int \sin x \cos^4 x \, dx, \ u = \cos x$

⑦ Figure 10.22 shows the curve $y = xe^{-2x^2}$.

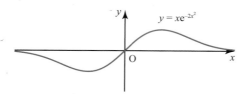

Figure 10.22

　(i) Find the coordinates of the turning points of the curve, leaving your answers in terms of e, and verify their nature.

　(ii) Find the area of the region bounded by the curve, the x-axis, and the lines $x = -1$ and $x = 1$. Give your answer to 3 significant figures.

PS ⑧ Using either a suitable substitution or recognition, find the following integrals.

　(i) $\int \sin x \ e^{\cos x} \, dx$

　(ii) $\int e^x \left(e^x - 5\right)^7 \, dx$

⑨ Figure 10.23 shows the curve $y = xe^{x^2}$.

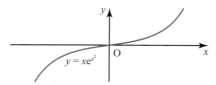

Figure 10.23

　(i) Find the coordinates of the points of intersection of the straight line $y = 3x$ and the curve $y = xe^{x^2}$.

　(ii) Find the total area of the regions bounded by the curve and the line.

⑩ (i) Use the double angle formula to express $\cos^2 x$ in terms of $\cos 2x$.

　(ii) Hence find $\int \cos^2 x \, dx$.

　(iii) Use a similar approach to find $\int \sin^2 x \, dx$.

⑪ (i) Show that $\sin^3 x \equiv \sin x - \sin x \cos^2 x$.

　(ii) Hence find $\int \sin^3 x \, dx$.

4 Integration involving the natural logarithmic function

Discussion points

→ Differentiate $\ln 2x$, $\ln 5x$ and $\ln 9x$ with respect to x.

→ What do your results tell you about different possible answers for the integral $\int \frac{1}{x} \, dx, \ x > 0$?

→ How can one integral have more than one answer?

→ Why is this integral restricted to positive values of x?

In these discussion points, you have seen that $\int \dfrac{1}{x}\,dx = \ln x + c$ for $x > 0$, where the restriction to $x > 0$ is because logarithms are undefined for negative numbers.

In Figure 10.24, region A is bounded by the curve $y = \dfrac{1}{x}$ and the x-axis between $x = -b$ and $x = -a$.

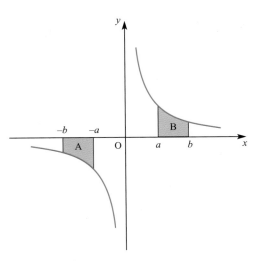

The area of region A is clearly a real area, and it must be possible to evaluate it. Region A is below the x-axis, and therefore the result by integration will be negative.

It looks as though the area of region A is the same as that of region B. Here is a proof that this is indeed the case.

Figure 10.24

$$\text{Area of region A} = -\int_{-b}^{-a} \frac{1}{x}\,dx.$$

> The minus sign at the front is because the area is below the x-axis, and so the result of the integration will be negative.

Substituting $u = -x$:

$$x = -b \Rightarrow u = b$$
$$x = -a \Rightarrow u = a$$

$$\frac{du}{dx} = -1 \Rightarrow dx = -du.$$

$$\text{Area of region A} = -\int_{b}^{a} \frac{1}{-u}(-du)$$

$$= -\int_{b}^{a} \frac{1}{u}\,du$$

$$= -(\ln a - \ln b)$$

$$= \ln b - \ln a$$

$$= \text{area of region B.}$$

Therefore the restriction that $x > 0$ may be dropped and the modulus sign introduced, so that the integral is written

$$\int \frac{1}{x}\,dx = \ln |x| + c$$

Example 10.17

Find the exact value of $\displaystyle\int_{5}^{7} \frac{1}{4 - x}\,dx$.

Solution

$$\int_{5}^{7} \frac{1}{4 - x}\,dx = -\int_{5}^{7} \frac{-1}{4 - x}\,dx$$

> This makes the top line the derivative of the bottom line, so you can integrate by recognition. Alternatively, you could use the substitution $u = 4 - x$.

$$= -\Big[\ln |4 - x|\Big]_{5}^{7}$$

$$= -\Big(\ln |-3| - \ln |-1|\Big)$$

$$= -\big((\ln 3) - (\ln 1)\big)$$

$$= -\ln 3$$

> Because you were asked for the exact value, leave your answer as a logarithm.

The arbitrary constant

When the result of integration is in the form of natural logarithms, it is often helpful to have the arbitrary constant written as a natural logarithm as well, $c = \ln k$, where $k > 0$. The rules of logarithms then allow you to write the integral as a single term, as shown below.

$$\int \frac{1}{x}\, dx = \ln|x| + c$$

> Since $c = \ln k$.

$$= \ln|x| + \ln k$$

$$= \ln(|x| \times k)$$

> Using the log rule $\log a + \log b = \log ab$

$$= \ln(|x| \times |k|)$$

$$= \ln|kx|$$

> As $k > 0$, $k = |k|$

Discussion points

→ Why is the restriction $k > 0$ needed?
→ Can every value of c be written in the form $\ln k$, $k > 0$?

The result above then becomes

$$\int \frac{1}{x}\, dx = \ln|kx| \text{ where } k > 0$$

Integrals of the form $\int \frac{f'(x)}{f(x)}\, dx$

Example 10.18

Find $\int \frac{x^2}{1 - x^3}\, dx$, $1 - x^3 > 0$.

Solution

Use the substitution $u = 1 - x^3$.

$$\frac{du}{dx} = -3x^2 \Rightarrow dx = \frac{1}{-3x^2}\, du$$

$$\int \frac{x^2}{1 - x^3}\, dx = \int \frac{x^2}{1 - x^3} \times \frac{1}{-3x^2}\, du$$

$$= \int -\frac{1}{3(1 - x^3)}\, du$$

$$= \int -\frac{1}{3u}\, du$$

$$= -\tfrac{1}{3}\int \frac{1}{u}\, du$$

$$= -\tfrac{1}{3}\ln|u| + c$$

$$= -\tfrac{1}{3}\ln|1 - x^3| + c$$

The integral in Example 10.18 is of the 'reverse chain rule' type. It can be written as

$$\int \frac{x^2}{1-x^3}\, dx = -\frac{1}{3}\int \frac{-3x^2}{1-x^3}\, dx$$

and so is of the form $\int \frac{f'(x)}{f(x)}\, dx$ because $-3x^2$ is the derivative of $1-x^3$.

Integrals of this form can be found by using the substitution $u = f(x)$.

Alternatively, the integral can be found by recognition using

$$\int \frac{f'(x)}{f(x)}\, dx = \ln |f(x)| + c$$

Prior knowledge

You need to know how to find partial fractions. This is covered in Chapter 7.

Using partial fractions

Many integrals leading to natural logarithmic functions need first to be expressed in suitable form, for example using partial fractions.

Example 10.19

(i)   Express $\dfrac{x-1}{(x+3)(x+1)}$ in partial fractions.

(ii) Find $\int \dfrac{x-1}{(x+3)(x+1)}\, dx$.

(iii) Express your answer to (ii) as a single logarithm.

Solution

(i)   Let $\dfrac{x-1}{(x+3)(x+1)} = \dfrac{A}{x+3} + \dfrac{B}{x+1}$

$$\dfrac{x-1}{(x+3)(x+1)} = \dfrac{A(x+1)+B(x+3)}{(x+3)(x+1)}$$

Then $x-1 = A(x+1)+B(x+3)$ ← Equating numerators.

Substituting $x=-1$:

$$2B = -2 \Rightarrow B = -1$$

Substituting $x=-3$:

$$-2A = -4 \Rightarrow A = 2$$

So $\dfrac{x-1}{(x+3)(x+1)} = \dfrac{2}{x+3} - \dfrac{1}{x+1}$

(ii) $\int \dfrac{x-1}{(x+3)(x+1)}\, dx = \int \left(\dfrac{2}{x+3} - \dfrac{1}{x+1}\right) dx$ Using partial fractions from (i).

$$= 2\ln|x+3| - \ln|x+1| + c$$

Replacing c with $\ln k$.

(iii) $2\ln|x+3| - \ln|x+1| + c = 2\ln|x+3| - \ln|x+1| + \ln k$

Since $(x+3)^2$ must be greater than or equal to zero, the modulus signs aren't necessary.

$$= \ln(x+3)^2 - \ln|x+1| + \ln k$$

$$= \ln\left(\dfrac{k(x+3)^2}{|x+1|}\right)$$

Combining terms using the laws of logarithms.

Example 10.20

Find $\int \dfrac{x+3}{(x+1)^2}\, dx$.

Solution

Let $\dfrac{x+3}{(x+1)^2} = \dfrac{A}{(x+1)} + \dfrac{B}{(x+1)^2}$ ← Partial fractions with a repeated factor.

$$= \dfrac{A(x+1)+B}{(x+1)^2}$$

Then $\quad x + 3 = A(x+1) + B$

Substituting $x = -1$:

$$2 = B$$

Comparing coefficients of x:

$$1 = A$$

So $\quad \dfrac{x+3}{(x+1)^2} = \dfrac{1}{(x+1)} + \dfrac{2}{(x+1)^2}$

$$\int \dfrac{x+3}{(x+1)^2}\, dx = \int \left(\dfrac{1}{(x+1)} + \dfrac{2}{(x+1)^2} \right) dx$$

$$= \ln|x+1| - 2(x+1)^{-1} + c$$

$$= \ln|x+1| - \dfrac{2}{x+1} + c$$

Notice that the second term is a power of −2 rather than a power of −1, so integrating it does not result in a logarithmic function.

ACTIVITY 10.7

In Example 10.20, instead of using partial fractions with a repeated factor,

$\dfrac{x+3}{(x+1)^2}$ can be expressed as $\dfrac{x+1+2}{(x+1)^2} = \dfrac{x+1}{(x+1)^2} + \dfrac{2}{(x+1)^2}$.

Show that this gives the same result.

Example 10.21

Show that $\int_0^1 \dfrac{x-3}{(x+2)(x+1)}\,dx = \ln\left(\dfrac{243}{512}\right)$.

Solution

Let $\qquad \dfrac{x-3}{(x+2)(x+1)} = \dfrac{A}{x+2} + \dfrac{B}{x+1}$

> Writing in partial fractions.

$$= \dfrac{A(x+1) + B(x+2)}{(x+2)(x+1)}$$

Then: $\qquad x - 3 = A(x+1) + B(x+2)$

Substituting $x = -2$:

$$-5 = -A \Rightarrow A = 5$$

Substituting $x = -1$:

$$-4 = B$$

So $\qquad \dfrac{x-3}{(x+2)(x+1)} = \dfrac{5}{x+2} - \dfrac{4}{x+1}$

$$\int_0^1 \dfrac{x-3}{(x+2)(x+1)}\,dx = \int_0^1 \left(\dfrac{5}{x+2} - \dfrac{4}{x+1}\right) dx$$

$$= \Big[5\ln|x+2| - 4\ln|x+1|\Big]_0^1$$

> Integrating each term separately.

$$= (5\ln 3 - 4\ln 2) - (5\ln 2 - 4\ln 1)$$

$$= (\ln 243 - \ln 16) - (\ln 32 - 0)$$

> Using rules of logarithms.

$$= \ln\left(\dfrac{243}{512}\right)$$

> Rewriting as a single natural logarithm using rules of logarithms.

Exercise 10.4

① Integrate the following functions with respect to x, where $x > 0$ in each case.

(i) $\dfrac{2}{x}$ (ii) $\dfrac{1}{2x}$ (iii) $\dfrac{1}{2x+7}$

② Find the following integrals.

(i) $\displaystyle\int \dfrac{1}{x-4}\,dx$ (ii) $\displaystyle\int \dfrac{1}{2-x}\,dx$

(iii) $\displaystyle\int \dfrac{1}{2x-7}\,dx$

③ Evaluate each of the following, giving your answer as a single logarithm.

(i) $\displaystyle\int_2^6 \dfrac{1}{x+2}\,dx$ (ii) $\displaystyle\int_{-7}^{-5} \dfrac{1}{x+1}\,dx$

(iii) $\displaystyle\int_0^1 \dfrac{2}{3-x}\,dx$

④ Use the substitution $u = x^2 + 1$ to find

$\displaystyle\int \dfrac{2x}{x^2+1}\,dx$.

⑤ Use the substitution $u = 2 + x^2$ to find $\displaystyle\int_{-1}^0 \dfrac{2x}{2+x^2}\,dx$, giving your answer as a single logarithm.

⑥ Find the following integrals.

(i) $\displaystyle\int \dfrac{3x^2}{x^3-4}\,dx$ (ii) $\displaystyle\int \dfrac{x}{2-x^2}\,dx$

⑦ Evaluate each of the following, giving your answer as a single logarithm.

(i) $\displaystyle\int_1^2 \dfrac{x^2}{x^3-2}\,dx$ (ii) $\displaystyle\int_{-1}^0 \dfrac{2x^3}{1+x^4}\,dx$

⑧ Find the following integrals.

(i) $\displaystyle\int \dfrac{1}{(1-x)(3x-2)}\,dx$ (ii) $\displaystyle\int \dfrac{1}{x^2(1-x)}\,dx$

(iii) $\displaystyle\int \dfrac{3x+3}{(x-1)(2x+1)}\,dx$

⑨ Evaluate the following definite integrals, giving your answers in exact form.

(i) $\displaystyle\int_0^1 \frac{1}{(x+1)(x+3)}\,dx$

(ii) $\displaystyle\int_2^4 \frac{7x-2}{(x-1)^2(2x+3)}\,dx$

(iii) $\displaystyle\int_0^2 \frac{5x+1}{(x+2)(2x+1)^2}\,dx$

⑩ Sketch the curve $y = \dfrac{3}{2x} + 1$ for positive values of x, showing clearly on your sketch any asymptotes to the curve. Find, in exact form, the area of the region enclosed by the curve, the x-axis and the lines $x = 1$ and $x = 3$, and indicate this region on your sketch.

⑪ Given that $y = \dfrac{3}{1-2x}$

(i) write down the equations of the asymptotes of the curve

(ii) sketch the curve

(iii) find the exact value of the area bounded by the curve, the x-axis and the lines $x = -2$ and $x = -1$.

⑫ (i) Express $\dfrac{3}{(1+x)(1-2x)}$ in partial fractions.

(ii) Find $\displaystyle\int_0^{0.1} \frac{3}{(1+x)(1-2x)}\,dx$ to 5 decimal places.

⑬ Figure 10.25 shows the curve $y = \dfrac{e^x}{1+e^x}$.

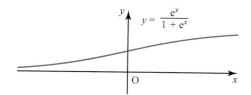

Figure 10.25

(i) Write down the coordinates of the point where the curve crosses the y-axis.

(ii) Describe the behaviour of y as $x \to \infty$ and as $x \to -\infty$.

(iii) Find the area of the region bounded by the curve, the coordinate axes and the line $x = 1$, giving your answer to 3 significant figures.

⑭ Find the following integrals.

(i) $\displaystyle\int \frac{1}{x^2-1}\,dx$ (ii) $\displaystyle\int \frac{x}{x^2-1}\,dx$

(iii) $\displaystyle\int \frac{x^2}{x^2-1}\,dx$

⑮ Figure 10.26 shows the curve $y = \dfrac{\sin x}{2+\cos x}$, $0 \leqslant x \leqslant 2\pi$.

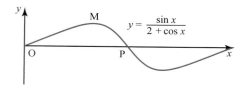

Figure 10.26

(i) Find the coordinates of the point M and use calculus to verify that it is a maximum point.

(ii) The area under the curve between $x = 0$ and $x = \pi$ can be approximated by the area of the triangle OMP.

Find this area and explain why this will give an underestimate.

(iii) Hence estimate the total area of the region bounded by the curve and the x-axis for $0 \leqslant x \leqslant 2\pi$.

(iv) Find the exact value of the total area bounded by the curve and the x-axis and compare this with your answer to (iii).

⑯ (i) Find $\displaystyle\int \frac{1}{x^2}\,dx$.

Selena tries to evaluate $\displaystyle\int_{-4}^2 \frac{1}{x^2}\,dx$. Her work is shown here.

$$\int_{-4}^2 \frac{1}{x^2}\,dx = \left[-\frac{1}{x}\right]_{-4}^2 = -\frac{1}{2} + \frac{1}{4} = -\frac{1}{4}$$

Selena says, 'This cannot be right because $\dfrac{1}{x^2}$ is positive for all values of x.'

(ii) Sketch the curve $y = \dfrac{1}{x^2}$ and explain why Selena's answer cannot represent the area under the curve between $x = -4$ and $x = 2$.

(iii) Is it possible to evaluate this area? Explain your answer.

(iv) Selena has broken a general rule for definite integration. State this rule in your own words.

5 Further integration by substitution

You already know how to use integration by substitution in cases where the process is the reverse of the chain rule, and the integration can alternatively be done by recognition. However, the method of substitution is used for other cases as well.

Suppose you want to integrate $\dfrac{x}{2x-1}$.

The numerator is not the derivative of the denominator, and therefore the integral is not of the form $\int\left(\dfrac{f'(x)}{f(x)}\right)dx$, and cannot be done by recognition.

However it can still be integrated by using the substitution $u = 2x - 1$.

First use the substitution to replace dx, in the normal way.

$$\frac{du}{dx} = 2 \implies dx = \tfrac{1}{2}du$$

$$\int \frac{x}{2x-1}\,dx = \int \frac{x}{2x-1} \times \tfrac{1}{2}du$$

$$= \tfrac{1}{2}\int \frac{x}{2x-1}\,du$$

> In these cases, there is often nothing that cancels out.

$$2x - 1 = u \implies x = \frac{u+1}{2}$$

> You must then rewrite both $2x - 1$ and x in terms of u, and replace all the x terms.

$$\tfrac{1}{2}\int \frac{x}{2x-1}\,du = \tfrac{1}{2}\int \frac{\left(\dfrac{u+1}{2}\right)}{u}\,du$$

$$= \tfrac{1}{4}\int \frac{u+1}{u}\,du$$

> Simplifying, and splitting the fraction into two separate terms, then gives you a function of u that you are able to integrate.

$$= \tfrac{1}{4}\int \left(1 + \frac{1}{u}\right)du$$

$$= \tfrac{1}{4}\left(u + \ln|u|\right) + c$$

Finally you must remember to write your final answer in terms of x.

$$\int \frac{x}{2x-1}\,dx = \tfrac{1}{4}(2x-1) + \tfrac{1}{4}\ln|2x-1| + c$$

Example 10.22

By using a suitable substitution, find $\displaystyle\int x\sqrt{x-2}\ dx$.

Solution

Let $u = x - 2$.

> By making the substitution $u = x - 2$, you can turn this integral into one you can do.

$$\frac{du}{dx} = 1 \qquad \implies dx = du$$

$$u = x - 2 \implies x = u + 2$$

$$\int x\sqrt{x-2}\ dx = \int x\sqrt{x-2}\ du$$

> Because $dx = du$.

$$= \int (u+2)\sqrt{u}\ du$$

$$= \int \left(u^{\frac{3}{2}} + 2u^{\frac{1}{2}}\right)du$$

$$= \tfrac{2}{5}u^{\frac{5}{2}} + \tfrac{4}{3}u^{\frac{3}{2}} + c$$

$$= \tfrac{2}{5}(x-2)^{\frac{5}{2}} + \tfrac{4}{3}(x-2)^{\frac{3}{2}} + c$$

> Rewrite in terms of x.

ACTIVITY 10.8

Show that the solution in Example 10.22 can be written as

$\frac{2}{15}(x-2)^{\frac{3}{2}}(3x+4)+c$.

ACTIVITY 10.9

For each of the following integrals
(a) write down a suitable substitution to use to perform the integration,
(b) use your substitution to integrate the function and
(c) check your answer by differentiation.

(i) $\int \frac{x}{(1+x)^2}\,dx$ (ii) $\int x\sqrt{2+x}\,dx$

Exercise 10.5

① Find the following integrals using the given substitution.

(i) $\int \left(\frac{x}{x+1}\right)^2 dx$, $u = x + 1$

(ii) $\int x\sqrt{2x+1}\,dx$, $u = 2x + 1$

② Evaluate the following definite integrals using the given substitution.

(i) $\int_2^3 \frac{x}{(x-1)^3}\,dx$, $u = x - 1$

(ii) $\int_{-1}^1 6x(3x-1)^3\,dx$, $u = 3x - 1$

③ For each of the following functions
(a) write down a suitable substitution to use to integrate the function with respect to x
(b) integrate the function.

(i) $\frac{x}{\sqrt{1-x}}$ (ii) $5x\sqrt{x+4}$

(iii) $\frac{-x}{2x+1}$

④ A curve has gradient function

$\frac{dy}{dx} = \frac{4x}{2x+1}$, $x \neq -\frac{1}{2}$.

(i) Use the substitution $u = 2x + 1$ to find the general equation of the family of solution curves.

(ii) Find the equation of the particular curve for which $y = 3 - \ln 5$ when $x = 2$.

⑤ Figure 10.27 shows the curve $y = \frac{x}{(x-1)^2}$.

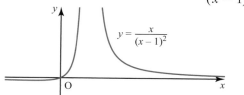

Figure 10.27

(i) The curve has a vertical asymptote at $x = a$. Write down the value of a.

(ii) Find the coordinates of the turning point of the curve and determine its nature.

(iii) Find the area of the region bounded by the curve, the x-axis and the line $x = -1$.

⑥ Find the integral $\int \frac{\ln x}{x}(1 + \ln x)\,dx$ using the substitution $u = 1 + \ln x$.

⑦ Figure 10.28 shows the curve $y = \frac{x}{\sqrt[3]{(x+3)}}$.

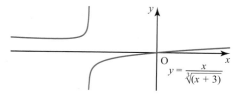

Figure 10.28

(i) The curve has a vertical asymptote at $x = k$. Write down the value of k.

(ii) Find the coordinates of the turning point of the curve, giving the y coordinate to 3 significant figures, and determine its nature.

(iii) Find, to 3 significant figures, the area of the region bounded by the curve, the x-axis and the line $x = -2$.

PS (8) The curve with gradient function $\dfrac{\mathrm{d}y}{\mathrm{d}x} = x(2x + 1)\sqrt{2x + 1}$ passes through the point $\left(0, \dfrac{69}{35}\right)$.

Find the equation of the curve.

(9) Using the substitution $u = 1 - \sin x$, evaluate

$$\int_{0.1}^{0.2} \sin 2x \sqrt{1 - \sin x} \ \mathrm{d}x,$$ giving your answer to 3 significant figures.

(10) (i) Use the substitution $x = 2\tan u$ to show that $\displaystyle\int \frac{1}{4 + x^2} \ \mathrm{d}x$ can be written as $\displaystyle\int \frac{1}{2} \ \mathrm{d}u$.

(ii) Hence find the exact value of

$$\int_0^2 \frac{1}{4 + x^2} \ \mathrm{d}x.$$

(11) (i) Use the substitution $x = 3\sin u$ to show that $\displaystyle\int \frac{1}{\sqrt{9 - x^2}} \ \mathrm{d}x$ can be written as $\displaystyle\int 1 \ \mathrm{d}u$.

(ii) Hence find the exact value of

$$\int_0^{\frac{3}{2}} \frac{1}{\sqrt{9 - x^2}} \ \mathrm{d}x.$$

6 Integration by parts

Suppose you want to integrate the function $x\cos x$. The function to be integrated is clearly a product of two simpler functions, x and $\cos x$, so your first thought may be to look for a substitution to enable you to perform the integration. However, there are some functions which are products but which cannot be integrated by substitution. This is one of them. You need a new technique to integrate such functions.

The technique of **integration by parts** is based on the reverse of the product rule. You can see from the working below that in the case of $\int x\cos x \,\mathrm{d}x$, you can start by differentiating $x\sin x$.

Differentiate the function $x\sin x$ by the product rule:

$$\frac{\mathrm{d}(x\sin x)}{\mathrm{d}x} = x\cos x + \sin x$$

Now integrate both sides with respect to x:

$$x\sin x = \int x\cos x \,\mathrm{d}x + \int \sin x \,\mathrm{d}x$$

Integrating $\dfrac{\mathrm{d}(x\sin x)}{\mathrm{d}x}$ 'undoes' the differentiation.

Rearranging this gives

$$\int x\cos x \,\mathrm{d}x = x\sin x - \int \sin x \,\mathrm{d}x$$
$$= x\sin x + \cos x + c$$

The method above of reversing the product rule and rearranging, can be generalised into the method of integration by parts, which can be used even when you are not told the function to differentiate (as you were in Example 10.22 and in Activity 10.10).

The general result for integration by parts

Using the product rule to differentiate the function uv:

$$\frac{\mathrm{d}(uv)}{\mathrm{d}x} = u\frac{\mathrm{d}v}{\mathrm{d}x} + v\frac{\mathrm{d}u}{\mathrm{d}x}$$

Integrating both sides:

$$uv = \int u\frac{\mathrm{d}v}{\mathrm{d}x} \ \mathrm{d}x + \int v\frac{\mathrm{d}u}{\mathrm{d}x} \ \mathrm{d}x$$

ACTIVITY 10.10

(i) By first differentiating $x\cos x$, find

$$\int x\sin x \,\mathrm{d}x.$$

(ii) By first differentiating xe^{2x}, find
$$\int 2xe^{2x} \,\mathrm{d}x.$$

Rearranging gives

$$\int u \frac{\mathrm{d}v}{\mathrm{d}x}\,\mathrm{d}x = uv - \int v \frac{\mathrm{d}u}{\mathrm{d}x}\,\mathrm{d}x$$

This is the formula you use when you need to integrate by parts.

To integrate $x\cos x$ using this formula, you split it into the two functions, x and $\cos x$. To fit the formula, you call one of these functions u and the other $\frac{\mathrm{d}v}{\mathrm{d}x}$. When deciding which is which, you should consider the following.

- To work out the right-hand side of the formula, you need to find v. Therefore, you must be able to integrate $\frac{\mathrm{d}v}{\mathrm{d}x}$.
- When you apply the formula, you will need to be able to integrate the function $v\frac{\mathrm{d}u}{\mathrm{d}x}$. Therefore u should be a function which becomes simpler when you differentiate.

Example 10.23

Find $\int x\mathrm{e}^x\,\mathrm{d}x$.

Solution

> x becomes simpler when you differentiate, but e^x stays as e^x. So choose $u = x$.

$$u = x \Rightarrow \frac{\mathrm{d}u}{\mathrm{d}x} = 1$$

$$\frac{\mathrm{d}v}{\mathrm{d}x} = \mathrm{e}^x \Rightarrow v = \mathrm{e}^x$$

Substituting into $\int u \frac{\mathrm{d}v}{\mathrm{d}x}\,\mathrm{d}x = uv - \int v \frac{\mathrm{d}u}{\mathrm{d}x}\,\mathrm{d}x$ gives

$$\int x\mathrm{e}^x\,\mathrm{d}x = x\mathrm{e}^x - \int \mathrm{e}^x\,\mathrm{d}x$$

$$= x\mathrm{e}^x - \mathrm{e}^x + c$$

ACTIVITY 10.11

Anna wants to integrate $\int x\sin x\,\mathrm{d}x$ by parts. She chooses $u = \sin x$, $\frac{\mathrm{d}v}{\mathrm{d}x} = x$.

(i) Apply the integration by parts formula, using Anna's choices for u and $\frac{\mathrm{d}v}{\mathrm{d}x}$.
(ii) Explain how you know that Anna has made a bad choice.

Example 10.24

Find $\int x\ln x\,\mathrm{d}x$.

Note

You might have expected to take $u = x$ in Example 10.24, as this is a function which gets simpler when differentiated. However, in this case, since you cannot integrate $\ln x$ easily, you need to take $u = \ln x$.

Solution

You are not able to integrate $\ln x$, so you cannot make this $\frac{\mathrm{d}v}{\mathrm{d}x}$. Therefore $\ln x$ must be u.

$$u = \ln x \Rightarrow \frac{\mathrm{d}u}{\mathrm{d}x} = \frac{1}{x}$$

$$\frac{\mathrm{d}v}{\mathrm{d}x} = x \Rightarrow v = \frac{x^2}{2}$$

Substituting into $\int u\dfrac{dv}{dx}\,dx = uv - \int v\dfrac{du}{dx}\,dx$ gives

$$\int x\ln x\,dx = \frac{x^2}{2}\ln x - \int\frac{x^2}{2}\times\frac{1}{x}\,dx$$

$$= \frac{x^2\ln x}{2} - \int\frac{x}{2}\,dx$$

$$= \frac{x^2\ln x}{2} - \frac{x^2}{4} + c$$

Definite integration by parts

When you use the method of integration by parts on a definite integral, you must remember that the term uv has already been integrated, and so should be written in square brackets with limits, and evaluated accordingly.

$$\int_a^b u\frac{dv}{dx}\,dx = \left[uv\right]_a^b - \int_a^b v\frac{du}{dx}\,dx$$

Example 10.25

Evaluate $\displaystyle\int_0^2 xe^{-x}\,dx$.

Solution

$$u = x \quad\Rightarrow\quad \frac{du}{dx} = 1$$

$$\frac{dv}{dx} = e^{-x} \quad\Rightarrow\quad v = -e^{-x}$$

Substituting into $\int_a^b u\dfrac{dv}{dx}\,dx = \left[uv\right]_a^b - \int_a^b v\dfrac{du}{dx}\,dx$ gives

$$\int_0^2 xe^{-x}\,dx = \left[-xe^{-x}\right]_0^2 - \int_0^2\left(-e^{-x}\right)\,dx$$

$$= \left[-xe^{-x}\right]_0^2 + \int_0^2 e^{-x}\,dx$$

$$= \left[-xe^{-x}\right]_0^2 + \left[-e^{-x}\right]_0^2$$

$$= \left(-2e^{-2} - 0\right) + \left(-e^{-2} + 1\right)$$

$$= 1 - 3e^{-2}$$

Using integration by parts twice

Sometimes it is necessary to use integration by parts twice (or more) to integrate a function successfully.

Example 10.26

Find $\int x^2 e^x \, dx$.

Solution

$$u = x^2 \Rightarrow \frac{du}{dx} = 2x$$

$$\frac{dv}{dx} = e^x \Rightarrow v = e^x$$

> You are able to integrate either function, so choose u to be the function that becomes simpler when you differentiate.

Substituting into $\int u \dfrac{dv}{dx} dx = uv - \int v \dfrac{du}{dx} dx$ gives

$$\int x^2 e^x \, dx = x^2 e^x - \int 2x e^x \, dx$$

To find $\int 2x e^x \, dx$ use integration by parts again:

$$u = 2x \Rightarrow \frac{du}{dx} = 2$$

$$\frac{dv}{dx} = e^x \Rightarrow v = e^x$$

Substituting into $\int u \dfrac{dv}{dx} dx = uv - \int v \dfrac{du}{dx} dx$ gives

$$\int 2x e^x \, dx = 2x e^x - \int 2 e^x \, dx$$

$$= 2x e^x - 2e^x + c$$

So $\int x^2 e^x \, dx = x^2 e^x - \left(2x e^x - 2e^x\right) + c$

$$= x^2 e^x - 2x e^x + 2e^x + c$$

Using integration by parts to integrate $\ln x$

At first glance, you would think that you cannot integrate $\ln x$ by parts because it is not the product of two functions. However, by writing $\ln x$ as the product of 1 and $\ln x$, you can still use this method.

$$\int \ln x \, dx = \int 1 \times \ln x \, dx$$

$$u = \ln x \Rightarrow \frac{du}{dx} = \frac{1}{x}$$

$$\frac{dv}{dx} = 1 \Rightarrow v = x$$

> You cannot integrate $\ln x$, so you must make it u and not $\dfrac{dv}{dx}$.

Substituting into $\int u \dfrac{dv}{dx} dx = uv - \int v \dfrac{du}{dx} dx$ gives

$$\int \ln x \, dx = x \ln x - \int x \times \frac{1}{x} \, dx$$

$$= x \ln x - \int 1 \, dx$$

$$= x \ln x - x + c$$

Exercise 10.6

① Find $\int x\cos 2x\,dx$ by using the integration by parts formula,

$$\int u\frac{dv}{dx}\,dx = uv - \int v\frac{du}{dx}\,dx,$$

with $u = x$ and $\frac{dv}{dx} = \cos 2x$.

② Evaluate $\int_{-1}^{1}(2x-1)\,e^x\,dx$ by using the integration by parts formula,

$$\int_a^b u\frac{dv}{dx}\,dx = [uv]_a^b - \int_a^b v\frac{du}{dx}\,dx,$$

with $u = 2x - 1$ and $\frac{dv}{dx} = e^x$.

③ Use integration by parts to integrate the following functions with respect to x.

(i) xe^{2x} (ii) $x\ln 3x$

④ Evaluate the following definite integrals.

(i) $\int_0^1 xe^{-x}\,dx$

(ii) $\int_0^{\pi}(x+1)\sin x\,dx$

⑤ Find $\int 2x(x-2)^4\,dx$

(i)   by using integration by parts

(ii)   by using the substitution $u = x - 2$.

⑥ Figure 10.29 shows the curve $y = (2-x)e^{-x}$.

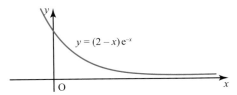

Figure 10.29

(i)   Find the coordinates of the points where the graph of $y = (2-x)e^{-x}$ cuts the x and y-axes.

(ii)   Use integration by parts to find the area of the region between the x-axis, the y-axis and the graph $y = (2-x)e^{-x}$.

PS ⑦ (i)   Sketch the graph of $y = x\sin x$ from $x = 0$ to $x = \pi$.

(ii)   Find the area of the region bounded by the curve and the x-axis.

PS ⑧ Find the area bounded by the curve $y = \ln 3x$, the x-axis and the line $x = 1$.

⑨ (i)   Use integration by parts to find the area of the blue region in Figure 10.30.

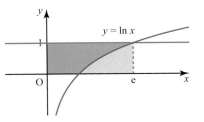

Figure 10.30

(ii)   Show that the area of the red region can be written as $\int_0^1 e^y\,dy$, and find this area in exact form.

(iii)   Add together your answers to (i) and (ii) and explain your answer.

PS ⑩ Find the following indefinite integrals.

(i) $\int x^2\sin 2x\,dx$

(ii) $\int x^2 e^{-x}\,dx$

⑪ Given the curve $y = x^2\ln x$, $x > 0$

(i)   find the coordinates of the point where the curve crosses the x-axis

(ii)   find the exact coordinates of the turning point of the curve and determine its nature

(iii)   sketch the curve

(iv)   find the exact area of the region bounded by the curve, the x-axis and the line $x = 2$.

⑫ Given that $I = \int e^x\sin x\,dx$

(i)   use integration by parts to show that
$$I = e^x\sin x - e^x\cos x - I$$

(ii)   hence find $\int e^x\sin x\,dx$.

You have now met several different methods of integration, some of which involve substitution. So when you need the answer to an integral, you have to select the best method, and you often have to decide on an appropriate substitution.

In this exercise there are 30 questions; having studied this chapter, you should be able to do all of them. However, they are in random order and so you will have to make decisions for yourself about the best method and what substitutions to make (if any).

① $\displaystyle\int xe^{-3x}\,dx$

② $\displaystyle\int (x-2)(x-3)\,dx$

③ $\displaystyle\int (2x+5)^{10}\,dx$

④ $\displaystyle\int xe^{x^2}\,dx$

⑤ $\displaystyle\int \cos x(1+\sin x)^3\,dx$

⑥ $\displaystyle\int \frac{1}{(3x-4)^5}\,dx$

⑦ $\displaystyle\int x\cos 3x\,dx$

⑧ $\displaystyle\int \frac{1}{x^2-4}\,dx$

⑨ $\displaystyle\int \frac{x}{x-2}\,dx$

⑩ $\displaystyle\int (x+3)^2\,dx$

⑪ $\displaystyle\int (x^3+4)(x^3-4)\,dx$

⑫ $\displaystyle\int \ln 3x\,dx$

⑬ $\displaystyle\int \sec^2 x\tan x\,dx$

⑭ $\displaystyle\int \frac{3x^2+1}{x^3+x}\,dx$

⑮ $\displaystyle\int \cos xe^{\sin x}\,dx$

⑯ $\displaystyle\int \sin x\cos x\,dx$

⑰ $\displaystyle\int (1+\tan^2 x)\,dx$

⑱ $\displaystyle\int \frac{(x-9)}{(x-3)(x-5)}\,dx$

⑲ $\displaystyle\int \sqrt{9x+4}\,dx$

⑳ $\displaystyle\int (1+\sec^2 x)\,dx$

㉑ $\displaystyle\int \frac{x^5-1}{x-1}\,dx$

㉒ $\displaystyle\int \left[(\sin x+\cos x)^2+(\sin x-\cos x)^2\right]dx$

㉓ $\displaystyle\int x^2e^{2x}\,dx$

㉔ $\displaystyle\int \frac{\sin 2x}{\sin x}\,dx$

㉕ $\displaystyle\int x(x+1)^2\,dx$

㉖ $\displaystyle\int x^3\ln x\,dx$

㉗ $\displaystyle\int \frac{x+x^{-1}}{x^2+2\ln x}\,dx$

㉘ $\displaystyle\int \frac{x^2}{x+1}\,dx$

㉙ $\displaystyle\int \frac{\cos x+\sin x}{\cos x}\,dx$

㉚ $\displaystyle\int \left(\frac{e^x+e^{-x}}{2}\right)^2\,dx$

LEARNING OUTCOMES

When you have completed this chapter, you should be able to:

➤ integrate, e^{kx}, $\dfrac{1}{x}$, $\sin kx$, $\cos kx$, and related sums, differences and constant multiples

➤ use definite integrals to find the area between two curves

➤ understand and use integration as the limit of a sum

➤ carry out simple cases of integration by substitution including:
  ○ finding a suitable substitution that leads to a function that can be integrated
  ○ understanding it as the inverse process of the chain rule

➤ carry out simple cases of integration by parts including:
  ○ more than one application of the method
  ○ understanding it as the inverse process of the product rule

➤ integrate using partial fractions that are linear in the denominator.

KEY POINTS

1. The definite integral $\int_a^b y\,dx$ is defined as $\displaystyle\lim_{\delta x \to 0}\left(\sum_{x=a}^{x=b} y\,\delta x\right)$.

2. The area between a curve and the y-axis between $y = a$ and $y = b$ is given by $\int_a^b x\,dy$.

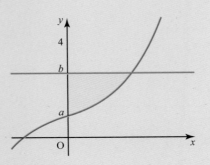

Figure 10.31

3. The area between two curves can be found by subtracting the equations of the curves and integrating between the limits.

4. Many functions can be integrated through knowledge of differentiation, for example

$$\int e^x\,dx = e^x + c$$

$$\int \sin x\,dx = -\cos x + c$$

$$\int \cos x\,dx = \sin x + c$$

$$\int \frac{1}{x}\,dx = \ln|x| + c = \ln|kx|$$

5. Substitution is often used to change a non-standard integral into a standard one.

6. In some cases, integration by substitution is the reverse of the chain rule. In these cases, the integration can also be done by recognition.

7. One important type of integral that can be done by recognition is those of the form $\dfrac{f'(x)}{f(x)}$, for which

$$\int \left(\frac{f'(x)}{f(x)}\right)dx = \ln|f(x)| + c = \ln|kf(x)|.$$

8. Some products may be integrated by parts using the formula

$$\int u\frac{dv}{dx}\,dx = uv - \int v\frac{du}{dx}\,dx.$$

FUTURE USES

You will use all these integration techniques in Chapter 13.

① $\left(1+\dfrac{x}{2}\right)^{-3}$ is expanded in a series of ascending powers of x.

Which of the following shows the full set of values for which it is valid?

A $-\dfrac{1}{2} < x < \dfrac{1}{2}$

B $-2 < x < 2$

C $-3 \leqslant x \leqslant 2$

D $-\dfrac{1}{2} \leqslant x \leqslant \dfrac{1}{2}$

② The solutions to $\cos 2\theta + \cos\theta = 0$ are the same as the solutions to which of the following:

A $2\cos^2\theta - \cos\theta + 1 = 0$

B $2\cos^2\theta + \cos\theta + 1 = 0$

C $2\cos^2\theta - \cos\theta - 1 = 0$

D $2\cos^2\theta + \cos\theta - 1 = 0$

③ Differentiate y^2 with respect to x. Which of the following is your answer?

A $y^2 \dfrac{\mathrm{d}y}{\mathrm{d}x}$

B $2y\dfrac{\mathrm{d}y}{\mathrm{d}x}$

C $y\dfrac{\mathrm{d}y}{\mathrm{d}x}$

D $2\dfrac{\mathrm{d}y}{\mathrm{d}x}$

④ Which of the following substitutions would be suitable for finding $\displaystyle\int \frac{x}{\sqrt{x-5}}\,\mathrm{d}x$?

A $u=\sqrt{x}$

B $u =\dfrac{1}{x}$

C $u = x - 5$

D $u = x + 5$

PS ① Figure 1 shows the graph of $y = \dfrac{1}{x}$. The two shaded regions have the same area.

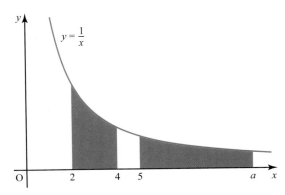

Figure 1

Find the value of a. [4 marks]

PS ② Figure 2 shows two curves. One is $y = 4 \sin x - 3 \cos x$ and the other is $y = 5 \sin x$.

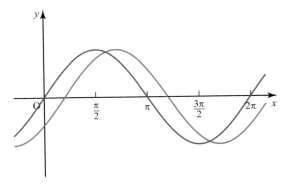

Figure 2

(i) Identify which curve is which, justifying your answer. [1 mark]

(ii) Describe precisely the translation that maps $f(x) = 4 \sin x - 3 \cos x$ on to $g(x) = 5 \sin x$. [4 marks]

③ (i) Write down the first 6 terms of the binomial expansion of $(1 - x)^{-2}$, in their simplest form, in increasing powers of x. [3 marks]

(ii) By substituting $x = 0.1$, show that the decimal expansion of $\dfrac{100}{81}$ begins $1.234\,56\ldots$. [2 marks]

P PS ④ Figure 3 shows a triangle OAB inscribed in a circle of radius r, centre O. M is the midpoint of AB. Angle AOM is α radians.

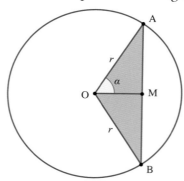

Figure 3

By considering the area of triangle OAB in two different ways, prove that $\sin 2\alpha = 2 \sin \alpha \cos \alpha$.　　　　　　　　　　　　　　　　[5 marks]

P ⑤ (i) Prove the identity $\sec^2 \theta + \operatorname{cosec}^2 \theta \equiv \dfrac{4}{\sin^2 2\theta}$.　　　　[3 marks]

　　 (ii) Hence or otherwise find all the roots of the equation $\sec^2 \theta + \operatorname{cosec}^2 \theta = 4$ in the range $0 < \theta < 2\pi$.　　　　[3 marks]

⑥ Figure 4 shows the curve with equation $x^2 y + 2y - 4x = 0$.

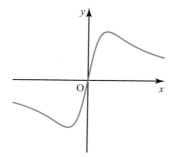

Figure 4

　　 (i) Show that $\dfrac{dy}{dx} = \dfrac{4 - 2xy}{2 + x^2}$.　　　　　　　　　　　　[3 marks]

　　 (ii) Hence find the coordinates of the maximum point on the curve. [4 marks]

⑦ (i) Write $\dfrac{x - 2}{(x + 1)(x - 3)}$ as partial fractions.　　　　　　[3 marks]

Figure 5 shows the curve with equation $y = \dfrac{x - 2}{(x + 1)(x - 3)}$.

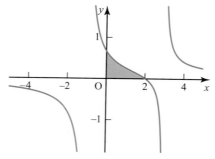

Figure 5

　　 (ii) Find the area of the shaded region enclosed by the curve and the axes.　　　　　　　　　　　　　　　　　　　　[5 marks]

⑧ Figure 6 shows the curve $y = x \sin x$ and the line $y = x$. At the point P, the line $y = x$ is a tangent to $y = x \sin x$.

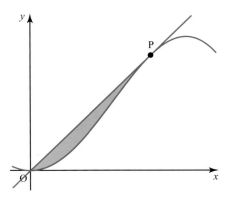

Figure 6

(i) Verify that P is the point with coordinates $\left(\dfrac{\pi}{2}, \dfrac{\pi}{2}\right)$. [4 marks]

(ii) Find the exact value of the area of the shaded region. [5 marks]

M **T** ⑨ (i) By using your calculator, or otherwise, draw the graph of

$$y = \frac{6e^x}{5 + e^x} \text{ for } x \geqslant 0.$$ [1 mark]

A particular social media platform defines a 'regular user' as someone who accesses the platform at least once every day. It currently has one million regular users. It models its future number of regular users, P (measured in millions), by the equation $P = \dfrac{6e^t}{5 + e^t}$, where t is measured in months.

(ii) According to this model, find the value of t when the platform is experiencing its fastest rate of growth. [7 marks]

(iii) Prove that according to this model P never reaches a value of 7. [3 marks]

Review: Coordinate geometry

1 Line segments

For a line segment AB between the points A (x_1, y_1) and B (x_2, y_2) then

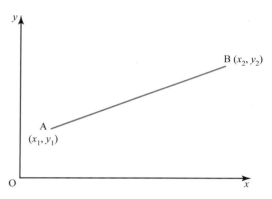

Figure R.1

- the gradient of AB is $\dfrac{y_2 - y_1}{x_2 - x_1}$

- the midpoint is $\left(\dfrac{x_1 + x_2}{2}, \dfrac{y_1 + y_2}{2} \right)$

> The midpoint of two values is their mean point.

- the distance AB is $\sqrt{(x_2 - x_1)^2 + (y_2 - y_1)^2}$.

> Using Pythagoras' theorem.

Parallel and perpendicular lines

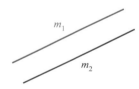
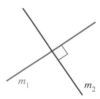

Figure R.2

For perpendicular lines, $m_1 = -\dfrac{1}{m_2}$ and likewise, $m_2 = -\dfrac{1}{m_1}$.

> m_1 and m_2 are the negative reciprocals of each other.

The equation of a straight line

The general equation of a straight line with y intercept $(0, c)$ and gradient m is

$$y = mx + c.$$

The gradient, m, of the line joining (x_1, y_1) and (x, y) is given by

$$m = \frac{y - y_1}{x - x_1}$$

$$\Rightarrow y - y_1 = m(x - x_1).$$

> This is a very useful form of the equation of a straight line.

Example R.1

The line L_1 passes through the points A $(0, 2)$ and B $(4, 10)$.

(i) The line L_2 is perpendicular to L_1 and passes through the midpoint of AB. Find the equation of the line L_2. Give your answer in the form $ax + by + c = 0$.

(ii) Find the area of the triangle bounded by the lines L_1 and L_2 and the x-axis.

Solution

(i)

Two perpendicular lines will only be at right angles in a diagram if you use the same scale for both axes.

Always start by drawing a diagram.

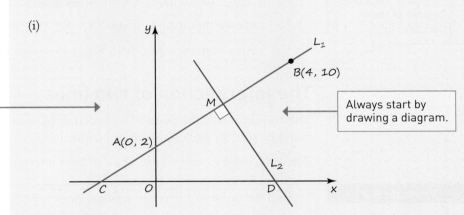

Figure R.3

Gradient of $L_1 = m_1 = \dfrac{y_B - y_A}{x_B - x_A}$

Gradient is difference in y coordinates divided by difference in x coordinates. It doesn't matter which point you use first, as long as you are consistent.

$= \dfrac{10 - 2}{4 - 0}$

$= 2$

The gradient of the line perpendicular to L_1 is the negative reciprocal of the gradient of L_1.

$\Rightarrow$ gradient of L_2 is $-\frac{1}{2}$.

Check: $2 \times -\frac{1}{2} = -1$ ✓

Midpoint M of AB $= \left(\dfrac{x_A + x_B}{2}, \dfrac{y_A + y_B}{2} \right)$

$= \left(\dfrac{0 + 4}{2}, \dfrac{2 + 10}{2} \right)$

$= (2, 6)$

So L_2 passes through M $(2, 6)$ and has gradient $-\frac{1}{2}$.
Therefore its equation is

$$y - y_1 = m(x - x_1)$$

It is usually easier to use this form of the equation of a straight line.

$$\Rightarrow \quad y - 6 = -\frac{1}{2}(x - 2)$$

$$\Rightarrow \quad 2y - 12 = -x + 2$$

Make sure you give your answer in the correct form. In this case the question asks for the form $ax + by + c = 0$.

$$\Rightarrow 2y + x - 14 = 0$$

Multiply both sides by 2 to clear the fraction.

From part (i), L_1 has gradient 2 and y intercept $(0, 2)$.

Discussion point

→ Use a different method to calculate the area of the triangle.

(ii) M has coordinates $(2, 6)$ so the vertical height of the triangle is 6.

The base is the distance CD.

C is the point where the line L_1 crosses the x-axis.

Use the formula:
area of a triangle $= \frac{1}{2} \times$ base $\times$ height.

Line L_1 has equation $y = 2x + 2$.

When $y = 0$ then $2x = -2 \Rightarrow x = -1$.

D is the point where the line L_2, $2y + x - 14 = 0$, crosses the x-axis.

When $y = 0$ then $x - 14 = 0 \Rightarrow x = 14$.

Hence the distance CD is $14 + 1 = 15$.

C is 1 unit to the left of O and D is 14 units right of O.

Area of triangle $= \frac{1}{2} \times 15 \times 6 = 45$ square units.

The intersection of two lines

The coordinates of the point of intersection of any two lines (or curves) can be found by solving their equations simultaneously.

You often need to find where a pair of lines intersect in order to solve problems.

Example R.2

Find the point of intersection of the lines $y = 2x - 1$ and $2y - 3x = 4$.

Solution

$$2(2x - 1) - 3x = 4$$

Substitute $y = 2x - 1$ into $2y - 3x = 4$.

$$4x - 2 - 3x = 4$$
$$x = 6$$
$$y = 11$$

Substitute $x = 6$ into either of the original equations to find y.

The point of intersection is $(6, 11)$.

Exercise R.1

① For each of the following pairs of points A and B calculate
  (a) the gradient of AB
  (b) the midpoint of AB
  (c) the exact distance AB.

  (i) A $(5, 2)$ and B $(3, 4)$

  (ii) A $(2, 5)$ and B $(4, 3)$

  (iii) A $(-5, 2)$ and B $(3, -4)$

  (iv) A $(-2, 5)$ and B $(-4, 3)$

② Find the gradient of each of the following lines.
  (i) $y = 3x + 5$ (ii) $2x + y = 5$
  (iii) $2x + 3y = 5$ (iv) $2y - 5x - 3 = 0$

③ Find the equation of each of the following lines.
  (i) gradient 2, passing through $(0, 3)$
  (ii) gradient $\frac{1}{2}$, passing through $(3, 0)$
  (iii) gradient $-\frac{1}{2}$, passing through $(3, -3)$
  (iv) passing through $(1, 2)$ and $(-3, 0)$
  (v) passing through $(-1, 2)$ and $(-3, 3)$

④ Find the coordinates of the point of intersection of each of the following pairs of lines.
  (i) $y = 2x - 1$ and $y = 3x - 4$
  (ii) $2x + 3y = 5$ and $y = 2x + 7$
  (iii) $y - 2x = 6$ and $2y + x - 7 = 0$
  (iv) $4x - 2y - 5 = 0$ and $3y - 2x + 3 = 0$

⑤ The point $(3, -5)$ lies on the line $2x + 3y + k = 0$, where k is a constant. Find the value of k.

⑥ (i) Prove that the points A $(1, 4)$, B $(5, 7)$ and C $(2, 11)$ form a right-angled triangle.

 (ii) Find the area of the triangle.

⑦ (i) Prove that the points A $(0, 4)$, B $(4, 2)$, C $(5, -1)$ and D $(-3, 3)$ form an isosceles trapezium.

> In an isosceles trapezium the two non-parallel sides are equal in length.

 (ii) Find the coordinates of the point where the diagonals of the trapezium intersect.

⑧ The line $3x + 4y = 15$ cuts the axes at the points A and B. Find the distance AB.

⑨ The line L has equation $kx + 3y + 8 = 0$. The point $(2, -4)$ lies on the line L.

 (i) Find the value of k.

 (ii) Find the equation of the line perpendicular to L that passes through the point on L where $x = 5$. Give your answer in the form $ax + by + c = 0$.

⑩ A and B have coordinates $(-5, 1)$ and $(1, 5)$ respectively. Find the equation of the perpendicular bisector of AB.

⑪ The line L is parallel to $2x + 5y = 1$ and passes through the point $(-1, 2)$. Find the coordinates of the points of intersection of L with the axes.

⑫ Find the area of the triangle bounded by the lines $3y = 2x + 3$, $2y + 3x = 15$ and $5y + x = 5$.

⑬ The point A has coordinates $(3, -5)$. The point B lies on the line $y = 2x - 5$. The distance AB is $\sqrt{17}$. Find the possible coordinates of B.

⑭ A and B are the points with coordinates $(-1, 6)$ and $(5, 2)$ respectively. The line L is the perpendicular bisector of AB. Find the area of the triangle bounded by the line L, the y-axis and the line through AB.

2 Circles

The equation of a circle

The general equation of a circle with centre (a, b) and radius r is

$$(x - a)^2 + (y - b)^2 = r^2$$

Example R.3

Find the centre and radius of the circle $x^2 + y^2 + 4x - 8y - 5 = 0$.

Complete the square on the terms involving x …

Note

The expanded equation in Example R.3 highlights some of the important characteristics of the equation of a circle. In particular:

- the coefficients of x^2 and y^2 are equal
- there is no xy term.

Solution

You need to rewrite the equation so it is in the form $(x - a)^2 + (y - b)^2 = r^2$.

$$x^2 + 4x + y^2 - 8y - 5 = 0$$
$$(x + 2)^2 - 4 + (y - 4)^2 - 16 - 5 = 0$$
$$(x + 2)^2 + (y - 4)^2 = 25$$

… then complete the square on the terms involving y.

Comparing this with the general equation for a circle with radius r and centre (a, b),

$$(x - a)^2 + (y - b)^2 = r^2,$$

you have $a = -2$, $b = 4$ and $r = 5$, so the centre is $(-2, 4)$ and the radius is 5.

Circle theorems

The circle theorems illustrated here are often useful when solving problems.

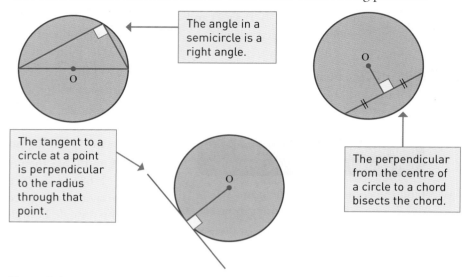

The angle in a semicircle is a right angle.

The tangent to a circle at a point is perpendicular to the radius through that point.

The perpendicular from the centre of a circle to a chord bisects the chord.

Figure R.4

Example R.4

A circle passes through the points $(1, 9)$, $(7, 9)$ and $(7, 5)$.
Find

(i)   the coordinates of the centre of the circle

(ii)   the equation of the circle.

Solution

(i)

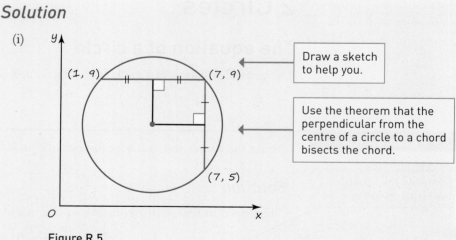

Draw a sketch to help you.

Use the theorem that the perpendicular from the centre of a circle to a chord bisects the chord.

Figure R.5

So the centre is at $\left(\dfrac{1+7}{2}, \dfrac{9+5}{2}\right) = (4, 7)$.

(ii)   The radius of a circle is the distance between the centre and any point on the circumference.

To find the equation of a circle you need the centre and the square of the radius.

$$r^2 = (7-4)^2 + (9-7)^2$$
$$= 3^2 + 2^2$$
$$= 13$$
$$\Rightarrow (x-4)^2 + (y-7)^2 = 13$$

Notice that the word 'curve' is sometimes used to describe a straight line. So this includes the intersection of a curve and a straight line.

→The intersection of two curves

The same principles apply to finding the intersection of two curves, but it is only in simple cases that it is possible to solve the equations simultaneously using algebra (rather than a numerical or graphical method).

Example R.5

Find the coordinates of the points of intersection of the circle $(x - 5)^2 + (y + 2)^2 = 16$ and the line $y = 7 - x$.

Solution

Substituting $y = 7 - x$ into $(x - 5)^2 + (y + 2)^2 = 16$ gives

$$(x - 5)^2 + (7 - x + 2)^2 = 16$$

$$\Rightarrow \qquad (x - 5)^2 + (9 - x)^2 = 16$$

$$\Rightarrow x^2 - 10x + 25 + 81 - 18x + x^2 = 16 \;\longleftarrow \boxed{\text{Multiply out the brackets.}}$$

$$\Rightarrow \qquad 2x^2 - 28x + 90 = 0 \;\longleftarrow \boxed{\text{Simplify.}}$$

$$\Rightarrow \qquad x^2 - 14x + 45 = 0$$

$$\Rightarrow \qquad (x - 5)(x - 9) = 0 \;\longleftarrow \boxed{\text{Factorise.}}$$

So $\quad x = 5$ and $y = 2$

or $\quad x = 9$ and $y = -2$. $\;\longleftarrow \boxed{\text{Don't forget to use } y = 7 - x \text{ to find the } y \text{ coordinates.}}$

So the coordinates are $(5, 2)$ and $(9, -2)$.

Exercise R.2

① Find the centre and radius of each of the following circles.

(i) $(x - 3)^2 + (y - 1)^2 = 25$

(ii) $(x - 3)^2 + (y + 1)^2 = 5$

(iii) $(x + 3)^2 + (y + 1)^2 = 25$

(iv) $(x - 1)^2 + (y + 3)^2 = 5$

② Find the equation of the circle with centre $(5, -2)$ and radius 4.

③ Find the equation of the circle with centre $(-3, 4)$ and radius 9.

Give your answer in the form $x^2 + ax + y^2 + by + c = 0$.

④ The points A and B have coordinates $(-2, 5)$ and $(4, 7)$ respectively. AB forms the diameter of a circle.

Find the equation of the circle.

⑤ (i) The equation of a circle, C, is

$x^2 - 8x + y^2 + 2y + 7 = 0$.

Show that the equation of C can be written in the form

$(x + a)^2 + (y + b)^2 = c$.

(ii) Find the centre and radius of the circle.

⑥ Find the coordinates of the points of intersection of the circle $(x - 5)^2 + (y + 2)^2 = 9$ and the line $y = x - 10$.

⑦ The line $y = 2x + 3$ intersects the circle $(x + 1)^2 + (y - 2)^2 = 2$ at the points A and B. The distance AB is $a\sqrt{b}$ where b is an integer.

Find the value of a and the value of b.

⑧ Find the coordinates of the points of intersection of the circle $(x + 2)^2 + (y - 1)^2 = 15$ and the line $2x + y = 2$.

Give your answers in surd form.

⑨ The line $y = x + k$ forms a tangent to the circle $(x + 2)^2 + (y - 1)^2 = 50$.

Find the possible values of k.

⑩ The points P $(-2, -1)$, Q $(8, -9)$ and R $(7, 0)$ lie on the circumference of a circle, C.

(i) Show that PQ forms the diameter of the circle.

(ii) Find the equation of the circle, C.

(iii) Find the exact coordinates of the points where C crosses the axes.

⑪ The points A $(3, 6)$ and B $(5, 2)$ lie on the circumference of a circle.

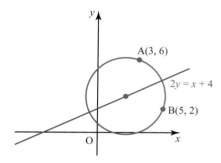

Figure R.6

(i) Show that the centre of the circle lies on the line $2y = x + 4$.

(ii) Given that the centre of the circle also lies on the line $y = 3$, find the equation of the circle.

⑫ The equation of the circle shown in Figure R.7 is $(x - 3)^2 + (y - 4)^2 = 20$.

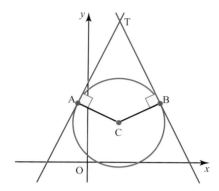

Figure R.7

The circle intersects the line $y = 6$ at the points A and B. The tangents at A and B meet at T.

Find the perimeter of the quadrilateral ACBT. Give your answer in the form $a\sqrt{b}$ where a and b are integers.

⑬ Find the coordinates of the point of intersection of the circles

$$x^2 + y^2 - 8x + 2y + 13 = 0$$
$$x^2 + (y - 2)^2 = 9$$

Find also the distance between their centres.

KEY POINTS

1 The equation of a straight line is $y = mx + c$.

2 When the points A and B have coordinates (x_1, y_1) and (x_2, y_2) respectively, then

   ■ the gradient of AB is $\dfrac{y_2 - y_1}{x_2 - x_1}$

   ■ the midpoint of AB is $\left(\dfrac{x_1 + x_2}{2}, \dfrac{y_1 + y_2}{2} \right)$

   ■ the distance AB between the two points is $\sqrt{(x_2 - x_1)^2 + (y_2 - y_1)^2}$.

3 Two lines are parallel when their gradients are equal.

4 Two lines are perpendicular when the product of their gradients is −1.

5 The equation of a straight line may take any of the following forms:

   ■ line parallel to the y-axis: $x = a$

   ■ line parallel to the x-axis: $y = b$

   ■ line through the origin with gradient m: $y = mx$

   ■ line through $(0, c)$ with gradient m: $y = mx + c$

   ■ line through (x_1, y_1) with gradient m: $y - y_1 = m(x - x_1)$

6 The equation of a circle

   ■ with centre $(0, 0)$ and radius r is $x^2 + y^2 = r^2$

   ■ with centre (a, b) and radius r is $(x - a)^2 + (y - b)^2 = r^2$.

7 **Circle theorems**

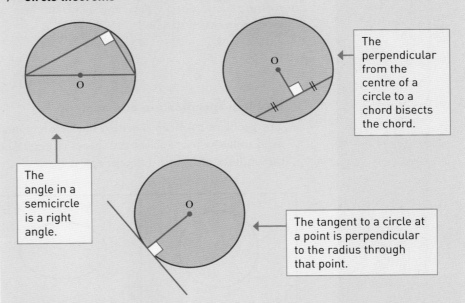

The angle in a semicircle is a right angle.

The perpendicular from the centre of a circle to a chord bisects the chord.

The tangent to a circle at a point is perpendicular to the radius through that point.

Figure R.8

8 To find the points of intersection of two curves (or lines), you solve their equations simultaneously.

Eggs

Figures 1 and 2 show a male robin and a nest containing a robin's egg. The egg has the distinctive shape of many birds' eggs.

Figure 1 A male robin

Figure 2 A robin's egg

Sally is starting a business reproducing replica eggs made out of wood. She has a machine that will produce eggs of any shape but first the equation of the desired shape has to be entered. Sally will then paint the eggs by hand. She wants to know the equation of a typical egg shape and how it can be adapted to allow for differences between species.

1 Problem specification and analysis

The problem has been specified but before you can make any progress you need to know more about the shapes of eggs. The diagrams in Figure 3 show the outlines of several different egg shapes.

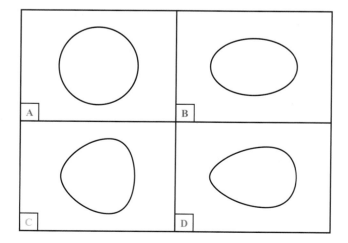

Figure 3

Shape A is a sphere and this is the most basic shape. So use the equation of a circle, centre $(0, 0)$ with radius 1 as a starting point. Spherical eggs are laid by many animals, for example turtles, but not by birds. So the equation needs to be adapted; two parameters, denoted here by a and k, are involved.

In **shape B** the sphere has been made longer and thinner. This is a feature of many eggs allowing the animal to have a narrower oviduct. It is particularly pronounced in fast flying birds. The equation of this shape can be obtained by applying stretches of scale factors a and $\frac{1}{a}$ in the x and y directions.

In **shape C** the egg is asymmetrical. This is a feature of almost all birds' eggs; it means that if the egg is disturbed, it rolls around in a circle rather than rolling out of the nesting area. To achieve this transformation the right hand side of the equation for y is multiplied by a factor of $(1 + kx)$ where k is a positive number less than 1. Some birds, for example owls and ostriches, have eggs like this.

In shape D both transformations are applied at the same time. This gives the common egg shape and for suitable values of the parameters, a and k, a full range of birds' egg shapes can be obtained.

2 **Information collection**

Notice that you will need a graphical calculator or graphing software.

You are advised to work in three stages.

Start by identifying the equation for shape B. This will involve the parameter a and you will need to find suitable values to give it to produce a variety of elongated but symmetrical shapes.

Then go back to the equation for the circle and apply the transformation needed for shape C. This involves the parameter k and you will need to find suitable values for it.

Then work with an equation that incorporates both transformations, giving shape D.

3 **Processing and representation**

A good way to present your work is to set up a display of, say, about 12 egg shapes, stating the different pairs of values for the parameters a and k that you have used.

You may be able to make this into a poster.

4 **Interpretion**

Your display will include the egg shapes for a variety of birds. To complete the task, match up your shapes with those typical of different species; you can easily find the information you need online.

Sally will need this information to paint realistic patterns on her eggs for different bird species.

11 Parametric equations

Fairground rides are scary because the ride follows a complicated path.

→ How could you find an equation for the path that this ride takes its riders on? Figure 11.1 shows a simplified version of the ride in the picture.

(a) (b)

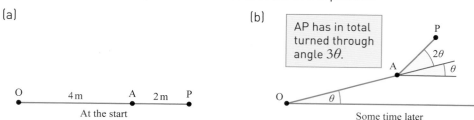

Figure 11.1

The rider's chair is on the end of a rod AP of length 2 m which is rotating about A. The rod OA is 4 m long and is itself rotating about O. The gearing of the mechanism ensures that the rod AP rotates twice as fast relative to OA as the rod OA does. This is illustrated by the angles marked on Figure 11.1(b), at a time when OA has rotated through an angle θ.

At this time, the coordinates of the point P, taking O as the origin (Figure 11.2), are given by

$$x = 4\cos\theta + 2\cos 3\theta$$

$$y = 4\sin\theta + 2\sin 3\theta$$

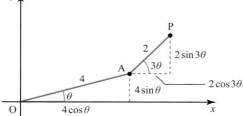

Figure 11.2

These two equations are called **parametric equations** of the curve. They do not give the relationship between x and y directly in the form $y = f(x)$ but use a third variable, θ, to do so. This third variable is called the **parameter**.

To plot the curve, you need to substitute values of θ and find the corresponding values of x and y.

Thus $\theta = 0° \implies x = 4 + 2 = 6$

$y = 0 + 0 = 0$ Point $(6, 0)$

$\theta = 30° \implies x = 4 \times 0.866 + 0 = 3.464$

$y = 4 \times 0.5 + 2 \times 1 = 4$ Point $(3.46, 4)$

and so on.

Joining points found in this way reveals the curve to have the shape shown in Figure 11.3.

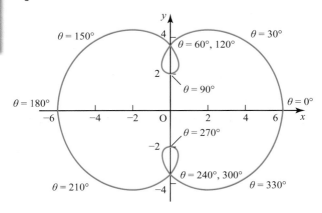

Figure 11.3

1 Graphs from parametric equations

Prior knowledge

You need to be able to sketch the graph of a curve (see the review section in Chapter 4).

Parametric equations are very useful in situations such as this, where an otherwise complicated equation may be expressed more simply in terms of a parameter. In fact, there are some curves which can be given by parametric equations but cannot be written as Cartesian equations (in terms of x and y only).

Example 11.1 is based on a simpler curve.

Example 11.1

A curve has the parametric equations $x = 2t$, $y = \dfrac{36}{t^2}$.

(i)   Find the coordinates of the points corresponding to $t = 1, 2, 3, -1, -2$ and -3.

(ii)   Plot the points you have found and join them to give the curve.

(iii)   Explain what happens as $t \to 0$ and as $t \to \infty$.

Solution

(i)

> Substituting t into $x = 2t$ gives the x coordinates ...

> ... and substituting t into $y = \dfrac{36}{t^2}$ gives the y coordinates.

t	-3	-2	-1	1	2	3
x	-6	-4	-2	2	4	6
y	4	9	36	36	9	4

The points required are $(-6, 4)$, $(-4, 9)$, $(-2, 36)$, $(2, 36)$, $(4, 9)$ and $(6, 4)$.

(ii)   The curve is shown in Figure 11.4.

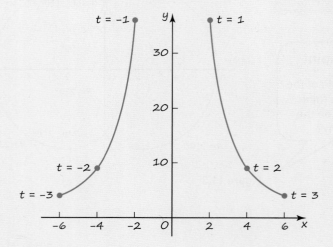

Figure 11.4

(iii)   As $t \to 0$, $x \to 0$ and $y \to \infty$. So the y-axis is an asymptote for the curve.

As $t \to \infty$, $x \to \infty$ and $y \to 0$. So the x-axis is an asymptote for the curve.

Example 11.2

A curve has the parametric equations $x = t^2$, $y = t^3 - t$.

(i) Find the coordinates of the points corresponding to values of t from -2 to $+2$ at half-unit intervals.

(ii) Sketch the curve for $-2 \leqslant t \leqslant 2$.

(iii) Are there any values of x for which the curve is undefined?

Solution

(i)

t	-2	-1.5	-1	-0.5	0	0.5	1	1.5	2
x	4	2.25	1	0.25	0	0.25	1	2.25	4
y	-6	-1.875	0	0.375	0	-0.375	0	1.875	6

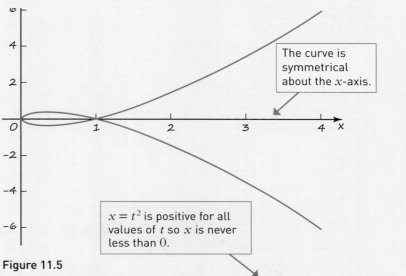

The curve is symmetrical about the x-axis.

$x = t^2$ is positive for all values of t so x is never less than 0.

Figure 11.5

(iii) The curve in Figure 11.5 is undefined for $x < 0$.

TECHNOLOGY

You can use a graphical calculator or graphing software to sketch parametric curves but, as with Cartesian curves, you need to be careful when choosing the range.

Exercise 11.1

In this exercise you should sketch the curves by hand. Use a graphical calculator or graphing software to check your results.

① A curve has parametric equations $x = 3 \cos \theta$, $y = 3 \sin \theta$. Use graphing software to sketch the graph.

② A curve has parametric equations $x = 2t$, $y = t^2$. Find the Cartesian coordinates of the point where $t = -1$

③ A curve has the parametric equations $x = 3t + 1$, $y = 2t^2 - t$.
  (i) Find the coordinates of the point with parameter
    (a) $t = 1$ (b) $t = 2$ (c) $t = -1$.
  (ii) What is the value of the parameter t at the point $(-8, 21)$?

④ A curve has the parametric equations $x = t^3$, $y = 3t^2$. Find the coordinates of the point where
  (i) $t = 1$ (ii) $x = 27$ (iii) $y = 12$.

⑤ For each of the following curves
  (a) copy and complete the table of values

t	x	y
-2		
-1.5		
-1		
-0.5		
0		
0.5		
1		
1.5		
2		

(b)   sketch the curve.

(i) $x = 2t$ (ii) $x = t^2$

$y = t^2$ $y = t^3$

(iii) $x = t^2$

$y = t^2 - t$

⑥ For each of the following pairs of parametric equations

(a)   copy and complete the table of values

θ	x	y
0°		
30°		
60°		
90°		
120°		
150°		
180°		
210°		
240°		
270°		
300°		
330°		
360°		

(b)   sketch the curve

(c)   state the values of x and y for which the curve is defined.

(i) $x = \cos 2\theta$

$y = \sin^2 \theta$

(ii) $x = \sin^2 \theta$

$y = 1 + 2\sin \theta$

(iii) $x = 2\sin^2 \theta$

$y = 3\cos \theta$

⑦ A curve has the parametric equations $x = t^2$, $y = t^4$.

(i)   Find the coordinates of the points corresponding to $t = -2$ to $t = 2$ at half-unit intervals.

(ii)   Sketch the curve for $-2 \leqslant t \leqslant 2$.

(iii)   Why is it not quite accurate to say this curve has equation $y = x^2$?

⑧ A curve has the parametric equations $x = 2\,\mathrm{cosec}\,\theta$, $y = 2\cot\theta$.

(i)   Sketch the curve.

(ii)   For which values of x is the curve undefined?

⑨ For each of the following pairs of parametric equations

(a)   sketch the curve

(b)   state the equations of any asymptotes.

(i) $x = \tan\theta$ (ii) $x = \dfrac{t}{1+t}$

$y = \tan 2\theta$ $y = \dfrac{t}{1-t}$

⑩ A curve has the parametric equations $x = e^t$, $y = \sin t$, where t is in radians.

(i)   Find, to 2 decimal places, the coordinates of the points corresponding to values of t from -2 to $+2$ at half-unit intervals.

(ii)   What can you say about the values of x for which the curve is defined?

(iii)   Sketch the curve for $-2 \leqslant t \leqslant 2$.

(iv)   Predict how this graph would continue if all values of t were considered (that is, $t < -2$ and $t > 2$).

⑪ A student is investigating the trajectory of a golf ball being hit over level ground. At first she ignores air resistance, and this leads her to an initial model given by $x = 40t$, $y = 30t - 5t^2$, where x and y are the horizontal and vertical distances in metres from where the ball is hit, and t is the time in seconds.

(i)   Plot the trajectory on graph paper for $t = 0, 1, 2, \ldots$, until the ball hits the ground again.

(ii)   How far does the ball travel horizontally before bouncing, according to this model?

The student then decides to make an allowance for air resistance to the horizontal motion and proposes the model $x = 40t - t^2$, $y = 30t - 5t^2$.

(iii)   Plot the trajectory according to this model using the same axes as in part (i).

(iv)   By how much does this model reduce the horizontal distance the ball travels before bouncing?

PS ⑫ The path traced out by a marked point on the rim of a wheel of radius a when the wheel is rolled along a flat surface is called a cycloid.

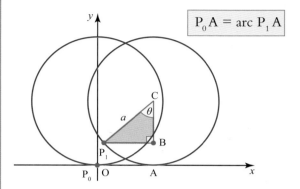

$$P_0A = \text{arc } P_1A$$

Figure 11.6

Figure 11.6 shows the wheel in its initial position, when the lowest point on the rim is P_0, and when it has rotated through an angle θ (radians). In this position, the point P_0 has moved to P_1 with parametric equations given by

$$x = OA - P_1B = a\theta - a\sin\theta$$
$$y = AC - BC = a - a\cos\theta.$$

(i) Find the coordinates of the points corresponding to values of θ from 0 to 6π at intervals of $\frac{\pi}{3}$.

(ii) Sketch the curve for $0 \leqslant \theta \leqslant 6\pi$.

(iii) What do you notice about the curve?

PS ⑬ The curve with parametric equations

$$x = a\cos^3\theta$$
$$y = a\sin^3\theta$$

is called an astroid.

(i) Sketch the curve.

(ii) On the same diagram sketch the curve

$$x = a\cos^n\theta$$
$$y = a\sin^n\theta$$

for $n = 1, 2, 3, 4, 5, 6$.
What happens if $n = 0$?

(iii) What can you say regarding the shape and position of the curve when $n \geqslant 7$ and

(a) n is even

(b) n is odd?

2 Finding the equation by eliminating the parameter

Prior knowledge

You need to be able to solve simultaneous equations – see Review: Algebra (2).

For some pairs of parametric equations, it is possible to eliminate the parameter and obtain the Cartesian equation for the curve. This is usually done by making the parameter the subject of one of the equations, and substituting this expression into the other.

Example 11.3

Eliminate t from the equations $x = t^3 - 2t^2$, $y = \frac{t}{2}$.

Solution

$$y = \frac{t}{2} \Rightarrow t = 2y$$

> Make t the subject of one equation ...

Substituting this in the equation $x = t^3 - 2t^2$ gives

$$x = (2y)^3 - 2(2y)^2$$

> ... then substitute into the other equation.

So $x = 8y^3 - 8y^2$.

Sometimes it is not straightforward to make the parameter the subject of one of the equations. In the next example you will see two different methods of eliminating the parameter.

Example 11.4

The parametric equations of a curve are $x = t + \dfrac{1}{t}$, $y = t - \dfrac{1}{t}$.

Eliminate the parameter by

(i) first finding $x + y$

(ii) first squaring x and y.

Solution

(i) Adding the two equations gives

$$x + y = 2t \quad \text{or} \quad t = \frac{x + y}{2}.$$

Substituting for t in the first equation (it could be either one) gives

> The parameter t has been eliminated, but the equation is not in its neatest form.

$$x = \frac{x + y}{2} + \frac{2}{x + y}.$$

Multiplying by $2(x + y)$ to eliminate the fractions:

$$2x(x + y) = (x + y)^2 + 4$$

$$\Rightarrow \quad 2x^2 + 2xy = x^2 + 2xy + y^2 + 4$$

$$\Rightarrow \quad x^2 - y^2 = 4.$$

(ii) Squaring gives

$$x^2 = t^2 + 2 + \frac{1}{t^2}$$

$$y^2 = t^2 - 2 + \frac{1}{t^2}.$$

Subtracting gives

$$x^2 - y^2 = 4.$$

Using graphing software to sketch $x^2 - y^2 = 4$ gives the curve shown in Figure 11.7.

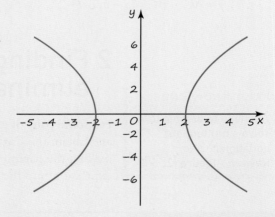

Figure 11.7

Prior knowledge

You need to be able to use trigonometric identities. These are covered in Chapters 6 and 8.

Trigonometric parametric equations

When trigonometric functions are used in parametric equations, you can use trigonometric identities to help you eliminate the parameter.

The next examples illustrate this.

Example 11.5

Eliminate θ from $x = 4\cos\theta$, $y = 3\sin\theta$.

Solution

The identity which connects $\cos\theta$ and $\sin\theta$ is

$$\cos^2\theta + \sin^2\theta = 1 \qquad \textcircled{1}$$

$$x = 4\cos\theta \implies \cos\theta = \frac{x}{4}$$

$$y = 3\sin\theta \implies \sin\theta = \frac{y}{3}.$$

Substituting these in $\textcircled{1}$ gives

$$\left(\frac{x}{4}\right)^2 + \left(\frac{y}{3}\right)^2 = 1.$$

> This looks similar to the equation of a circle, $x^2 + y^2 = 1$.

This is usually written as

$$\frac{x^2}{16} + \frac{y^2}{9} = 1.$$

> This curve is called an ellipse.

Discussion point

➔ What transformations map the unit circle $x^2 + y^2 = 1$ on to this ellipse?

To find where the curve crosses the axes, substitute in $x = 0$ and then $y = 0$:

When $y = 0$, $x^2 = 16 \implies x = \pm 4$.

When $x = 0$, $y^2 = 9 \implies y = \pm 3$.

The curve is shown in Figure 11.8.

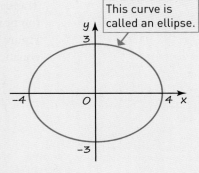

Figure 11.8

The double angle formulae giving $\cos 2\theta$ in terms of either $\sin\theta$ or $\cos\theta$ are also useful when converting from parametric to Cartesian equations. Remember:

$$\cos 2\theta = 1 - 2\sin^2\theta \quad \text{and} \quad \cos 2\theta = 2\cos^2\theta - 1.$$

Example 11.6

Eliminate θ from $y = \cos 2\theta$, $x = \sin\theta + 2$.

Discussion point

Use graphing software to sketch these curves.

(i) $y = \cos 2\theta$,
$x = \sin\theta + 2$

(ii) $y = 1 - 2(x - 2)^2$

➔ Why are they not exactly the same?

Solution

The relationship between $\cos 2\theta$ and $\sin\theta$ is

$$\cos 2\theta = 1 - 2\sin^2\theta.$$

Now $x - 2 = \sin\theta$

so $\qquad y = 1 - 2(x - 2)^2.$

The parametric equation of a circle

The circle with centre $(0, 0)$

Prior knowledge

You need to know the Cartesian equation of a circle – see Review: Coordinate geometry.

The circle with centre $(0, 0)$ and radius 4 units has the equation $x^2 + y^2 = 16$.

Alternatively, using the triangle OAB and the angle θ in Figure 11.9, you can write the equations

$$x = 4\cos\theta$$

$$y = 4\sin\theta.$$

Generalising, a circle with centre $(0, 0)$ and radius r has the parametric equations

$$x = r\cos\theta$$

$$y = r\sin\theta.$$

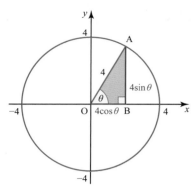

Figure 11.9

The circle with centre (a, b)

Translating the centre of the circle to the point (a, b) gives the circle in Figure 11.10 with the parametric equations

$$x = a + r\cos\theta$$

$$y = b + r\sin\theta.$$

The general point P has coordinates $(a + r\cos\theta, b + r\sin\theta)$.

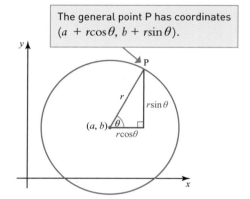

Figure 11.10

Converting from a Cartesian equation to parametric equations

Discussion points

→ How would your answers to Activity 11.1 change if the parametric equations described the position of an object at time t?

→ What happens when t is replaced by $2t$?

! Activity 11.1 showed that the parametric equations for a curve are not unique and also that the shape of the curve depends on the range of values chosen for t.

ACTIVITY 11.1

(i) Match together the parametric equations that describe the same curve.

$x = 2 + 5\sin t$ $y = 3 + 5\cos t$	$x = 5\cos t$ $y = 5\sin t$	$x = 2 + 5\cos 2t$ $y = 3 + 5\sin 2t$
$x = 5\cos 2t$ $y = 5\sin 2t$	$x = 3 + 5\sin t$ $y = 2 + 5\cos t$	$x = 2 + 5\cos t$ $y = 3 + 5\cos t$
$x = 3 + 5\cos\frac{1}{2}t$ $y = 2 + 5\sin\frac{1}{2}t$	$x = 2 + 5\sin\frac{1}{2}t$ $y = 3 + 5\sin\frac{1}{2}t$	$x = 5\cos\frac{1}{2}t$ $y = 5\sin\frac{1}{2}t$
$x = 2 + 5\sin t$ $y = 3 + 5\sin t$	$x = 2 + 5\cos t$ $y = 3 + 5\sin t$	$x = 2 - 5\cos t$ $y = 3 - 5\sin t$

(ii) For each set of parametric equations you found in part (i), investigate the curve for

(a) $0 \leqslant t \leqslant \dfrac{\pi}{2}$ (b) $0 \leqslant t \leqslant \pi$ (c) $0 \leqslant t \leqslant 2\pi$.

What effect does changing the interval for the parameter have on the curve?

Example 11.7

(i) Find parametric equations to describe the curve $y = \dfrac{2}{1 + x^2}$ for $1 \leqslant x \leqslant 4$ when

    (a) $x = t$ (b) $x = e^t$.

(ii) Explain why you can't use (b) to describe the curve for all values of x.

Solution

(i)   (a)   Substituting $x = t$ into $y = \dfrac{2}{1 + x^2}$ gives $y = \dfrac{2}{1 + t^2}$.

        Since $x = t$ and $1 \leqslant x \leqslant 4$ then t is also between 1 and 4, so the parametric equations are

> If you use a value of t outside of this range you will get a different section of the curve.

$$x = t, \quad y = \dfrac{2}{1 + t^2} \text{ for } 1 \leqslant t \leqslant 4.$$

      (b)   Substituting $x = e^t$ into $y = \dfrac{2}{1 + x^2}$ gives

$$y = \dfrac{2}{1 + \left(e^t\right)^2} = \dfrac{2}{1 + e^{2t}}.$$

> Make t the subject so you can find the values of the parameter corresponding to $0 \leqslant x \leqslant 4$.

        Since $x = e^t$ then $t = \ln x$.

        When $x = 1$ then $t = \ln 1 = 0$.
        When $x = 4$ then $t = \ln 4$.

        So the parametric equations are

$$x = e^t, \quad y = \dfrac{2}{1 + e^{2t}} \text{ for } 0 \leqslant t \leqslant \ln 4.$$

(ii)   Since $t = \ln x$ the parameter is only defined for $x > 0$.

T Discussion point

Use graphing software to sketch the following curves.

(i) $x = t,\ y = \dfrac{2}{1 + t^2}$

(ii) $x = e^t,\ y = \dfrac{2}{1 + e^{2t}}$

➜ What do you notice?

Exercise 11.2

① Match together these Cartesian and parametric equations.

$xy = 9$

$x = 1 - t$
$y = t^2$

$x = 3t^2$
$y = 6t$

$xy = 3$

$4y = 4 - x^2$

$y = (1 - x)^2$

$x = 2t$
$y = 1 - t^2$

$x = 3t$
$y = \dfrac{3}{t}$

$x = \dfrac{1}{t}$
$y = 3t$

$y^2 = 12x$

② Find the Cartesian equation of each of these curves from question 3 in Exercise 11.1.

  (i) $x = 2t$
        $y = t^2$

  (ii) $x = t^2$
         $y = t^3$

  (iii) $x = t^2$
          $y = t^2 - t$

③ Find parametric equations to describe the line $y = 5 - 2x$ for $0 \leqslant x \leqslant 4$ when

  (i) $x = 2t$ (ii) $x = u - 1$

④ A circle has equation
$(x - 3)^2 + (y + 1)^2 = 16$.
Find parametric equations to describe the circle given that

  (i) $x = 3 + 4\cos t$.

  (ii) $x = 3 - 4\sin t$.

⑤ Find the Cartesian equation of each of these curves from question 4 in Exercise 11.1. Write down the values of x and y for which your equation is valid.

  (i) $x = \cos 2\theta$ (ii) $x = \sin^2 \theta$
        $y = \sin^2 \theta$ $y = 1 + 2\sin \theta$

  (iii) $x = 2\sin^2 \theta$
          $y = 3\cos \theta$

⑥ Sketch the circles given by the following equations. Write down the Cartesian equation of each curve.

   (i) $x = 5\cos\theta$
       $y = 5\sin\theta$

   (ii) $x = 3\cos\theta$
       $y = 3\sin\theta$

   (iii) $x = 4 + 3\cos\theta$
       $y = 1 + 3\sin\theta$

   (iv) $x = 2\cos\theta - 1$
       $y = 3 + 2\sin\theta$

⑦ (i) Sketch both of these curves on the same axes.

     (a) $x = t, y = \dfrac{1}{t}$

     (b) $x = 4t, y = \dfrac{4}{t}$

   (ii) Find the Cartesian equation of each curve.

   (iii) Comment on the relationship between them.

⑧ Figure 11.11 shows the graph of $y = 2x + 1$.

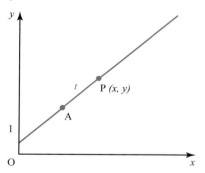

Figure 11.11

   (i) Show that the point A (3, 7) lies on $y = 2x + 1$.

   (ii) Write down

     (a) the position vector of A

     (b) a vector in the same direction as the line.

A point P (x, y) lies on $y = 2x + 1$ at a distance t from A (3, 7).

   (iii) Find the position vector of P in terms of t.

   (iv) Hence write down the parametric equations of the line. Use t as the parameter.

⑨ A curve has parametric equations $x = (t + 1)^2$, $y = t - 1$.

   (i) Sketch the curve for $-4 \leqslant t \leqslant 4$.

   (ii) State the equation of the line of symmetry of the curve.

   (iii) By eliminating the parameter, find the Cartesian equation of the curve.

PS ⑩ Find the Cartesian equation of each of these curves from questions 6 and 7 in Exercise 11.1.

   (i) $x = 2\operatorname{cosec}\theta$ (ii) $x = \tan\theta$
      $y = 2\cot\theta$ $y = \tan 2\theta$

⑪ Figure 11.12 shows the circle with equation $(x - 4)^2 + (y - 6)^2 = 25$.
C is the centre of the circle and CP is a radius of the circle which is parallel to the x-axis.
A is a variable point on the circle and t is the angle shown.

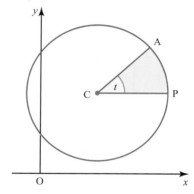

Figure 11.12

   (i) Give the coordinates of C and P.

   (ii) Using t as the parameter, find the equation of the circle in parametric form.

The point X has parameter $t = \dfrac{\pi}{3}$ and the point Y has parameter $t = \dfrac{5\pi}{6}$.

   (iii) (a) Find, in exact form, the coordinates of X and the length of XY.

       (b) Does XY form a diameter of the circle?

3 Parametric differentiation

To differentiate a function which is defined in terms of a parameter t, you need to use the chain rule:

$$\frac{\mathrm{d}y}{\mathrm{d}x} = \frac{\mathrm{d}y}{\mathrm{d}t} \times \frac{\mathrm{d}t}{\mathrm{d}x}.$$

Since

$$\frac{\mathrm{d}t}{\mathrm{d}x} = \frac{1}{\frac{\mathrm{d}x}{\mathrm{d}t}}$$

it follows that

$$\frac{\mathrm{d}y}{\mathrm{d}x} = \frac{\frac{\mathrm{d}y}{\mathrm{d}t}}{\frac{\mathrm{d}x}{\mathrm{d}t}} \qquad \text{provided that } \frac{\mathrm{d}x}{\mathrm{d}t} \neq 0.$$

> **Prior knowledge**
>
> You need to understand the chain rule (covered in Chapter 5) and to be able to differentiate a variety of functions including those involving trigonometric functions (covered in Chapter 9). You also need to be able to find the equation of the tangent or normal to a curve (see the review section in Chapter 5).

Example 11.8

A curve has the parametric equations $x = t^2$, $y = 2t$.

(i)   Find $\frac{\mathrm{d}y}{\mathrm{d}x}$ in terms of the parameter t.

(ii) Find the equation of the tangent to the curve at the general point $(t^2, 2t)$.

(iii) Find the equation of the tangent at the point where $t = 3$.

(iv) Eliminate the parameter to find the Cartesian equation of the curve. Hence sketch the curve and the tangent at the point where $t = 3$.

Solution

(i) $x = t^2 \implies \dfrac{\mathrm{d}x}{\mathrm{d}t} = 2t$

$y = 2t \implies \dfrac{\mathrm{d}y}{\mathrm{d}t} = 2$

$\dfrac{\mathrm{d}y}{\mathrm{d}x} = \dfrac{\frac{\mathrm{d}y}{\mathrm{d}t}}{\frac{\mathrm{d}x}{\mathrm{d}t}} = \dfrac{2}{2t} = \dfrac{1}{t}$

(ii) Using $y - y_1 = m(x - x_1)$ and taking the point (x_1, y_1) as $(t^2, 2t)$, the equation of the tangent at the point $(t^2, 2t)$ is

$$y - 2t = \frac{1}{t}(x - t^2)$$

$$\implies \quad ty - 2t^2 = x - t^2$$

$$\implies x - ty + t^2 = 0.$$

(iii) Substituting $t = 3$ into this equation gives the equation of the tangent at the point where $t = 3$.

The tangent is $x - 3y + 9 = 0$.

(iv) Eliminating t from $x = t^2$, $y = 2t$ gives

$$x = \left(\frac{y}{2}\right)^2 \quad \text{or} \quad y^2 = 4x.$$

This is a parabola with the x-axis as its line of symmetry.

The point where $t = 3$ has coordinates $(9, 6)$.

The tangent $x - 3y + 9 = 0$ crosses the axes at $(0, 3)$ and $(-9, 0)$. ➔

The curve is shown in Figure 11.13.

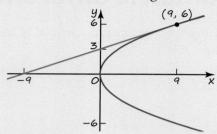

Figure 11.13

Example 11.9

A curve has parametric equations $x = 4\cos\theta$, $y = 3\sin\theta$.

(i)   Find $\dfrac{dy}{dx}$ at the point with parameter θ.

(ii)   Find the equation of the normal at the general point $(4\cos\theta, 3\sin\theta)$.

(iii) Find the equation of the normal at the point where $\theta = \dfrac{\pi}{4}$.

Solution

(i) $x = 4\cos\theta \Rightarrow \dfrac{dx}{d\theta} = -4\sin\theta$

$y = 3\sin\theta \Rightarrow \dfrac{dy}{d\theta} = 3\cos\theta$

$$\frac{dy}{dx} = \frac{\dfrac{dy}{d\theta}}{\dfrac{dx}{d\theta}} = \frac{3\cos\theta}{-4\sin\theta}$$

$$= -\frac{3\cos\theta}{4\sin\theta}$$

This is the same as $-\frac{3}{4}\cot\theta$.

(ii)   The tangent and normal are perpendicular, so the gradient of the normal is

Find the negative reciprocal.

$$\frac{-1}{\dfrac{dy}{dx}} \quad \text{which is} \quad \frac{4\sin\theta}{3\cos\theta} = \frac{4}{3}\tan\theta.$$

Using $y - y_1 = m(x - x_1)$ and (x_1, y_1) as $(4\cos\theta, 3\sin\theta)$.

The equation of the normal at the point $(4\cos\theta, 3\sin\theta)$ is

$$y - 3\sin\theta = \frac{4}{3}\tan\theta(x - 4\cos\theta)$$

$$\Rightarrow \quad 3y - 9\sin\theta = 4x\tan\theta - 16\tan\theta\cos\theta$$

$$\Rightarrow \quad 3y - 9\sin\theta = 4x\tan\theta - 16\sin\theta$$

$$\Rightarrow \quad\quad\quad 3y = 4x\tan\theta - 7\sin\theta$$

$\tan\theta\cos\theta = \dfrac{\sin\theta}{\cos\theta}\cos\theta$

$= \sin\theta$

(iii) Substitute $\theta = \dfrac{\pi}{4}$ into $3y = 4x\tan\theta - 7\sin\theta$:

$$3y = 4x \times 1 - 7\frac{\sqrt{2}}{2}$$

Multiply both sides by 2:

$$6y = 8x - 7\sqrt{2}.$$

Remember $\tan\dfrac{\pi}{4} = 1$

and $\sin\dfrac{\pi}{4} = \dfrac{\sqrt{2}}{2}$.

Turning points

When the equation of a curve is given parametrically, you can distinguish between turning points by considering the sign of $\frac{dy}{dx}$.

Example 11.10

The path of a projectile at time t seconds is modelled by the parametric equations

$$x = 20\sqrt{3}t$$

$$y = 20t - 5t^2$$

where x and y are measured in metres.

Find the maximum height, y metres, reached by the projectile and justify that this height is a maximum.

Solution

$$x = 20\sqrt{3}t \quad \Rightarrow \frac{dx}{dt} = 20\sqrt{3}$$

$$y = 20t - 5t^2 \Rightarrow \frac{dy}{dt} = 20 - 10t$$

$$\frac{dy}{dx} = \frac{\dfrac{dy}{dt}}{\dfrac{dx}{dt}} = \frac{20 - 10t}{20\sqrt{3}}$$

> ❗ You must be careful to ensure that you take points which are to the left and right of the turning point, i.e. have x coordinates smaller and larger than those at the turning point. These will not necessarily be points whose parameters are smaller and larger than those at the turning point.

Turning points occur when $\frac{dy}{dx} = 0$

so $\quad 20 - 10t = 0 \Rightarrow t = 2$.

At $t = 2$: $\quad x = 20\sqrt{3} \times 2 = 40\sqrt{3}$

$$y = 20 \times 2 - 5 \times 2^2 = 20.$$

To justify that this is a maximum you need to look at the sign of $\frac{dy}{dx}$ either side of $x = 40\sqrt{3} = 69.3$.

At $t = 1.9$: $\quad x = 65.8$ (to the left); $\frac{dy}{dx} = 0.0288\ldots$ (positive).

At $t = 2.1$: $\quad x = 72.7$ (to the right); $\frac{dy}{dx} = -0.0288\ldots$ (negative).

Therefore the maximum height of the projectile is 20 metres at $t = 2$ seconds.

Exercise 11.3

① The equation of the tangent to a curve is $y - 4tx + 8t^2 = 0$. What is the equation at the point where $t = \frac{1}{2}$?

② For each of the following curves, find

    (a) $\dfrac{dx}{dt}$ (b) $\dfrac{dy}{dt}$ (c) $\dfrac{dy}{dx}$.

  (i) $x = 3t^2$
      $y = 2t^3$

  (ii) $x = 4t - 1$
      $y = t^4$

  (iii) $x = t + \dfrac{1}{t}$

      $y = t - \dfrac{1}{t}$

③ A curve has the parametric equations $x = (t + 1)^2$, $y = (t - 1)^2$. Find

  (i) (a) $\dfrac{dx}{dt}$ (b) $\dfrac{dy}{dt}$ (c) $\dfrac{dy}{dx}$

  (ii) (a) the coordinates of the point where $t = 3$

      (b) the gradient of the curve at $t = 3$

      (c) the equation of the tangent at $t = 3$.

④ For each of the following curves, find $\dfrac{dy}{dx}$ in terms of the parameter.

(i) $x = \theta - \cos\theta$
$y = \theta + \sin\theta$

(ii) $x = 3\cos\theta$
$y = 2\sin\theta$

(iii) $x = \theta\sin\theta + \cos\theta$
$y = \theta\cos\theta - \sin\theta$

⑤ A curve has the parametric equations
$x = \dfrac{t}{1+t}, \ y = \dfrac{t}{1-t}.$

(i) (a) Find $\dfrac{dy}{dx}$ in terms of the parameter t.

(b) Find $\dfrac{dy}{dx}$ when $t = 2$.

(ii) (a) Find the Cartesian equation of the curve.

(b) Find $\dfrac{dy}{dx}$ in terms of x.

(iii) Show that both of your expressions for the gradient have the same value at the point where $t = 2$.

⑥ A ball is thrown from the top of a tower. The trajectory of the ball at time t seconds is modelled by the parametric equations

$$x = 20t, \ y = 50 + 20t - 5t^2$$

where x and y are measured in metres.

(i) Find the position of the ball when $t = 0.5$ seconds.

(ii) Find the maximum height, y metres, reached by the ball.

(iii) At what time does the ball land? How far from the base of the tower does the ball land?

(iv) Illustrate the trajectory of the ball on a graph.

⑦ Figure 11.14 shows part of the curve with the parametric equations

$x = \dfrac{3t}{1+t^3}$

$y = \dfrac{3t^2}{1+t^3}$

Figure 11.14

(i) Find $\dfrac{dy}{dx}$ in terms of the parameter t.

(ii) Find the exact coordinates of the maximum point of the curve.

(iii) The curve intersects the line $y = x$ at the origin and at point P.
Find the coordinates of P and the distance OP. What is the significance of the point P?

⑧ A curve has the parametric equations
$x = e^{2t} + 1, \ y = e^t$.

(i) Find $\dfrac{dy}{dx}$ in terms of the parameter t.

(ii) Find the equation of the tangent at the point where

(a) $t = 0$ (b) $t = 1$.

⑨ For each of the following curves

(a) find the turning points of the curve and distinguish between them

(b) draw the graph of the curve.

(i) $x = 8t + 1$
$y = 4t^2 - 3$

(ii) $x = 3t$
$y = \dfrac{3}{1+t^2}$

⑩ A curve has the parametric equations
$x = \tan\theta, \ y = \tan 2\theta$. Find

(i) the value of $\dfrac{dy}{dx}$ when $\theta = \dfrac{\pi}{6}$

(ii) the equation of the tangent to the curve at the point where $\theta = \dfrac{\pi}{6}$

(iii) the equation of the normal to the curve at the point where $\theta = \dfrac{\pi}{6}$.

PS ⑪ A curve has the parametric equations
$x = t^2, \ y = 1 - \dfrac{1}{2t}$ for $t > 0$. Find

(i) the coordinates of the point P where the curve cuts the x-axis

(ii) the gradient of the curve at this point

(iii) the equation of the tangent to the curve at P

(iv) the coordinates of the point where the tangent cuts the y-axis.

⑫ A curve has the parametric equations
$x = at^2, \ y = 2at$, where a is constant. Find

(i) the equation of the tangent to the curve at the point with parameter t

(ii) the equation of the normal to the curve at the point with parameter t

(iii) the coordinates of the points where the normal cuts the x and y axes.

⑬ A curve has the parametric equations
$x = \cos\theta, \ y = \cos 2\theta$.

(i) Show that $\dfrac{dy}{dx} = 4x$.

(ii) Find the coordinates of the stationary point and identify its nature.

⑭ The parametric equations of a curve are $x = at$, $y = \dfrac{b}{t}$, where a and b are constant. Find in terms of a, b and t

(i) $\dfrac{dy}{dx}$

(ii) the equation of the tangent to the curve at the general point $\left(at, \dfrac{b}{t}\right)$

(iii) the coordinates of the points X and Y where the tangent cuts the x and y axes.

PS

(iv) Show that the area of triangle OXY is constant, where O is the origin.

⑮ A curve has parametric equations $x = t^3$, $y = (t^2 - 1)^2$.

(i) Prove algebraically that no point on the curve is below the x-axis.

(ii) Find the coordinates of the points where the curve touches the x-axis.

(iii) Investigate the behaviour of the curve where it crosses the y-axis.

(iv) Sketch the curve.

PS ⑯ A particle P moves in a plane so that at time t its coordinates are given by $x = 4\cos t$, $y = 3\sin t$. Find

(i) $\dfrac{dy}{dx}$ in terms of t

(ii) the equation of the tangent to its path at time t

(iii) the values of t for which the particle is travelling parallel to the line $x + y = 0$.

⑰ A circle has parametric equations $x = 3 + 2\cos\theta$, $y = 3 + 2\sin\theta$.

(i) Find the equation of the tangent at the point with parameter θ.

(ii) Show that this tangent will pass through the origin provided that $\sin\theta + \cos\theta = -\dfrac{2}{3}$.

(iii) By writing $\sin\theta + \cos\theta$ in the form $R\sin(\theta + \alpha)$, solve the equation $\sin\theta + \cos\theta = -\dfrac{2}{3}$ for $0 \leqslant \theta \leqslant 2\pi$.

(iv) Illustrate the circle and tangents on a sketch, showing clearly the values of θ which you found in part (iii).

PS ⑱ The parametric equations of the circle with centre $(2, 5)$ and radius 3 units are $x = 2 + 3\cos\theta$, $y = 5 + 3\sin\theta$.

(i) Find the gradient of the circle at the point with parameter θ.

(ii) Find the equation of the normal to the circle at this point.

(iii) Show that the normal at any point on the circle passes through the centre.

> This is an alternative proof of the result 'the tangent and radius are perpendicular'.

LEARNING OUTCOMES

When you have completed this chapter, you should be able to:
➤ understand and use the parametric equations of curves
➤ convert between Cartesian and parametric forms
➤ use parametric equations for modelling in a variety of contexts
➤ differentiate simple functions and relations defined parametrically, for first derivative only
➤ use trigonometric functions to solve kinematics problems in context.

KEY POINTS

1 In parametric equations the relationship between two variables is expressed by writing both of them in terms of a third variable or **parameter**.

2 To draw a graph from parametric equations, plot the points on the curve given by different values of the parameter.

3 Eliminating the parameter gives the Cartesian equation of the curve.

4 The parametric equations of circles:
- centre $(0, 0)$ and radius r $x = r\cos\theta$, $y = r\sin\theta$
- centre (a, b) and radius r $x = a + r\cos\theta$, $y = b + r\sin\theta$

5 $\dfrac{dy}{dx} = \dfrac{\frac{dy}{dt}}{\frac{dx}{dt}}$ provided that $\dfrac{dx}{dt} \neq 0$.

12 Vectors

→ Is it possible to sail faster than the wind? How?

1 Vectors

A quantity which has size (magnitude) only is called a **scalar**. The mass of the sailing boat (80 kg) is an example of a scalar.

A quantity which has both size (magnitude) and direction is called a **vector**. The velocity of the sailing boat is an example of a vector, as it has size (e.g. 8 knots ≈ 15 km h^{-1}) and direction (e.g. south-east, or on a course of 135°).

If the boat is out at sea, you might also want to include a component in the vertical direction to describe it rising and falling in the swell. In that case, you would need a three-dimensional vector. This chapter shows how you can use vectors in three dimensions.

Notation

You can use an arrow to represent a vector:
- the length of the arrow represents the magnitude of the vector
- the direction is the angle made with the positive x-axis.

> Vectors have magnitude (size) and direction.

A two-dimensional vector can be represented in **magnitude–direction** form (Figure 12.1) or in **component form**.

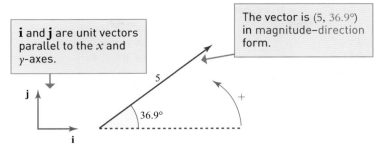

Figure 12.1

Using trigonometry,

$$5\cos 36.9° = 4$$

$$5\sin 36.9° = 3.$$

So $(5, 36.9°) = 4\mathbf{i} + 3\mathbf{j}$ or $\begin{pmatrix} 4 \\ 3 \end{pmatrix}$. ⟵ The vector is now in component form.

In general, for all values of θ:

$$(r, \theta) \Rightarrow \begin{pmatrix} r\cos\theta \\ r\sin\theta \end{pmatrix} = (r\cos\theta)\mathbf{i} + (r\sin\theta)\mathbf{j}.$$

Equal vectors and parallel vectors

In textbooks, a vector may be printed in **bold**, for example **a** or **OA**, or with an arrow above it, $\overrightarrow{OA}$.

When you write a vector by hand, it is usual to underline it, for example $\underline{a}$ or $\underline{OA}$, or to put an arrow above it, as in $\overrightarrow{OA}$.

Equal vectors have the same magnitude and direction – the actual location of the vector doesn't matter. Both the red lines in Figure 12.2 represent the vector **a**.

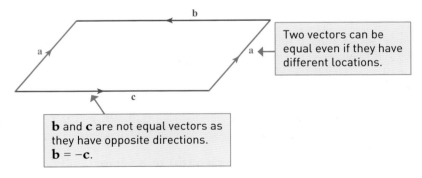

Figure 12.2

The **negative of a vector** has the same magnitude but the **opposite direction**.

You can multiply a vector by a scalar to form a **parallel vector**: the vectors **a** and **b** are parallel when **a** = k**b** for some scalar (number) k.

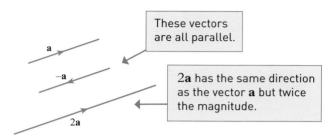

These vectors are all parallel.

2**a** has the same direction as the vector **a** but twice the magnitude.

Figure 12.3

Working with vectors

Points and position vectors

In two dimensions

A point has two coordinates, usually called x and y.

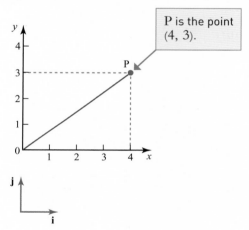

P is the point (4, 3).

Figure 12.4

The position vector of point P in Figure 12.4, $\overrightarrow{OP}$,

is $\begin{pmatrix} 4 \\ 3 \end{pmatrix}$ or 4**i** + 3**j**.

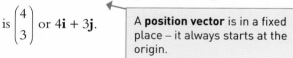

A **position vector** is in a fixed place – it always starts at the origin.

In three dimensions

A point has three coordinates, usually called x, y and z.

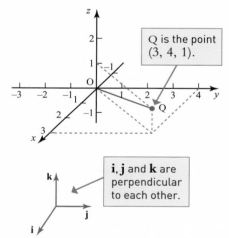

Q is the point (3, 4, 1).

i, **j** and **k** are perpendicular to each other.

Figure 12.5

The position vector of point Q in Figure 12.5, $\overrightarrow{OQ}$,

is $\begin{pmatrix} 3 \\ 4 \\ 1 \end{pmatrix}$ or 3**i** + 4**j** + **k**.

k is the unit vector parallel to the z axis.

Magnitude (or length) of a vector

You can use Pythagoras' theorem to find the magnitude (sometimes called the modulus) of a vector.

In two dimensions

$$\left| \overrightarrow{OP} \right| = \sqrt{4^2 + 3^2}$$
$$= \sqrt{25}$$
$$= 5$$

In three dimensions

$$\left| \overrightarrow{OQ} \right| = \sqrt{3^2 + 4^2 + 1^2}$$
$$= \sqrt{26}$$

| | are modulus signs. You say, 'mod OQ'.

Direction of a vector

In two dimensions

You can use trigonometry to find the direction of a vector in two dimensions.

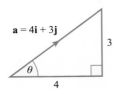

Figure 12.6

For the two-dimensional vector **a** in Figure 12.6,

$$\tan\theta = \frac{3}{4}$$

> So this vector is (5, 36.9°) – see page 267.

$$\Rightarrow \quad \theta = 36.9° \text{ to the } \mathbf{i} \text{ direction.}$$

In three dimensions

It is hard to define the direction of a vector in three dimensions so vectors are left in component form.

Vector arithmetic

The **resultant** of two or more vectors is found by adding those vectors together. You can add two vectors by adding the components.

You can also multiply a vector by a scalar (a number)

In two dimensions

For $\mathbf{a} = 2\mathbf{i} - 3\mathbf{j}$ and $\mathbf{b} = -\mathbf{i} + 2\mathbf{j}$,

$$\mathbf{a} + \mathbf{b} = (2\mathbf{i} - 3\mathbf{j}) + (-\mathbf{i} + 2\mathbf{j})$$

> The **resultant** of **a** and **b** is $\mathbf{i} - \mathbf{j}$.

$$= \mathbf{i} - \mathbf{j}$$
$$\mathbf{a} - \mathbf{b} = 2\mathbf{i} - 3\mathbf{j} - (-\mathbf{i} + 2\mathbf{j})$$
$$= 2\mathbf{i} - 3\mathbf{j} + \mathbf{i} - 2\mathbf{j}$$

> You could use column notation instead.

$$= 3\mathbf{i} - 5\mathbf{j}$$

In three dimensions

For $\mathbf{p} = \begin{pmatrix} 2 \\ 0 \\ -1 \end{pmatrix}$ and $\mathbf{q} = \begin{pmatrix} 1 \\ 2 \\ 4 \end{pmatrix}$,

$$\mathbf{p} + 3\mathbf{q} = \begin{pmatrix} 2 \\ 0 \\ -1 \end{pmatrix} + 3\begin{pmatrix} 1 \\ 2 \\ 4 \end{pmatrix}$$

> You could use **i**, **j**, **k** notation instead.

$$= \begin{pmatrix} 2 \\ 0 \\ -1 \end{pmatrix} + \begin{pmatrix} 3 \\ 6 \\ 12 \end{pmatrix}$$
$$= \begin{pmatrix} 5 \\ 6 \\ 11 \end{pmatrix}$$

Unit vectors

A unit vector has magnitude 1.

To find the unit vector in the same direction as a given vector, divide that vector by its magnitude.

> **i**, **j** and **k** are unit vectors as they have magnitude 1.

In two dimensions

For $\mathbf{a} = 2\mathbf{i} - 3\mathbf{j}$

$$|\mathbf{a}| = \sqrt{2^2 + (-3)^2} = \sqrt{13}$$
$$\hat{\mathbf{a}} = \frac{2}{\sqrt{13}}\mathbf{i} - \frac{3}{\sqrt{13}}\mathbf{j}$$

> The unit vector in the direction of a vector **a** is denoted by $\hat{\mathbf{a}}$.

In three dimensions

For $\mathbf{q} = \mathbf{i} + 2\mathbf{j} + 4\mathbf{j}$

$$|\mathbf{q}| = \sqrt{1^2 + 2^2 + 4^2} = \sqrt{21}$$
$$\hat{\mathbf{q}} = \frac{1}{\sqrt{21}}\mathbf{i} + \frac{2}{\sqrt{21}}\mathbf{j} + \frac{4}{\sqrt{21}}\mathbf{k}$$

The vector of a line joining two points

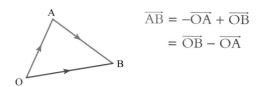

$$\overrightarrow{AB} = -\overrightarrow{OA} + \overrightarrow{OB}$$
$$= \overrightarrow{OB} - \overrightarrow{OA}$$

Figure 12.7

In two dimensions

For $\overrightarrow{OA} = \begin{pmatrix} 1 \\ -3 \end{pmatrix}$

and $\overrightarrow{OB} = \begin{pmatrix} 4 \\ -1 \end{pmatrix}$

$$\overrightarrow{AB} = \begin{pmatrix} 4 \\ -1 \end{pmatrix} - \begin{pmatrix} 1 \\ -3 \end{pmatrix}$$
$$= \begin{pmatrix} 3 \\ 2 \end{pmatrix}$$

> You could use **i, j, k** notation instead.

In three dimensions

For $\overrightarrow{OA} = \mathbf{i} - 3\mathbf{j} + 2\mathbf{k}$

and $\overrightarrow{OB} = 4\mathbf{i} - \mathbf{j} + 4\mathbf{k}$

$$\overrightarrow{AB} = (4\mathbf{i} - \mathbf{j} + 4\mathbf{k}) - (\mathbf{i} - 3\mathbf{j} + 2\mathbf{k})$$
$$= 3\mathbf{i} + 2\mathbf{j} + 2\mathbf{k}$$

> You could use column notation instead.

Discussion point

→ Give some examples of vector quantities you meet in mechanics.

Using vectors in mechanics

In mechanics, it is often useful to add vectors together. The sum of two or more vectors is called the **resultant**.

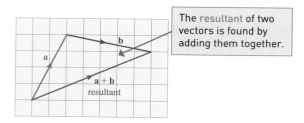

> The resultant of two vectors is found by adding them together.

Figure 12.8

For example, if Peter walks 2 km east and 4 km north followed by 5 km east and 1 km south his overall displacement in kilometres is

$$\begin{pmatrix} 2 \\ 4 \end{pmatrix} + \begin{pmatrix} 5 \\ -1 \end{pmatrix} = \begin{pmatrix} 7 \\ 3 \end{pmatrix}$$

You can also use vectors to represent the forces acting on an object. The resultant of the forces is the single force that has the same effect as all the forces acting on the object. When the resultant force is zero, the object is **in equilibrium**.

> The effects of all the individual forces cancel each other out.

Example 12.1

The unit vectors **i**, **j** and **k** act in the directions shown in Figure 12.9.

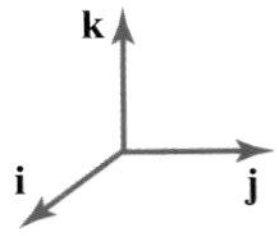

Figure 12.9

Forces **p** and **q** are given, in newtons, by $\mathbf{p} = 4\mathbf{i} - 6\mathbf{j} - 7\mathbf{k}$ and $\mathbf{q} = 6\mathbf{i} - 9\mathbf{j} + 2\mathbf{k}$.

(i) Find the resultant force, $\mathbf{p} + \mathbf{q}$, and show that it is parallel to $2\mathbf{i} - 3\mathbf{j} - \mathbf{k}$.

(ii) A particle is in equilibrium under the forces $3\mathbf{p}$, $a\mathbf{q}$ and $b\mathbf{k}$. Show that $a = -2$ and find the value of b.

Solution

You might find it easier to work with column vectors.

(i) $\mathbf{p} + \mathbf{q} = \begin{pmatrix} 4 \\ -6 \\ -7 \end{pmatrix} + \begin{pmatrix} 6 \\ -9 \\ 2 \end{pmatrix} = \begin{pmatrix} 10 \\ -15 \\ -5 \end{pmatrix}$

$\begin{pmatrix} 10 \\ -15 \\ -5 \end{pmatrix}$ and $\begin{pmatrix} 2 \\ -3 \\ -1 \end{pmatrix}$ are parallel if $\begin{pmatrix} 10 \\ -15 \\ -5 \end{pmatrix} = \lambda \begin{pmatrix} 2 \\ -3 \\ -1 \end{pmatrix}$ for some scalar λ.

Equating the **i** components gives

$10 = 2\lambda$ and so $\lambda = 5$.

Use the other two components to check this:

$5 \times (-3) = -15$ ✓

$5 \times (-1) = -5$ ✓

Since $\begin{pmatrix} 10 \\ -15 \\ -5 \end{pmatrix} = 5 \times \begin{pmatrix} 2 \\ -3 \\ -1 \end{pmatrix}$ then the vectors are parallel.

(ii) 'In equilibrium' means the resultant force is zero, so

$3\begin{pmatrix} 4 \\ -6 \\ -7 \end{pmatrix} + a\begin{pmatrix} 6 \\ -9 \\ 2 \end{pmatrix} + b\begin{pmatrix} 0 \\ 0 \\ 1 \end{pmatrix} = \begin{pmatrix} 0 \\ 0 \\ 0 \end{pmatrix}$

$\Rightarrow \begin{pmatrix} 12 \\ -18 \\ -21 \end{pmatrix} + \begin{pmatrix} 6a \\ -9a \\ 2a \end{pmatrix} + \begin{pmatrix} 0 \\ 0 \\ b \end{pmatrix} = \begin{pmatrix} 0 \\ 0 \\ 0 \end{pmatrix}$

Read across the top line of both the vectors …

So $12 + 6a = 0 \Rightarrow a = -2$ ← … and then the middle line.

and $-18 - 9a = 0 \Rightarrow a = -2$ as required.

Use the bottom line to find b.

To find b: $-21 + 2 \times (-2) + b = 0 \Rightarrow b = 25$.

Vectors

Exercise 12.1

① Write each vector in Figure 12.10 in terms of the unit vectors **i** and **j**.

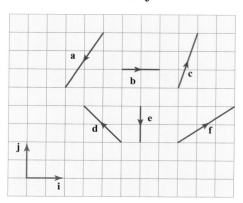

Figure 12.10

② Figure 12.11 shows several vectors.

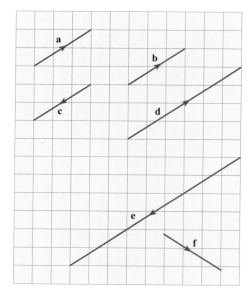

Figure 12.11

(i) Write each of these vectors in terms of the vector **a**.
Which vector cannot be written in terms of **a**?

(ii) Draw diagrams to show each of the following.

(a) **a** + **f**   (b) **a** − **f**

(c) **2c** + **f**   (d) **a** + **f** + **c**

③ Write each of the following vectors in magnitude–direction form.

(i) $5\mathbf{i} + 12\mathbf{j}$ (ii) $\begin{pmatrix} 12 \\ 5 \end{pmatrix}$ (iii) $5\mathbf{i} - 12\mathbf{j}$

(iv) $\begin{pmatrix} -5 \\ 12 \end{pmatrix}$ (v) $\begin{pmatrix} -12 \\ -5 \end{pmatrix}$

④ Find the magnitude of each of the following vectors.

(i) $3\mathbf{j} + 4\mathbf{k}$ (ii) $-4\mathbf{i} + 2\mathbf{j} + \mathbf{k}$

(iii) $\begin{pmatrix} 2 \\ 0 \\ 1 \end{pmatrix}$ (iv) $\begin{pmatrix} 3 \\ 2 \\ -1 \end{pmatrix}$

⑤ The vectors **a**, **b** and **c** are given by

$\mathbf{a} = 3\mathbf{i} - \mathbf{k}, \quad \mathbf{b} = \mathbf{i} - 2\mathbf{j} + 3\mathbf{k}, \quad \mathbf{c} = -3\mathbf{i} - \mathbf{j}.$

Find, in component form, each of the following vectors.

(i) **a** + **b** + **c**   (ii) **b** − **a**

(iii) **b** + **c**   (iv) 2**a** + **b** − 3**c**

(v) **c** − 2(**a** − **b**) (vi) 3(**a** − **b**) + 2(**b** + **c**)

⑥ (i) Show that $\frac{1}{3}\mathbf{i} - \frac{2}{3}\mathbf{j} + \frac{2}{3}\mathbf{k}$ is a unit vector.

(ii) Find unit vectors in the direction of

(a) $2\mathbf{i} - 6\mathbf{j} + 3\mathbf{k}$ (b) $\mathbf{i} + \mathbf{j} + \mathbf{k}$

⑦ Match together the parallel vectors.

$\mathbf{a} = \begin{pmatrix} 6 \\ -2 \\ 4 \end{pmatrix}$ $\mathbf{b} = \mathbf{i} + 4\mathbf{j} + 9\mathbf{k}$

$\mathbf{c} = \begin{pmatrix} 1 \\ 2 \\ 3 \end{pmatrix}$ $\mathbf{d} = 3\mathbf{i} - \mathbf{j} + 2\mathbf{k}$

$\mathbf{e} = \begin{pmatrix} 3 \\ -1 \\ 2 \end{pmatrix}$ $\mathbf{f} = \begin{pmatrix} -3 \\ 1 \\ -2 \end{pmatrix}$

$\mathbf{g} = -9\mathbf{i} + 3\mathbf{j} - 6\mathbf{k}$ $\mathbf{h} = \begin{pmatrix} -0.2 \\ -0.4 \\ -0.6 \end{pmatrix}$

⑧ Given that the vectors $\begin{pmatrix} x \\ -6 \\ 4 \end{pmatrix}$ and $\begin{pmatrix} -1 \\ y \\ -2 \end{pmatrix}$ are parallel, find the value of x and the value of y.

⑨ The vectors **r** and **s** are defined as
$$\mathbf{r} = a\mathbf{i} + (a + b)\mathbf{j} \quad \text{and}$$
$$\mathbf{s} = (6 - b)\mathbf{i} - (2a + 3)\mathbf{j}.$$
Given that $\mathbf{r} = \mathbf{s}$, find the values of a and b.

⑩ The force **p**, given in newtons, $-4\mathbf{i} + 2\mathbf{j} + k(2\mathbf{i} - 3\mathbf{j})$ acts in direction $2\mathbf{i} - 5\mathbf{j}$. Find the value of k and the magnitude of **p**.

⑪ The position vectors of the points A, B and C are $\mathbf{a} = \mathbf{i} + \mathbf{j} - 2\mathbf{k}$, $\mathbf{b} = 6\mathbf{i} - 3\mathbf{j} + \mathbf{k}$ and $\mathbf{c} = -2\mathbf{i} + 2\mathbf{j}$ respectively.

  (i)   Find $\overrightarrow{AC}$, $\overrightarrow{AB}$ and $\overrightarrow{BC}$.

  (ii) Show that $\left|\mathbf{a} + \mathbf{b} - \mathbf{c}\right|$ is not equal to $\left|\mathbf{a}\right| + \left|\mathbf{b}\right| - \left|\mathbf{c}\right|$.

⑫ Given forces $\mathbf{F}_1$, $\mathbf{F}_2$ and $\mathbf{F}_3$, where
$$\mathbf{F}_1 = (3\mathbf{i} - \mathbf{j})\,\text{N}, \ \mathbf{F}_2 = (3\mathbf{i} + 2\mathbf{j})\,\text{N} \quad \text{and}$$
$$\mathbf{F}_3 = (-2\mathbf{i} + 4\mathbf{j})\,\text{N},$$
find

  (i)   the resultant of $\mathbf{F}_1$ and $\mathbf{F}_2$

  (ii)   the resultant of $\mathbf{F}_1$, $\mathbf{F}_2$ and $\mathbf{F}_3$

  (iii) the magnitude and direction of $\mathbf{F}_3$

  (iv) the magnitude and direction of $(\mathbf{F}_1 + \mathbf{F}_3)$.

⑬ The resultant of the forces $(5\mathbf{i} + 7\mathbf{j})\,\text{N}$, $(a\mathbf{i} - 3\mathbf{j})\,\text{N}$ and $(4\mathbf{i} + b\mathbf{j})\,\text{N}$ is $(3\mathbf{i} - \mathbf{j})\,\text{N}$. Find the value of a and the value of b.

⑭ The following systems of forces act on an object that is in equilibrium. Find the value of a and the value of b in each case.

  (i) $(\mathbf{i} - \mathbf{j})\,\text{N}$, $(2\mathbf{i} + 3\mathbf{j})\,\text{N}$ and $(a\mathbf{i} + b\mathbf{j})\,\text{N}$

  (ii) $(4\mathbf{i} - 3\mathbf{j})\,\text{N}$, $(2\mathbf{i} - 3\mathbf{j})\,\text{N}$ and $(a\mathbf{i} + b\mathbf{j})\,\text{N}$

  (iii) $(\mathbf{i} + 2\mathbf{j})\,\text{N}$, $(-\mathbf{i} - 3\mathbf{j})\,\text{N}$ and $(a\mathbf{i} + b\mathbf{j})\,\text{N}$

⑮ In this question the origin is taken to be at a harbour and the unit vectors **i** and **j** to have lengths of 1 km in the directions east and north respectively.

A cargo vessel leaves the harbour and its position vector t hours later is given by
$$\mathbf{r}_1 = 12t\mathbf{i} + 16t\mathbf{j}.$$
A fishing boat is trawling nearby and its position at time t is given by
$$\mathbf{r}_2 = (10 - 3t)\mathbf{i} + (8 + 4t)\mathbf{j}.$$

  (i)   How far apart are the two boats when the cargo vessel leaves harbour?

  (ii)   How fast is each boat travelling?

  (iii) What happens?

2 Using vectors to solve problems

You can use vectors in two and three dimensions to solve geometry problems.

| **Example 12.2** | Figure 12.12 shows triangle AOB. C is a point on AB and divides it in the ratio $2 : 3$.

 Find $\overrightarrow{OC}$ in terms of the vectors **a** and **b**. |

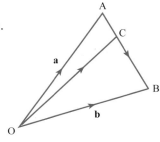

Figure 12.12

> Imagine going on a vector 'walk' – walking from O to C is the same as walking from O to A and then from A to C.

Solution

$$\overrightarrow{OC} = \overrightarrow{OA} + \overrightarrow{AC}$$
$$= \overrightarrow{OA} + \tfrac{2}{5}\overrightarrow{AB}$$

> Since C divides AB in the ratio 2 : 3, C is $\tfrac{2}{5}$ of the way along AB from A.

$$\overrightarrow{OA} = \mathbf{a} \quad \text{and} \quad \overrightarrow{AB} = \mathbf{b} - \mathbf{a}$$

so

$$\overrightarrow{OC} = \mathbf{a} + \tfrac{2}{5}(\mathbf{b} - \mathbf{a})$$
$$= \mathbf{a} + \tfrac{2}{5}\mathbf{b} - \tfrac{2}{5}\mathbf{a}$$
$$= \tfrac{3}{5}\mathbf{a} + \tfrac{2}{5}\mathbf{b}$$

Example 12.3

Relative to an origin O, the position vectors of the points A, B and C are given by

$$\overrightarrow{OA} = \begin{pmatrix} -2 \\ 3 \\ -2 \end{pmatrix}, \quad \overrightarrow{OB} = \begin{pmatrix} 1 \\ 9 \\ 10 \end{pmatrix} \quad \text{and} \quad \overrightarrow{OC} = \begin{pmatrix} 13 \\ 5 \\ 9 \end{pmatrix}.$$

Use vectors to prove that angle ABC is 90°.

Solution

If angle ABC is 90° then triangle ABC is right-angled

> Pythagoras' theorem.

$$\Leftrightarrow \left|\overrightarrow{AB}\right|^2 + \left|\overrightarrow{BC}\right|^2 = \left|\overrightarrow{AC}\right|^2.$$

$$\overrightarrow{AB} = \overrightarrow{OB} - \overrightarrow{OA} = \begin{pmatrix} 1 \\ 9 \\ 10 \end{pmatrix} - \begin{pmatrix} -2 \\ 3 \\ -2 \end{pmatrix} = \begin{pmatrix} 3 \\ 6 \\ 12 \end{pmatrix}$$

$$\Rightarrow \left|\overrightarrow{AB}\right|^2 = 3^2 + 6^2 + 12^2 = 189$$

$$\overrightarrow{BC} = \overrightarrow{OC} - \overrightarrow{OB} = \begin{pmatrix} 13 \\ 5 \\ 9 \end{pmatrix} - \begin{pmatrix} 1 \\ 9 \\ 10 \end{pmatrix} = \begin{pmatrix} 12 \\ -4 \\ -1 \end{pmatrix}$$

$$\Rightarrow \left|\overrightarrow{BC}\right|^2 = 12^2 + (-4)^2 + (-1)^2 = 161$$

$$\overrightarrow{AC} = \overrightarrow{OC} - \overrightarrow{OA} = \begin{pmatrix} 13 \\ 5 \\ 9 \end{pmatrix} - \begin{pmatrix} -2 \\ 3 \\ -2 \end{pmatrix} = \begin{pmatrix} 15 \\ 2 \\ 11 \end{pmatrix}$$

$$\Rightarrow \left|\overrightarrow{AC}\right|^2 = 15^2 + 2^2 + 11^2 = 350$$

$$\left|\overrightarrow{AB}\right|^2 + \left|\overrightarrow{BC}\right|^2 = 189 + 161$$
$$= 350$$
$$= \left|\overrightarrow{AC}\right|^2$$

> **Discussion point**
> Find the position vector of a fourth point D such that ABCD forms a parallelogram.
> → How many different answers are there?

That is,

$$\left|\overrightarrow{AB}\right|^2 + \left|\overrightarrow{BC}\right|^2 = \left|\overrightarrow{AC}\right|^2$$

so triangle ABC is right-angled and hence angle ABC is 90°, as required.

Exercise 12.2

① (i) ABCD in Figure 12.13 is a rectangle.

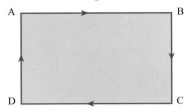

Figure 12.13

Show that $\overrightarrow{AB} + \overrightarrow{BC} + \overrightarrow{CD} + \overrightarrow{DA} = 0$.

(ii) Does the same result hold for any quadrilateral? Justify your answer.

② In the regular hexagon in Figure 12.14, $\overrightarrow{OA} = \mathbf{a}$, $\overrightarrow{OB} = \mathbf{b}$ and $\overrightarrow{OC} = \mathbf{c}$.

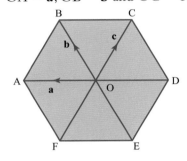

Figure 12.14

(i) Express the following vectors in terms of $\mathbf{a}$, $\mathbf{b}$ and $\mathbf{c}$.

(a) $\overrightarrow{EB}$ (b) $\overrightarrow{AC}$ (c) $\overrightarrow{AE}$ (d) $\overrightarrow{ED}$

(ii) Prove that $\mathbf{a} + \mathbf{c} - \mathbf{b} = 0$.

③ In the parallelogram in Figure 12.15, $\overrightarrow{OA} = \mathbf{a}$, $\overrightarrow{OC} = \mathbf{c}$ and M is the midpoint of AB.

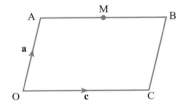

Figure 12.15

(i) Express the following vectors in terms of $\mathbf{a}$ and $\mathbf{c}$.

(a) $\overrightarrow{CB}$ (b) $\overrightarrow{OB}$ (c) $\overrightarrow{AC}$ (d) $\overrightarrow{CA}$
(e) $\overrightarrow{BO}$ (f) $\overrightarrow{AM}$ (g) $\overrightarrow{OM}$ (h) $\overrightarrow{MC}$

(ii) P is the midpoint of OB and Q is the midpoint of AC.
Find $\overrightarrow{OP}$ and $\overrightarrow{OQ}$. What theorem does your result prove?

④ In the cuboid in Figure 12.16, $\overrightarrow{OA} = \mathbf{p}$, $\overrightarrow{OE} = \mathbf{q}$ and $\overrightarrow{OG} = \mathbf{r}$.

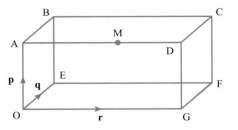

Figure 12.16

(i) Express the following vectors in terms of $\mathbf{p}$, $\mathbf{q}$ and $\mathbf{r}$.

(a) $\overrightarrow{GF}$ (b) $\overrightarrow{CF}$ (c) $\overrightarrow{OB}$
(d) $\overrightarrow{OD}$ (e) $\overrightarrow{OC}$

(ii) The point M divides AD in the ratio $3:2$.
Find $\overrightarrow{OM}$ in terms of $\mathbf{p}$, $\mathbf{q}$ and $\mathbf{r}$.

(iii) Use vectors to prove that OC and BG bisect each other.

⑤ A quadrilateral has vertices A, B, C and D at the points $(1, 2, 4)$, $(3, 5, 9)$, $(2, 9, 15)$ and $(-2, 3, 5)$ respectively. Use vectors to find out what type of quadrilateral ABCD is.

⑥ Relative to an origin O, the position vectors of the points A, B and C are given by

$$\overrightarrow{OA} = \begin{pmatrix} 2 \\ 1 \\ 3 \end{pmatrix}, \overrightarrow{OB} = \begin{pmatrix} -2 \\ 4 \\ 3 \end{pmatrix} \text{ and } \overrightarrow{OC} = \begin{pmatrix} -1 \\ 2 \\ 1 \end{pmatrix}.$$

Find the perimeter of triangle ABC.

⑦ In Figure 12.17, $\overrightarrow{OA} = \mathbf{a}$ and $\overrightarrow{OB} = \mathbf{b}$.
$\overrightarrow{OA} : \overrightarrow{OX} = \overrightarrow{OB} : \overrightarrow{OY} = 3:2$ and
$\overrightarrow{OA} : \overrightarrow{OP} = \overrightarrow{OB} : \overrightarrow{OQ} = 2:3$.

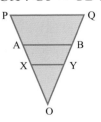

Figure 12.17

(i) Find $\overrightarrow{PQ}$ in terms of $\mathbf{a}$ and $\mathbf{b}$.

(ii) Prove that the vectors $\overrightarrow{AB}$, $\overrightarrow{XY}$ and $\overrightarrow{PQ}$ are parallel.

(iii) Find $\overrightarrow{PQ} : \overrightarrow{XY}$.

⑧ Relative to an origin O, the position vectors of the points A, B, C and D are given by

$$\overrightarrow{OA} = \begin{pmatrix} 3 \\ 1 \\ 5 \end{pmatrix}, \quad \overrightarrow{OB} = \begin{pmatrix} 5 \\ 5 \\ 13 \end{pmatrix},$$

$$\overrightarrow{OC} = \begin{pmatrix} 8 \\ 2 \\ 7 \end{pmatrix} \text{ and } \overrightarrow{OD} = \begin{pmatrix} 6 \\ -2 \\ -1 \end{pmatrix}.$$

Use vectors to prove that ABCD is a parallelogram.

⑨ Figure 12.18 shows a trapezium where $\overrightarrow{OA} = \mathbf{p}$, $\overrightarrow{OC} = \mathbf{q}$ and $\overrightarrow{AB} = 2\mathbf{q}$.
D and E are the midpoints of AC and OB respectively.

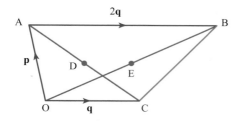

Figure 12.18

Use vectors to prove that $\overrightarrow{DE} = \frac{1}{2}\mathbf{q}$.

⑩ Relative to an origin O, the position vectors of the points A, B and C are given by

$$\overrightarrow{OA} = \begin{pmatrix} -2 \\ 3 \\ 5 \end{pmatrix}, \quad \overrightarrow{OB} = \begin{pmatrix} 0 \\ 7 \\ 3 \end{pmatrix}$$

$$\text{and } \overrightarrow{OC} = \begin{pmatrix} 3 \\ 8 \\ 8 \end{pmatrix}.$$

(i) Find $\overrightarrow{OD}$ such that ABCD is a parallelogram with sides AB, BC, CD and DA.

(ii) Prove that ABCD is, in fact, a rectangle.

(iii) X is the centre of the rectangle ABCD.
Find $\overrightarrow{OX}$.

⑪ In Figure 12.19, $\overrightarrow{OA} = \mathbf{a}$, $\overrightarrow{OB} = \mathbf{b}$, $\overrightarrow{BD} = \frac{1}{2}\mathbf{a}$ and the point C divides AB in the ratio 2:1.

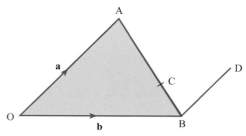

Figure 12.19

Prove that OCD is a straight line and find the ratio OC:OD.

⑫ Relative to an origin O, the points P and Q have position vectors

$$\overrightarrow{OP} = \begin{pmatrix} 2 \\ 6 \\ 4 \end{pmatrix} \quad \text{and} \quad \overrightarrow{OQ} = \begin{pmatrix} -1 \\ 2 \\ -3 \end{pmatrix}.$$

(i) The point M is such that $\overrightarrow{PM} = 3\overrightarrow{MQ}$. Find the vector $\overrightarrow{OM}$.

(ii) The point N lies on PQ such that PN:NQ is 2:3. Find the vector $\overrightarrow{ON}$.

⑬ Relative to an origin O, the position vectors of the points P and Q given by
$\overrightarrow{OP} = 3\mathbf{i} + \mathbf{j} + 4\mathbf{k}$ and
$\overrightarrow{OQ} = \mathbf{i} + a\mathbf{j} - 2\mathbf{k}$.

Find the values of a for which the magnitude of PQ is 7.

⑭ Relative to an origin O, the points A and B have position vectors $\mathbf{a}$ and $\mathbf{b}$.

Prove that when a point C divides AB in the ratio $s:t$ then

$$\overrightarrow{OC} = \frac{t}{s+t}\mathbf{a} + \frac{s}{s+t}\mathbf{b}.$$

⑮ Relative to an origin O, the position vectors of the points A, B, C and D are given by

$$\overrightarrow{OA} = \begin{pmatrix} 1 \\ 0 \\ 0 \end{pmatrix}, \quad \overrightarrow{OB} = \begin{pmatrix} 0 \\ 1 \\ 0 \end{pmatrix}$$

$$\overrightarrow{OC} = \begin{pmatrix} 0 \\ 0 \\ 1 \end{pmatrix} \quad \text{and} \quad \overrightarrow{OD} = \begin{pmatrix} 0 \\ 0 \\ -1 \end{pmatrix}.$$

Is ABCD a quadrilateral? Justify your answer.

LEARNING OUTCOMES

When you have completed this chapter, you should be able to:

➤ use vectors in three dimensions

➤ calculate the magnitude of a vector in three dimensions

➤ perform the algebraic operations of vector addition and multiplication by scalars

➤ understand the geometrical interpretations of vector addition and multiplication by scalars

➤ use vectors to solve problems in pure mathematics

➤ use vectors to solve problems in context such as forces.

KEY POINTS

1 A vector quantity has magnitude and direction.

2 A scalar quantity has magnitude only.

3 Vectors are typeset in **bold**, **a** or **OA**, or in the form $\overrightarrow{OA}$.
   They are handwritten either in the underlined form $\underline{a}$, or as $\overrightarrow{OA}$.

4 Unit vectors in the x, y and z directions are denoted by **i**, **j** and **k**, respectively.

5 A vector in two dimensions may be specified in
   - magnitude–direction form, as (r, θ)
   - component form, as $a\mathbf{i} + b\mathbf{j}$ or $\begin{pmatrix} a \\ b \end{pmatrix}$.

6 A vector in three dimensions is usually written in component form, as
   $a\mathbf{i} + b\mathbf{j} + c\mathbf{k}$ or $\begin{pmatrix} a \\ b \\ c \end{pmatrix}$.

7 The resultant of two (or more) vectors is found by the sum of the vectors.
   A resultant vector is usually denoted by a double-headed arrow.

8 The position vector $\overrightarrow{OP}$ of a point P is the vector joining the origin, O, to P.

9 The vector $\overrightarrow{AB}$ is **b** − **a**, where **a** and **b** are the position vectors of A and B.

10 The length (or modulus or magnitude) of the vector **r** is written as r or as $|\mathbf{r}|$.
   $\mathbf{r} = a\mathbf{i} + b\mathbf{j} + c\mathbf{k} \Rightarrow |\mathbf{r}| = \sqrt{a^2 + b^2 + c^2}$

11 A unit vector in the *same* direction as $\mathbf{r} = a\mathbf{i} + b\mathbf{j}$ is
   $\dfrac{a}{\sqrt{a^2 + b^2}}\mathbf{i} + \dfrac{b}{\sqrt{a^2 + b^2}}\mathbf{j}$.

FUTURE USES

■ You will use vectors to represent displacement, velocity and acceleration in Chapter 18.

■ You will use vectors to represent force in Chapter 19.

■ If you study Further Mathematics, you will learn about how vectors can be used to solve problems involving lines and planes.

13 Differential equations

➔ How long do you have to wait for a typical cup of coffee to be drinkable?

➔ How long does it take to go cold?

➔ What do the words 'drinkable' and 'cold' mean in this context?

Newton's law of cooling states that the rate of change of the temperature of an object is proportional to the difference between the object's temperature and the temperature of its surroundings.

This leads to the equation

$$\frac{dT}{dt} = -k(T - T_0),$$ where T is the temperature of the object at time t, T_0 is the temperature of the surroundings, and k is a constant of proportionality.

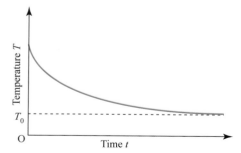

Figure 13.1

This is an example of a **differential equation**.

To be able to predict the temperature of the object at different times, you need to solve the differential equation. In this chapter, you will learn how to solve problems like this involving rates of change.

1 First order differential equations

A **differential equation** is an equation involving at least one derivative. The **order** of the differential equation is the order of the highest derivative occurring in the equation.

Discussion point

➜ What is the order of each of the following differential equations?

(i) $\dfrac{dy}{dx} = e^x + 3$

(ii) $3\dfrac{dy}{dx} - 4x\dfrac{d^3y}{dx^3} = y^5$

(iii) $4x^2 = 3x\dfrac{dy}{dx} - 2y$

(iv) $\dfrac{d^2y}{dx^2} = 4x^2$

The solution to a differential equation that contains one or more arbitrary constants is called a **general solution**, and represents a family of solution curves. If you are given additional information, you can find a **particular solution**, which represents one specific member of the family of solution curves.

Because an arbitrary constant comes into your solution each time you integrate, the number of arbitrary constants in a general solution will be the same as the order of the differential equation.

The simplest differential equations are first order, of the form $\dfrac{dy}{dx} = f(x)$; they can be solved by simply integrating with respect to x. You have actually already solved some differential equations of this type, when you have used the gradient function of a curve to find the general equation of the family of solution curves, and in some cases the equation of a particular curve from this family.

Example 13.1

(i) Find the general solution of the differential equation $\dfrac{dy}{dx} = 2x$.

(ii) Find the particular solution that passes through the point $(1, 2)$.

(iii) Sketch several members of the family of solution curves, and indicate the particular solution from (ii) on your sketch.

Solution

(i) $\dfrac{dy}{dx} = 2x$

$\Rightarrow\quad y = \int 2x\,dx$ ⟵ Integrate both sides of the equation.

$\Rightarrow\quad y = x^2 + c$

The general solution is $y = x^2 + c$.

(ii)   When $x = 1, y = 2$

$\Rightarrow 2 = 1^2 + c$ ◀— | Substitute values into the general solution.

$\Rightarrow c = 1$

The particular solution is $y = x^2 + 1$.

(iii)   Figure 13.2 shows a set of solution curves. The particular solution in (ii) is the red curve.

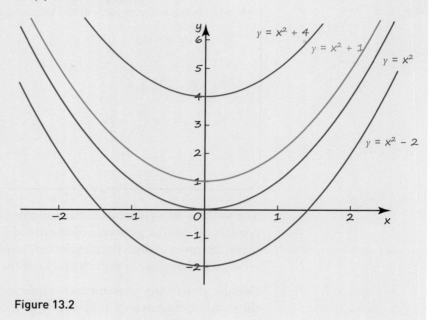

Figure 13.2

Forming differential equations

Differential equations are used to model rates of change. They open up new opportunities to you, extending the situations that you can work with in both pure and applied mathematics, but first you need to become familiar with the vocabulary and notation you will need.

For example, the rate of change of temperature, T, with respect to distance, x, is written as $\frac{dT}{dx}$.

| 'Rate of change' by itself means with respect to time, t.

The rate of change of velocity is written as $\frac{dv}{dt}$. This is also sometimes written as $\dot{v}$; by convention a dot above a variable indicates its rate of change with respect to time.

Some of the situations you meet in this chapter involve motion along a straight line, and so you will need to know the meanings of the associated terms.

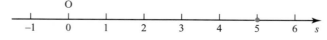

Figure 13.3

The **position** of an object (+5 in Figure 13.3) is its distance (or displacement) from the origin O in the direction you have chosen to define as being positive.

The term 'speed' is sometimes used informally to mean velocity along a line.

The rate of change of position of the object with respect to time is its **velocity**, v, and this can take positive or negative values according to whether the object is moving away from the origin or towards it.

$$v = \frac{dr}{dt}$$

The rate of change of an object's velocity with respect to time is called its **acceleration**, a.

$$a = \frac{dv}{dt}$$

Velocity and acceleration are vector quantities but in one-dimensional motion there is no choice in direction, only in sense (i.e. whether positive or negative). Consequently, as you may already have noticed, the conventional bold type for vectors is not used in this chapter.

Example 13.2

An object is moving through a liquid so that its acceleration is proportional to its velocity at any given instant. When it has a velocity of $5\,\mathrm{m\,s^{-1}}$, the velocity is decreasing at a rate of $1\,\mathrm{m\,s^{-2}}$.

Find the differential equation to model this situation.

Solution

$$\frac{dv}{dt} \propto v$$

This means 'the rate of change of the velocity is proportional to the velocity'.

$$\Rightarrow \frac{dv}{dt} = kv$$

When $v = 5$, $\frac{dv}{dt} = -1$, so

Use the additional information given to find the value of k.

$$-1 = 5k$$

Notice that because the velocity is decreasing, the constant of proportionality, k, turns out to be negative.

$$\Rightarrow k = -\frac{1}{5}$$

Therefore the differential equation is $\frac{dv}{dt} = -\frac{1}{5}v$.

Example 13.3

A model is proposed in which the temperature within a star decreases with respect to the distance, x, from the centre of the star at a rate which is inversely proportional to the square of this distance.

(i)   Express this model as a differential equation.

(ii)   Find the general solution of the differential equation.

(iii)   What happens when $x = 0$?

→

Solution

(i) In this example the rate of change of temperature is not with respect to time but with respect to the distance from the centre of the star. If T represents the temperature of the star and x the distance from the centre of the star, the rate of change of temperature with respect to distance may be written as $\dfrac{dT}{dx}$.

So $\dfrac{dT}{dx} \propto \dfrac{1}{x^2}$

$\Rightarrow \dfrac{dT}{dx} = \dfrac{k}{x^2}$ ← Because the rate of change is decreasing, not increasing, k must be a negative constant.

(ii) $\dfrac{dT}{dx} = \dfrac{k}{x^2}$

$\Rightarrow T = -\dfrac{k}{x} + c$

(iii) When $x = 0$ (at the centre of the star) the model predicts that the temperature is infinitely large. So the model must break down near the centre of the star, otherwise it would be infinitely hot there.

Exercise 13.1

① Given the differential equation $\dfrac{dy}{dx} = 3x^2$

　(i) find the general solution and sketch the family of solution curves

　(ii) find the particular solution for which $y = -1$ when $x = 1$, and indicate this solution on your sketch.

② The differential equation $\dfrac{dv}{dt} = 5v^2$ models the motion of a particle, where v is the velocity of the particle in $m\,s^{-1}$ and t is the time in seconds.

Explain the meaning of $\dfrac{dv}{dt}$ and what the differential equation tells you about the motion of the particle.

PS ③ The rate of increase in the number of bacteria in a colony, N, is proportional to the number of bacteria present.

Form a differential equation to model this situation.

④ State which is the odd one out among the following and give your reasons.

　(i) The rate of change of y with respect to x.

　(ii) $\dot{y}$

　(iii) $\dfrac{dy}{dx}$

　(iv) The gradient of the curve in a graph of y against x.

⑤ Given the differential equation $\dfrac{dy}{dx} = 2x - 4$

　(i) find the general solution and sketch the family of solution curves

　(ii) find the particular solution for which $y = 0$ when $x = 2$, and indicate this solution on your sketch.

⑥ After a major advertising campaign, an engineering company finds that its profits are increasing at a rate proportional to the square root of the profits at any given time.

Form a differential equation to model this situation.

⑦ A moving object has velocity $v\,\mathrm{m\,s^{-1}}$.
When $v \geqslant 0$, the acceleration of the object
is inversely proportional to the square root of
its velocity. When the velocity is $4\,\mathrm{m\,s^{-1}}$, the
acceleration is $2\,\mathrm{m\,s^{-2}}$.

Form a differential equation to model this
situation.

⑧ A poker which is 80 cm long has one end in
a fire. The temperature of the poker decreases
with respect to the distance from that end at
a rate proportional to that distance. Halfway
along the poker, the temperature is decreasing
at a rate of $10°\mathrm{C\,cm^{-1}}$.

Form a differential equation to model this
situation.

⑨ Given the differential equation $\dfrac{\mathrm{d}y}{\mathrm{d}x} = -4\mathrm{e}^{-2x}$

(i) find the general solution and sketch three
members of the family of solution curves

(ii) find the equation of the particular solution
which passes through the point $(\ln 2, 0)$.

⑩ A cup of tea cools at a rate proportional to the
difference between the temperature of the tea
and that of the surrounding air. Initially, the tea
is at a temperature of $95°\mathrm{C}$ and it is cooling
at a rate of $0.5°\mathrm{C\,s^{-1}}$. The surrounding air is
at $15°\mathrm{C}$.
Model this situation as a differential equation.

⑪ A bonfire is held in a field. It burns a circle of
grass of radius 8 metres. After the fire is over,
the grass grows back from the circumference of
the circle inwards. The radius, r m, of the circle
without any grass decreases at a rate proportional
to the square root of the time, t weeks, since the
bonfire. One week after the bonfire, the grass is
growing back at a rate of 1.5 metres per week.

(i) Form a differential equation to model this
situation.

(ii) Solve the differential equation.

(iii) Find how long it takes the grass to grow
back completely.

⑫ The acceleration of a particle is inversely
proportional to $t + 2$, where t is the time,
measured in seconds. The particle starts from
rest with an initial acceleration of $1\,\mathrm{m\,s^{-2}}$.

(i) Find an expression for $\dfrac{\mathrm{d}v}{\mathrm{d}t}$ to model the
motion.

(ii) Show that the velocity of the particle after
$6\,\mathrm{s}$ is $(6 \ln 2)\,\mathrm{m\,s^{-1}}$.

⑬ The mass of a pumpkin t weeks after it first forms
is M kg. The rate of change of M is modelled by
the quadratic graph in Figure 13.4.

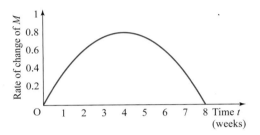

Figure 13.4

(i) Represent this information in the form of
a differential equation.

(ii) Find the greatest mass of the pumpkin.

⑭ A spherical balloon is allowed to deflate.
The rate at which air is leaving the balloon
is proportional to the volume V of air left in
the balloon. When the radius of the balloon
is 15 cm, air is leaving at a rate of $8\,\mathrm{cm^3\,s^{-1}}$.
Show that
$$\frac{\mathrm{d}V}{\mathrm{d}t} = -\frac{2V}{1125\pi}.$$

⑮ Liquid is being poured into a large vertical
circular cylinder of radius 40 cm at a constant
rate of $2000\,\mathrm{cm^3\,s^{-1}}$, and is leaking out of a
hole in the base at a rate proportional to the
square root of the height of the liquid already
in the cylinder. Show that the rate of change of
the height of the liquid in the cylinder can be
modelled by
$$\frac{\mathrm{d}h}{\mathrm{d}t} = \frac{5}{4\pi} - c\sqrt{h}.$$

(16) The height of a tree is h metres when the tree is t years old. For the first 10 years of the life of the tree, $\frac{dh}{dt} = 0.5$. For the rest of the tree's life, its rate of growth is inversely proportional to its age.

(i) Describe the growth of the tree during its first 10 years. What is its height when it is 10 years old?

There is no sudden change in its rate of growth when the tree is exactly 10 years old.

(ii) Form a differential equation for the tree's rate of growth for $t > 10$, i.e. for the rest of the tree's life.

(iii) Solve the differential equation you obtained in part (ii) to show that for $t > 10$,

$$h = 5 + 5\ln\left(\frac{t}{10}\right).$$

The tree dies when its height is 15 m.

(iv) How old is the tree when it dies?

(v) Sketch the graph of h against t for the lifetime of the tree.

2 Solving differential equations by separating the variables

It is not difficult to solve a differential equation like $\frac{dy}{dx} = 3x^2 - 2$, because the right-hand side is a function of x only. So long as the function can be integrated, the equation can be solved.

Now look at the differential equation $\frac{dy}{dx} = y^2$.

This cannot be solved directly by integration, because the right-hand side is a function of y. Instead, a method called **separating the variables** is used. This is shown in the next example.

Example 13.4

(i) Find the general solution of the differential equation $\frac{dy}{dx} = y^2$, and sketch three members of the family of solution curves.

(ii) Find the particular solution for which $y = 1$ when $x = 0$.

Solution

(i)
$$\frac{1}{y^2}\frac{dy}{dx} = 1$$ — Start by rewriting the equation so that the right-hand side does not involve y.

$$\Rightarrow \int \frac{1}{y^2}\frac{dy}{dx}\,dx = \int 1\,dx$$ — Integrate both sides with respect to x.

$$\Rightarrow \int \frac{1}{y^2}\,dy = \int 1\,dx$$ — Replace $\frac{dy}{dx}\,dx$ on the left-hand side with dy.

$$\Rightarrow -y^{-1} = x + c$$

$$\Rightarrow y = -\frac{1}{x + c}$$ — Rearrange to make y the subject.

This is the general solution. Three members of the family of solution curves are shown in Figure 13.5.

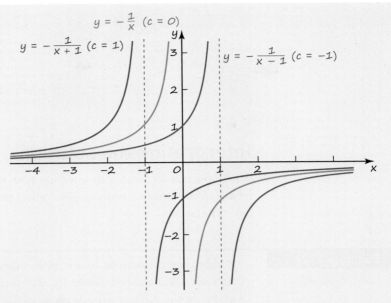

Figure 13.5

(ii) $y = 1$ when $x = 0$, so

$$1 = -\frac{1}{c} \longleftarrow \boxed{\text{Substitute values into } y = -\dfrac{1}{x + c}}$$

$$\Rightarrow c = -1$$

The particular solution is $y = -\dfrac{1}{x - 1}$ or $y = \dfrac{1}{1 - x}$.

This is the blue curve in Figure 13.5.

Note that the first part of this process, going from $\dfrac{dy}{dx} = y^2$ to $\displaystyle\int \dfrac{1}{y^2}\, dy = \int x\, dx$,

is usually done in one step. You can think of this process as rearranging $\dfrac{dy}{dx}$ as

though it were a fraction, to end up with all the y terms on one side and the x terms on the other, and then inserting integration signs (dx and dy must both end up in the top line).

In the next example, the function on the right-hand side of the differential equation is a function of both x and y.

Example 13.5

Find, for $y > 0$, the general solution of the differential equation $\dfrac{dy}{dx} = xy$.

Solution

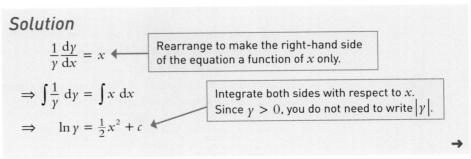

$$\frac{1}{y}\frac{dy}{dx} = x \longleftarrow \boxed{\begin{array}{l}\text{Rearrange to make the right-hand side} \\ \text{of the equation a function of } x \text{ only.}\end{array}}$$

$$\Rightarrow \int \frac{1}{y}\, dy = \int x\, dx \qquad \boxed{\begin{array}{l}\text{Integrate both sides with respect to } x. \\ \text{Since } y > 0, \text{ you do not need to write } |y|.\end{array}}$$

$$\Rightarrow \quad \ln y = \tfrac{1}{2}x^2 + c$$

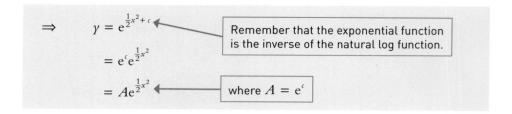

$$\Rightarrow \qquad y = e^{\frac{1}{2}x^2 + c}$$

Remember that the exponential function is the inverse of the natural log function.

$$= e^c e^{\frac{1}{2}x^2}$$

$$= A e^{\frac{1}{2}x^2}$$

where $A = e^c$

Interpreting the solution

It is important to be able to interpret how the solution of a differential equation relates to the original problem, including recognising that the model used may have limitations.

Example 13.6

Rabbits are introduced to a remote island and the size of the population increases. A suggested model for the number of rabbits, N, after t years, is given by the differential equation

$$\frac{dN}{dt} = kN$$

where $k > 0$.

(i)   Find the general solution for N in terms of t and k.

(ii) Find the particular solution for which $N = 10$ when $t = 0$.

(iii) What will happen to the number of rabbits when t becomes very large? Why is this not a realistic model for an actual population of rabbits? What would you expect to happen to the graph of N against t in a real population of rabbits as t becomes very large?

Solution

(i) $$\int \frac{1}{N} \, dN = \int k \, dt$$

$$\Rightarrow \qquad \ln N = kt + c$$

Since the number of rabbits must be positive, writing $|N|$ is not necessary.

$$\Rightarrow \qquad N = A e^{kt}$$

(ii) $N = 10$ when $t = 0$

$$\Rightarrow 10 = A e^0$$

$$\Rightarrow A = 10$$

So the particular solution is $N = 10 e^{kt}$.

(iii) As t becomes very large, $N \to \infty$. This is not a realistic model because there would be limitations on how large an actual population of rabbits would get, due to factors such as food and predators. In a real population, the graph of N against t would flatten out as t became very large.

> **Note**
>
> The differential equation $\dfrac{\mathrm{d}N}{\mathrm{d}t} = kN(1 - cN)$ is commonly used to model
>
> population growth. It has the general solution $N = \dfrac{Ae^{kt}}{1 + Ace^{kt}}$, which has
>
> the shape shown in Figure 13.6 for $t \geqslant 0$.
>
>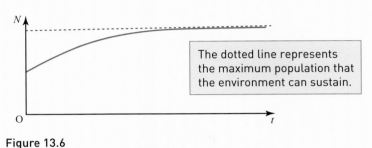
>
> The dotted line represents the maximum population that the environment can sustain.
>
> **Figure 13.6**

Exercise 13.2

① Make y the subject of $\ln y = \frac{1}{3} x^3, y > 0$

② Find the general solution of each of the following differential equations by separating the variables, expressing y in terms of x.

  (i) $\dfrac{\mathrm{d}y}{\mathrm{d}x} = \dfrac{x}{y^2}$

  (ii) $\dfrac{\mathrm{d}y}{\mathrm{d}x} = xy^2$

③ Find the particular solutions of each of the following differential equations.

  (i) $\dfrac{\mathrm{d}y}{\mathrm{d}x} = xe^{-y}$, given that $y = 0$
    when $x = 0$

  (ii) $\dfrac{\mathrm{d}y}{\mathrm{d}x} = y^2 \sin x$, given that $y = 1$
    when $x = 0$

④ A mathematical model for the number of bacteria, n, in a culture states that n is increasing at a rate proportional to the number present. At 10:00 there are 5000 bacteria and at 10:30 there are 7000.

  At what time, to the nearest minute, does the model predict 10 000 bacteria?

PS ⑤ An object is moving so that its velocity $v \left(= \dfrac{\mathrm{d}s}{\mathrm{d}t}\right)$ is inversely proportional to its displacement s from a fixed point. Its velocity is $1\,\mathrm{m\,s}^{-1}$ when its displacement is $2\,\mathrm{m}$.

Form a differential equation to model the situation. Find the general solution of your differential equation.

⑥ A cold liquid is standing in a warm room. The temperature of the liquid is $\theta°\mathrm{C}$, where $\theta < 20$; it obeys the differential equation

$$\frac{\mathrm{d}\theta}{\mathrm{d}t} = 2(20 - \theta)$$

where the time t is measured in minutes.

  (i) Find the general solution of this differential equation.

  (ii) Find the particular solution for which $\theta = 5$ when $t = 0$.

  (iii) For this particular solution, sketch the graph of temperature against time.

  (iv) In this case, how long, to the nearest second, does the liquid take to reach a temperature of $18°\mathrm{C}$?

  (v) What happens to the temperature of the liquid in the long term?

⑦ Given that

$$\frac{\mathrm{d}y}{\mathrm{d}x} = \frac{x\left(y^2 + 1\right)}{y\left(x^2 + 1\right)}$$

and that $y = 2$ when $x = 1$, find y as a function of x.

⑧ (i) Express $\dfrac{2x-1}{(x-1)(2x-3)}$ in partial fractions.

(ii) Given that $x \geqslant 2$, find the general solution of the differential equation

$$(2x-3)(x-1)\frac{\mathrm{d}y}{\mathrm{d}x} = (2x-1)y.$$

⑨ Water is stored in a tank, with a tap 5 cm above the base of the tank. When the tap is turned on, the flow of water out of the tank is modelled by the differential equation

$$\frac{\mathrm{d}h}{\mathrm{d}t} = -3\sqrt{(h-5)}$$

where h cm is the height of water in the tank, and t is the time in minutes. Initially the height of water in the tank is 105 cm.

(i) Find an expression for h in terms of t.

(ii) Explain what happens when $h = 5$.

(iii) Find, to the nearest second, how long it takes the depth of water in the tank to fall to a height of 40 cm.

⑩ A patch of oil pollution in the sea is approximately circular in shape. When first seen, its radius was 100 m and was increasing at a rate of 0.5 m per minute. At a time t minutes later, its radius is r metres. An expert believes that, if the patch is untreated, its radius will increase at a rate which is proportional to $\dfrac{1}{r^2}$.

(i) Write down a differential equation for this situation, using a constant of proportionality, k.

(ii) Using the initial conditions, find the value of k. Hence calculate the expert's prediction of the radius of the oil patch after 2 hours.

The expert thinks that if the oil patch is treated with chemicals, then its radius will increase at a rate which is proportional to $\dfrac{1}{r^2(2+t)}$.

(iii) Write down a differential equation for this new situation and, using the same initial conditions as before, find the value of the new constant of proportionality.

(iv) Calculate the expert's prediction of the radius of the treated oil patch after 2 hours.

PS ⑪ The acceleration of an object is inversely proportional to its velocity at any given time, and the direction of motion is taken to be positive. When the velocity is $1\,\mathrm{m\,s^{-1}}$, the acceleration is $3\,\mathrm{m\,s^{-2}}$.

(i) Find a differential equation to model this situation.

(ii) Find the particular solution to this differential equation for which the initial velocity is $2\,\mathrm{m\,s^{-1}}$.

(iii) In this case, how long does the object take to reach a velocity of $8\,\mathrm{m\,s^{-1}}$?

⑫ To control the pests inside a large greenhouse, 600 ladybirds are introduced. After t days there are P ladybirds in the greenhouse. In a simple model, P is assumed to be a continuous variable satisfying the differential equation

$$\frac{\mathrm{d}P}{\mathrm{d}t} = kP, \text{ where } k \text{ is a constant.}$$

(i) Solve the differential equation, with initial condition $P = 600$ when $t = 0$, to express P in terms of k and t.

Observations of the number of ladybirds (estimated to the nearest hundred) were as shown in Table 13.1.

Table 13.1

t	0	150	250
P	600	1200	1500

(ii) Show that $P = 1200$ when $t = 150$ implies that $k \approx 0.00462$. Show that this is not consistent with the observed value when $t = 250$.

In a refined model, allowing for seasonal variations, it is assumed that P satisfies the differential equation

$$\frac{\mathrm{d}P}{\mathrm{d}t} = P(0.005 - 0.008\cos(0.02t))$$

with initial condition $P = 600$ when $t = 0$.

(iii) Solve this differential equation to express P in terms of t, and comment on how well this fits with the data given in Table 13.1.

(iv) Show that, according to the refined model, the number of ladybirds will decrease initially, and find the smallest number of ladybirds in the greenhouse.

⑬ The relationship between the price of a commodity, p, and demand for the commodity, q, is modelled by the differential equation

$$\frac{\mathrm{d}q}{\mathrm{d}p} = -\eta \frac{q}{p}$$

where η is called the elasticity, and is a constant for a given commodity in a particular set of conditions.

(i) Find the general solution for q in terms of p.

(ii) When a particular retailer increases the price of a DVD from £15 to £20, the demand falls from 100 a month to 80 a month. For this case

(a) calculate the value of the elasticity

(b) find the particular solution of the differential equation

(c) sketch the graph of demand against price.

PS ⑭ The rate of increase of a population is modelled as being directly proportional to the size of the population, P.

(i) Form a differential equation to describe this situation.

(ii) Given that the initial population is P_0, and the initial rate of increase of the population per day is twice the initial population, solve the differential equation, to find P in terms of P_0 and t.

(iii) Find the time taken, to the nearest minute, for the population to double.

(iv) In an improved model, the population growth is modelled by the differential equation

$$\frac{\mathrm{d}P}{\mathrm{d}t} = kP\cos kt$$

where P is the population, t is the time measured in days and k is a positive constant. Given that the initial population is again P_0 and the initial rate of increase of the population per day is again twice the initial population, solve the second differential equation to find P in terms of P_0 and t.

(v) Find the time taken, to the nearest minute, for the population to double, using this new model.

(vi) For each of the models, describe how the population varies with time.

⑮ (i) Show that the differential equation

$$\frac{\mathrm{d}N}{\mathrm{d}t} = kN(1 - cN) \text{ has the general}$$

solution $N = \dfrac{Ae^{kt}}{1 + Ace^{kt}}$.

(ii) Under this model, what is the limiting value of the population size?

LEARNING OUTCOMES

When you have completed this chapter, you should be able to:
- ➤ construct simple differential equations in pure mathematics and in context
- ➤ evaluate the analytical solution of simple first order differential equations with separable variables including finding particular solutions
- ➤ separate variables when it requires factorisation involving a common factor
- ➤ interpret the solution of a differential equation in the context of solving a problem:
  - ○ identifying limitations of the solution
  - ○ identifying links to kinematics.

KEY POINTS

1 Differential equations are used to model rates of change.

2 $\dfrac{\mathrm{d}y}{\mathrm{d}x}$ is the rate of change of y with respect to x.

   $\dfrac{\mathrm{d}z}{\mathrm{d}t}$ is the rate of change of z with respect to t.

   If the words 'with respect to ...' are omitted, the change is assumed to be with respect to time.

3 A differential equation involves derivatives such as $\dfrac{\mathrm{d}y}{\mathrm{d}x}$ and $\dfrac{\mathrm{d}^2 y}{\mathrm{d}x^2}$.

4 A first order differential equation involves a first derivative only.

5 Some first order differential equations may be solved by direct integration.

6 Some first order differential equations may be solved by separating the variables.

7 A general solution is one in which the constant of integration is left in the solution, and a particular solution is one in which additional information is used to calculate the constant of integration.

8 A general solution may be represented by a family of curves, and a particular solution by a particular member of that family.

FUTURE USES

■ If you study Further Mathematics you will learn to model further situations using differential equations, and to solve a wider range of differential equations.

14 Numerical methods

It is the true nature of mankind to learn from his mistakes.
Fred Hoyle (1915–2001)

A golfer doesn't often hit a ball into the hole at the first attempt! Instead, he or she will try to hit the ball as close to the hole as possible. After that, successive attempts will usually be closer and closer to the hole, until the ball finally lands in the hole.

→ Think of some other situations where you need to make a rough approximation for your first attempt, and then gradually improve your attempts.

1 Solving equations numerically

ACTIVITY 14.1

Which of the following equations can be solved algebraically, and which cannot?
For the equations that can be solved algebraically, find the exact roots. ◄

Remember that exact roots may be given in terms of irrational numbers like π, e or $\sqrt{2}$.

For the equations that cannot be solved algebraically, use a graphical calculator or graphing software to find approximate roots.

(i) $x^2 - 4x + 3 = 0$ (ii) $x^2 + 10x + 8 = 0$ (iii) $x^5 - 5x + 3 = 0$

(iv) $x^3 - x = 0$ (v) $e^x - 4x = 0$

Although you could not solve the equations $x^5 - 5x + 3 = 0$ and $e^x - 4x = 0$ algebraically, you were still able to find approximate values for the roots, and, by zooming in more closely, you could have increased the accuracy of your approximations.

Figure 14.1 shows the graph of $y = e^x - 4x$.

From the graph, you can see that there is one root between 0 and 1, and another between 2 and 3.

Zooming in more closely shows that the first root is between 0 and 0.5, and the second between 2 and 2.5.

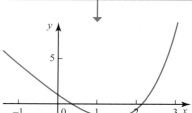
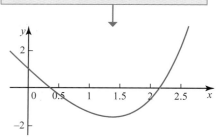

Figure 14.1

Discussion points

Here is the same graph, zoomed in further (Figure 14.2).

→ What can you say about each of the roots now?

→ From Figure 14.2, can you state the roots correct to 1 decimal place?

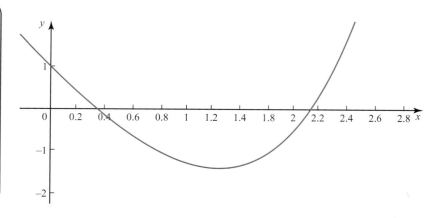

Figure 14.2

A **root** of an equation is a value which satisfies the equation. An equation may have more than one root. For the equation $x^2 + x - 6 = 0$, $x = 2$ is a root and so is $x = -3$.

The **solution** of an equation is the set of all the roots. So the solution of the equation $x^2 + x - 6 = 0$ is $x = 2$ and $x = -3$.

An **analytical** method for solving an equation is an algebraic method which gives exact values for the roots of the equation.

A **numerical method** for solving an equation does not give exact values for the roots of the equation, but it can usually find an approximate value to any required degree of accuracy.

Numerical methods permit you to solve many mathematical problems that cannot be solved analytically. Methods like these do not give you an exact answer, but by repeated application or refinement, they can usually give you a solution to any degree of accuracy that you require.

■ Only use numerical methods when algebraic methods are not available. If you can solve an equation algebraically (e.g. a quadratic equation), that is usually the best method to use.

■ Before starting to use a calculator or computer software, always start by drawing a sketch graph of the function whose equation you are trying to solve. This will show you how many roots the equation has and their approximate positions. It will also warn you of possible difficulties with particular methods. When using a graphing calculator or graphing software ensure that the range of values of x is sufficiently large to find all the roots.

■ Always give a statement about the accuracy of an answer (e.g. to 5 decimal places, or $\pm 0.000\,005$). An answer obtained by a numerical method is worthless without this.

■ The fact that at some point your calculator display reads, say, 1.6764705882 does not mean that all these figures are valid.

■ Your statement about the accuracy must be obtained from within the numerical method itself. Usually you find a sequence of estimates of ever-increasing accuracy.

■ Remember that the most suitable method for one equation may not be the most suitable for another.

Change of sign methods

Suppose you are looking for the roots of the equation $f(x) = 0$. This means that you want the values of x for which the graph of $y = f(x)$ crosses the x-axis.

As the curve crosses the x-axis, $f(x)$ changes sign, so provided that $f(x)$ is a continuous function (its graph has no asymptotes or other breaks in it), once you have located an interval in which $f(x)$ changes sign, you know that that interval must contain a root (Figure 14.3 overleaf).

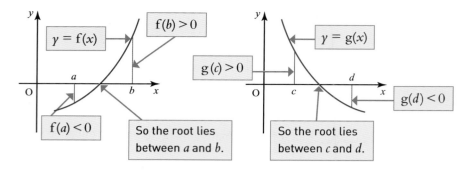

Figure 14.3

! Notice that $|g(d)| < |g(c)|$, but this does not mean that the root is closer to d than c.

Example 14.1

(i) Show that the equation $e^x - 4x = 0$ has a root in the interval $[2.1, 2.2]$.

(ii) How would you use decimal search to find the root to 2 decimal places?

Solution

> The equation is already in the required form of $f(x) = 0$.

(i) $f(x) = e^x - 4x$

> Evaluate the function at the beginning and end of the interval

$f(2.1) = -0.23... < 0$

$f(2.2) = 0.22... > 0$

> Don't be tempted to leave out this final sentence; it's an important part of your answer

Since one of the values is positive and one is negative, and the function is continuous, $e^x - 4x = 0$ has a root in the interval $[2.1, 2.2]$.

(ii) You would need to try the values $f(2.11), f(2.12), f(2.13)$, etc., until you reach a change of sign. This would tell you that the root is in the interval $[2.15, 2.16]$. You would then repeat the process by finding $f(2.151)$, $f(2.152)$ etc., until you reach a change of sign. This would show that the root is closer to 2.15 than to 2.16. So the root is 2.15 to 2 d.p.

Discussion points

➜ Is decimal search an efficient method for solving an equation
  (i) if you are programming a computer?
  (ii) if you are doing it yourself on a calculator?
➜ If you are doing it yourself on a calculator, is it possible to speed up the process?

Another change of sign method is called **interval bisection**. In this case, after finding one interval containing the root, the interval is divided into two equal parts – it is bisected.

TECHNOLOGY

The Table feature on a scientific calculator is useful for finding the values of a function for different values of x.

So to find the root of the equation $e^x - 4x = 0$ that is in the interval $[2, 3]$ (with $f(2) < 0$ and $f(3) > 0$), you would start by taking the midpoint of the interval, 2.5.

$f(2.5) = 2.18... > 0$ so the root is in the interval $[2, 2.5]$.

Now take the midpoint of this interval, 2.25.

$f(2.25) = 0.48... > 0$ so the root is in the interval $[2, 2.25]$.

The midpoint of this interval is 2.125.

$$f(2.125) = -0.12... < 0 \quad \text{so the root is in the interval } [2.125, 2.25].$$

TECHNOLOGY

Create a spreadsheet like the one in Figure 14.4. Use your spreadsheet to find the root correct to 8 decimal places.

Discussion point

→ What advantages and disadvantages does interval bisection have, compared with decimal search?

Discussion points

Figure 14.4 shows a spreadsheet that has been used to carry out interval bisection to find the root of $e^x - 4x = 0$ that is in the interval [2, 3].

→ After how many iterations can you state the root correct to 1 decimal place? 2 decimal places? 3 decimal places?

	Home	Insert	Page Layout	Formulas	Data	Review	
		fx					
	A	B	C	D	E	F	G
1	Iteration	x=a	x=b	x=m	f(a)	f(b)	f(m)
2	1	2	3	2.5	−0.610943901	8.085536923	2.182493961
3	2	2	2.5	2.25	−0.610943901	2.182493961	0.487735836
4	3	2	2.25	2.125	−0.610943901	0.487735836	−0.127102512
5	4	2.125	2.25	2.1875	−0.127102512	0.487735836	0.162902981
6	5	2.125	2.1875	2.15625	−0.127102512	0.162902981	0.013681785
7	6	2.125	2.15625	2.140625	−0.127102512	0.013681785	−0.057748562
8	7	2.140625	2.15625	2.1484375	−0.057748562	0.013681785	−0.02229497
9	8	2.1484375	2.15625	2.15234375	−0.02229497	0.013681785	−0.004372244
10	9	2.15234375	2.15625	2.154296875	−0.004372244	0.013681785	0.004638326
11	10	2.15234375	2.154296875	2.153320313	−0.004372244	0.004638326	0.000128934
12	11	2.15234375	2.153320313	2.152832031	−0.004372244	0.000128934	−0.002122681

Figure 14.4

Change of sign methods have the great advantage that they automatically provide bounds (the two ends of the interval) within which a root lies, so the maximum possible error in a result is known. Knowing that a root lies in the interval [0.61, 0.62] means that you can take the root as 0.615 with a maximum error of ±0.005.

Determining the accuracy of your answer is an essential part of any numerical method. As in this case, it must come out of the method itself; just rounding your answer is not enough.

Problems with change of sign methods

There are a number of situations which can cause problems for change of sign methods if they are applied blindly, for example by entering the equation into computer software without prior thought. In all cases you can avoid problems by first drawing a sketch graph, so that you know what dangers to look out for.

The curve touches the *x*-axis

In this case there is no change of sign, so change of sign methods will not work (see Figure 14.5).

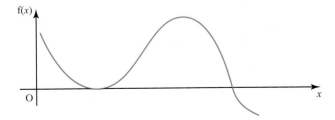

Figure 14.5

There are several roots close together

Where there are several roots close together, it is easy to miss a pair of them.

The equation

$$f(x) = x^3 - 1.65x^2 + 0.815x - 0.105 = 0$$

has roots at 0.2, 0.7 and 0.75. A sketch of the curve of $f(x)$ is shown in Figure 14.6.

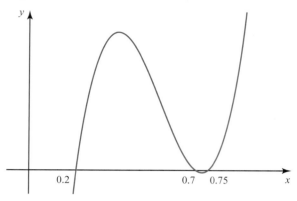

Figure 14.6

In this case $f(0) < 0$ and $f(1) > 0$, so you know there is a root between 0 and 1. A decimal search would show that $f(0.2) = 0$, so that 0.2 is a root. You would be unlikely to search further in this interval.

Interval bisection gives $f(0.5) > 0$, so you would search the interval $[0, 0.5]$ and eventually arrive at the root 0.2, unaware of the existence of those at 0.7 and 0.75.

There is a discontinuity in f(x)

The curve $y = \dfrac{1}{x - 2.7}$ has a discontinuity at $x = 2.7$, as shown by the asymptote in Figure 14.7.

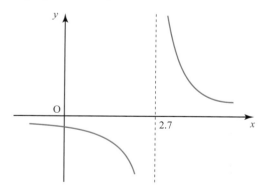

Figure 14.7

The equation $\dfrac{1}{x - 2.7} = 0$ has no root, but all change of sign methods will converge on a false root at $x = 2.7$.

It is noticeable if you look at the size of $f(x)$ as well as its sign because it will get larger as you approach the false root.

Fixed point iteration

An **iteration** or **iterative process** is a method of generating a sequence of numbers by repeating the same process over and over again. If the sequence of numbers approaches a limiting value, then you say that they **converge** to that number. For example, if you start with any positive number and repeatedly take the square root, this sequence converges to 1.

Using fixed point iteration to approximate a root of an equation starts with an estimate of the value of x, not, as with the change of sign method, with an interval in which x must lie. You then use iteration to improve this approximation, using a process that converges to the real value of x.

Figure 14.8 shows the curve $y = x^5 - 5x + 3$. From the graph, you can see that the equation $x^5 - 5x + 3 = 0$ has three real roots.

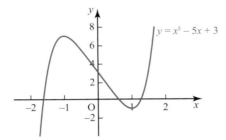

$y = x^5 - 5x + 3$

Figure 14.8

The equation $x^5 - 5x + 3 = 0$ can be rewritten in a number of different ways. One of these is $x = \dfrac{x^5 + 3}{5}$.

So the roots of $x^5 - 5x + 3 = 0$ are the same as the roots of the equation $x = \dfrac{x^5 + 3}{5}$.

These roots are also the x coordinates of the intersection points of the graphs $y = x$ and $y = \dfrac{x^5 + 3}{5}$.

These two graphs are shown in Figure 14.9.

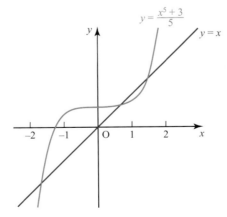

$y = \dfrac{x^5 + 3}{5}$

$y = x$

Figure 14.9

This provides the basis for the iterative formula $x_{n+1} = \dfrac{x_n^5 + 3}{5}$.

Taking $x = 1$ as a starting point to find the root in the interval $[0, 1]$, successive approximations are shown in the spreadsheet in Figure 14.10.

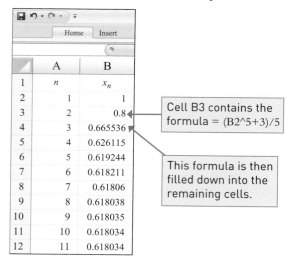

	A	B
1	n	x_n
2	1	1
3	2	0.8
4	3	0.665536
5	4	0.626115
6	5	0.619244
7	6	0.618211
8	7	0.61806
9	8	0.618038
10	9	0.618035
11	10	0.618034
12	11	0.618034

Cell B3 contains the formula = (B2^5+3)/5

This formula is then filled down into the remaining cells.

Figure 14.10

You can see that in this case the iteration has converged quite rapidly to the root for which you were looking.

It looks likely that the root is 0.618 to 3 d.p.; you cannot be sure without using a change of sign method. To verify that the root is $x = 0.618$ correct to 3 decimal places, you must show that the root lies in the interval $[0.6175, 0.6185]$ because all values in this interval round to 0.618.

> This check is an essential part of this method of solving equations; it is not an optional extra.

In general, an equation $f(x) = 0$, can be rearranged into the form $x = g(x)$, which gives the iterative formula $x_{n+1} = g(x_n)$. Starting with your first estimate for the root of x_1, if this iterative formula converges to the root, then $x_2 = g(x_1)$ will be a better approximation, and $x_3 = g(x_2)$ even better. By doing as many iterations as necessary, you can find an approximation to the root to whatever degree of accuracy is required.

T **ACTIVITY 14.2**

Set up a spreadsheet like the one shown in Figure 14.10. Change the starting value in cell B2 to try to find the other two roots of the equation, one in the interval $[-2, -1]$ and one in the interval $[1, 2]$. What happens?

In Activity 14.2 you will have seen that this method does not always converge to the root that you are looking for. The iterations may **diverge** (i.e. they get numerically larger and larger) or they may converge to a different root. However, often using a different rearrangement may be successful.

Example 14.2

You are given that $f(x) = x^5 - 5x + 3$.

(i) Show that the equation $f(x) = 0$ has a root in the interval $[1, 2]$.

(ii) Show that the equation $f(x) = 0$ can be rewritten in the form $x = \sqrt[5]{5x - 3}$.

(iii) Starting with an estimate for the solution of $x_1 = 1$, use the iterative formula $x_{n+1} = \sqrt[5]{5x_n - 3}$ to find the root in the interval $[1, 2]$ correct to three decimal places.

Solution

(i) $f(1) = -1 < 0$

$f(2) = 25 > 0$

Since one value is positive and one is negative, and $f(x)$ is a continuous function, the equation $f(x) = 0$ has a solution in the interval $[1, 2]$.

(ii) $x^5 - 5x + 3 = 0$

$$\Rightarrow x^5 = 5x - 3$$

$$\Rightarrow x = \sqrt[5]{5x - 3}$$

(iii) $x_1 = 1 \Rightarrow x_2 = \sqrt[5]{5x_1 - 3} = 1.148\,698\,35\ldots$

$x_3 = 1.223\,659\,91\ldots$

$x_4 = 1.255\,404\,16\ldots$

$x_5 = 1.267\,931\,64\ldots$

$x_6 = 1.272\,742\,08\ldots$

$x_7 = 1.274\,570\,08\ldots$

$x_8 = 1.275\,261\,99\ldots$

$x_9 = 1.275\,523\,49\ldots$

$x_{10} = 1.275\,622\,26\ldots$

It looks as if the root is 1.276 correct to 3 decimal places.

$f(1.2755) = -0.0015\ldots < 0$

$f(1.2765) = 0.0067\ldots > 0$

Check this by showing that the root lies in the interval $[1.2755, 1.2765]$

so the root is 1.276 correct to 3 decimal places.

TECHNOLOGY

Most calculators allow you to carry out iterations like this very quickly and easily using the 'answer' key for x_n in your iterative formula. Find out how to do this on your calculator.

T

ACTIVITY 14.3

What intervals do you need to use with the change of sign method to show that the following roots of equations are correct to the given degree of accuracy?

(i)   0.5 to 1 decimal place

(ii) −0.369 to 3 decimal places

(iii) 21.6342 to 4 decimal places.

Staircase and cobweb diagrams

The iteration process is easiest to understand if you think about the graph. Rewriting the equation $f(x) = 0$ in the form $x = g(x)$ means that instead of looking for points where the graph of $y = f(x)$ crosses the x-axis, you are now finding the points of intersection of the curve $y = g(x)$ and the line $y = x$ (Table 14.1 and Figure 14.11, overleaf).

Table 14.1

What you do	What it looks like on the graph
Take your initial estimate of x, x_1	Take a starting point on the x-axis
Find the corresponding value of $g(x_1)$	Move vertically to the curve $y = g(x_1)$
Take this value $g(x_1)$ as the new value of x, i.e. $x_2 = g(x_1)$	Move horizontally to the line $y = x$
Find the value of $g(x_2)$ and so on	Move vertically to the curve

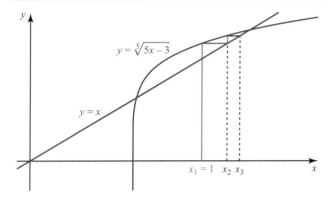

Figure 14.11

Figure 14.11 is called a **staircase diagram**, because the successive steps look like a staircase approaching the root. However, sometimes the successive values oscillate about the root, producing a **cobweb diagram**. You can see a cobweb diagram in Example 14.3.

Example 14.3

The curve $y = x^3 + 4x - 3$ intersects the x-axis at the point where $x = k$.

(i) Show that k lies in the interval $[0.5, 1]$.

(ii) Show that the equation $x^3 + 4x - 3 = 0$ can be rearranged into the form $x = \dfrac{3 - x^3}{4}$.

(iii) Use the iterative formula $x_{n+1} = \dfrac{3 - x_n^3}{4}$ with $x_1 = 0.5$ to find x_2, x_3 and x_4.

(iv) Draw a diagram to show how convergence takes place, indicating the positions of x_1, x_2, x_3 and x_4.

Solution

(i) k is the solution of the equation $x^3 + 4x - 3 = 0$.

Using the change of sign method with $f(x) = x^3 + 4x - 3$:

$$f(0.5) = -0.875 < 0$$

$$f(1) = 2 > 0$$

Since one value is positive and one is negative and the function is continuous, k lies in the interval $[0.5, 1]$ ◀

> Remember to include the final sentence of explanation.

(ii)
$$x^3 + 4x - 3 = 0$$
$$\Rightarrow \qquad 4x = 3 - x^3$$
$$\Rightarrow \qquad x = \frac{3 - x^3}{4}$$

(iii) $x_2 = 0.71875$

$x_3 = 0.65717315\ldots$

$x_4 = 0.67904558\ldots$

(iv)

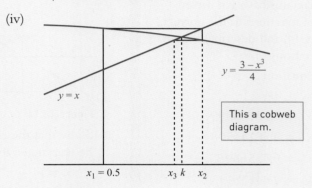

This a cobweb diagram.

Figure 14.12

> ! If you write down rounded figures for the iterations, remember not to use the rounded figure for the next iteration but keep the working on your calculator.

Problems with the fixed point iteration method

You have already seen that sometimes a particular rearrangement will result in the iterations diverging, or converging to the wrong root.

Discussion point

→ If a numerical method finds a root of an equation, but not the one you were looking for, is it a failure of the method?

What is happening geometrically when this happens?

The equation $xe^x + x^2 - 1 = 0$ has a solution of $x = 0.48$ correct to 2 decimal places. The equation can be rearranged into the form $x = \dfrac{1 - xe^x}{x}$.

However, using the iterative formula $x_{n+1} = \dfrac{1 - x_n e^{x_n}}{x_n}$ with a starting point of $x_1 = 0.5$ produces $x_2 = 0.35$, $x_3 = 1.43$, $x_4 = -3.46$, …, with repeated iterations giving values further from the solution.

Discussion point

The cobweb diagram in Figure 14.13 shows the positions of x_1, x_2 and x_3 for the above iterative formula.

→ What feature of the curve $y = \dfrac{1 - xe^x}{x}$ causes the iteration to diverge?

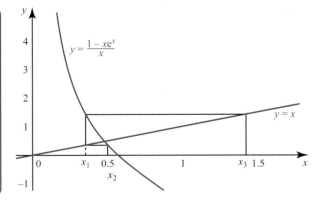

Figure 14.13

A particular rearrangement of the equation $f(x) = 0$ into the form $x = g(x)$ will give an iteration formula that converges to a root α of the equation if the gradient of the curve $y = g(x)$ is not too steep near $x = \alpha$.

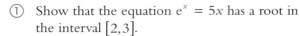

Exercise 14.1

① Show that the equation $e^x = 5x$ has a root in the interval $[2, 3]$.

PS ② Abdul has carried out iterations of the form $x_{n+1} = g(x_n)$ on four equations to try to find a root, α. His work is shown below. He has also drawn diagrams to illustrate each set of iterations, but the diagrams have got mixed up. Match each set of iterations to the correct diagram in Figure 14.14, and describe what is happening in the iterations.

Equation 1	Equation 2
$x_1 = 1$	$x_1 = 0.4$
$x_2 = 0.5714285 7\ldots$	$x_2 = 0.8210609 9\ldots$
$x_3 = 0.4552269 8\ldots$	$x_3 = 0.5814450 8\ldots$
$x_4 = 0.4420482 0\ldots$	$x_4 = 0.7356698 3\ldots$
$x_5 = 0.4409113 0\ldots$	$x_5 = 0.6413814 0\ldots$

Equation 3	Equation 4
$x_1 = 2.5$	$x_1 = 1$
$x_2 = 2.6607142 8\ldots$	$x_2 = 2.8$
$x_3 = 3.1194660 1\ldots$	$x_3 = -1.3094$
$x_4 = 4.7651052 0\ldots$	$x_4 = 3.5375876 3\ldots$
$x_5 = 15.8853662 3\ldots$	$x_5 = -5.8542466 70\ldots$

(a)
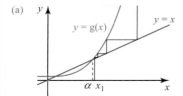

(b)
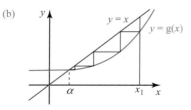

(c)
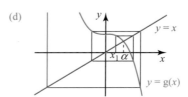

(d)
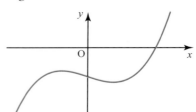

Figure 14.14

③ Show that the equation $x^4 + x - e^x = 0$ can be rearranged into each of the following forms:

(i) $x = e^x - x^4$ (ii) $x = \ln(x^4 + x)$

(iii) $x = \sqrt[4]{e^x - x}$ (iv) $x = \sqrt{\dfrac{e^x - x}{x^2}}$

④ The curve $y = x^3 - x - 2$ is shown in Figure 14.15.

Figure 14.15

(i) Show that the root of the equation $x^3 - x - 2 = 0$ is between $x = 1$ and $x = 2$.

(ii) Show that the equation can be rearranged into the form $x = \sqrt[3]{x + 2}$.

(iii) With a starting value of $x_1 = 1$, use the iterative formula $x_{n+1} = \sqrt[3]{x_n + 2}$ four times to get improved approximations for the root.

(iv) Show that your final answer is accurate to 3 decimal places.

⑤ (i) Show that the equation $e^x - x - 2 = 0$ has a root in the interval $[1,2]$.

(ii) Show that this equation may be written in the form $x = \ln(x + 2)$.

(iii) Use the iterative formula $x_{n+1} = \ln(x_n + 2)$ with a starting value of $x_1 = 2$ to find two further approximations for the root.

(iv) Sketch on the same axes the graphs of $y = x$ and $y = \ln(x + 2)$ for $0 \le x \le 3$.

(v) On your sketch, draw a diagram, indicating the positions of x_1, x_2 and x_3.

PS ⑥ (i) Show that the equation $6 - x = \ln x$ has a root between $x = 4$ and $x = 5$.

(ii) Determine whether the root is closer to $x = 4$ or to $x = 5$.

⑦ (i) On the same axes, sketch the curves $y = 3^x$ and $y = 9 - x^2$ for $-3 \le x \le 3$, and state the number of roots of the equation $3^x = 9 - x^2$.

(ii) The equation is rearranged into the form $x = \sqrt{9 - 3^x}$. Show, by trying starting numbers of

(a) $x = 2$ (b) $x = 1$

that the iterative formula $x_{n+1} = \sqrt{9 - 3^{x_n}}$ does not converge to the positive root.

(iii) Show that the equation $3^x = 9 - x^2$ can also be rearranged into the form
$$x = \frac{\ln(9 - x^2)}{\ln 3}.$$

(iv) Use the iterative formula
$$x_{n+1} = \frac{\ln(9 - x_n^2)}{\ln 3}$$ and a starting value of $x_1 = 2$ to find the positive root correct to 4 d.p.

⑧ (i) On the same axes, sketch the line $y = x$ and the curve $y = \sin x + 1$ for $0 \le x \le \pi$.

(ii) Use the iterative formula $x_{n+1} = \sin x_n + 1$ with $x_1 = 1$ to find values for x_2, x_3 and x_4.

(iii) On your answer to (i), draw a diagram to show how convergence takes place indicating the positions of x_1, x_2, x_3 and x_4.

(iv) At this stage you know the values of x_1, x_2, x_3 and x_4. With what accuracy can you now state the value of the root?

(v) Carry out enough further iterations to find the root correct to 2 d.p.

⑨ The equation $x^3 + 3x^2 - e^x = 0$ can be rearranged into each of the following forms:

(i) $x = \sqrt{\dfrac{e^x - x^3}{3}}$

(ii) $x = \ln(x^3 + 3x^2)$

(iii) $x = \sqrt[3]{e^x - 3x^2}$

(iv) $x = \sqrt{\dfrac{e^x}{x^2 + 3x}}$

Use a spreadsheet to perform iterations using each of these four rearrangements. Your spreadsheet might look something like that in Figure 14.16.

Use your spreadsheet with different starting numbers to investigate the different rearrangements. Comment on your results.

⑩ (i) Sketch the curves $y = e^x$ and $y = x^2 + 2$ on the same graph, and use your sketch to explain why the equation $e^x - x^2 - 2 = 0$ has only one root.

PS (ii) By rearranging the equation into the form $x = g(x)$ to find a suitable iterative formula, find the root of the equation accurate to 2 decimal places.

(iii) Prove that your answer is correct to this degree of accuracy.

A	B	C	D	E
Iteration	Rearrangement 1	Rearrangement 2	Rearrangement 3	Rearrangement 4
Start	1	1	1	1
1	0.756809494	1.386294361	−0.093906057	0.679570457
2	0.752328439	2.131753281	0.294637746	0.78904557

Figure 14.16

⑪ (i) Show that if the sequence defined by the iterative formula $x_{n+1} = \frac{1}{2}\left(\dfrac{k}{x_n} + x_n\right)$ converges, then it will converge to $\sqrt{k}$.

(ii) Use this formula, with a sensible starting value, to find the value of $\sqrt{10}$ to 4 decimal places.

PS ⑫ Use iteration to find the non-zero root of the equation $x^2 = \ln(x + 1)$ correct to 3 decimal places.

2 The Newton–Raphson method

Another fixed-point iteration method that starts with an estimate of the root and improves on this is called the **Newton–Raphson method**.

If you have an equation $f(x) = 0$ and you have an estimate x_1 for the value of a root, then you can get a better approximation, x_2, for the root by drawing in the tangent to the curve at x_1 and working out where this crosses the x-axis. This is shown in Figure 14.17.

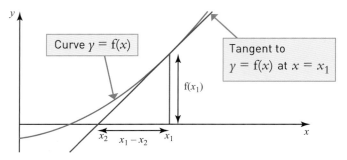

Figure 14.17

Using the right-angled triangle, the gradient of the tangent can be written as $\dfrac{f(x_1)}{x_1 - x_2}$. The gradient of the tangent is also the gradient of the curve at the point $x = x_1$, which is given by $f'(x_1)$.

Therefore $f'(x_1) = \dfrac{f(x_1)}{x_1 - x_2}$

$$\Rightarrow x_1 - x_2 = \frac{f(x_1)}{f'(x_1)}$$

$$\Rightarrow \quad x_2 = x_1 - \frac{f(x_1)}{f'(x_1)}$$

Repeating this process gives the Newton–Raphson iterative formula.

$$x_{n+1} = x_n - \frac{f(x_n)}{f'(x_n)}$$

Example 14.4

You are given that $f(x) = x^3 + 2x - 6$.

(i)   Show that the equation $f(x) = 0$ has a root lying between $x = 1$ and $x = 2$.

(ii)   Taking $x = 1$ as the first approximation to the root use the Newton–Raphson method to find three further approximations to the root.

(iii)   Show that your final approximation is accurate to 2 decimal places.

Solution

(i) $f(1) = -3 < 0$

$f(2) = 6 > 0$

Since one value is positive and the other is negative, and the function is continuous, the root must lie between $x = 1$ and $x = 2$.

(ii) $f(x) = x^3 + 2x - 6$

$f'(x) = 3x^2 + 2$

The Newton–Raphson iterative formula is therefore

$$x_{n+1} = x_{1n} - \frac{x_n^3 + 2x_n - 6}{3x_n^2 + 2}.$$

$x_1 = 1$

$x_2 = 1.6$

$x_3 = 1.466115\ldots$

$x_4 = 1.456215\ldots$

(iii)   To show that the root is 1.46 accurate to 2 decimal places, you must show that it lies in the interval $[1.455, 1.465]$.

$f(1.455) = -0.0097\ldots < 0$

$f(1.465) = 0.074\ldots > 0$

Since one value is positive and the other is negative, and the function is continuous, the root must lie between $x = 1.455$ and $x = 1.465$. Therefore it is equal to 1.46, correct to 2 decimal places.

> **TECHNOLOGY**
>
> Remember to use your calculator efficiently by using the ANS key to perform the iterations quickly.

Problems with the Newton–Raphson method

If your initial value is close enough to the root, then the Newton–Raphson method will almost always give convergence to it. The rate of convergence to the root using the method depends on the choice of initial value and the shape of the curve in the neighbourhood of the root. If the initial value is not close to the root, then the iteration may converge to another root or diverge.

In the following figures, the second approximation, x_2, is further from the root α than the initial approximation, x_1.

1   Poor choice of initial value: x_1 is too far away from α (for example, Figure 14.18).

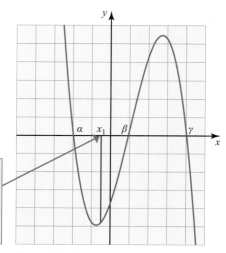

Although the starting point x_1 lies between the roots α and β, the iteration actually finds the third root, γ.

Figure 14.18

Discussion point

→ What happens when x_1 is at a stationary point? Draw a diagram to illustrate your answer.

2   $f'(x_1)$ is too small, usually because x_1 is close to a stationary point (as shown in Figure 14.19).

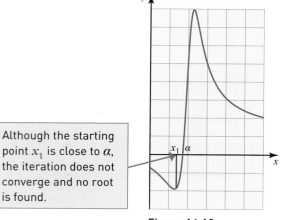

Although the starting point x_1 is close to α, the iteration does not converge and no root is found.

Figure 14.19

The Newton–Raphson method may break down if the function

■   is not defined over the whole of $\mathbb{R}$, because the tangent may meet the x-axis at a point outside the domain

■   is discontinuous.

Exercise 14.2

① The graph of $y = f(x)$, where $f(x) = x^3 - x - 3$ is shown in Figure 14.20.

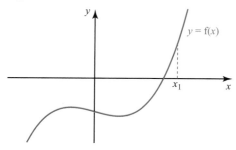

Figure 14.20

(i)   Find $f'(x)$.

(ii) Write down the Newton–Raphson iterative formula for the equation $f(x) = 0$.

(iii) Taking the first approximation to the root as $x_1 = 2$, apply the Newton–Raphson method once to find a second approximation, x_2.

(iv) Copy Figure 14.20 and show graphically how the Newton–Raphson method finds this second approximation.

② The equation $x^3 - 6x^2 + 12x - 11 = 0$ has one root, α.

   (i) Show that α lies between $x = 3$ and $x = 4$.

   (ii) Use the Newton–Raphson method twice with a starting value of $x_1 = 3.5$ to find two further approximations to the root.

   (iii) Show that your answer is correct to 3 s.f.

③ You are given that $f(x) = e^{2x} - 15x - 2$.

   (i) Show that the equation $f(x) = 0$ has a root between $x = 1.5$ and $x = 1.7$.

   (ii) Taking 1.6 as your first approximation, apply the Newton–Raphson method once to find a second approximation.

   (iii) Show that your answer is correct to 3 significant figures.

④ You are given that $f(x) = 2x^3 + 5x + 2$.

   (i) Find $f'(x)$ and hence show that the equation $f(x) = 0$ has only one root.

   (ii) Show that the root lies in the interval $[-1,0]$.

   (iii) Use the Newton–Raphson method twice with a starting value of $x_1 = -0.5$ to find two further approximations, x_2 and x_3, to the root.

   (iv) Show that your value for x_3 gives the root correct to 3 decimal places.

⑤ (i) Find the turning points of the curve $y = x^3 - 3x + 3$, and identify their nature. Hence sketch the curve.

   (ii) Using $x = -2$ as your first approximation to the root of $x^3 - 3x + 3 = 0$, apply the Newton–Raphson method once to find an improved approximation.

   (iii) Indicate on your sketch the positions of the first and second approximations, and the process by which the second approximation has been found.

   (iv) Show that your second approximation does not give the root correct to 2 d.p. but does give it correct to 1 d.p.

⑥ (i) Show that the equation $x^4 - 7x^3 + 1 = 0$ has a root in the interval $[0, 1]$.

   (ii) Use the Newton–Raphson method repeatedly to find this root correct to 2 d.p., starting with $x_1 = 1$.

   (iii) Explain why $x_1 = 0$ is not a suitable starting point.

PS ⑦ Use a graphical method to find an initial approximation to the smallest positive root of the equation $\tan x = x + 1$, and then apply the Newton–Raphson method to find the root correct to 3 s.f. Show that your answer is correct to this degree of accuracy

PS ⑧ Use a graphical method to find a first approximation to the root of the equation $\ln x = \cos x$, and then apply the Newton–Raphson method to find the root correct to 3 d.p. Show that your answer is correct to this degree of accuracy.

⑨ (i) Sketch the curves $y = e^x$ and $y = \dfrac{4}{x}$. At how many points do they intersect?

   (ii) Sketch the graph of the function $y = \dfrac{4}{x} - e^x$ for all values of x.

   (iii) Use the Newton–Raphson method to find the value of x where the curve $y = \dfrac{4}{x} - e^x$ crosses the x-axis, correct to 3 d.p., taking $x_1 = 2$.

   (iv) Explain what happens if you use a starting value of $x_1 = 3$.

PS ⑩ A geometric series has first term equal to 3 and common ratio x. The sum of the first twelve terms is equal to 750. By using the Newton–Raphson method with starting value $x_1 = 1.5$ with an appropriate equation, find the value of the common ratio correct to 5 d.p. and confirm that your answer is correct to 5 d.p.

3 Numerical integration

Numerical methods are not only used to solve equations. They can also be used for differentiation, integration, solving differential equations and many other applications.

There are many functions that cannot be integrated algebraically, and numerical methods can be used to find approximate values for definite integrals. They can also be used to estimate the area under a curve if you do not know the function in algebraic form, but just have a set of points (perhaps derived from an experiment).

In the same way that when using numerical methods to approximate the root of an equation it is important to remember that a numerical answer without any estimate of its accuracy, or error bounds, is valueless.

Note

You should not use a numerical method when an algebraic technique is available to you. Numerical methods should be used only when other methods do not work.

The trapezium rule

In this chapter, one numerical method of integration is introduced: the trapezium rule. As an illustration of the rule, it is used to find the area under the curve $y = \sqrt{5x - x^2}$ for values of x between 0 and 4.

It is in fact possible to integrate this function algebraically, but not using the techniques that you have met so far.

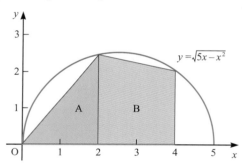

Figure 14.21

Figure 14.21 shows the area approximated by two trapezia of equal width.

Remember the formula for the area of a trapezium is area $= \frac{1}{2}h(a + b)$, where a and b are the lengths of the parallel sides and h is the distance between them.

In the cases of the trapezia A and B, the parallel sides are vertical. The left-hand side of trapezium A has zero height, and so the trapezium is also a triangle.

When $x = 0$ $\Rightarrow y = \sqrt{0} = 0$

when $x = 2$ $\Rightarrow y = \sqrt{6} = 2.4495$ (to 4 d.p.)

when $x = 4$ $\Rightarrow y = \sqrt{4} = 2$

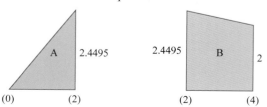

Figure 14.22

The area of trapezium A $= \frac{1}{2} \times 2 \times (0 + 2.4495) = 2.4495$

The area of trapezium B $= \frac{1}{2} \times 2 \times (2.4495 + 2) = \underline{4.4495}$

Total   6.8990

For greater accuracy you can use four trapezia, P, Q, R and S, each of width 1 unit as shown in Figure 14.23. The area is estimated in just the same way.

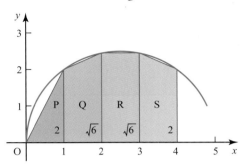

Figure 14.23

Discussion point

If you have had your wits about you, you may have found a method of finding this area without using calculus at all.

→ How can this be done? How close are your estimates?

ACTIVITY 14.4

1 Calculate the sum of the trapezia P, Q, R and S.
2 Use graphing software to calculate the area using progressively more strips and observe the convergence. **T**
3 Are your estimates too big or too small? Explain your answer.

The procedure

In Activity 14.4 above, the sum of the areas of the four trapezia P, Q, R and S can be written as

$$\frac{1}{2} \times 1 \times (0 + 2) + \frac{1}{2} \times 1 \times (2 + \sqrt{6}) + \frac{1}{2} \times 1 \times (\sqrt{6} + \sqrt{6})$$
$$+ \frac{1}{2} \times 1 \times (\sqrt{6} + 2) + \frac{1}{2} \times 1 \times (2 + 0)$$

and this can be written as

This is the strip width: 1

$$\frac{1}{2} \times 1 \times \left[0 + 2(2 + \sqrt{6} + \sqrt{6}) + 2 \right]$$

These are the heights of the ends of the whole area: 0 and 2

These are the heights of the intermediate vertical lines.

This is often stated in words as

area $\approx \frac{1}{2} \times$ strip width $\times$ [ends + twice middles]

or in symbols, for n strips of width h

$$A \approx \frac{1}{2} h \left[(y_0 + y_n) + 2(y_1 + y_2 + \cdots + y_{n-1}) \right].$$

This is called the **trapezium rule** (see Figure 14.24).

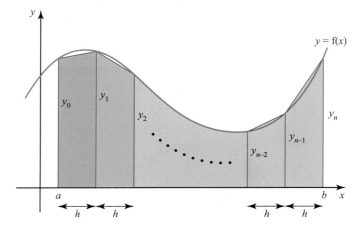

Figure 14.24

Too big or too small?

When the curve is concave upwards (Figure 14.25), all the 'tops' of the trapezia are above the curve, and the trapezium rule gives an overestimate for the area under the curve.

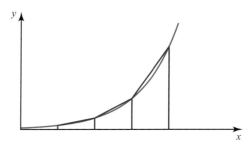

Figure 14.25

When the curve is concave downwards (Figure 14.26), the trapezia are underneath the curve; the trapezium rule underestimates the area.

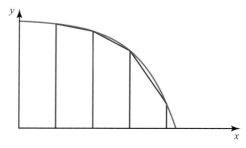

Figure 14.26

If part of the curve is concave upwards, and part is concave downwards, you cannot usually tell if the trapezium rule gives an overestimate or an underestimate for the area.

Example 14.5

(i) Use the trapezium rule with four strips to find an approximation for $\int_0^1 (3^x + 1)\, dx$, giving your answer to 3 s.f.

(ii) By sketching the curve, explain whether your answer is an overestimate or an underestimate.

(iii) You are given the following results from using the trapezium rule to find further approximations to the integral.

8 strips	2.823 338 52…
16 strips	2.821 193 63…
32 strips	2.820 627 26…

What can you say about the true value of the integral?

Solution

(i) $h = \dfrac{b - a}{n} = \dfrac{1 - 0}{4} = 0.25$

$x_0 = 0 \quad \Rightarrow y_0 = 3^0 + 1 = 2$

$x_1 = 0.25 \Rightarrow y_1 = 3^{0.25} + 1 = 2.3161$

$x_2 = 0.5 \;\; \Rightarrow y_2 = 3^{0.5} + 1 = 2.7321$

$x_3 = 0.75 \Rightarrow y_3 = 3^{0.75} + 1 = 3.2795$ ◄── To work out y values, substitute x values into the equation of the curve.

$x_4 = 1 \quad\;\; \Rightarrow y_4 = 3^1 + 1 = 4$

$\int_0^1 (3^x + 1)\, dx \approx \dfrac{0.25}{2}\big(2 + 4 + 2(2.3161 + 2.7321 + 3.2795)\big)$

$\approx 2.83 \quad$ (3 s.f.)

(ii) This is an overestimate, since the curve $y = 3^x + 1$ is concave upwards.

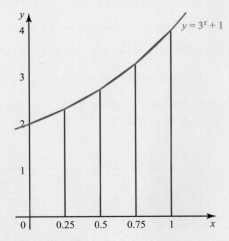

Figure 14.27

(iii) The geometry of the figure shows that all the approximations are overestimates. So the smallest of them, 2.820 627 26, is an upper bound for the true value. The pattern of the convergence indicates that 2.8 is certainly a lower bound and 2.82 may possibly be but more strips would be needed to state this with confidence.

Discussion point

This method is valid if your curve is either increasing or decreasing between a and b.

→ What problem might arise if your curve has a turning point between a and b?

Using rectangles to find bounds for the area under a curve

You can also find an estimate for the area underneath a curve by dividing the area under the curve into rectangular strips of width h and using the sum of the areas of these rectangles.

Using rectangles that have smaller y values than the curve (except at one point) as their heights will give you an underestimate for the area under the curve (Figure 14.28). This gives a **lower bound** for the area under the curve.

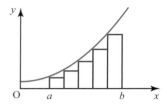

Figure 14.28

Using rectangles that have greater y values than the curve (except at one point) as their heights will give you an overestimate for the area under the curve (Figure 14.29). This gives a **upper bound** for the area under the curve.

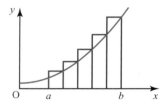

Figure 14.29

Increasing the number of rectangles allows you to improve these bounds, and home in on the true value of the area.

Example 14.6

You are given that $f(x) = \dfrac{1}{x^3 + 1}$, $x > -1$.

(i)   Show that $f(x)$ is a decreasing function.

(ii)   Copy and complete Table 14.2 for $y = f(x)$, giving your answers to 4 decimal places.

Table 14.2

x_r	2	2.25	2.5	2.75	3
y_r					

(iii) Using the sums of the areas of four rectangles, form an inequality for the value of $\int_2^3 f(x)\,dx$, giving your bounds to 3 d.p.

(iv) Give the value of $\int_2^3 f(x)\,dx$ to as great a degree of accuracy as possible from your answer to (iii), and explain how you could refine this method to enable you to give $\int_2^3 f(x)\,dx$ to a greater degree of accuracy.

Solution

(i) $$f(x) = \left(x^3 + 1\right)^{-1}$$

$$\Rightarrow f'(x) = -\left(x^3 + 1\right)^{-2} \times 3x^2 \longleftarrow \boxed{\text{Using the chain rule}}$$

$$= -\frac{3x^2}{\left(x^3 + 1\right)^2}$$

$x^2 \geqslant 0$ since it is a perfect square, and $\left(x^3 + 1\right)^2 > 0$ since it is also a perfect square and $x \neq -1$.

Therefore $f'(x) = -\dfrac{3x^2}{\left(x^3 + 1\right)^2} \leqslant 0$, and so $f(x)$ is a decreasing function.

Substitute the x values into the equation
$y = \dfrac{1}{x^3 + 1}$.

(ii)

x_r	2	2.25	2.5	2.75	3
y_r	0.1111	0.0807	0.0602	0.0459	0.0357

The difference between the x values.

(iii) The width of the rectangles is 0.25.

To get the lower bound:

area $\approx 0.25 \times 0.0807 + 0.25 \times 0.0602 + 0.25 \times 0.0459$
$\quad\quad + 0.25 \times 0.0357$

$\Rightarrow$ area ≈ 0.056

To get the upper bound:

area $\approx 0.25 \times 0.1111 + 0.25 \times 0.0807 + 0.25 \times 0.0602$
$\quad\quad + 0.25 \times 0.0459$

$\Rightarrow$ area ≈ 0.074

Therefore $0.056 < \int_2^3 f(x)\,dx < 0.074$.

You are not certain of the second decimal place because it could be either 0.06 or 0.07.

(iv) From (iii), $\int_2^3 f(x)\,dx = 0.1$ to 1 d.p.

To get a more accurate answer, divide the area into more rectangles.

① Table 14.3 shows values of x and y for a curve $y = f(x)$. Figure 14.30 shows how the area under the curve can be estimated using three trapezia.

Table 14.3

x	0	3	6	9
y	0	5.2	8.0	10.1

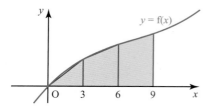

Figure 14.30

Use the trapezium rule to estimate $\int_0^9 f(x)\,dx$, giving your answer to the nearest integer.

② Use the trapezium rule to find an approximation to $\int_0^1 \dfrac{1}{1 + x^3}\,dx$

(i) using two strips

(ii) using four strips.

③ Sketch the graph of $y = \dfrac{1}{x + 1}$, $x > -1$.

Use the trapezium rule with 4 strips to estimate the value of $\int_2^4 \dfrac{1}{x + 1}\,dx$, and state with a reason whether your value is an underestimate or an overestimate.

④ Sketch the graph of $y = \sqrt{x - 2}$, $x \geqslant 2$. Use the trapezium rule with 5 strips to estimate the value of $\int_3^4 \sqrt{x - 2}\,dx$, giving your answer to 3 significant figures, and state with a reason whether your value is an underestimate or an overestimate.

⑤ The speed v in $\mathrm{m\,s^{-1}}$ of a train is given at time t seconds in Table 14.4.

Table 14.4

t	0	10	20	30	40	50	60
v	0	5.0	6.7	8.2	9.5	10.6	11.6

The distance that the train has travelled is given by the area under the graph of the speed (vertical axis) against time (horizontal axis).

(i) Estimate the distance the train travels in this 1 minute period.

(ii) Give two reasons why your method cannot give a very accurate answer.

⑥ (i) Copy and complete Table 14.5 for the curve $y = f(x)$ where

$$f(x) = \frac{1}{1 + x^2}.$$

Table 14.5

x	0	0.2	0.4	0.6	0.8	1.0
y						

(ii) Using the method of finding the sum of the area of a series of rectangles in each case, find an upper bound and a lower bound for the value of the area under the curve between $x = 0$ and $x = 1$.

(iii) State the value of $\int_0^1 \dfrac{1}{1 + x^2}\,dx$ to as great a degree of accuracy as you can from your answers to (ii).

(iv) Explain how you could refine this method to find the value of $\int_0^1 \dfrac{1}{1 + x^2}\,dx$ to a greater degree of accuracy.

⑦ (i) Use the trapezium rule with 6 strips to find an estimate for $\int_2^{2.3} \ln x\,dx$, giving your answer to 7 s.f.

(ii) Find the exact value of $\int_2^{2.3} \ln x\,dx$, and hence find the percentage error in your answer to (i) giving your answer to 2 s.f.

⑧ By using the trapezium rule to provide one bound, and the sum of the areas of a series of rectangles to provide the other, find $\int_2^3 \dfrac{1}{\sqrt{1 + x^2}}\,dx$ correct to 2 d.p.

⑨ In statistics, the equation of the Normal distribution curve for a distribution with mean 0 and standard deviation 1 is given by

$$\phi(x) = \frac{1}{\sqrt{2\pi}} e^{-\frac{1}{2}x^2}.$$

(i) Using graphing software or a spreadsheet, use the trapezium rule to estimate the value of $\int_{-1}^{1} \phi(x)\,dx$ to 3 decimal places. How many strips did you use?

(ii) Using graphing software or a spreadsheet, use the sum of rectangles to find upper and lower bounds for $\int_{-1}^{1} \phi(x)\,dx$. How many rectangles did you use to give the value of $\int_{-1}^{1} \phi(x)\,dx$ correct to 3 decimal places?

LEARNING OUTCOMES

When you have completed this chapter, you should be able to:

➤ locate roots of $f(x) = 0$ by considering changes of sign of $f(x)$ in an interval of x on which $f(x)$ is sufficiently well-behaved

➤ understand how change of sign methods can fail

➤ solve equations approximately using simple iterative methods; be able to draw associated cobweb and staircase diagrams.

➤ solve equations using the Newton–Raphson method and other recurrence relations of the form $x_{n+1} = g(x_n)$

➤ understand how such methods can fail

➤ understand and use numerical integration of functions, including the use of the trapezium rule and estimating the approximate area under a curve and limits that it must lie between

➤ use numerical methods to solve problems in context.

KEY POINTS

1 When $f(x)$ is a continuous function, if $f(a)$ and $f(b)$ have opposite signs, there will be at least one root of $f(x) = 0$ in the interval $[a, b]$.

2 Fixed point iteration may be used to solve an equation $f(x) = 0$ by either of the following methods.
   - Rearranging the equation $f(x) = 0$ into the form $x = g(x)$ and using the iteration $x_{n+1} = g(x_n)$.
   - The Newton–Raphson method using the iteration $x_{n+1} = x_n - \dfrac{f(x_n)}{f'(x_n)}$.

3 You can use the trapezium rule, with n strips of width h, to find an approximate value for a definite integral:

$$\int_a^b y \,dx \approx \frac{1}{2}h\big((y_0 + y_n) + 2(y_1 + y_2 + \ldots + y_{n-1})\big)$$

Using more strips increases the accuracy of the approximation.

4 If the curve is concave upwards, the trapezium rule gives an overestimate. If the curve is concave downwards, the trapezium rule gives an underestimate.

5 By using the sum of the areas of a series of rectangles, upper and lower bounds for the area under a curve can be found.

FUTURE USES

This work is developed further in the Numerical Methods option in Further Mathematics.

ⓣ Numerical integration

Vesna is carrying out numerical integration, approximating the area under a curve by rectangles. She makes the following statements.

1   Using rectangles drawn from the left hand ends of the intervals is a poor method. Using the right hand ends is no better. They always give you extreme estimates for the area under the curve. You can see it in my diagrams (Figures 1 and 2).

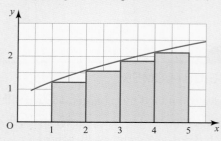

Figure 1 Low estimate in this case

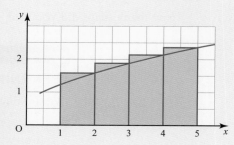

Figure 2 **H**igh estimate in this case

2   Taking the average of the left hand and right hand rectangles is better. The answer is always the same as you get from the trapezium rule.

3   An even better method is to draw the rectangles at the midpoints of the intervals as shown in Figure 3. For any number of intervals that will be more accurate.

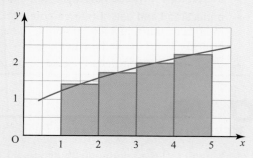

Figure 3 Best estimate

Investigate whether Vesna's statements are true.

1   **Problem specification and analysis**

Vesna has made three statements. Look at them carefully and decide how you are going to proceed.

■   Which of them are little more than common sense and which need real work?

■   How you are going to investigate those that need work?

   –   What technology will you use?

   –   What examples will you choose to work with?

   –   How many different examples do you expect to use?

   –   To what level of accuracy do you expect to work?

■ How you are going to report the outcomes?

– Will you be happy just to say 'true' or 'false', or do you expect to make statements such as 'It is usually true but there are exceptions such as …'?

– How much explanation do you expect to give?

2 **Information collection**

You may be able to make some comments on Vesna's statements just by thinking about them but you will also need to carry out some investigations of your own. This will require the use of technology.

It will be helpful, at least in some cases, to choose functions which you know how to integrate, allowing you to know the answer to which a numerical method should be converging.

However, the whole point of using a numerical method is to find an answer when an analytical method is not available to you. It may be that there is one that you don't know or it may be that one just does not exist. So you should also use at least one example where you will only know the answer (to your chosen level of accuracy) when you have completed your work.

Do not be content to work with just one or two types of functions. Try a variety of functions but always start with a sketch of the curve of the function. You can of course use graphing software to obtain this.

3 **Processing and representation**

The previous stage will probably result in you having a lot of information. Now you need to sort through it and to organise it in a systematic way that allows you to comment on Vesna's three statements.

Where you can explain your results using algebra, then you should do so. In other cases you may present them as experimental outcomes.

4 **Interpretion**

The method illustrated in Figure 3 is called the midpoint rule. Much of your work on this task will have been focused on the midpoint rule, and you need to comment on whether this is a good method for numerical integration. In order to do so you will need to explain what you mean by 'good' and what the desirable features are in such a method.

There are other methods of numerical integration and you may choose to conclude by saying something about them.

① A curve has parametric equations $x = 1 + 2t$, $y = 8t^2$. Which of the following is its Cartesian equation?

A $y = 2x^2 - 1$

B $y = 2(x - 1)^2$

C $y = 2x^2 - 2$

D $y = 2(x + 1)^2$

② Which of the following is **not** parallel to $-2\mathbf{i} + 5\mathbf{j} + \mathbf{k}$?

A $2\mathbf{i} - 5\mathbf{j} - \mathbf{k}$

B $-4\mathbf{i} + 10\mathbf{j} + 2\mathbf{k}$

C $-\mathbf{i} + \dfrac{5}{2}\mathbf{j} - \dfrac{1}{2}\mathbf{k}$

D $-20\mathbf{i} + 50\mathbf{j} + 10\mathbf{k}$

③ Which of the following rearrangements is **not** equivalent to $x^5 - 4x + 2 = 0$

A $x = \sqrt[5]{4x - 2}$

B $x = \dfrac{4x - 2}{x^4}$

C $x = \sqrt{\dfrac{4x - 2}{x}}$

D $x = \dfrac{x^5 + 2}{4}$

④ You are solving an equation using fixed point iteration. You think that the root you are searching for is 0.901 to 3 decimal places. Which two values should you check to verify this?

A $[0.9005, 0.9015]$

B $[0.900, 0.902]$

C $[0.905, 0.915]$

D $[0.9015, 0.9025]$

M ① In a chemical reaction, the rate of change of the mass, m grams, of a substance at time t minutes is inversely proportional to the square root of m. Initially, the mass is 9 grams and is increasing at a rate of 10 grams per minute.

Formulate a differential equation between m and t. [3 marks]

② Forces $\mathbf{F}_1 = (2a\mathbf{i} + 3b\mathbf{j})\,\mathrm{N}$, $\mathbf{F}_2 = (-b\mathbf{i} + a\mathbf{j})\,\mathrm{N}$ and $\mathbf{F}_3 = (10\mathbf{i} - 2\mathbf{j})\,\mathrm{N}$ are in equilibrium, so that $\mathbf{F}_1 + \mathbf{F}_2 + \mathbf{F}_3 = 0$. Find a and b. [4 marks]

③ The triangle ABC has coordinates A $(1, 2, 0)$, B $(-1, 3, -1)$ and C$(3, 0, -6)$.

(i) Find the column vectors $\overrightarrow{AB}$, $\overrightarrow{AC}$ and $\overrightarrow{BC}$. [2 marks]

(ii) Prove that triangle ABC is right angled. [3 marks]

P ④ In Figure 1, ABCD is a parallelogram. D divides CB in the ratio $1:2$ and E divides OD in the ratio $3:1$.

$\overrightarrow{OA} = \mathbf{u}$ and $\overrightarrow{OC} = \mathbf{v}$.

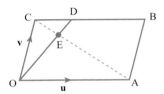

Figure 1

(i) Find the vectors $\overrightarrow{OE}$ and $\overrightarrow{CE}$ in terms of $\mathbf{u}$ and $\mathbf{v}$. [3 marks]

(ii) Hence prove that the point E lies on the diagonal AC. [2 marks]

T ⑤ (i) Use the Newton–Raphson method to find a root of $f(x) = 0$, where $f(x) = 4\sin 2x - x^3$. Use a starting value $x_0 = 1$, and give the results of the first 4 iterations correct to 5 decimal places. Hence suggest the value of the root to 3 significant figures. [4 marks]

(ii) Verify that your root is indeed correct to 3 significant figures. [1 mark]

T ⑥ The spreadsheet (Figure 2) is used to identify a root of the equation $x^5 - 2x + 3 = 0$.

	A	B	C	D	E	F	G
	Clipboard			Font			Alignment
B2			$f\!x$	=A2^5−2*A2+3			
1	a=	−2	h=	0.1			
2	−2	−25					
3	−1.9	−17.961					
4	−1.8	−12.2957					
5	−1.7	−7.79857					
6	−1.6	−4.28576					
7	−1.5	−1.59375					
8	−1.4	0.42176					
9	−1.3	1.88707					
10	−1.2	2.91168					
11	−1.1	3.58949					
12	−1	4					
13							

Figure 2

(i) Identify an interval which contains the root to 1 decimal place, justifying your answer. [2 marks]

(ii) Use your calculator to find the interval containing the root correct to 2 decimal places. [2 marks]

(iii) Identify the root correct to 2 decimal places. [2 marks]

(T) (7) Ben and Carrie are attempting to solve the equation $x^3 - 2x - 5 = 0$ using fixed point iteration.

Ben rearranges the equation to get an iterative formula of the form $x_{n+1} = ax_n^3 + b$.

(i) Find the constants a and b. [1 mark]

(ii) Using an initial value $x_0 = 1$, find x_1, x_2 and x_3. Comment on your results. [2 marks]

Carrie rearranges the equation to get an iterative formula of the form $x_{n+1} = \sqrt[3]{c\,x_n + d}$.

(iii) Find the constants c and d. [1 mark]

(iv) Use this iterative formula, together with an initial value $x_0 = 1$, to find x_1, x_2 and x_3. [2 marks]

(v) Verify that x_3, when rounded, gives a root of the equation correct to 2 decimal places. [2 marks]

(8) Figure 3 shows a curve with parametric equations $x = 1 + 2t$, $y = 2t + t^2$ for $-2 \leqslant t \leqslant 2$.

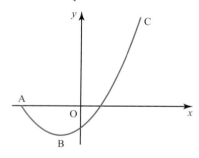

Figure 3

(i) Find the coordinates of A and C. [4 marks]

(ii) Find $\dfrac{dy}{dx}$ in terms of t. Hence find the coordinates of the turning point B of the curve. [5 marks]

(iii) Find the Cartesian equation of the curve in the form $y = ax^2 + bx + c$. [3 marks]

(M) (PS) (9) At the moment he opens his parachute, a sky diver is moving vertically downwards at a speed of 10 metres per second. His speed, $v\,\mathrm{m\,s^{-1}}$, t seconds after this is modelled by the differential equation $\dfrac{dv}{dt} = -\dfrac{1}{3}kv(v - 3)$, where k is a positive constant.

(i) Show that $v = \dfrac{3}{1 - 0.7\,\mathrm{e}^{-kt}}$. [10 marks]

(ii) Hence find the terminal velocity of the sky diver. [2 marks]

Review: Working with data

1 Statistical problem solving

Statistics provides a powerful set of tools for solving problems. While many of the techniques are specific to statistics they are nonetheless typically carried out within the standard cycle.

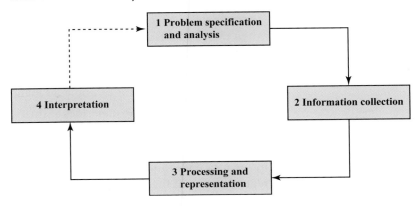

Figure R.1

This chapter reviews the techniques that are used in **Information collection** and **Processing and representation**. The questions at the end also involve elements of the other two stages.

Information collection

The information needed in statistics is usually in the form of data so this stage is also called **data collection**.

> Notice that 'data' is a plural word, so that you should say, 'The data are …', rather than 'The data is …'.

Data collection often requires you to take a sample, a set of items which are drawn from the relevant population and should be representative of it.

Here is a check list of questions to ask yourself when you are taking a sample.

- Are the data relevant to the problem?
- Are the data unbiased?
- Is there any danger that the act of collection will distort the data?
- Is the person collecting the data suitable?
- Is the sample of a suitable size?
- Is a suitable sampling procedure being followed?

> **Discussion point**
> → Give examples of cases where the answers to these questions are 'No'.

The last of the questions above asks about sampling procedures. There are many of these and here are some that you are likely to come across.

Random sampling is a general description in which every item in the population has a non-zero probability of being selected. The term is often restricted to cases where that probability is equal for every item.

> *Data by itself is useless. Data is only useful if you apply it.*
>
> Todd Park (1973–)

329

Simple random sampling is a stronger term than random sampling. It refers to a situation in which every possible sample of a given size is equally likely to be selected. This can only happen if the selections are independent of each other. This is the ideal form of sampling but practical considerations mean that it is not always a possibility.

Stratified sampling is used when there are clear groups, or strata, within the population and items are chosen from each of them. In proportional stratified sampling the numbers chosen from the various groups are proportional to their sizes.

Cluster sampling is typically used when the population is in groups, or clusters, some of which may be more accessible than others. A sample of badgers in the UK might, for example, involve a number of animals from several setts located at different places round the country.

Systematic sampling is carried out when there is a list of the population. Items are selected at regular intervals through the list. This is a good method for when the population data are stored on a spreadsheet.

Quota sampling is often used in market and social surveys. The sampling is carried out by a number of people each of whom has a quota of different types of people to interview, for example 30 men, 30 women and 40 children.

Opportunity sampling describes a situation when a sample is easily available, for example the delegates at a conference. It can be quite unreliable.

Self-selecting sampling occurs when the participants volunteer to take part in the data collection exercise. A self-selected sample may well not be very representative of the population.

Processing and representation

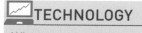

TECHNOLOGY

When working with a large set of data, a spreadsheet or statistics package is useful for all these processes.

At the start of this stage you have a set of raw data; by the end you have worked them into forms that will allow people to see the information that this set contains, with particular emphasis on the problem in hand. Three processes are particularly important.

- Cleaning the data. This involves checking outliers, errors and missing data, and formatting the data, often using a spreadsheet or statistics package.
- Presenting the data using suitable diagrams.
- Calculating summary measures.

Describing data

The data items you collect are often values of **variables** or of **random variables.** The height of an adult human is a variable because it varies from one person to another; because it does so in an unpredictable manner it is a random variable. Rather than repeatedly using the phrase 'the height of an adult person in metres' it is usual to use an upper case letter like X to represent it. Particular values of a random variable are denoted by a lower case letter; often (but not always) the same letter is used. So if the random variable X is 'the height of an adult in metres', for a person with height 1.83 m, you could say $x = 1.83$.

The number of times that a particular value of a random variable occurs is called its **frequency**.

When there are many possible values of the variable, it is convenient to allocate the data to classes. An example of the use of **grouped data** is the way people are allocated to age groups.

The pattern in which the values of a variable occur is called its **distribution**. This is often displayed in a diagram with the variable on the horizontal scale and

a measure of frequency or probability on the vertical scale. If the diagram has one peak the distribution is **unimodal**; if the peak is to the left of the middle the distribution has **positive skew** and if it is to the right there is **negative skew**. If the distribution has two distinct peaks it is **bimodal**.

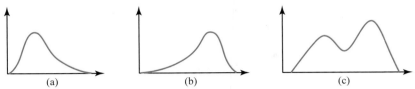

Figure R.2 (a) Positive skew (b) Negative skew (c) A bimodal distribution

A data item which is far away from the rest is called an **outlier**. An outlier may be a mistake, for example a faulty reading from an experiment, or it may be telling you something really important about the situation you are investigating. When you are cleaning your data it is essential to look at any outliers and decide which of these is the case, and so whether to reject or accept them.

The data you collect can be of a number of different types. You always need to know what type of data you are working with as this will affect the ways you can display them and what summary measures you can use.

Categorical (or qualitative) data

These data come in classes or categories, like types of bird or makes of car. Categorical data are also called *qualitative*, particularly if they can be described without using numbers.

Common displays for categorical data are pictograms, dot plots, tallies, pie charts and bar charts. A summary measure for the most typical item of categorical data is the modal class.

Categorical data may be contrasted with **numerical (or quantitative) data** which are defined in some way by numbers. Examples of numerical data are the times people take to run a race, the numbers of trucks in freight trains and the birth weights of babies.

Ranked data

These data are given by their position within a group rather than by measurements or scores. For example the competitors in a competition could be given their positions as 1st, 2nd, 3rd ... .

Drawing a stem-and-leaf diagram can be helpful when ranking data.

The median divides the data into two groups, those with high ranks and those with low ranks. The **lower quartile** and the **upper quartile** do the same for these two groups so between them the two quartiles and the median divide the data into four equal sized groups according to their ranks. These three measures are sometimes denoted by Q_1, Q_2 and Q_3. These values, with the highest and lowest value can be used to create a box plot (or box-and-whisker diagram).

Figure R.3

Identifying outliers

There are two common tests.

- Is the item more than 2 standard deviations from the mean?
- Is the item more than 1.5 × the interquartile range beyond the nearer quartile?

Note

A pie chart is used for showing proportions of a total.

There should be gaps between the bars in a bar chart.

The median is a typical middle value and so is sometimes called an **average**. More formally it is a **measure of central tendency**. It usually provides a good representative value.

The median is easy to work out if the data are stored on a spreadsheet since that will do the ranking for you. Notice that extreme values have little, if any, effect on the median. It is described as resistant to outliers. It is often useful when some data values are missing but can be estimated.

Interquartile range and semi-interquartile range are measures of spread for ranked data, as is the range.

Percentiles tell you where specified percentages of the data are. For instance, to say that 'K is the 60th percentile' means that 60% of the data values are less than K. The median is therefore the 50th percentile and the lower quartile is the 25th percentile.

Discrete numerical data

These data can take certain particular numerical values but not those in between. The number of spots on a ladybird (2, 3, 4, …) the number of goals a football team scores in a match (0, 1, 2, 3, …) and shoe sizes in the UK $\left(1, 1\frac{1}{2}, 2, 2\frac{1}{2}, \ldots\right)$ are all examples of discrete variables. If there are many possible values it is common to group discrete data.

Commonly used displays for discrete data include a vertical line chart and a stem-and-leaf diagram.

A frequency table can be useful in recording, sorting and displaying discrete numerical data.

Summary measures for discrete numerical data include

- Central tendency: mean, weighted mean, mode, median, mid-range, modal class (if the data are grouped)
- Spread: range, interquartile range, semi-interquartile range, standard deviation.

Continuous numerical data

These data can take any appropriate value if measured accurately enough.

Distance, mass, temperature and speed are all continuous variables. You cannot list all the possible values.

If you are working with continuous data you will always need to group them. This includes two special cases.

- The variable is actually discrete but the intervals between values are very small. For example cost in £ is a discrete variable with steps of £0.01 (i.e. 1 penny) but this is so small that the variable may be regarded as continuous.

- The underlying variable is continuous but the measurements of it are rounded (for example, to the nearest mm), making your data discrete. All measurements of continuous variables are rounded and providing the rounding is not too coarse, the data should normally be treated as continuous. A special case of rounding occurs with people's age; this is a continuous variable but is usually rounded down to the nearest completed year.

Remember

In frequency charts and histograms the values of the variables go at the ends of the bars. In a bar chart the labels are in the middle.

A frequency chart and a histogram are the commonest ways of displaying continuous data. Both have a continuous horizontal scale covering the range of values of the variable. Both have vertical bars.

- In a frequency chart, frequency is represented by the height of a bar. The vertical scale is frequency.

- In a histogram, frequency is represented by the area of a bar. The vertical scale is frequency density.

Remember

In frequency charts and histograms the values of the variables go at the ends of the bars. In a bar chart the labels are in the middle.

Discussion point

→ Look at the frequency chart and histogram in Figure R.4 and Figure R.5. They show the length, in seconds, of passages of bird song recorded one morning. What is the same about them and what is different?

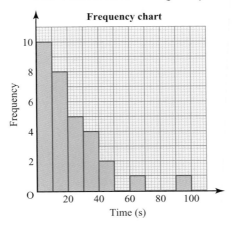

Figure R.4

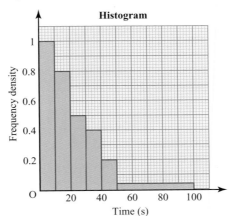

Figure R.5

Remember

If you are using a frequency chart the class intervals should all be equal. For a histogram they do not have to be equal. So if you have continuous data grouped into classes of unequal width, you should expect to use a histogram.

The summary measures used for continuous data are the same as those for discrete data.

Numerical data can also be displayed on a **cumulative frequency curve**. To draw a cumulative frequency curve you plot the cumulative frequency (vertical axis) against the upper boundary of each class interval (horizontal axis). Then you join the points with a smooth curve. This lends itself to using the median, quartiles and other percentiles as summary measures.

Remember

It is usual in a case like this, based on grouped data, to estimate the median as the value of term $\frac{n}{2}$ and the quartiles as the values of terms $\frac{n}{4}$ and $\frac{3n}{4}$.

Bivariate data

These data cover two variables, such as the people's height and weight.

The usual form of display for bivariate data is a scatter diagram.

Note

Multivariate data cover more variables.

When you are working with bivariate data you are likely to be interested in the relationship between the two variables, how this can be shown on a scatter diagram and how it can be quantified. These topics are covered in Chapter 17.

Standard deviation, a new measure of spread

You have met range and interquartile range as measures of spread. In this section a different and very important measure, standard deviation, is introduced. The next example and the subsequent work introduce you to the meaning of standard deviation.

Karen and Jake are keen dancers and enter competitons where they are given marks out of 10. The producer of a television dancing programme wants to choose one of them to take part.

These are their scores for their last ten competitions.

| Karen | 4 | 3 | 6 | 8 | 5 | 6 | 3 | 9 | 6 | 10 |
| Jake | 6 | 6 | 6 | 5 | 5 | 7 | 6 | 5 | 6 | 8 |

Who do you think should be chosen?

> ### Note
>
> As part of the explanation, the text shows how to work out the value of the standard deviation of a data set by hand. However, some people would often just enter the data into their calculators and read off the answer.

Solution

The means of their scores should give an indication of their overall performance. They both have a total score of 60 points so have the same mean of 6.0 points per competition.

So in this case the mean does not help the producer decide between the two dancers. However, the spread of their points could give an indication of their reliability.

Start by finding how far each score is from the mean. This is the **deviation**, $(x - \bar{x})$. Table R.1 shows Karen's data.

Table R.1

	Karen's points										Total, Σ
Points, x	4	3	6	8	5	6	3	9	6	10	60
Deviation $(x - \bar{x})$	−2	−3	0	+2	−1	0	−3	+3	0	+4	0

The total of the deviations is zero.

Instead you want the absolute value of the deviation.

Whether positive or negative, it is counted as positive.

It is denoted by $|x - \bar{x}|$.

You can see this in Table R.2.

Table R.2

	Karen's points										Total, Σ
Points, x	4	3	6	8	5	6	3	9	6	10	60
Absolute deviation, $|x - \bar{x}|$	+2	+3	0	+2	+1	0	+3	+3	0	+4	18

The next step is to find the means of their absolution deviations.

> ### Note
>
> The reason the total is zero is because of the definition of the mean. The sum of the deviations above it must be equal to the sum of those below it.

For Karen it is $\frac{18}{10} = 1.8$. For Jake the mean absolute deviation works out to be $\frac{6}{10} = 0.6$. ◄ | Check Jake's mean absolute deviation for yourself. |

So there is a greater spread in Karen's scores.

Who should be chosen?

If the producer wants a dancer with a steady reliable performance, the choice should probably be Jake. On the other hand if the producer wants someone who has good days and bad days but is very good at her best, then it should be Karen.

Mean absolute deviation is given by $\dfrac{\sum |x - \bar{x}|}{n}$. It is an acceptable measure of spread but it is not widely used because it is difficult to work with. Instead the thinking behind it is taken further with **standard deviation** which is more important mathematically and consequently is very widely used.

To work out the mean absolute deviation, you had to treat all deviations as if they were positive. Another way to get rid of the unwanted negative signs is to square the deviations. For Karen's data this could be written as in Table R.3.

Table R.3

	Karen's points										Total, Σ
Points, x	4	3	6	8	5	6	3	9	6	10	60
Squared deviation $(x - \bar{x})^2$	4	9	0	4	1	0	9	9	0	16	52

So the mean of the squared deviations is $\frac{52}{10} = 5.2$.

This is not a particular easy measure to interpret but if you take its square root and so get the root mean squared deviation, this can then be compared with the actual data values. In this example the root mean squared deviation for Karen's data is $\sqrt{5.2} = 2.28\ldots$ It is a measure of how much the value of a typical item of data might be above or below the mean.

It is good practice to use standard notation to explain your working, and this is given below.

Notation

The sum of the squared deviations is denoted by S_{xx}.

So $\quad S_{xx} = \displaystyle\sum_{i=1}^{i=n} (x_i - \bar{x})^2$.

This is often written more simply as $S_{xx} = \sum (x - \bar{x})^2$.

An equivalent form is $S_{xx} = \sum x^2 - n\bar{x}^2$.

When the data are given using frequencies, these formulae become

$S_{xx} = \displaystyle\sum_{i=1}^{i=n} f_i(x_i - \bar{x})^2 \ $ or $\ S_{xx} = \sum f(x - \bar{x})^2 = \sum fx^2 - n\bar{x}^2 \ $ respectively.

Mean squared deviation is given by $\dfrac{S_{xx}}{n}$

and root mean squared deviation (rmsd) is $\sqrt{\dfrac{S_{xx}}{n}}$.

You are now one step away from finding **variance** and **standard deviation**.

The calculation of the *mean squared deviation* and the *root mean squared deviation* involves dividing the sum of squares, S_{xx}, by n. If, instead, S_{xx} is divided by $(n - 1)$ you obtain the variance and the standard deviation. For sample data, they are denoted by s^2 and s.

Variance $\quad s^2 = \dfrac{S_{xx}}{n - 1}$

Standard deviation $\quad s = \sqrt{\dfrac{S_{xx}}{n - 1}}$

For Karen's data, $S_{xx} = 26$ and $n = 10$.

So the variance, $s^2 = \dfrac{26}{10 - 1} = 2.888\ldots$ and the standard deviation

$s = \sqrt{2.888\ldots} = 1.699\ldots$, (or 1.700 to 3 d.p.).

Standard deviation is by far the most important measure of spread in statistics. It is used for both discrete and continuous data and with ungrouped and grouped data.

Note

Find the root mean squared deviation for Jake's data for yourself.

Remember

$\sum x = n\sum \dfrac{x}{n} = n\bar{x}$

$\sum \bar{x}^2 = n\bar{x}^2$

Note

The reason for dividing by $(n - 1)$ rather than by n is that when you come to work out the n deviations of a data set from the mean there are only $(n - 1)$ independent variables.

TECHNOLOGY

Check whether your calculator gives the standard deviation or the root mean squared deviation or whether it offers you a choice.

Discussion point

→ What are the variance and standard deviation for Jake's data?

① A conservationist spends a night observing bats. Afterwards he draws these four diagrams to show his results. Three of them are correct but one is not. Which one is wrong?

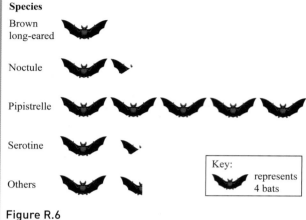

Species

Brown long-eared

Noctule

Pipistrelle

Serotine

Others

Key:

represents 4 bats

Figure R.6

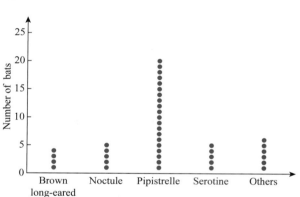

Figure R.7

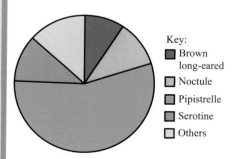

Key:
- ◼ Brown long-eared
- ◻ Noctule
- ◼ Pipistrelle
- ◼ Serotine
- ◻ Others

Figure R.8

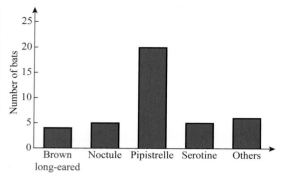

Figure R.9

② This back-to-back stem-and-leaf diagram (Figure R.10) shows the mean GDP per person in Western and Eastern European countries, in thousands of US$.

Western Europe		Eastern Europe
	0	4 7 8 8 8
9 1	1	1 2 4 6 8
8 4 4 3 3	2	0 1 2 3 5 6 8
8 8 7 7 6 6 1 0 0	3	
6 5 3 3 1 1 1 1 0	4	
7 5 4	5	
6 1	6	
	7	
9 0	88	

Key 3 | 7 = 37 000 US$

Figure R.10

(i) What is the main point that the diagram tells you?

(ii) Find the modes for
  (a) Western Europe
  (b) Eastern Europe.

(iii) Which are the modal groups for
  (a) Western Europe
  (b) Eastern Europe
  (c) Europe as a whole?

(iv) Which is the more representative of each region, its mode or its modal group?

(v) Compare the medians for Western and Eastern Europe.

(vi) The mean of the figures represented in the stem-and-leaf diagram for Western Europe is 40 875 US$.

(a) Calculate the equivalent figure for Eastern Europe.

(b) Explain why these figures do not represent the mean GDP per person living in those two regions and state what sort of means would give that information.

(vii) The mean per capita income for the UK is 37 300 US$. What is the rank of the UK among European countries?

③ (i) The mean height of a sample of sunflowers is 162.3 cm and the standard deviation is 15.2 cm. Calculate the mean and standard deviation of the height of the plants in metres.

(ii) The mean temperature of soup served in a restaurant is 89°F. The standard deviation of the temperatures is 1.1°F. Calculate the mean and standard deviation of the temperatures in degrees Celsius. The formula to convert from f degrees Fahrenheit to c degrees Celsius is

$$c = \frac{5}{9}(f - 32)$$

Questions 4 and 5 use the data in Table R.4, giving how many countries won different numbers of gold medals at the 2016 Olympic Games.

The standard deviation of the number of gold medals per country is 4.79.

④ (i) Describe the distribution of the number of gold medals per country.

(ii) Find the median and the quartiles of the data.

(iii) Explain why a box-and-whisker plot is not an appropriate way to display these data.

(iv) Consider other possible ways of displaying the data and comment on the problems associated with them.

Note

The five countries with the most medals were: USA (46), Great Britain (27), China (26), Russia (19) and Germany (17).

Table R.4

Gold medals	Countries
0	148
1	21
2	11
3	6
4	3
5	2
6	2
7	2
8	4
9	1
10	0
11	0
12	2
17	1
19	1
26	1
27	1
46	1

⑤ Six statements are given here. Decide whether each of them is
- definitely a TRUE conclusion from the data
- definitely a FALSE conclusion from the data
- UNCERTAIN because it is impossible to say from the information given.

A   Over half the gold medals went to just 4 countries.

B   The mean number of gold medals per country was 1.5, correct to 1 decimal place.

C   Over 80% of countries obtained fewer than the mean number of gold medals per country.

D   It was the smaller countries that did not get any gold medals.

E   Seven countries can be identified as outliers because the number of gold medals they won was more than 2 standard deviations greater than the mean.

F   The mode and midrange are both zero.

⑥ As part of a study, a doctor keeps a record of the duration of the pregnancies of 200 patients and displays the data on this histogram (Figure R.11).

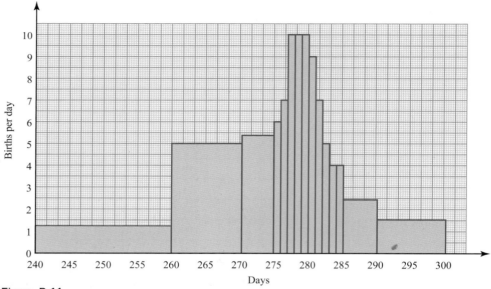

Figure R.11

(i) Show that this study covers 200 women.

The 'due date' of a mother to be is set at 280 days. Babies born before 260 days are described as 'pre-term'. Those born after 287 days are 'post-term'.

(ii) Find the percentage of the births in this study that resulted in babies that were

   (a) born on their due date

   (b) pre-term

   (c) post-term.

(iii) Describe the distribution shown in the histogram. Explain why, given the context, you would expect this to be the case.

(iv) The vertical scale is labelled 'Births per day'. Explain how this corresponds to the usual 'Frequency density' for a histogram.

⑦ (i) Summary measures for a data set are

$n = 207$, $\sum x = 309$ and $\sum x^2 = 5183$.

Calculate the standard deviation.

(ii) Show that $\sum (x - \overline{x})^2 = \sum x^2 - n\overline{x}^2$ and so the two formulae for S_{xx} are equivalent.

(iii) Are there any situations where one method might be preferred to the other?

Working with a large data set

🖥 **TECHNOLOGY**

Questions 8–11 are based on the data that Robin, a reporter from *The Avonford Star*, collected in order to investigate cycling accidents. He collected data for 93 accidents covering 13 fields. The full data set is available as a spreadsheet at www.hoddereducation.co.uk/AQAMathsYear2. It can also be found on pages 521–522 at the back of this book. You are expected to use a spreadsheet or statistical software when answering most of the questions.

Ⓣ ⑧ What does 'cleaning' a data set mean? Identify cases where you would need to clean the cycling accidents data set.

Ⓣ ⑨ Looking at the cycling accidents data spreadsheet, decide for which fields it is sensible to calculate some form of average. Where appropriate, say which form of average is the best one to choose for each field.

Ⓣ ⑩ Robin wants to write case studies of a sample of 12 of the cyclists. What sort of sample would you advise him to select?

Ⓣ ⑪ How would you compare the risks of serious accidents to young people and to older people?

KEY POINTS

1 Statistical problem solving uses the standard 4-stage problem solving cycle
   - problem specification and analysis
   - information collection
   - processing and representation
   - interpretation.

   The information collected is usually in the form of data.

2 Data collection often involves taking a sample. Types of samples include the following:
   - simple random
   - stratified
   - cluster
   - systematic
   - quota
   - opportunity
   - self-selected.

3 Processing and representation involves
   - cleaning the data
   - selecting, calculating and interpreting summary measures that are appropriate for the type of data and the context
   - selecting, presenting and interpreting display diagrams that are appropriate for the type of data and the context.

4 Typically summary measures and displays are given in Table R.5.

Table R.5

Type of data	Typical display	Summary measures	
		Central tendency	Spread
Categorical	Bar chart, Pie chart, Dot plot, Tally, Pictogram	Modal class	
Ranked	Stem-and-leaf diagram Boxplot	Median	
Discrete numerical (ungrouped)	Stem-and-leaf diagram Vertical line chart	Men, Mode, Median, Weighted mean	Range Interquartile range Semi-interquartile range
Discrete numerical (grouped)	Bar chart (groups as categories)	Mean, Modal class, Weighted mean	Range Standard deviation
Continuous	Frequency chart Histogram	Mean, Median Modal class	Range Standard deviation
	Cumulative frequency curve	Median	Interquartile range Semi-interquartile range

5 For a data set, $x_1, x_2, \ldots x_n$

Mean $\bar{x} = \dfrac{\sum x_i}{n}$

Sum of squared deviations: $S_{xx} = \sum \left(x_i - \bar{x} \right)^2 = \sum x_i^2 - n\bar{x}^2$

Variance: $s^2 = \dfrac{S_{xx}}{n - 1}$

Standard deviation: $s = \sqrt{\text{variance}} = \sqrt{\dfrac{S_{xx}}{n - 1}}$

Trains

The details of this situation are fictitious, but they are based on a real problem.

Here is one of many letters of complaint that a Train Operating Company receives about a particular train.

Figure 1

I am writing to complain about the overcrowding on your 1723 train. I use this train regularly to get home from work but the 28 minute journey is the most uncomfortable time of my day.

If you thought of your passengers you would add extra coaches to the train but you don't. Everyone on the train says it is because you put profits before passengers.

We are all hoping you lose the franchise when it comes up next year.

What should the company do about the situation?

1 **Problem specification and analysis**

Letters of complaint like this are not good news for the company and the danger of losing the franchise is serious. They review the situation.

- The train mentioned in the letter runs between five stations (A to E) each weekday from Monday to Friday. The timings are the same each day.

  A 1618   **B** 1648   **C** 1723   **D** 1751   **E** 1827

  The letter is complaining about the journey from C to D.

- The train has 4 coaches with a total of exactly 200 seats. It is possible to add one more coach with an additional 58 seats but this would have to be for the whole journey every day. If all the stages are overcrowded this would be an obvious solution to the problem, but if that is not the case it might be a costly thing to do. The company has a rule of thumb that a train is running profitably if at least 50% of its seats are occupied.

- A slow stopping train leaves C at 1715 and is almost empty when it arrives at D at 1829. It is run by a different company.

  A fast train, also run by a different company, leaves C at 1728 and goes through D without stopping even though a platform is available. This train overtakes the 1723 while it is stationary at D.

So, as a first step, the company decides to investigate how many passengers its train carries at each stage between A and E on a typical day.

2 Information collection

Devise an information collection strategy for the company. It should include advice on

- the number of observations to be made, i.e. the sample size
- when the observations should be made
- who should collect the information.

Your report should consider any possible causes of inaccuracy or bias, and how they might relate to the timescale for the work to be carried out.

3 Processing and representation

The company actually uses the train manager on each train to collect data on the number of people on the train at the various stages during two consecutive weeks. The data are given in Table 1.

Table 1

Day \ Stage	A to B	B to C	C to D	D to E
Monday	108	124	240	110
Tuesday	85	126	248	116
Wednesday	90	90	90	90
Thursday	76	108	Too crowded to count	–
Friday	42	87	198	98
Monday	121	156	252	108
Tuesday	91	110	238	128
Wednesday	75	98	212	12
Thursday	72	84	204	106
Friday	48	79	201	86

Clean these data, explaining any decisions you make.

Then present them in a way that will be as helpful as possible for the company's decision makers. This will include using one or more suitable data displays.

4 Interpretation

Based on all the available information, state what you would recommend the company to do.

Probability

→ How does probability affect the outcome in a court of law in

(i) a civil case

(ii) a criminal case?

Review: Probability

The probability of an event

Probability is a number between 0 and 1 which measures likelihood. A probability of 0 means impossibility and 1 means certainty. Probability may be estimated experimentally or theoretically.

The probability of an event occurring is estimated by repeating an experiment and noting how many times the event occurs. This gives the approximate probability of an event which cannot be calculated theoretically. The more often the experiment is repeated, the more confident you can be in the result obtained.

As an example, the probability of a drawing pin landing pin-up is estimated by throwing it up in the air 10 times. It lands pin-up 3 times so the probability of this event is estimated to be $\frac{3}{10}$ or 0.3. This is not, however, a figure that you would feel very confident about. So the experiment is repeated to give a total of 100 throws. It lands pin-up 28 times so the probability is now estimated to be 0.28. The experiment is then repeated to obtain a total of 1000 throws, of which 291 are pin-up. You now have three estimates of the probability; with increasing confidence they are 0.3, 0.28 and 0.291.

In some situations, like tossing a coin, throwing dice or selecting a playing card from a pack at random, all the possible outcomes are equally likely, or equiprobable. In such cases, you can calculate the theoretical probability of an event.

For example, the probability of getting a total score of 9 when adding the scores on two dice is calculated from the **sample space diagram** which shows all the possible outcomes from throwing two dice. (You may find it helpful to think of them as one red and one blue, as indicated in the diagram.)

Figure 15.1

Scores	1	2	3	4	5	6
1	2	3	4	5	6	7
2	3	4	5	6	7	8
3	4	5	6	7	8	9
4	5	6	7	8	9	10
5	6	7	8	9	10	11
6	7	8	9	10	11	12

As the dice are different the order of the outcomes is important. '4 and 5' is not the same as '5 and 4'.

$$P(9) = \frac{4}{36} \text{ where}$$

- 4 is the number of ways you could get a total score of 9. They are 3 and 6; 4 and 5; 5 and 4; 6 and 3.

- 36 is the total number of possible outcomes.

This is expressed as the following rule for an event A.

$$P(A) = \frac{n(A)}{n(\mathscr{E})}$$

where $n(A)$ is the number of ways that event A can occur and

$n(\mathscr{E})$ is the total number of ways that the possible events can occur.

The probability of the **complement** of an event A, that is the event not happening, is given by

$$P(A') = 1 - P(A)$$

where the notation A' is used to denote the complementary event 'not A'.

The probability of two events from a single experiment

You can show the outcomes of two events resulting from a single experiment on a Venn diagram like that in Figure 15.2 (overleaf) for A and B. You can then use it to work out the probability of any pair of outcomes for the two events. This is usually referred to as the **probability of two events**.

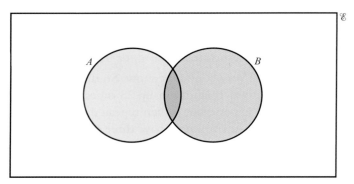

Figure 15.2

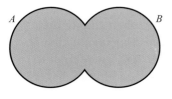

Figure 15.3 ($A \cup B$)

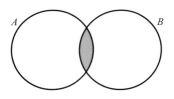

Figure 15.4 ($A \cap B$)

This shows you that for any two events, A and B,

$$P(A \cup B) = P(A) + P(B) - P(A \cap B).$$

$P(A \cup B)$ is the probability of the **union** of two events, as shown in Figure 15.3. This is more commonly read as 'the probability of A **or** B occurring'.

$P(A \cap B)$ is the probability of the **intersection** of two events (Figure 15.4). This is more commonly read as 'the probability of A **and** B occurring'.

Where the events are **mutually exclusive** (i.e. they cannot both occur) then this is still true, but the probability of $P(A \cap B)$ is zero so the equation simplifies to: $P(A \cup B) = P(A) + P(B)$.

In this case the Venn diagram looks like Figure 15.5.

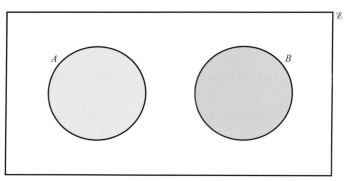

Figure 15.5 $P(A \cap B) = 0$ and so $P(A \cup B) = P(A) + P(B)$

Example 15.1

There are 15 goldfish and 30 tench in a tank at a shop. There are 8 male fish and 3 of these are goldfish; the other fish are all female.

A fish is caught at random in a net.

(i)   Identify the experiment, the events and their possible outcomes, in this situation.

(ii) Using G for 'The fish is a goldfish' and M for 'The fish is male', draw a labelled Venn diagram to illustrate the situation, and write the correct number in each region.

(iii) Find the probability that the fish is
(a) a male goldfish   (b) a male or a goldfish   (c) a female tench.

(iv) Show that $P(G) + P(M) - P(G \cap M) = P(G \cup M)$.

Solution

(i)

Experiment	Events	Outcomes

Selecting a fish at random

Type of fish → Goldfish, Tench

Sex of fish → Male, Female

(ii)

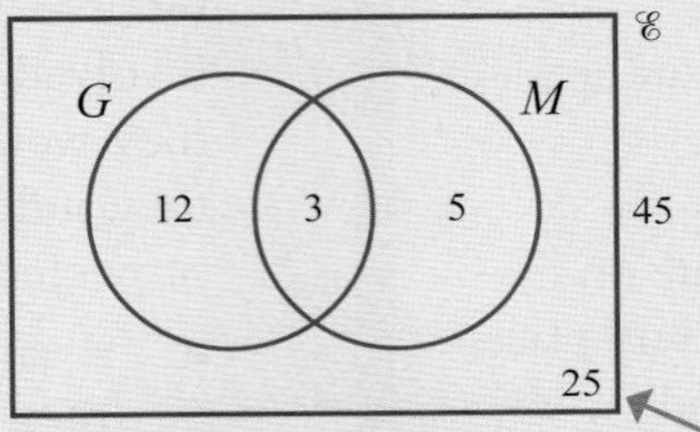

Figure 15.6

A female tench is not a goldfish (G') and ($\cap$) not a male (M').

(iii) (a) $P(G \cap M) = \frac{3}{45}$

(b) $P(G \cup M) = \frac{20}{45}$

(c) $P(G' \cap M') = \frac{25}{45}$

(iv) $P(G) + P(M) - P(G \cap M) = P(G \cup M)$

$\frac{15}{45} + \frac{8}{45} - \frac{3}{45} = \frac{20}{45}$

as required.

The expected frequency of an event

In the previous example, the probability of catching a male goldfish was $\frac{3}{45}$ or $\frac{1}{15}$.

If the fish was put back and the experiment was repeated 30 times, you would expect to get a male goldfish on $30 \times \frac{1}{15}$ or 2 occasions. The number 2 (in this case) is the **expected frequency** or the **expectation**.

Review exercise

① Aled has three 2p coins, two 10p coins and one £1 coin in his pocket. He selects one coin at random. Assuming that all the coins are equally likely to be selected, find the probability that it is worth

(i) 10p   (ii) less than 10p

(iii) more than 10p   (iv) less than £2

(v) £2 or more.

② Figure 15.7 shows the cards in a standard pack.

	2	3	4	5	6	7	8	9	10	J	Q	K	A
♠	*	*	*	*	*	*	*	*	*	*	*	*	*
♥	*	*	*	*	*	*	*	*	*	*	*	*	*
♦	*	*	*	*	*	*	*	*	*	*	*	*	*
♣	*	*	*	*	*	*	*	*	*	*	*	*	*

Figure 15.7

Two cards are chosen at random from a standard pack. The first one is not replaced before the second one is chosen.

Say whether the following statements are TRUE or FALSE. If FALSE, explain why.

(i)   Figure 15.7 is an example of a sample space diagram.

(ii)   The probability that the two cards are those shown by blue stars (★) is $\frac{1}{2652}$.

(iii)   The probability that both cards are the Queen of Spades is 0.

(iv)   The probability that the first card chosen is a 4 or a heart is $\frac{17}{52}$.

(v)   The probability that both cards are spades is $\left(\frac{1}{4}\right)^2 = \frac{1}{16}$.

(vi)   The probability that the total score on the two cards is 15 is 0.0362 (to 4 decimal places).

③   Given that P(A) = 0.8, P(B) = 0.6 and P($A \cap B$) = 0.5, find P($A \cup B$).

④   Given that P(C) = 0.6, P(D) = 0.7, and that C and D are independent, find P($C \cap D$).

⑤   Given that P(E') = 0.44, P(F') = 0.28 and P($E \cup F$) = 1, find P($E \cap F$).

⑥   A poultry keeper sets a night vision camera to record what predators visit her garden at night during the month of June one year. She records the results and illustrates them on the Venn diagram in Figure 15.8, using the letters B and F to denote the following events for any night.

■ B: at least one badger visits her garden

■ F: at least one fox visits her garden

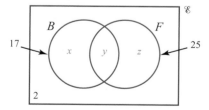

Figure 15.8

(i)   State the values of the numbers x, y and z on the Venn diagram.

(ii)   Find P(B), P(F), P($B \cup F$) and P($B \cap F$).

(iii)   Show that
P($B \cup F$) = P(B) + P(F) − P($B \cap F$)

(iv)   State, with an explanation, whether the events B and F are mutually exclusive.

⑦   In a year cohort the probability that a randomly chosen student is studying French is 0.84, the probability that a randomly chosen student is studying German is 0.65, and 8% of the cohort study neither French nor German. Find the probability that a randomly chosen student is studying both French and German.

⑧   Given that P(A) = 0.9 and P(B) = 0.7, find the smallest possible value of P($A \cap B$).

Working with a large data set

⑨   In the Avonford cycling accidents data set, information is available on 85 cyclists involved in accidents regarding whether they were wearing helmets and whether they suffered from concussion (actual or suspected).

■   Event C is that an individual cyclist suffered from concussion.

■   Event H is that an individual cyclist was wearing a helmet.

■   22 cyclists suffered from concussion.

■   55 cyclists were wearing helmets.

■   13 cyclists were both wearing helmets and suffered from concussion.

(i)   Copy this Venn diagram and place the correct numbers in the four different regions.

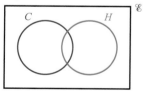

Figure 15.9

One of the cyclists is selected at random.

(ii)   Write down (a) P(C')   (b) P($C \cap H$)

(iii)   Verify that for these data
P($C \cup H$) = P(C) + P(H) − P($C \cap H$)

(iv)   Show that P($C' \cap H'$) = 1 − P($C \cup H$)

(v)   Compare the percentages of those wearing helmets who suffered from concussion and those not wearing helmets who suffered from concussion.

Comment on whether this is a reliable comparison.

1 The probability of events from two experiments

This story describes two pieces of good fortune on the same day. Veronica said the probability was about $\frac{1}{1000\,000}$. What was it really?

The two events resulted from two different experiments, the raffle draw and the programme draw. This is not the same as two events based on the outcomes from a single experiment. Consequently, this situation is different from looking at two outcomes from a single experiment (like a playing card chosen at random from a pack being a heart or being an ace).

The total sales of raffle tickets were 1245 and of programmes 324. The draws were conducted fairly, that is each number had an equal chance of being selected. Table 15.1 sets out the two experiments and their corresponding events with associated probabilities.

Table 15.1

Experiments	**Events (and estimated probabilities)**
Raffle draw	Winning with a single ticket $\frac{1}{1245}$
	Not winning with a single ticket $\frac{1244}{1245}$
Programme draw	Winning with a single programme $\frac{1}{324}$
	Not winning with a single programme $\frac{323}{324}$

> If you want to prove that events A and B are independent, you should show that they satisfy this equation.

In situations like this, the possible outcomes resulting from the different experiments are often shown on a tree diagram.

If the events A and B are independent then $P(A \cap B) = P(A) \times P(B)$.

Multiplying probabilities together can often give incorrect answers when the events are not independent.

Example 15.2

Find, in advance of the results of the two draws, the probability that Veronica would do the following:

(i) win both draws,

(ii) fail to win either draw,

(iii) win exactly one of the two draws. →

Solution

The possible results are shown on the tree diagram in Figure 15.10.

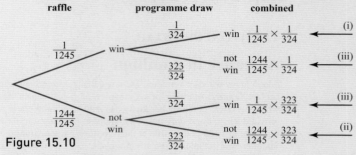

Figure 15.10

(i) The probability that Veronica wins both $= \frac{1}{1245} \times \frac{1}{324} = \frac{1}{403\,380}$

This is not quite Veronica's 'one in a million' but it is not very far off it. In fact the 'miracle' is even less unlikely than this calculation suggests. Although the probability that Veronica herself wins both draws is $\frac{1}{403\,380}$, the probability that any one person wins both is the probability that the one person who won the first draw also wins the second, which is simply $\frac{1}{324}$. Given the number of village fetes in the country, this probability suggests that such so-called 'miracles' are going to happen quite often somewhere.

(ii) The probability that Veronica wins neither $= \frac{1244}{1245} \times \frac{323}{324} = \frac{401\,812}{403\,380}$

This of course is much the most likely outcome.

> **Wins raffle but not programme draw.**

> **Wins programme draw but not raffle.**

(iii) The probability that Veronica wins one but not the other is given by

$$= \frac{1}{1245} \times \frac{323}{324} + \frac{1244}{1245} \times \frac{1}{324} = \frac{1567}{403\,380}$$

Look again at the structure of the tree diagram in Figure 15.10.

There are two experiments, the raffle draw and the programme draw. These are considered as 'first, then' experiments and set out 'first' on the left and 'then' on the right. The rest of the layout falls into place, with the different outcomes or events appearing as branches. In this example there are two branches at each stage; sometimes there may be three or more. Notice that for a given situation the component probabilities sum to 1.

> **Notice that this check was made easier by not cancelling common factors in the fractions.**

$$= \frac{1}{403\,380} + \frac{323}{403\,380} + \frac{1224}{403\,380} + \frac{401\,812}{403\,380} = \frac{403\,380}{403\,380} = 1$$

Example 15.3

Some friends buy a six-pack of crisps. Two of the bags are salted potato crisps (S), the rest are fruit crisps (F). They decide to allocate the bags by lucky dip. Find the probability that

(i) the first two bags chosen are the same as each other

(ii) the first two bags chosen are different from each other.

> **Note**
>
> P(F, S) means the probability of drawing a fruit crisps bag (F) on the first dip and a salted potato crisps bag (S) on the second.

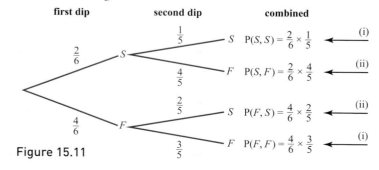

Figure 15.11

Solution

(i) The probability that the first two bags chosen are the same as each other $= P(S, S) + P(F, F)$

$$= \frac{2}{6} \times \frac{1}{5} + \frac{4}{6} \times \frac{3}{5}$$

$$= \frac{2}{30} + \frac{12}{30}$$

$$= \frac{14}{30}$$

(ii) The probability that the first two bags chosen are different from each other $= P(S, F) + P(F, S)$

$$= \frac{2}{6} \times \frac{4}{5} + \frac{4}{6} \times \frac{2}{5}$$

$$= \frac{8}{30} + \frac{8}{30}$$

$$= \frac{16}{30}$$

> **Note**
>
> The answer to part (ii) on the right hinges on the fact that two orderings (S then F, and F then S) are possible for the same combined event (that the two bags selected include one salted potato crisps and one fruit crisps bag).

The probabilities changed between the first dip and the second dip. This is because the outcome of the second dip is dependent on the outcome of the first one (with fewer bags remaining to choose from).

By contrast, the outcomes of the two experiments involved in tossing a coin twice are independent, and so the probability of getting a head on the second toss remains unchanged at 0.5, whatever the outcome of the first toss.

Although you may find it helpful to think about combined events in terms of how they would be represented on a tree diagram, you may not always actually draw them in this way. If there are several experiments and perhaps more than two possible outcomes from each, drawing a tree diagram can be very time-consuming.

Example 15.4

A roulette wheel has 37 numbers on it (0, 1, 2, 3, ..., 36); they are all equally likely to come up. The odds offered by most casinos for a single number are 35–1, meaning that a player who bets £1 on the number that comes up wins £35 and gets the £1 back. A roulette player has a 'lucky' number of 17. He bets on it 40 times without winning.

(i) Does this mean that his 'lucky' number is actually unlucky?

(ii) He bets £1 each time. How much does he expect to win or lose in 40 rolls?

Solution

(i) The probability that a given number (e.g. 17) comes up on any roll is $\frac{1}{37}$.

On one roll, $P(\text{win}) = \frac{1}{37}$ $P(\text{not win}) = 1 - \frac{1}{37} = \frac{36}{37}$

On 40 rolls, $P(\text{no wins}) = \frac{36}{37} \times \frac{36}{37} \times \frac{36}{37} \times \cdots \times \frac{36}{37} = \left(\frac{36}{37}\right)^{40}$

$$= 0.334$$

So no wins in 40 rolls is not particularly unlikely.

➜

(ii)   So on one roll the expectation of his winnings is

$$£\left(35 \times \tfrac{1}{37} - 1 \times \tfrac{36}{37}\right) = -2.7\text{p or a 2.7p loss.}$$

On 40 rolls his expected loss is $40 \times 2.7\text{p} = £1.08$.

Example 15.5

THE AVONFORD STAR

Is this justice?

In 2011, local man David Starr was sentenced to 12 years imprisonment for armed robbery solely on the basis of an identification parade. He was one of 12 people in the parade and was picked out by one witness but not by three others. Many Avonford people who knew

David well believe he was incapable of such a crime. We in the *Star* are now adding our voice to the clamour for a review of his case. How conclusive, we ask, is this sort of evidence, or, to put it another way, how likely is it that a mistake has been made?

Investigate how likely it is that David Starr really did commit the robbery. In this you should assume that all the four witnesses at the identity parade selected one person and that their selections were done entirely at random.

Solution

In this situation you need to assess the probability of an innocent individual being picked out by chance alone.

Assume that David Starr was innocent.

Since the witnesses were selecting in a purely random way, the probability was $\tfrac{1}{12}$ of selecting any person and $\tfrac{11}{12}$ of not selecting that person.

So the probability that none of the four witnesses chose David Starr is $\tfrac{11}{12} \times \tfrac{11}{12} \times \tfrac{11}{12} \times \tfrac{11}{12}$.

The probability that at least one of the witnesses selected him is $1 -$ the probability that none of them did.

P(at least one selection) = 1 − P(no selections)

$$= 1 - \tfrac{11}{12} \times \tfrac{11}{12} \times \tfrac{11}{12} \times \tfrac{11}{12}$$
$$= 1 - 0.706 = 0.294$$

In other words, there is about a 30% chance of an innocent person being chosen in this way by at least one of the witnesses.

Discussion point

→ Discuss the assumptions underlying the calculation in this example.

The same newspaper article concluded:

Is 30% really the sort of figure we have in mind when judges use the phrase 'beyond reasonable doubt'? Because if it is, many innocent people will be condemned to a life behind bars.

Exercise 15.1

①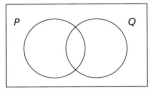

P(P) = 0.7, P(P∪Q)′ = 0.1, P(Q′) = 0.4. Work out P(P∩Q).

② The probability of a pregnant woman giving birth to a girl is about 0.49.

Draw a tree diagram showing the possible outcomes if she has two babies (not twins). From the tree diagram, calculate the probability that

(i) the babies are both girls

(ii) the babies are the same sex

(iii) the second baby is of different sex to the first.

③ In a certain district of a large city, the probability of a household suffering a break-in in a particular year is 0.07 and the probability of its car being stolen is 0.12.

Assuming these two events are independent of each other, draw a tree diagram showing the possible outcomes for a particular year. Calculate, for a randomly selected household with one car, the probability that

(i) the household is a victim of both crimes during that year

(ii) the household suffers only one of these misfortunes during that year

(iii) the household suffers at least one of these misfortunes during that year.

④ There are 12 people at an identification parade; one of them is guilty and the others are innocent. Three witnesses are called to identify the guilty person.

(i) Assuming they make their choice purely by random selection, draw a tree diagram showing the possible outcomes labelling them Correct or Wrong.

(ii) From the tree diagram, calculate the probability that

(a) all three witnesses select the accused person

(b) none of the witnesses selects the accused person

(c) at least two of the witnesses select the accused person.

(iii) Suppose now that the composition of people in the identification parade is changed. Two witnesses now know that the guilty person is one out of four individuals in the parade. The third chooses at random. Draw a new tree diagram and calculate the probability that

(a) all three witnesses select the accused person

(b) none of the witnesses selects the accused person

(c) at least two of the witnesses select the accused person.

⑤ Ruth drives her car to work – provided she can get it to start! When she remembers to put the car in the garage the night before, it starts next morning with a probability of 0.95. When she forgets to put the car away, it starts next morning with a probability of 0.75. She remembers to garage her car 90% of the time. What is the probability that Ruth drives her car to work on a randomly chosen day?

⑥ All the Jacks, Queens and Kings are removed from a pack of playing cards. Giving the Ace a value of 1, this leaves a pack of 40 cards consisting of four suits of cards numbered 1 to 10. The cards are well shuffled and one is drawn and noted. This card is not returned to the pack and a second card is drawn.

Find the probability that

(i) both cards are even

(ii) at least one card is odd

(iii) both cards are of the same suit

(iv) only one of the cards has a value greater than 7.

⑦ Three dice are thrown. Find the probability of obtaining

(i) at least two 6s

(ii) no 6s

(iii) different scores on all the dice.

⑧ Explain the flaw in this argument about throwing dice and rewrite it as a valid statement.

The probability of getting a 6 on one throw = $\frac{1}{6}$. Therefore the probability of getting at least one 6 in six throws is

$\frac{1}{6} + \frac{1}{6} + \frac{1}{6} + \frac{1}{6} + \frac{1}{6} + \frac{1}{6} = 1$ *so it is a certainty.*

⑨ In a Donkey Derby event, there are three races. There are six donkeys entered for the first race, four for the second and three the third. Sheila places a bet on one donkey in each race. She knows nothing about donkeys and chooses each donkey at random.

Find the probability that she backs at least one winner.

⑩ The probability of someone catching flu in a particular winter when they have been given the flu vaccine is 0.1. Without the vaccine, the probability of catching flu is 0.4. If 30% of the population has been given the vaccine, what is the probability that a person chosen at random from the population will catch flu over that winter?

⑪ In a game, players pay an entry fee of £1. They then throw dice, one at a time, until one shows 6. If only one throw is needed, the player receives £3. If two throws are needed, the player receives £2. If three throws are needed, the player gets his or her entry fee back; otherwise the £1 is lost.

Find the probability that a player

(i) receives £3   (ii) receives £2

(iii) gets the £1 entry fee back

(iv) loses the entry fee

Working with a large data set

⑫ This question is based on the Avonford cycling accidents data set. The relevant data are unclear or missing for 5 of the cyclists and so are excluded from this question, leaving data for a total of 88 cyclists.

■ 44 of the cycling accidents involved a motor vehicle (car, lorry, bus etc); the others did not.

■ After an accident involving a motor vehicle, 24 of the cyclists spent at least one night in hospital. The others involved in such accidents did not spend a night in hospital.

■ The other accidents did not involve a motor vehicle, Out of the cyclists involved in those accidents, 35 did not spend a night in hospital.

Having an accident involving a motor vehicle is referred to as event M. Spending at least one night in hospital is event H.

(i) Copy this tree diagram and write the missing probabilities on the branches.

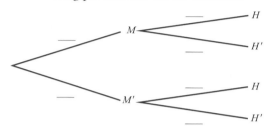

Figure 15.12

(ii) Find the probability that a person selected at random from the data set

(a) had an accident involving a motor vehicle and spent at least one night in hospital

(b) did not spend a night in hospital.

(iii) Copy this table and fill in the cells with the correct numbers of cyclists.

Table 15.2

	H	H'	Total
M			
M'			
Total			88

(iv) Robin says 'The data tell me that accidents involving motor vehicles tend to be more serious'.

Explain why he might come to this conclusion.

2 Conditional probability

What is the probability that somebody chosen at random will die of a heart attack in the next 12 months? To answer this question, you need to develop a probability model. One model would be based on saying that, since there are about 150 000 deaths per year from diseases of the circulatory system among the 64 000 000 population of the UK,

$$\text{Probability} = \frac{\text{Number of deaths from circulatory diseases per year in UK}}{\text{Total population of UK}}$$

$$= \frac{150\,000}{64\,000\,000} = 0.0023$$

However, if you think about it, you will probably realise that this is rather a meaningless figure. For a start, young people are much less at risk than those in or beyond middle age.

So you might wish to refine your model so that it gives two answers:

$$P_1 = \frac{\text{Deaths from circulatory diseases among over-55s}}{\text{Population of over-55s}}$$

$$P_2 = \frac{\text{Deaths from circulatory diseases among under-55s}}{\text{Population of under-55s}}$$

Typically only 5000 of the deaths would be among the under-55s leaving (on the basis of these figures) 145 000 among the over-55s.

About 18 000 000 people in the UK are over 55, and 46 000 000 under 55. This gives

'Over–55s' are those who have had their 55th birthday.

$$P_1 = \frac{\text{Deaths from circulatory diseases among over-55s}}{\text{Population of over-55s}}$$

$$= \frac{145\,000}{18\,000\,000} = 0.0081$$

$$P_2 = \frac{\text{Deaths from circulatory diseases among under-55s}}{\text{Population of under-55s}}$$

$$= \frac{5000}{46\,000\,000} = 0.00011$$

So, according to this model, somebody in the older group is over 70 times more likely to die of a heart attack than somebody in the younger group. Putting them both together as an average figure resulted in a figure that was representative of neither group. (The figures used in this section are approximated from those supplied by the British Heart Foundation who had themselves used a number of official sources.)

But why stop there? You could, if you had the figures, divide the population up into 10-year, 5-year, or even 1-year intervals. That would certainly improve the accuracy of this model, but there are also more factors that you might wish to take into account, such as the following.

■ Is the person overweight?
■ Does the person smoke?
■ Does the person take regular exercise?

The more conditions you build in, the more accurate will be the estimate of the probability obtained from the model.

You can see how the conditions are brought in by looking at P_1:

$$P_1 = \frac{\text{Deaths from circulatory diseases among over-55s}}{\text{Population of over-55s}}$$

$$= \frac{145\,000}{18\,000\,000} = 0.0081$$

You would write this in symbols as follows:

Event G: Somebody selected at random is over 55.

Event H: Somebody selected at random dies from circulatory diseases.

The probability of someone dying from circulatory diseases given that he or she is over 55 is given by the conditional probability $P(H\,|\,G)$ where

> $P(H\,|\,G)$ means the probability of event H occurring *given that* event G has occurred.

$$P(H\,|\,G) = \frac{n(H \cap G)}{n(G)}$$

$$= \frac{\dfrac{n(H \cap G)}{n(\mathscr{E})}}{\dfrac{n(G)}{n(\mathscr{E})}}$$

$$= \frac{P(H \cap G)}{P(G)}$$

In general, for events A and B, $P(B\,|\,A) = \dfrac{P(B \cap A)}{P(A)}$

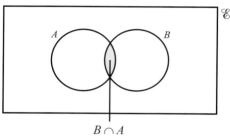

$B \cap A$

Figure 15.13

Conditional probability is used when your estimate of the probability of an event is altered by your knowledge of whether some other event has occurred. In this case the estimate of the probability of somebody dying from circulatory diseases, $P(H)$, is altered by a knowledge of whether the person is over 55 or not.

Thus, conditional probability addresses the question of whether one event is dependent on another one. It is very important in probability models. If the probability of event B is not affected by the occurrence of event A, you say that B is independent of A. If, on the other hand, the probability of event B is affected by the occurrence (or not) of event A, you say that B is dependent on A.

> A' is the event not–A, i.e. that A does not occur.

If A and B are independent, then $P(B\,|\,A) = P(B\,|\,A')$ and this is just $P(B)$.

If A and B are dependent, then $P(B\,|\,A) \neq P(B\,|\,A')$.

> The probability of both A and B occurring.

> The probability of B occurring, given that A has happened.

As you have already seen, the probability of a combined event is the product of the separate probabilities of each event, provided the question of dependence between the two events is properly dealt with. Specifically:

> The probability of A occurring.

For dependent events $P(A \cap B) = P(A) \times P(B\,|\,A)$.

When A and B are independent events, then, because $P(B\,|\,A) = P(B)$ this can be written as

For independent events $P(A \cap B) = P(A) \times P(B)$.

Example 15.6

A company is worried about the high turnover of its employees and decides to investigate whether they are more likely to stay if they are given training.

On 1 January one year the company employed 256 people. During that year a record was kept of who received training as well as who left the company. The results are summarised in this Table 15.3:

Table 15.3

	Still employed	Left company	Total
Given training	109	43	152
Not given training	60	44	104
Total	169	87	256

This is a two-way table.

Find the probability that a randomly selected employee

(i)   received training

(ii)   did not leave the company

(iii)   received training and did not leave the company

(iv)   did not leave the company, given that the person had received training

(v)   did not leave the company, given that the person had not received training.

Alternatively, you can say, 'given that the person had received training' limits the population to 152 people. 109 did not leave, so the probability is $\frac{109}{152} = 0.72$ (2 s.f.).

Discussion point

➜ How would you show that the event T is not independent of the event S?

Solution

Using the notation T: the employee received training

S: the employee stayed in the company

(i)   P(received training) $P(T) = \dfrac{n(T)}{n(\mathscr{E})} = \dfrac{152}{256}$

(ii)   P(did not leave) $P(S) = \dfrac{n(S)}{n(\mathscr{E})} = \dfrac{169}{256}$

(iii)   P(received training and did not leave) $P(T \cap S) = \dfrac{n(T \cap S)}{n(\mathscr{E})} = \dfrac{109}{256}$

(iv)   P(did not leave given that they received training)

$$P(S \mid T) = \frac{P(S \cap T)}{P(T)} = \frac{\frac{109}{256}}{\frac{152}{256}} = \frac{109}{152} = 0.72$$

(v)   P(did not leave given that they did not receive training)

$$P(S \mid T') = \frac{P(S \cap T')}{P(T')} = \frac{\frac{60}{256}}{\frac{104}{256}} = \frac{60}{104} = 0.58$$

Since $P(S \mid T)$ is not the same as $P(S \mid T')$, the event S is not independent of the event T. Each of S and T is dependent on the other, a conclusion which matches common sense. It is almost certainly true that training increases employees' job satisfaction and so makes them more likely to stay, but it is also probably true that the company is more likely to go to the expense of training the employees who seem less inclined to move on to other jobs.

In some situations you may find it helps to represent a problem such as this as a Venn diagram as shown in Figure 15.14.

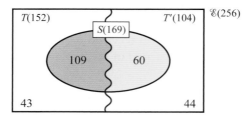

Figure 15.14

> **Discussion points**
>
> → What do the various numbers and letters represent?
>
> → Where is the region S'?
>
> → How are the numbers in Figure 15.14 related to the answers to parts (i) to (v) of Example 15.6?

In other situations it may be helpful to think of conditional probabilities in terms of tree diagrams. Conditional probabilities are needed when events are dependent, that is when the outcome of one trial affects the outcomes from a subsequent trial, so for dependent events, the probabilities of the second and any subsequent layers of a tree diagram are conditional.

Example 15.7

Rebecca is buying two goldfish from a pet shop. The shop's tank contains seven male fish and eight female fish but they all look the same (Figure 15.15).

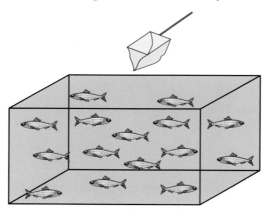

Figure 15.15

Find the probability that Rebecca's fish are
(i)   both the same sex
(ii)   both female
(iii)   both female given that they are the same sex.

Solution

The situation is shown on the tree diagram in Figure 15.16.

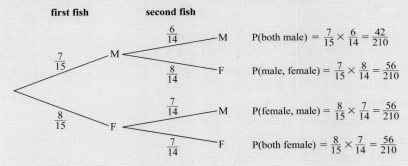

Figure 15.16

(i)   P(both the same sex) = P(both male) + P(both female)

$$= \frac{42}{210} + \frac{56}{210} = \frac{98}{210}$$

(ii)   P(both female) $= \frac{56}{210}$

(iii) P(both female | both the same sex)

$$= \frac{P(\text{both female and the same sex})}{P(\text{both the same sex})}$$

This is the same as P(both female).

$$= \frac{\frac{56}{210}}{\frac{98}{210}} = \frac{4}{7}$$

The idea in the last example can be expressed more generally using formal notation for any two dependent events, A and B. The tree diagram would be as in Figure 15.17.

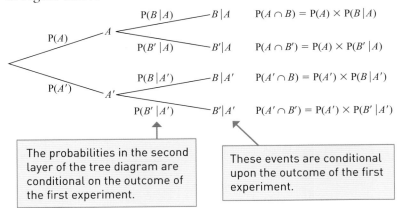

The probabilities in the second layer of the tree diagram are conditional on the outcome of the first experiment.

These events are conditional upon the outcome of the first experiment.

Figure 15.17

Discussion point

→ How were these results used in Example 15.7 about the goldfish?

The tree diagram shows you that

$$P(B) = P(A \cap B) + P(A' \cap B)$$

$$= P(A) \times P(B|A) + P(A') \times P(B|A')$$

and that $P(A \cap B) = P(A) \times P(B|A)$

$$\Rightarrow P(B|A) = \frac{P(A \cap B)}{P(A)}$$

Exercise 15.2

Those questions in this exercise where the context is drawn from real life involve working with mathematical models.

① In a school of 600 students, of whom 360 are girls, there are 320 hockey players, of whom 200 are girls. Among the hockey players there are 28 goalkeepers, 19 of them girls. Find the probability that

(i) a student chosen at random is a girl

(ii) a girl chosen at random plays hockey

(iii) a hockey player chosen at random is a girl

(iv) a student chosen at random is a goalkeeper

(v) a goalkeeper chosen at random is a boy

(vi) a male hockey player chosen at random is a goalkeeper

(vii) a hockey player chosen at random is a male goalkeeper

(viii) two students chosen at random are both goalkeepers

(ix) two students chosen at random are a male goalkeeper and a female goalkeeper

(x) two students chosen at random are one boy and one girl

② Table 15.4 below gives the numbers of offenders in England and Wales sentenced for indictable offences, by type of offence and type of sentence, in 2013.

(i) Find the probability that a randomly selected person indicted for an offence will be

(a) discharged

(b) given a caution

(c) fined

(d) sent to jail

(e) sent to jail for a motoring offence

(f) sent to jail given that the person has committed a motoring offence

(g) guilty of a motoring offence given that the person is sent to jail.

(ii) Criticise this statement:

Based on these figures nearly 1.5% of the country's prison population are there for motoring offences.

Table 15.4

Offence group	Penalty notices for disorder	Cautions	Proceedings	Convictions	Sentenced	Discharged	Fine	Custody	Total
Violence against the person	–	11 966	55 161	34 061	33 706	1 439	2 345	12 678	151 356
Sexual offences	–	1 447	9 791	5 512	5 489	79	124	3 259	2 571
Burglary	–	2 264	26 887	20 500	20 143	332	249	10 658	81 033
Robbery	–	160	10 595	7 377	7 368	21	16	4 634	30 171
Theft and handling stolen goods	25 331	32 006	118 919	105 721	105 033	22 287	16 276	22 148	447 721
Fraud and forgery	–	4 347	20 282	15 576	15 352	1 493	1 935	3 789	62 774
Criminal damage	–	3 445	3 445	5 394	5 298	1 066	698	834	20 180
Drug offences	15 374	37 515	63 522	57 218	56 518	9 801	20 909	8 617	269 474
Indictable motoring offences	–	–	3 193	2 788	2 813	35	182	1 083	10 094
Other indictable offences	–	4 933	51 901	37 811	37 846	2 871	10 967	9 344	155 673
Total indictable offences	40 705	98 083	363 696	291 958	289 566	39 424	53 701	77 044	1 254 177

③ 100 cars are entered for a roadworthiness test which is in two parts, mechanical and electrical. A car passes only if it passes both parts. Half the cars fail the electrical test and 62 pass the mechanical. 15 pass the electrical but fail the mechanical test.

Find the probability that a car chosen at random

(i)   passes overall

(ii) fails on one test only

(iii) given that it has failed, failed the mechanical test only.

④ Two dice are thrown. What is the probability that the total is

(i)   7

(ii) a prime number

(iii) 7, given that it is a prime number?

⑤ In a school of 400 students, 250 play a musical instrument and 100 sing in the choir. The probability that a student chosen at random neither plays a musical instrument nor sings in the choir is $\frac{1}{5}$.

(i)   How many students both sing in the choir and play a musical instrument?

(ii) Find the probability that a student chosen at random sings in the choir but does not play an instrument.

(iii) Find the probability that a member of the choir chosen at random does not play an instrument.

(iv) Find the probability that someone who does not play an instrument, chosen at random, is in the choir.

⑥ Quark hunting is a dangerous occupation. On a quark hunt, there is a probability of $\frac{1}{4}$ that the hunter is killed. The quark is twice as likely to be killed as the hunter. There is a probability of $\frac{1}{3}$ that both survive.

(i)   Copy and complete this table of probabilities.

Table 15.5

	Hunter dies	Hunter lives	Total
Quark dies			$\frac{1}{2}$
Quark lives		$\frac{1}{3}$	$\frac{1}{2}$
Total	$\frac{1}{4}$		

Find the probability that

(ii) both the hunter and the quark die

(iii) the hunter lives and the quark dies

(iv) the hunter lives, given that the quark dies.

⑦ There are 90 players in a tennis club. Of these, 23 are juniors, the rest are seniors. 34 of the seniors and 10 of the juniors are male. There are 8 juniors who are left-handed, 5 of whom are male. There are 18 left-handed players in total, 4 of whom are female seniors.

(i)   Represent this information in a Venn diagram.

(ii) What is the probability that

(a) a male player selected at random is left-handed?

(b) a left-handed player selected at random is a female junior?

(c) a player selected at random is either a junior or a female?

(d) a player selected at random is right-handed?

(e) a right-handed player selected at random is not a junior?

(f) a right-handed female player selected at random is a junior?

Working with a large data set

⑧ The Avonford cycling accidents data set covers 93 people. However there are 3 people whose sex is not given and 8 more for whom there is no information on whether they were wearing a helmet.

This question refers to the group of people for whom complete information is given.

In this group there are:

■ 54 people wearing a helmet

■ 26 females

■ 13 females wearing a helmet.

(i)   Copy this Venn diagram and write the number of people in each of the four regions.

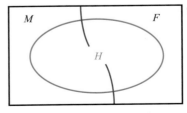

Key:
M = Male
F = Female
H = Wearing a helmet

Figure 15.18

A person is chosen at random from the group.

(ii) Find the probability that the person

(a) is female

(b) was wearing a helmet

(c) is female given that the person was wearing a helmet

(d) was wearing a helmet given that the person is female.

(iii) At an editorial meeting, the other journalists ask Robin 'Can we say in the paper whether males and females are equally likely to be wearing helmets? Is there any simple evidence we can quote?'

How might Robin respond to these questions?

LEARNING OUTCOMES

When you have completed this chapter, you should be able to:

➤ know what is meant by mutually exclusive events

➤ know what is meant by independent events

➤ understand and use conditional probability, including the use of:

○ tree diagrams

○ Venn diagrams

○ two-way tables

○ the formula $P(B|A) = \dfrac{P(B \cap A)}{P(A)}$

➤ know that in many situations probability provides a model for reality

➤ critique the assumptions made and consider the likely effect of more realistic assumptions.

KEY POINTS

1 The probability of an outcome A, $P(A) = \dfrac{n(A)}{n(\mathcal{E})}$

where $n(A)$ is the number of ways that A can occur and $n(\mathcal{E})$ is the total number of ways that all possible events can occur, all of which are equally likely.

2 The expected frequency of an event is given by:

expected frequency = P(event) × number of times the experiment is repeated

3 For any two events, A and B, of the same experiment,

$$P(A \cup B) = P(A) + P(B) - P(A \cap B)$$

4 Two events are mutually exclusive if they cannot both occur.

5 The regions representing mutually exclusive events on a Venn diagram do not overlap and so

$$P(A \cup B) = P(A) + P(B)$$

6 Where an experiment produces two or more mutually exclusive events, the probabilities of the separate events sum to 1.

7 $P(A) + P(A') = 1$

8 $P(B|A)$ means the probability of event B occurring given that event A has already occurred.

$$P(B|A) = \dfrac{P(A \cap B)}{P(A)}$$

9 The probability that event A and then event B occur, in that order, is

$$P(A) \times P(B|A).$$

10 If event B is independent of event A, $P(B \mid A) = P(B \mid A') = P(B)$.

11 If and only if events A and B are independent then $P(A \cap B) = P(A) \times P(B)$.

12 If events A and B are independent then A' and B' are independent, as are A and B', and A' and B.

16 Statistical distributions

> *An approximate answer to the right problem is worth a good deal more than an exact answer to an approximate problem.*
>
> John Tukey (1915–2000)

→ Contrast this photograph of the 2012 Olympics medal winners with people driving themselves to work on a typical morning. As a thought experiment, imagine choosing a commuter's car at random. Estimate the probability that it contains just 1 person. What about 2, 3, 4, 5, over 5 people? How could you make your estimates more accurate?

Review: The binomial distribution

The binomial distribution may be used to model situations in which:

- you are conducting trials on random samples of a certain size, denoted by n
- on each trial, the outcomes can be classified as either **success** or **failure**. ◀

In addition, the following modelling assumptions are needed if the binomial distribution is to be a good model and give reliable answers.

- The outcome of each trial is independent of the outcome of any other trial.

- The probability of success is the same on each trial.

> The probability of success is usually denoted by p and that of failure by q, so $p + q = 1$.

Note

In any given situation there will rarely if ever be any doubt as to whether the first two of these **conditions** are satisfied in a given situation. The third and fourth may be much more open to doubt, and it may be necessary to **assume** that they hold; they would then be **modelling assumptions**.

**TECHNOLOGY**

Your calculator should have a function to calculate binomial probabilities.

The probability that the number of successes, X, has the value r is given by

$$P(X = r) = \binom{n}{r} q^{n-r} p^r$$

> This notation means that the distribution is binomial, with n trials each with probability of success p.

For $B(n, p)$ the expectation of the number of successes is np.

Example 16.1

National statistics show that in about 60% of all car journeys only the driver is in the car.

10 cars are selected at random.

(i)   What is the expectation of the number of cars that contain only the driver?

(ii)   Find the probability that exactly 7 of them contain only the driver.

(iii)   Find the probability that fewer than 8 of them contain only the driver.

Note

Check for yourself that

$= \binom{10}{7}$ is 120.

You can also find this from cumulative binomial tables or, if you are a glutton for punishment, you can work out the individual probabilities for $r = 0, 1, \ldots$ up to 7 and add them all together. You could also find this by finding the probability of 8, 9 or 10 and subtracting that probability from 1.

Solution

(i)   Let X be the number of cars, in the sample of 10, that contain only the driver. The distribution is binomial with $n = 10$ and $p = 0.6$.

For the binomial distribution $E(X) = np$.

So the expectation is $10 \times 0.6 = 6$ cars.

> This is denoted by $B(10, 0.6)$.

(ii) $P(X = 7) = \binom{10}{7} \times (0.6)^7 \times (0.4)^3$

$= \dfrac{10!}{7!3!} \times 0.6^7 \times 0.4^3$

$= 0.215$ to 3 d.p.

(ii)   Fewer than 8 means 0 or 1 or … up to 7.

So you want the cumulative binomial probability for $n = 10$, $p = 0.6$ and $r = 7$.

By calculator, this is 0.833 to 3 d.p.

Review exercise

T ① A spinner with four equal sectors numbered 1, 2, 3 and 4 is rolled 100 times.

(i)   State the distribution of the number of 3s scored.

(ii)   What is the expected number of times a 3 is scored?

(iii)   What is the probability of scoring 3 exactly 20 times?

(iv)   What is the probability of scoring 3 at most 20 times?

② At a village fete a stall offers the chance to win a car. The game is to roll seven fair dice and if all seven show a 6 on the top face then the car is won.

(i)   Describe this as a binomial distribution, defining p.

(ii)   What is the probability of winning the car?

(iii)   How many tries would be expected before the car was won?

③ A quarter of Flapper fish have red spots, the rest have blue spots. A fisherman catches 10 Flapper fish.

(i)   What is the probability that

(a)   exactly 8 have blue spots

(b)   at least 8 have blue spots?

On another occasion, fishing at the same place, the fisherman catches 100 Flapper fish.

(ii)   What is the probability that exactly 80 of them have blue spots?

(iii)   Out of these 100 Flapper fish, what is the expected number with blue spots?

(iv) In fact the 100 Flapper fish all came from the same shoal. Suggest a reason why this might mean that the binomial distribution is not a suitable model.

④ The random variable $X \sim B(20, 0.3)$.

   (i) State

      (a) the number of trials

      (b) the probability of success in any trial

      (c) the probability of failure in any trial.

   (ii) Show that the probability of at least 10 successes is just under 5%.

   (iii) Find the values of the integer a such that $P(X < a) < 0.05$ and $P(X \le a) > 0.05$.

⑤ A company manufactures light bulbs in batches of 1000. The probability of a light bulb being faulty is known to be 0.01.

What is the probability that in a batch of light bulbs that there are less than 12 faulty light bulbs?

⑥ Fred travels to work by train each morning. The probability of his train being late is 0.15.

   (i) What is the probability that his train is late once in a 5-day working week?

   (ii) What is the probability that it is late at least once in a 5-day working week?

   (iii) Fred buys a season ticket which allows him to travel for four weeks. How many times would he expect the train to be late in this four week period?

   (iv) State any assumptions needed for your answer to part (i), and comment on whether they are likely to be reasonable.

Working with a large data set

⑦ (i) The random variable X has the distribution $B(22, 0.7)$. Find the probability that X takes the following values

      (a) 20   (b) 21   (c) 22   (d) 20, 21 or 22.

Over recent years the schools in Avonford have run a campaign called 'Safe cycling'. Among other things they emphasise the need to wear safety helmets. There is an immediate target of persuading 70% of cyclists to do so.

Table 16.1 shows whether those involved in accidents were wearing helmets, according to age group.

Table 16.1

	Under 15	15 to 29	Over 29
Wearing a helmet	20	19	15
Not wearing a helmet	2	8	19
No usable information	4	2	4

   (ii) The organiser of the campaign says 'Look at the under-15s. If only 70% of that age group were wearing helmets, then the chances that out of the 22 having accidents, 20 or more were wearing helmets is tiny. So we can be certain that we are exceeding our target.'

      Comment on this statement.

   (iii) Estimate the percentages of cyclists in the other two age groups wearing helmets.

      What do you conclude from these figures?

1 Discrete random variables

Discussion point

→ How would you collect information on the volume of traffic in the town centre?

THE AVONFORD STAR

Traffic chaos in town centre – car-share scheme to go ahead

In an attempt to reduce the volume of traffic coming into and going out of Avonford town centre at peak times Avonford Council are to promote a car-share scheme.

Council spokesman, Andrew Siantonas, told the *Star*, 'In a recent traffic survey we found that there were far too many cars with only the driver in the car. We need to take action now to reduce the number of vehicles travelling through the town centre in the early morning and late afternoon.

The Council have put aside a sum of money to fund a car-share scheme. Interested drivers will be put in touch with one another using a central database. An advertising campaign is to start in the next few weeks. In six months' time we will conduct another survey to measure the success of the scheme.'

Discussion point

→ What are the main features of this distribution? How would you illustrate such a distribution?

A traffic survey, at critical points around the town centre, was conducted at peak travelling times over a period of a working week. The survey involved 1000 cars. The number of people in each car was noted, with the following results.

Table 16.2

Number of people per car	1	2	3	4	5	>5
Frequency	561	242	148	39	11	0

The numbers of people per car are necessarily discrete. A discrete frequency distribution is best illustrated by a vertical line chart, as in Figure 16.1. This shows you that the distribution has positive skew, with the bulk of the data at the lower end of the distribution.

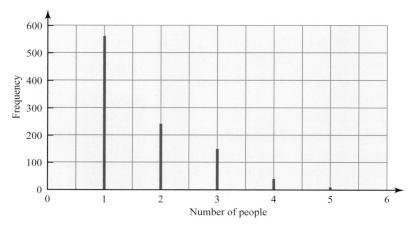

Figure 16.1

The survey involved 1000 cars. This is a large sample and so it is reasonable to use the results to estimate the probabilities of the various possible outcomes: 1, 2, 3, 4, 5 people per car. You divide each frequency by 1000 to obtain the relative frequency, or probability, of each outcome (number of people per car).

Table 16.3

Outcome (number of people)	1	2	3	4	5	>5
Probability (relative frequency)	0.561	0.242	0.148	0.039	0.011	0

A probability distribution provides a mathematical model to describe a particular situation. In statistics you are often looking for models to describe and explain the data you find in the real world. In this chapter you are introduced to some of the techniques for working with probability distributions. In the example above, the model describes the distribution of a discrete random variable.

In this case, the only possible values are all positive integers.

The variable is discrete since it can only take certain values. The number of people in a car is a random variable since the actual value varies from one car to another and can only be predicted with a given probability, i.e. the values occur at random.

Discrete random variables may take a finite or an infinite number of values. In this case the number is finite; in the survey the maximum number of people observed was five, but the maximum could be larger but only up to the seating capacity of the largest available car. It would be finite. A well-known example of a finite discrete random variable is the number of successes in a binomial distribution.

On the other hand, if you decided to throw a fair coin until you got a head, there is no theoretical maximum as each throw has the probability of 0.5, so each could land with tails. There is also the case where the maximum is so large that it can be considered to be infinite; one example of this would be the number of hits on a website in a given day.

The study of discrete random variables in this chapter is limited to finite cases.

Notation and conditions for a discrete random variable

- A random variable is denoted by an upper case letter, such as X, Y, or Z.

- The particular values that the random variable takes are denoted by lower case letters, such as x, y, z and r.

- In the case of a discrete variable these are sometimes given suffixes such as $r_1, r_2, r_3, \ldots$.

- Thus $P(X = r_1)$ means the probability that the random variable X takes a particular value r_1.

- The expression $P(X = r)$ is used to express a more general idea, as, for example, in a table heading.

- If a finite discrete random variable can take n distinct values $r_1, r_2, \ldots, r_n$, with associated probabilities $p_1, p_2, \ldots, p_n$, then the sum of the probabilities must equal 1.

- In that case $p_1 + p_2 + \cdots + p_n = 1$.

- This can be written more formally as $\displaystyle\sum_{k=1}^{n} p_k = \sum_{k=1}^{n} P(X = r_k) = 1$.

- If there is no ambiguity then $\displaystyle\sum_{k=1}^{n} P(X = r_k)$ is often abbreviated to $\displaystyle\sum P(X = r)$.

> **Note**
>
> The various outcomes cover all possibilities; they are exhaustive.

Diagrams of discrete random variables

Just as with frequency distributions for discrete data, the most appropriate diagram to illustrate a discrete random variable is a vertical line chart. Figure 16.2 shows the probability distribution of X, the number of people per car. Note that it is identical in shape to the corresponding frequency diagram. The important difference is that the vertical axis is now for probability rather than frequency.

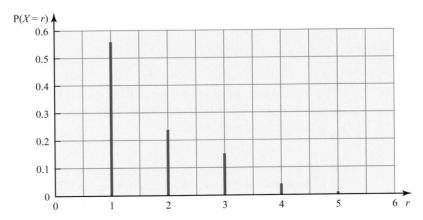

Figure 16.2

Example 16.2

Two tetrahedral dice, each with faces labelled 1, 2, 3 and 4, are thrown and the random variable X represents the sum of the numbers face-down on the dice.

(i)   Find the probability distribution of X.

(ii)   Illustrate the distribution and describe the shape of the distribution.

(iii)   What is the probability that any throw of the dice results in a value of X which is an odd number?

Solution

The table shows all the possible total scores when the two dice are thrown.

Table 16.4

	First throw			
+	1	2	3	4
Second throw 1	2	3	4	5
2	3	4	5	5
3	4	5	5	7
4	5	5	7	8

Table 16.5

Total score	2	3	4	5	6	7	8
$P(X = r)$	$\frac{1}{16}$	$\frac{2}{16}$	$\frac{3}{16}$	$\frac{4}{16}$	$\frac{3}{16}$	$\frac{2}{16}$	$\frac{1}{16}$

(i)   You can use the table to write down the probability distribution for X.

(ii)   The vertical line chart illustrates this distribution, which is symmetrical.

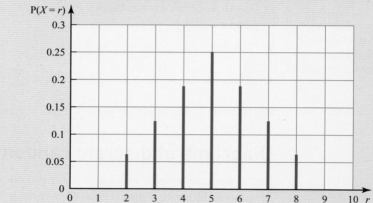

Figure 16.3

(iii)   The probability that X is an odd number

$$= P(X = 3) + P(X = 5) + P(X = 7)$$

$$= \frac{2}{16} + \frac{4}{16} + \frac{2}{16}$$

$$= \frac{1}{2}$$

Discussion point

→ What other situations give rise to a uniform distribution?

Notice that the score on one of these dice has a **uniform distribution.** The probability is the same for each value.

Table 16.6

Score	1	2	3	4
Probability	$\frac{1}{4}$	$\frac{1}{4}$	$\frac{1}{4}$	$\frac{1}{4}$

As well as defining a discrete random variable by tabulating the probability distribution, another effective way is to use an algebraic definition of the form $P(X = r) = f(r)$ for given values of r.

The following example illustrates how this may be used.

Example 16.3

The probability distribution of a random variable X is given by

$$P(X = r) = kr \quad \text{for } r = 1, 2, 3, 4$$
$$P(X = r) = 0 \quad \text{otherwise.}$$

(i)   Find the value of the constant k.

(ii)   Illustrate the distribution and describe the shape of the distribution.

(iii)   Two successive values of X are generated independently of each other. Find the probability that

    (a)   both values of X are the same

    (b)   the total of the two values of X is greater than 6.

Solution

(i)   Tabulating the probability distribution for X gives this table.

Table 16.7

r	1	2	3	4
$P(X = r)$	k	$2k$	$3k$	$4k$

Since X is a random variable,

$$\sum P(X = r) = 1$$
$$k + 2k + 3k + 4k = 1$$
$$10k = 1$$
$$k = 0.1$$

(ii)   Hence $P(X = r) = 0.1r$, for $r = 1, 2, 3, 4$, which gives the following probability distribution.

Table 16.8

r	1	2	3	4
$P(X = r)$	0.1	0.2	0.3	0.4

The vertical line chart in Figure 16.4 illustrates this distribution. It has negative skew.

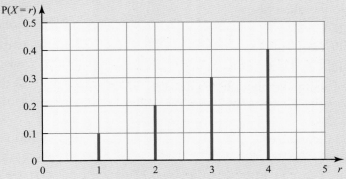

Figure 16.4

(iii) (a) P(both values of X are the same)

$$= P(1 \text{ and } 1 \text{ or } 2 \text{ and } 2 \text{ or } 3 \text{ and } 3 \text{ or } 4 \text{ and } 4)$$

$$= P(1 \text{ and } 1) + P(2 \text{ and } 2) + P(3 \text{ and } 3) + P(4 \text{ and } 4)$$

$$= P(X = 1) \times P(X = 1) + P(X = 2) \times P(X = 2)$$
$$\quad + P(X = 3) \times P(X = 3) + P(X = 4) \times P(X = 4)$$

$$= (0.1)^2 + (0.2)^2 + (0.3)^2 + (0.4)^2$$

$$= 0.01 + 0.04 + 0.09 + 0.16$$

$$= 0.3$$

(b) P(total of the two values is greater than 6)

$$= P(4 \text{ and } 3 \text{ or } 3 \text{ and } 4 \text{ or } 4 \text{ and } 4)$$

$$= P(4 \text{ and } 3) + P(3 \text{ and } 4) + P(4 \text{ and } 4)$$

$$= P(X = 4) \times P(X = 3) + P(X = 3) \times P(X = 4) + P(X = 4)$$
$$\quad \times P(X = 4)$$

$$= 0.4 \times 0.3 + 0.3 \times 0.4 + 0.4 \times 0.4$$

$$= 0.12 + 0.12 + 0.16$$

$$= 0.4$$

Exercise 16.1

Those questions in this exercise where the context is drawn from real life involve working with mathematical models.

① The probability distribution for a discrete random variable is given in this table.

r	0	1	2
$P(X = r)$	0.12	a	$3a$

Find the value of a.

② The random variable X is given by the sum of the scores when two ordinary dice are thrown.

(i) Find the probability distribution of X.

(ii) Illustrate the distribution and describe the shape of the distribution.

(iii) Find the values of

(a) $P(X > 8)$

(b) $P(X \text{ is even})$

(c) $P(|X - 7| < 3)$.

③ The probability distribution of a discrete random variable is given by $P(X = r) = k$ for $1 \leqslant r \leqslant 20$.

(i) Find the value of k.

(ii) Describe the distribution.

(iii) Find $(3 \leqslant X \leqslant 12)$.

④ The random variable Y is given by the absolute difference between the scores when two ordinary dice are thrown.

(i) Find the probability distribution of Y.

(ii) Illustrate the distribution and describe the shape of the distribution.

(iii) Find the values of

(a) $P(Y < 3)$ (b) $P(Y \text{ is odd})$.

⑤ The probability distribution of a discrete random variable X is given by

$$P(X = r) = \frac{kr}{8} \quad \text{for } r = 2, 4, 6, 8$$
$$P(X = r) = 0 \quad \text{otherwise.}$$

(i) Find the value of k and tabulate the probability distribution.

(ii) Two successive values of X are generated independently. Find the probability that

(a) the two values are equal

(b) the first value is greater than the second value.

⑥ A curiously shaped six-sided spinner produces scores, X, for which the probability distribution is given by

$$P(X = r) = \frac{k}{r} \quad \text{for } r = 1, 2, 3, 4, 5, 6$$
$$P(X = r) = 0 \quad \text{otherwise.}$$

(i) Find the value of k and illustrate the distribution.

(ii) Show that, when this spinner is used twice, the probability of obtaining two equal scores is very nearly $\frac{1}{4}$.

⑦ Three fair coins are tossed.

(i) By considering the set of possible outcomes, HHH, HHT, etc., tabulate the probability distribution for X, the number of heads occurring.

(ii) Illustrate the distribution and describe the shape of the distribution.

(iii) Find the probability that there are more heads than tails.

(iv) Without further calculation, state whether your answer to part would be the same if four fair coins were tossed. Give a reason for your answer.

⑧ Two tetrahedral dice, each with faces labelled 1, 2, 3 and 4, are thrown and the random variable X is the product of the numbers shown on the dice.

(i) Find the probability distribution of X.

(ii) What is the probability that any throw of the dice results in a value of X which is an odd number?

⑨ A sociologist is investigating the changing pattern of the numbers of children which women have in a country. She denotes the present number by the random variable X which she finds to have the following probability distribution.

Table 16.9

r	0	1	2	3	4	5+
$P(X = r)$	0.09	0.22	a	0.19	0.08	negligible

(i) Find the value of a.

She is keen to find an algebraic expression for the probability distribution and suggests the following model.

$$P(X = r) = k(r + 1)(5 - r)$$
for $r = 0, 1, 2, 3, 4, 5$
$$P(X = r) = 0 \quad \text{otherwise.}$$

(ii) Find the value of k for this model.

(iii) Compare the algebraic model with the probabilities she found, illustrating both distributions on one diagram. Do you think it is a good model?

⑩ In a game, each player throws three ordinary six-sided dice. The random variable X is the largest number showing on the dice, so for example, for scores of 2, 5 and 4, $X = 5$.

(i) Find the probability that $X = 1$.

(ii) Find $P(X \leqslant 2)$ and deduce that $P(X = 2) = \frac{7}{216}$.

(iii) Find $P(X \leqslant r)$ and so deduce $P(X = r)$, for $r = 3, 4, 5, 6$.

(iv) Illustrate and describe the probability distribution of X.

⑪ A box contains six black pens and four red pens. Three pens are taken at random from the box.

(i) By considering the selection of pens as sampling without replacement, illustrate the various outcomes on a probability tree diagram.

The random variable X represents the number of red pens obtained.

(ii) Find the probability distribution of X.

Working with a large data set

⑫ (i) All the people covered by the Avonford cycling accidents data set went to a hospital. The table below gives the frequency distribution of the numbers of nights spent in hospital by the 93 cyclists involved. A number is missing from one of the cells.

Table 16.10

Nights	0	1	2	3	4	5	>5	No data
Frequency	58	23		2	0	2	2	1

What is the missing value?

(ii) A model for the probability distribution for 0, 1, 2, 3, … nights is given by the terms of the sequence

$$(1 - p), p(1 - p), p^2(1 - p), p^3(1 - p), p^4(1 - p), \ldots, \text{ where } 0 \leqslant p < 1.$$

Prove that the sum to infinity of these terms is 1.

(iii) It is suggested that this model, with the value of p taken to be 0.35, would be suitable for Avonford. Find the frequency distribution that this model predicts for 92 people.

(iv) Compare the frequency distributions in parts (i) and (iii) and comment on any outliers.

(v) The model is based on the parameter p. State what p and $(1 - p)$ represent. Hence explain how the model has been set up.

2 The Normal distribution

Wilf Harris is clearly exceptionally tall, but how much so? Is he one in a hundred, or a thousand or even a million? To answer that question you need to know the distribution of heights of adult British men.

The first point that needs to be made is that height is a continuous variable and not a discrete one. If you measure accurately enough it can take any value.

This means that it does not really make sense to ask 'What is the probability that somebody chosen at random has height exactly 195 cm?' The answer is zero.

However, you can ask questions like 'What is the probability that somebody chosen at random has height between 194 cm and 196 cm?' and 'What is the probability that somebody chosen at random has height at least 195 cm?'. When the variable is continuous, you are concerned with a range of values rather than a single value.

Like many other naturally occurring variables, the heights of adult men may be modelled by a Normal distribution, shown in Figure 16.5. You will see that this has a distinctive bell-shaped curve and is symmetrical about its middle. The curve is continuous as height is a continuous variable.

In Figure 16.5, area represents probability so the shaded area to the right of 195 cm represents the probability that a randomly selected adult male is at least 195 cm tall.

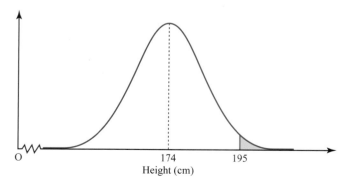

Figure 16.5

Before you can start to find this area, you must know the mean and standard deviation of the distribution, in this case about 174 cm and 7 cm respectively.

So Wilf's height is 195 cm − 174 cm = 21 cm above the mean, and that is

$$\frac{21}{7} = 3 \text{ standard deviations.}$$

The number of standard deviations beyond the mean, in this case 3, is denoted by the letter z. Thus the shaded area gives the probability of obtaining a value of $z \geqslant 3$.

Your calculator will provide this value. The answer is 0.0013.

The probability of a randomly selected adult male being 195 cm or over is 0.0013. Slightly more than one man in a thousand is at least as tall as Wilf.

It is also possible to find this area by using tables. You look up the value of $\Phi(z)$, in this case when $z = 3$, in a Normal distribution table of $\Phi(z)$ as shown in Figure 16.6, and then calculate $1 - \Phi(z)$.

TECHNOLOGY

Some calculators will give the answer $P(z \geqslant 3) = 0.0013$ directly. Make sure you can work out how to do this on your calculator.

Others only give the area to the left. This is the cumulative probability and is denoted by $\Phi(z)$. In that case the area to the right is given by $1 - \Phi(z)$.

The symbol Φ is the Greek letter phi. The equivalent in English is PH.

z	.00	.01	.02	.03	.04	.05	.06	.07	.08	.09	1	2	3	4	5	6	7	8	9
0.0	.5000	5040	5080	5120	5160	5199	5239	5279	5319	5359	4	8	12	16	20	24	28	32	36
0.1	.5398	5438	5478	5517	5557	5596	5636	5675	5714	5753	4	8	12	16	20	24	28	32	35
0.2	.5793	5832	5871	5910	5948	5987	6026	6064	6103	6141	4	8	12	15	19	23	27	31	35
0.3	.6179	6217	6255	6293	6331	6368	6406	6443	6480	6517	4	8	11	15	19	23	26	30	34
0.4	.6554	6591	6628	6664	6700	6736	6772	6808	6844	6879	4	7	11	14	18	22	25	29	32
0.5	.6915	6950	6985	7019	7054	7088	7123	7157	7190	7224	3	7	10	14	17	21	24	27	31
0.6	.7257	7291	7324	7357	7389	7422	7454	7486	7517	7549	3	6	10	13	16	19	23	26	29
0.7	.7580	7611	7642	7673	7704	7734	7764	7794	7823	7852	3	6	9	12	15	18	21	24	27
0.8	.7881	7910	7939	7967	7995	8023	8051	8078	8106	8133	3	6	8	11	14	17	19	22	25
3.0	.9987	9987	9988	9988	9988	9989	9989	9989	9990	9990									
3.1	.9990	9991	9991	9991	9992	9992	9992	9992	9993	9993				*differences*					
3.2	.9993	9993	9994	9994	9994	9994	9994	9995	9995	9995				*untrustworthy*					
3.3	.9995	9995	9996	9996	9996	9996	9996	9996	9996	9997									
3.4	.9997	9997	9997	9997	9997	9997	9997	9997	9997	9998									

Figure 16.6

This gives $\Phi(3) = 0.9987$, and so $1 - \Phi(3) = 0.0013$.

Note

It is often helpful to know that in a Normal distribution, roughly
- 68% of the values lie within ±1 standard deviation of the mean
- 95% of the values lie within ±2 standard deviations of the mean
- 99.75% of the values lie within ±3 standard deviations of the mean.

These figures are particularly useful when you are doing a rough estimate or checking that your answer is sensible.

Notation

The function $\Phi(z)$ gives the area under the Normal distribution curve to the left of the value z, that is the shaded area in Figure 16.7. It is the cumulative distribution function. The total area under the curve is 1, and the area given by $\Phi(z)$ represents the probability of a value smaller than z.

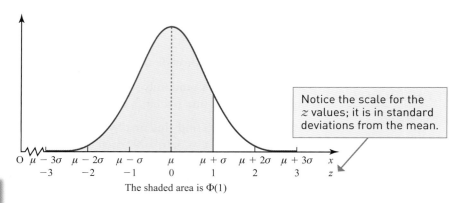

Notice the scale for the z values; it is in standard deviations from the mean.

The shaded area is $\Phi(1)$

Figure 16.7

Note

Notice how lower case letters, x and z, are used to indicate particular values of the random variables, whereas upper case letters, X and Z, are used to describe or name those variables.

If the variable X has mean μ and standard deviation σ then the random variable Z defined by

$$Z = \frac{X - \mu}{\sigma}$$

has mean 0 and variance 1.

Table 16.11

	Actual distribution, X	Standardised distribution, Z
Mean	μ	0
Standard deviation	σ	1
Particular value	x	$z = \dfrac{x - \mu}{\sigma}$

The Normal distribution function on your calculator is easy to use but you should always make a point of drawing a diagram and shading the region you are interested in.

Example 16.4

Assuming the distribution of the heights of adult men is Normal, with mean 174 cm and standard deviation 7 cm, find the probability that a randomly selected adult man is

(i)   under 185 cm

(ii)   over 185 cm

(iii)   over 180 cm

(iv)   between 180 cm and 185 cm

(v)   under 170 cm

giving your answers to 3 decimal places.

> Expect to use your calculator to find the values of $\Phi(z)$. You may find that your calculator allows you to obtain these values without transforming to the standardised Normal distribution, i.e. entering values of x rather than z.

Solution

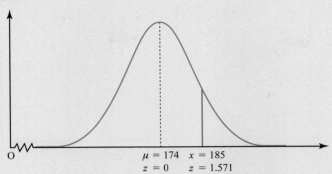

Figure 16.8

The mean height, $\mu = 174$.

The standard deviation, $\sigma = 7$.

(i) The probability that an adult man selected at random is under 185 cm.

The area required is that shaded in Figure 16.8.

$$x = 185 \text{ cm}$$

And so $\quad z = \dfrac{185 - 174}{7} = 1.571$

$$\Phi(1.571) = 0.9419$$

$$= 0.942 \quad (3 \text{ d.p.})$$

Answer: The probability that an adult man selected at random is under 185 cm is 0.942.

(ii) The probability that an adult man selected at random is over 185 cm.

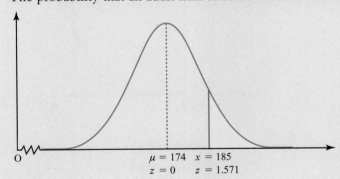

Figure 16.9

The area required is the complement of that for part (i).

$$\text{Probability} = 1 - \Phi(1.571)$$

$$= 1 - 0.9419$$

$$= 0.0581$$

$$= 0.058 \quad (3 \text{ d.p.})$$

Answer: the probability that an adult man selected at random is over 185 cm is 0.058. ➜

(iii)   The probability that an adult man selected at random is over 180 cm.

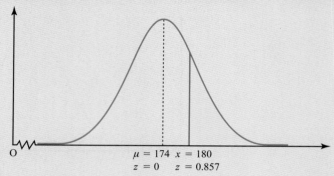

Figure 16.10

$x = 180$ and so $z = \dfrac{180 - 174}{7} = 0.857$

The area required $= 1 - \Phi(0.857)$

$= 1 - 0.8042$

$= 0.1958$

$= 0.196$ (3 d.p.)

Answer: the probability that and adult man selected at random is over 180 cm is 0.196.

(iv)   The probability that an adult man selected at random is between 180 cm and 185 cm.

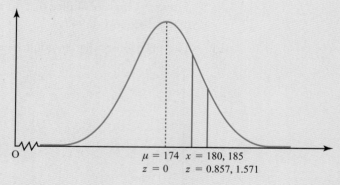

Figure 16.11

The required area is shown in Figure 16.11. It is

$\Phi(1.571) - \Phi(0.857) = 0.9419 - 0.8042$

$= 0.1377$

$= 0.138$ (3 d.p.)

Answer: the probability that an adult man selected at random is over 180 cm but less than 185 m is 0.138.

(v)   The probability that an adult man selected at random is under 170 cm.

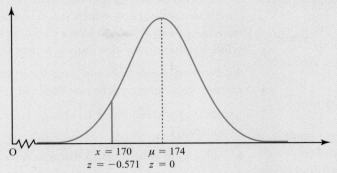

Figure 16.12

In this case $x = 170$

And so $z = \dfrac{170 - 174}{7} = -0.571$

However, when you come to look up $\Phi(-0.571)$, you will find that only positive values of z are given in your tables. You overcome this problem by using the symmetry of the Normal curve. The area you want in this case is that to the left of -0.571 and this is clearly just the same as that to the right of $+0.571$.

So $\Phi(-0.571) = 1 - \Phi(0.571)$

$= 1 - 0.7160$

$= 0.2840$

$= 0.284$ (3 d.p.)

Answer: the probability that an adult man selected at random is under 170 cm is 0.284.

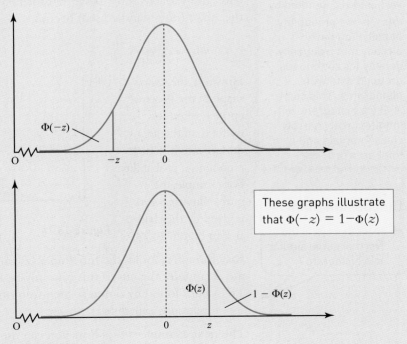

These graphs illustrate that $\Phi(-z) = 1 - \Phi(z)$

Figure 16.13

The Normal curve

$y_i = a + bx_i$ is an example of a linear transformation.

There are two symbols which you will need to distinguish between. The Normal curve is $y = \phi(z)$ (using a lower case Greek letter phi and it is also written φ) while the cumulative distribution is denoted by $\Phi(z)$ (using an upper case phi).

All Normal curves have the same basic shape, so that by scaling the two axes suitably you can always fit one Normal curve exactly on top of another one.

Sometimes you may wish to change your measurements, denoted by x_i, in various ways:

- changing your units so that x_i is now recorded as bx_i
- measuring from a different datum level so that x_i becomes $a + x_i$

or a combination of these two changes. Taken together these changes result in your data values being transformed from x_i to y_i where $y_i = a + bx_i$.

If your original distribution was Normal with mean $\bar{x}$, standard deviation s and variance s^2 then your new distribution will also be Normal. The new mean will be $\bar{y} = a + b\bar{x}$, the new standard deviation will be bs and the new variance will be b^2s^2.

The curve for the Normal distribution with mean μ and standard deviation σ (i.e. variance σ^2) is given by the function $\phi(x)$ in

$$\phi(x) = \frac{1}{\sigma\sqrt{2\pi}}\, e^{-\frac{1}{2}\left(\frac{x-\mu}{\sigma}\right)^2}$$

The notation $N(\mu, \sigma^2)$ is used to describe this distribution. The mean μ and standard deviation σ (or variance σ^2) are the two parameters used to define the distribution. Once you know their values, you know everything there is to know about the distribution. The standardised variable Z has mean 0 and variance 1, so its distribution is $N(0, 1)$.

Note

The variable on the vertical axis, denoted by $\phi(z)$, is the probability density. You have already met frequency density for continuous grouped data with histograms. This curve is like a histogram with infinitesimally narrow class intervals and a total area of 1.

After the variable X has been transformed to Z using $z = \dfrac{x - \mu}{\sigma}$, the form of the curve (now standardised) becomes

$$\phi(z) = \frac{1}{\sqrt{2\pi}}\, e^{-\frac{1}{2}z^2}$$

However, the exact shape of the Normal curve is often less useful than the area underneath it, which represents a probability. For example, the probability that $z \leqslant 2$ is given by the shaded area in Figure 16.14.

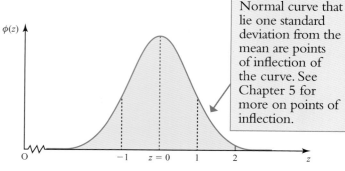

The points on the Normal curve that lie one standard deviation from the mean are points of inflection of the curve. See Chapter 5 for more on points of inflection.

Figure 16.14

Discussion point

→ In what ways is the Normal curve similar to a histogram?

Easy though it looks, the function $\phi(z)$ cannot be integrated algebraically to find the area under the curve; this can only be found by using a numerical method. The values found by doing so are given on your calculator or as a table and this area function is called $\Phi(z)$.

The next example shows how you can use experimental data to estimate the mean and standard deviation of a Normal distribution.

Example 16.5

Skilled operators make a particular component for an engine. The company believes that the time taken to make this component may be modelled by the Normal distribution. They time one of their operators, Sheila, over a long period.

They find that only 10% of the components take her over 90 minutes to make, and that 20% take her less than 70 minutes. Estimate the mean and standard deviation of the time Sheila takes.

> You are asked to infer the mean, μ, and standard deviation, σ, from the areas under different parts of the graph in Figures 16.15 and 16.16.

There are two pieces of information.

The first is that 10% take her 90 minutes or more. This means that the shaded area is 0.1.

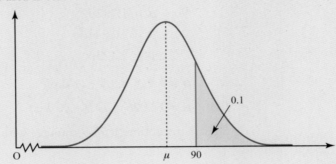

Figure 16.15

$$z = \frac{90 - \mu}{\sigma}$$

$$\Phi(z) = 1 - 0.1 = 0.9$$

The next step is to use the inverse Normal function on your calculator, $\Phi^{-1}(p) = z$

$$\Phi^{-1}(0.9) = 1.282$$

So $z = 1.282$

> **TECHNOLOGY**
>
> Make sure that you know how to do this on your calculator. (It is also possible to use tables to find the value of z.)

So $\dfrac{90 - \mu}{\sigma} = 1.282 \Rightarrow 90 - \mu = 1.282\sigma$

The second piece of information, that 20% of the components took Sheila under 70 minutes, is shown in Figure 16.16.

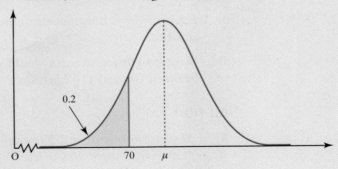

Figure 16.16

$$z = \frac{70 - \mu}{\sigma}$$

(z has a negative value in this case, the point being to the left of the mean.)

$$\Phi(z) = 0.2$$

and so, by symmetry

$$\Phi(-z) = 1 - 0.2 = 0.8.$$

Using the table or the inverse Normal function gives

$$-z = 0.8416 \quad \text{or} \quad z = -0.8416$$

This gives a second equation for μ and σ.

$$\frac{70 - \mu}{\sigma} = -0.8416 \Rightarrow 70 - \mu = -0.8416\sigma$$

The two simultaneous equations are now solved:

$$90 - \mu = 1.282\sigma$$
$$70 - \mu = -0.8416\sigma$$
$$20 = 2.123\sigma$$

$$\sigma = 9.418 = 9.4 \quad \text{(1 d.p.)}$$

$$\mu = 77.926 = 77.9 \quad \text{(1 d.p.)}$$

Sheila's mean time is 77.9 minutes with standard deviation 9.4 minutes.

Modelling discrete situations

Although the Normal distribution applies strictly to a continuous variable, it is also common to use it in situations where the variable is discrete providing that

- the distribution is approximately Normal; this requires that the steps in possible values are small compared to its standard deviation.

- **continuity corrections** are applied where appropriate.

The next example shows how this is done and the meaning of the term continuity correction is explained.

Example 16.6

The result of an Intelligence Quotient (IQ) test is an integer score, X. Tests are designed so that X has a mean value of 100 and standard deviation 15.

A large number of people have their IQs tested.

(i)   Explain why it is reasonable to use a Normal approximation in this situation.

(ii)   What proportion of them would you expect to have IQs measuring between 106 and 110 (inclusive)?

Solution

(i)   If you assume that an IQ test is measuring innate, natural intelligence (rather than the results of learning), then it is reasonable to assume a Normal distribution.

Although the random variable X is an integer and so discrete, the steps of 1 in its possible values are small compared with the standard deviation of 15. So it is reasonable to treat it as if it is continuous.

(ii)

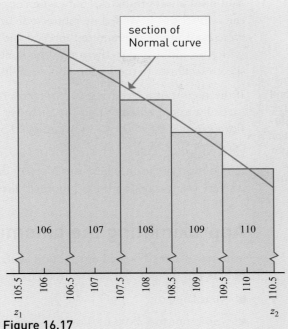

Figure 16.17

If you draw the probability distribution function for the discrete variable X it looks like the blue part of Figure 16.17. The area you require is the total of the five bars representing 106, 107, 108, 109 and 110.

You can see that the equivalent section of the Normal curve, outlined in red, does not run from 106 to 110 but from 105.5 to 110.5. When you change from the discrete scale to the continuous scale the numbers 106, 107, etc. no longer represent the whole intervals, just their centre points.

So the area you require under the Normal curve is given by $\Phi(z_2) - \Phi(z_1)$ where

$$z_1 = \frac{105.5 - 100}{15} \quad \text{and} \quad z_2 = \frac{110.5 - 100}{15}.$$

This is $\Phi(0.7000) - \Phi(0.3667) = 0.7580 - 0.6431 = 0.1149$

Answer: The proportion with IQs between 106 and 110 (inclusive) should be approximately 11.5%.

Continuity corrections

In Example 16.6, both end values needed to be adjusted to allow for the fact that a continuous distribution was being used to approximate a discrete one. The adjustments

$$106 \rightarrow 105.5 \quad \text{and} \quad 110 \rightarrow 110.5$$

are called **continuity corrections**. Whenever a discrete distribution is approximated by a continuous one a continuity correction may need to be used.

You must always think carefully when applying a continuity correction. Should the corrections be added or subtracted? In the previous example 106 and 110 were both inside the required interval and so any value (like 105.7 and 110.4) which would round to them must be included. It is often helpful to draw a sketch to illustrate the region you want, like the one in Figure 16.17.

It is common to write the interval between 19.5 and 30.5 as $19.5 \leq X < 30.5$ (as has been done here), but as the normal distribution is continuous it makes no difference whether you use $<$ or $\leq$, provided you are consistent.

If the region of interest is given in terms of inequalities, you should look carefully to see whether they are inclusive ($\leq$ or $\geq$) or exclusive ($<$ or $>$). For example, $20 \leq X \leq 30$ becomes $19.5 \leq X < 30.5$ whereas $20 < X < 30$ becomes $20.5 \leq X < 29.5$.

A common situation where the Normal distribution is used to approximate a discrete one occurs with the binomial distribution.

Approximating the binomial distribution

You may use the Normal distribution as an approximation for the binomial $B(n, p)$ (where n is the number of trials each having a probability p of success) when

- n is large

- np is not too close to 0 or n.

These are combined as the condition

- $0 \ll np \ll n$.

This condition ensures that, like the Normal, the binomial distribution is reasonably symmetrical and not skewed from either end. This is illustrated in Figure 16.18. The larger the value of n, the greater the range of values of p for which this is the case.

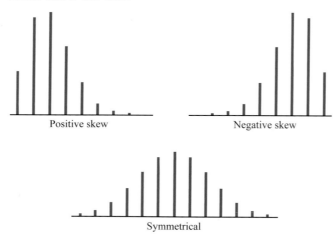

Figure 16.18

The parameters for the Normal distribution are then

Mean: $\mu = np$

Variance: $\sigma^2 = np(1 - p) = npq$

so that it can be denoted by $N(np, npq)$.

Example 16.7

> **Note**
>
> In theory you could expand $(0.43 + 0.57t)^{1700}$ and use the answer to find the probability of obtaining 0, 1, 2, 3, ... , 850 Labour voters in your sample of 1700. However, such a method would be impractical because of the work involved. Instead you can use the Normal approximation.

This is a true story. During voting at a by-election, an exit poll of 1700 voters indicated that 50% of people had voted for the Labour party candidate. When the votes were counted, it was found that he had in fact received 57% support.

Of the 1700 people interviewed 850 said they had voted Labour but 57% of 1700 is 969, a much higher number. What went wrong? Is it possible to be so far out just by being unlucky and asking the wrong people?

Solution

This situation can be modelled by the binomial distribution $B(1700, 0.57)$.

The conditions for a Normal approximation apply.

■ n is large. It is 1700.

■ np is not too close to 0 or n. The value of np is 969. So it is not near 0 nor is it near 1700.

The parameters for the Normal approximation are given by

Mean: $\mu = np = 1700 \times 0.57 = 969$

Variance: $\sigma^2 = np(1 - p) = 1700 \times 0.57 \times 0.43 = 416.67$

Standard deviation: $\sigma = \sqrt{416.67} = 20.4$

> Notice that the standard deviation of 20.4 is large compared with the steps of 1 in the possible number of voters.

So the approximating Normal distribution can be written as $N(969, 416.67)$.

The probability of selecting no more than 850 Labour voters is given by $\Phi(z)$ where

$$z = \frac{850.5 - 969}{20.4} = -5.8$$

> Notice the continuity correction with 850.5 rather than 850.

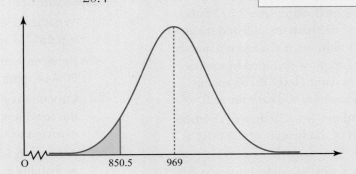

Figure 16.19

> **Discussion point**
>
> → What do you think went wrong with the exit poll?
>
> (Remember that this really did happen.)

This value of z corresponds to a probability of about 0.00001. So, allowing for an equivalent result in the tail above the mean, the probability of a result as extreme as this is 0.00002. It is clearly so unlikely that this was a consequence of random sampling that another explanation must be found.

Exercise 16.2

Those questions in this exercise where the context is drawn from real life involve working with mathematical models.

① The times taken to complete a skills test are distributed Normally with mean 75 seconds and standard deviation 15 seconds.

Find the probability that a person chosen at random

(i) took less than 80 seconds to complete the task

(ii) took more than 85 seconds to complete the task

(iii) took between 80 and 85 seconds to complete the task.

② The distribution of the heights of 18-year-old girls may be modelled by the Normal distribution with mean 162.5 cm and standard deviation 6 cm. Find the probability that the height of a randomly selected 18-year-old girl is

(i) under 168.5 cm

(ii) over 174.5 cm

(iii) between 168.5 and 174.5 cm.

③ The heights of the tides in a harbour have been recorded over many years and found to be Normally distributed with mean 2.512 fathoms above a mark on the harbour wall and standard deviation 1.201 fathoms. A change is made so that the heights are now recorded in metres above a different datum level, 0.755 metres lower than the mark on the harbour wall.

Given that 1 fathom is 1.829 metres, describe the distribution of the heights of the tides as now measured.

④ A pet shop has a tank of goldfish for sale. All the fish in the tank were hatched at the same time and their weights may be taken to be Normally distributed with mean 100 g and standard deviation 10 g. Melanie is buying a goldfish and is invited to catch the one she wants in a small net. In fact the fish are much too quick for her to be able to catch any particular one and the fish which she eventually nets is selected at random. Find the probability that its weight is

(i) over 115 g

(ii) under 105 g

(iii) between 105 and 115 g

⑤ When he makes instant coffee, Tony puts a spoonful of powder into a mug. The weight of coffee in grams on the spoon may be modelled by the Normal distribution with mean 5 g and standard deviation 1 g. If he uses more than 6.5 g Julia complains that it is too strong and if he uses less than 4 g she tells him it is too weak. Find the probability that he makes the coffee

(i) too strong

(ii) too weak

(iii) all right.

⑥ A machine is set to produce metal rods of length 20 cm, with standard deviation 0.8 cm. The lengths of rods are Normally distributed.

(i) Find the percentage of rods produced between 19 cm and 21 cm in length.

The machine is reset to be more consistent so that the percentage between 19 cm and 21 cm is increased to at least 95%.

(ii) Calculate the new standard deviation to 1 d.p.

⑦ An aptitude test for applicants for a senior management course has been designed to have a mean mark of 100 and a standard deviation of 15. The distribution of the marks is approximately Normal.

(i) What is the least mark needed to be in the top 35% of applicants taking this test?

(ii) Between which two marks will the middle 90% of applicants lie?

(iii) On one occasion 150 applicants take this test. How many of them would be expected to score 110 or over?

⑧ (i) A fair coin is tossed 10 times. Evaluate the probability that exactly half of the tosses result in heads.

(ii) The same coin is tossed 100 times. Use the Normal approximation to the binomial to estimate the probability that exactly half of the tosses result in heads.

(iii) Also estimate the probability that more than 60 of the tosses result in heads.

(iv) Explain why a continuity correction is made when using the Normal approximation to the binomial.

[MEI adapted]

⑨ When a butcher takes an order for a Christmas turkey, he asks the customer what weight in kilograms the bird should be. He then sends his order to a turkey farmer who supplies birds of about the requested weight. For any particular weight of bird ordered, the error in kilograms may be taken to be Normally distributed with mean $0\,\text{kg}$ and standard deviation $0.75\,\text{kg}$.

Mrs Jones orders a $10\,\text{kg}$ turkey from the butcher. Find the probability that the one she gets is

(i) over $12\,\text{kg}$

(ii) under $10\,\text{kg}$

(iii) within $0.5\,\text{kg}$ of the weight she actually ordered.

⑩ A biologist finds a nesting colony of a previously unknown sea bird on a remote island. She is able to take measurements on 100 of the eggs before replacing them in their nests. She records their weights, $w\,\text{g}$, in this frequency table.

Table 16.11

Weight, w	Frequency
$25 < w \leqslant 27$	2
$27 < w \leqslant 29$	13
$29 < w \leqslant 31$	35
$31 < w \leqslant 33$	33
$33 < w \leqslant 35$	17
$35 < w \leqslant 37$	0

(i) Find the mean and standard deviation of these data.

(ii) Assuming the weights of the eggs for this type of bird are Normally distributed and that their mean and standard deviation are the same as those of this sample, find how many eggs you would expect to be in each of these categories.

(iii) Do you think the assumption that the weights of the eggs are Normally distributed is reasonable?

⑪ A farmer knows the cut height of her daffodils can be taken to be Normally distributed with mean $40\,\text{cm}$ and standard deviation $10\,\text{cm}$.

(i) Find the probability that a daffodil chosen at random will be more than $45\,\text{cm}$ tall.

(ii) Find the probability that a daffodil is between 35 and $45\,\text{cm}$ tall.

(iii) The farmer can sell daffodils which are over $32\,\text{cm}$ tall. She has cut $10\,000$ daffodils – how many can she expect to be tall enough to sell?

⑫ During an advertising campaign, the manufacturers of Wolfitt (a dog food) claim that 60% of dog owners preferred to buy Wolfitt.

(i) Assuming that the manufacturers' claim is correct for the population of dog owners, calculate the probability that at least 6 of a random sample of 8 dog owners prefer to buy Wolfitt, using

(a) the binomial distribution

(b) a Normal approximation to the binomial.

(ii) Comment on the agreement, or disagreement, between your answers to part (i)(a) and (b). Would the agreement have been better if the proportion had been 80% instead of 60%?

(iii) Continuing to assume the manufacturers' figure of 60% is correct, use the Normal approximation to the binomial to estimate the probability that, of a random sample of 100 dog owners, the number preferring to buy Wolfitt is between 60 and 70 inclusive.

[MEI adapted]

Working with a large data set

⑬ The histogram in Figure 16.20 illustrates the ages of those involved in cycling accidents in the Avonford cycling accidents data set.

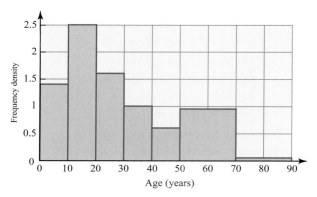

Figure 16.20

(i) Verify that the histogram represents 91 people.

(ii) State two features of this histogram which suggest that the data do not come from a Normal distribution.

To the nearest whole number, the mean age of the cyclists in the Avonford data set is 30 and the standard deviation is 20. Robin decides for comparison to draw the histogram based on a Normal distribution with the same mean and standard deviation as the Avonford data. Robin uses the same population size and the same class intervals.

(iii) Copy and complete Table 16.12 giving the frequency density for each of the intervals. Give your answers to 3 decimal places.

(iv) How many people would this histogram represent?

Explain why this is not 91.

(v) The data include 14 people in their 60s.

Robin thinks that there may be two different Normal populations of cyclists involved, one of young and working age people, and another of people who have just retired or are about to do so. Comment briefly on whether Robin's idea is consistent with the data.

Table 16.12

Age, a years	$0 \leqslant a < 10$	$10 \leqslant a < 20$	$20 \leqslant a < 30$	$30 \leqslant a < 40$	$40 \leqslant a < 50$	$50 \leqslant a < 70$	$70 \leqslant a < 90$
Frequency density	0.836		1.743				0.098

KEY POINTS

1 For a **discrete random variable**, X, which can only take the values $r_1, r_2, \dots, r_n$, with probabilities $p_1, p_2, \dots, p_n$ respectively:

   - $p_1 + p_2 + \dots + p_n = \sum_{k-1}^{n} P_k = \sum_{k=1}^{n} P(X = p_k) = 1$

2 The binomial distribution may be used to model situations subject to these conditions.
   - You are conducting trials on random samples of a certain size, n.
   - On each trial, the outcomes can be classified as either **success** or **failure**.
   - The probabilities of success and failure are denoted by p and q, and $p + q = 1$.

   In addition, the following modelling assumptions are needed if the binomial distribution is to be a good model and give reliable answers.
   - The probability of success in any trial is independent of the outcomes of previous trials
   - The probability of success is the same on each trial.

   The probability that the number of successes, X, has the value r, is given by

   $$P(X = r) = \binom{n}{r} q^{n-r} p^r$$

   For $B(n, p)$ the expectation of the number of successes is np.

3 The Normal distribution with mean μ and standard deviation σ is denoted by $N(\mu, \sigma^2)$.

4 This may be given in standardised form by using the transformation

   $$z = \frac{x - \mu}{\sigma}.$$

5 In the standardised form, $N(0, 1)$, the mean is 0, the standard deviation and the variance both 1.

6 The standardised Normal curve is given by

   $$\phi(z) = \frac{1}{\sqrt{2\pi}} e^{-\frac{1}{2}(z)^2}$$

7 The area to the left of the value z in Figure 16.21, representing the probability of a value less than z, is denoted by $\Phi(z)$, and can be read from a calculator or tables.

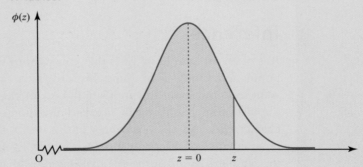

Figure 16.21

8 The Normal distribution may be used to approximate suitable discrete distributions but continuity corrections are then required.

9 The binomial distribution $B(n, p)$ may be approximated by $N(np, npq)$, provided n is large and p is not close to 0 or 1.

17 Statistical hypothesis testing

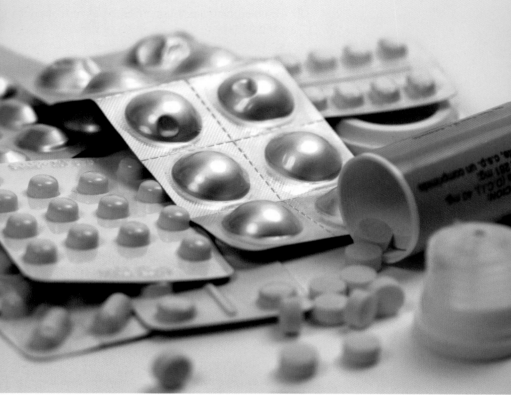

When we spend money on testing an item, we are buying confidence in its performance.

Tony Cutler (1954–)

➜ How can you be certain that any medicine you get from a pharmacy is safe?

Review

Inference

It is often the case in statistics that you want to find out information about a large population. One way of doing this is to take a sample and use it to infer what you want to know about the population. Usually, this is the value of a parameter of the population such as its mean, standard deviation or the proportion with the same characteristic. An obvious danger in this is that, for some reason, the sample is not representative of the population as a whole. Even if the procedure for collecting the sample is sound, it may be that the items that happen to be selected give an inaccurate impression.

A statistical hypothesis testing is a procedure that allows you to estimate how reliable your conclusions from a sample are likely to be.

Steps for conducting a hypothesis test

■ Establish the null and alternative hypotheses.

  ○ The null hypothesis, H_0, is the default position that nothing special has occurred. The parameter has the expected value.

  ○ The alternative hypothesis, H_1, is that there has been a change in the value of the parameter indicated by the null hypothesis.

  ○ The alternative hypothesis may be 1-tail or 2-tail according to whether the direction of change is or is not specified.

■ Decide on the significance level. This is the level at which you say there is enough evidence to reject the null hypothesis in favour of your alternative hypothesis.

■ Collect suitable data using a random sampling procedure that ensures the items are independent.

■ Use the data to determine the test statistic. This is the measure that will be used to decide whether the test is significant. In the case of a binomial test it is the number of successes.

■ Conduct the test doing the necessary calculations. Then:

  ○ either work out the p-value, the probability of a result at least as extreme as that obtained, and compare it with the significance level

  ○ or work out the critical value for the particular significance level and compare it with the test statistic

  ○ or work out the critical (or rejection) region for the particular significance level and determine whether the test statistic lies within it or outside it in the acceptance region. (The terms rejection and acceptance refer to the null hypothesis.)

Note

The significance level is the probability of rejecting the null hypothesis when it is actually true. This is called a **Type 1** error. Accepting the null hypothesis when it is false is called a **Type 2** error.

Typical ways of describing the outcome of a hypothesis test

■ Either:

  ○ the evidence is not strong enough to reject the null hypothesis in favour of the alternative hypothesis

  ○ there is not enough evidence to conclude that the value of the parameter has changed/increased/decreased.

■ Or:

  ○ there is sufficient evidence to reject the null hypothesis in favour of the alternative hypothesis

  ○ there is sufficient evidence to conclude that the value of the parameter has changed/increased/decreased.

■ You should then add a comment relating to the situation you are investigating.

Hypothesis testing for the proportion in a binomial distribution

The next two examples illustrate hypothesis tests for the proportion, p, of a binomial distribution.

Example 17.1

Sunita is a zoologist. She finds a colony of a previously unknown species of bat roosting in a cave in a remote region. There are 16 bats. Among other things, Sunita wants to know if males and females occur in equal proportions for this species. She observes that 5 of them are male and the rest female. She carries out a hypothesis test at the 10% significance level.

(i) State the null and alternative hypotheses for Sunita's test and state whether it is 1-tail or 2-tail.

(ii) Find the critical region for the test.

(iii) Illustrate the critical region on a bar chart.

(iv) State what conclusion can be drawn from the test.

Solution

So $q = 1 - p$ is the proportion of females.

(i) Let p be the proportion of males.

Null hypothesis, H_0: $p = 0.5$ males and females of this species occur in equal proportions.

Alternative hypothesis, H_1: $p \neq 0.5$ males and females of this species occur in different proportions.

The test is 2-tail because the alternative hypothesis is $p \neq 0.5$ and so is two sided. If the alternative hypothesis had been $p > 0.5$ or if it had been $p < 0.5$ then it would have been one sided and a 1-tail test would have been needed.

2-tail test

Significance level: 10%

(ii) Let X be the number of male bats in a sample of size 16. $X \sim B(16, p)$.

Since the test is 2-tail, the critical region should be up to 5% at each side.

On the assumption that the null hypothesis is true, and so $p = 0.5$, the probabilities for small values of X and the cumulative percentages are given in Table 17.1.

You would expect to use your calculator to do this work for you. Use the figures here to check that you know how to do this.

Table 17.1

X	$p(X)$	Cumulative % (1 d.p.)
0	0.000015 ...	0.0
1	0.000244 ...	0.0
2	0.001831...	0.2
3	0.008544 ...	1.1
4	0.027770 ...	3.8
5	0.066650 ...	10.5

This shows that the critical region for the left-hand tail is $X \leqslant 4$. Since the null hypothesis is $p = 0.5$, the equivalent critical region for the right-hand tail can be found by symmetry. It is $X \geqslant 12$.

(iii)   This is illustrated in the bar chart in Figure 17.1.

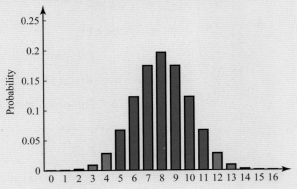

Figure 17.1

(iv)   Sunita observed the number of male bats to be 5.

Since $4 < 5 < 12$, this value of X is not in the critical region. So the evidence is not strong enough to reject the null hypothesis that the proportions of male and female bats of this species are the same in favour of the alternative hypothesis that they are different.

Although there were fewer male bats than females the difference is not great enough to be certain, even at the 10% significance level, that this is not just by chance.

Example 17.2

A driving instructor claims that 60% of his pupils pass their driving test at the first attempt.

(i)   Supposing this claim is true, find the probability that, of 20 pupils taking their first test

(a)   exactly 12 pass   (b)   more than 12 pass.

A local newspaper reporter suspects that the driving instructor is exaggerating his success rate and so she decides to carry out a statistical investigation.

(ii)   State the null and alternative hypotheses which she is testing, and whether the test is 1- or 2-tail.

(iii)   She contacts 20 of his pupils who have recently taken their first test and discovers that N passed. Given that she performs a 5% significance test and that she concludes that the driving instructor's claim was exaggerated, what are the possible values of N?

[MEI adapted]

Solution

This statement can be written as $X \sim \text{B}(20, 0.6)$.

The number of passes is modelled by a random variable X, with a binomial distribution for which $n = 20$ and $p = 0.6$.

(i)   (a)   (from calculator) $\text{P}(X = 12) = 0.180$

(b)   (from calculator) $\text{P}(X > 12) = 1 - \text{P}(X \leqslant 12) = 1 - 0.584 = 0.416$

3

(ii)   Let p be the probability that a randomly selected pupil passes the test at the first attempt.

Null hypothesis, H_0:

$p = 0.6$ The probability of passing first time is 0.6.

Alternative hypothesis, H_1:

$p < 0.6$ The probability of passing first time is less than 0.6.

1-tail test ←

> The reporter is only interested in whether the claim is exaggerated.

Significance level: 5%

T

(iii)   The binomial function on a calculator gives the following cumulative probabilities

$X = 7,$ Cumulative probability $= 0.0210$ (2.1%)

$X = 8,$ Cumulative probability $= 0.0565$ (5.65%)

> So
> 7 is in the critical region
> 8 is outside the critical region

So, at the 5% significance level, the critical region is $0 \leqslant X \leqslant 7$.

The possible values for the number of passes, N, are between 0 and 7 inclusive.

Review exercise

① A firm producing mugs has a quality control scheme in which a random sample of 10 mugs from each batch is inspected. For 50 such samples, the numbers of defective mugs are as follows.

Table 17.2

Number of defective mugs	Number of samples
0	5
1	13
2	15
3	12
4	4
5	1
6+	0

(i)   Find the mean and standard deviation of the number of defective mugs per sample.

(ii)   Show that a reasonable estimate for p, the probability that a mug is defective, is 0.2. Use this figure to calculate the probability that a randomly chosen sample will contain exactly two defective mugs. Comment on the agreement between this value and the observed data.

The management is not satisfied with 20% of mugs being defective and introduces a new process to reduce the proportion of defective mugs.

(iii)   A random sample of 20 mugs, produced by the new process, contains just one which is defective. Test, at the 5% level, whether it is reasonable to suppose that the proportion of defective mugs has been reduced, stating your null and alternative hypotheses clearly.

(iv)   What would the conclusion have been if the management had chosen to conduct the test at the 10% level?   [MEI]

② A leaflet from the Department of Health recently claimed that 70% of businesses operate a no smoking policy on their premises. A member of the public who believed the true figure to be lower than 70% rang a random sample of 19 businesses to ask whether or not they operated a no smoking policy. She then carried out a hypothesis test.

(i)   Write down the null and alternative hypotheses under test.

(ii)   Of the 19 businesses, k say that they do operate a no smoking policy. Use tables to write down the critical region for a 10% test.

(iii) A second person decided to carry out a similar test, also at the 10% level, but sampled only 4 businesses. Write down the critical region in this case.

(iv) Find, for each test, the probability that the null hypothesis is rejected if the true figure is 65%. Hence state which of the two tests is preferable and explain why.

[MEI]

③ In a certain country, 90% of letters are delivered the day after posting.

A resident posts eight letters on a certain day. Find the probability that

(i) all eight letters are delivered the next day

(ii) at least six letters are delivered the next day

(iii) exactly half the letters are delivered the next day.

It is later suspected that the service has deteriorated as a result of mechanisation. To test this, 17 letters are posted and it is found that only 13 of them arrive the next day. Let p denote the probability that, after mechanisation, a letter is delivered the next day.

(iv) Write down suitable null and alternative hypotheses for the value of p.

(v) Carry out the hypothesis test, at the 5% level of significance, stating your results clearly.

(vi) Write down the critical region for the test, giving a reason for your choice. [MEI]

④ For most small birds, the ratio of males to females may be expected to be about 1:1. In one ornithological study birds are trapped by setting fine-mesh nets. The trapped birds are counted and then released. The catch may be regarded as a random sample of the birds in the area.

The ornithologists want to test whether there are more male blackbirds than females.

(i) Assuming that the sex ratio of blackbirds is 1:1, find the probability that a random sample of 16 blackbirds contains

(a) 12 males

(b) at least 12 males.

(ii) State the null and alternative hypotheses the ornithologists should use.

In one sample of 16 blackbirds there are 12 males and 4 females.

(iii) Carry out a suitable test using these data at the 5% significance level, stating your conclusion clearly. Find the critical region for the test.

(iv) Another ornithologist points out that, because female birds spend more time sitting on the nest, females are less likely to be caught than males.

Explain how this would affect your conclusions. [MEI]

⑤ A seed supplier advertises that, on average, 80% of a certain type of seed will germinate. Suppose that 18 of these seeds, chosen at random, are planted.

(i) Find the probability that 17 or more seeds will germinate if

(a) the supplier's claim is correct

(b) the supplier is incorrect and 82% of the seeds, on average, germinate.

Mr Brewer is the advertising manager for the seed supplier. He thinks that the germination rate may be higher than 80% and he decides to carry out a hypothesis test at the 10% level of significance. He plants 18 seeds.

(ii) Write down the null and alternative hypotheses for Mr Brewer's test, explaining why the alternative hypothesis takes the form it does.

(iii) Find the critical region for Mr Brewer's test. Explain your reasoning.

(iv) Determine the probability that Mr Brewer will reach the wrong conclusion if

(a) the true germination rate is 80%

(b) the true germination rate is 82%.

[MEI]

⑥ Given that X has a binomial distribution in which $n = 15$ and $p = 0.5$, find the probability of each of the following events.

(i) $X = 4$ (ii) $X \leqslant 4$

(iii) $X = 4$ or $X = 11$ (iv) $X \leqslant 4$ or $X \geqslant 11$

A large company is considering introducing a new selection procedure for job applicants. The selection procedure is intended to result over a long period in equal numbers of men and women being offered jobs. The new procedure is tried with a random sample of applicants and 15 of them, 11 women and 4 men, are offered jobs.

(v)   Carry out a suitable test at the 5% level of significance to determine whether it is reasonable to suppose that the selection procedure is performing as intended. You should state the null and alternative hypotheses under test and explain carefully how you arrive at your conclusions.

(vi)   Suppose now that, of the 15 applicants offered jobs, w are women. Find all the values of w for which the selection procedure should be judged acceptable at the 5% level.   [MEI]

Working with a large data set

⑦ Sally conjectures that among children aged under 13, boys are more likely than girls to have cycling accidents.

(i)   In the Avonford cycling accidents data set, there are 13 boys and 7 girls under 13 (there is also one person whose sex is not given). Treating this group as a random sample, use these figures to test Sally's conjecture at the 10% significance level.

(ii)   In the complete data set, for all ages, there are 59 males, 31 females and 3 people whose sex is not given. Comment on Sally's conjecture in the light of these figures.

(iii)   Three suggestions are put forward to explain the disparity in the numbers between the sexes.

- Females are safer cyclists.
- More males than females cycle regularly.
- Males tend to cycle further, and so spend longer on their bicycles, than females.

Do the data support any of these suggestions?

1 Interpreting sample data using the Normal distribution

THE AVONFORD STAR

Avonford set to become greenhouse?

From our Science Correspondent Ama Williams

On a recent visit to Avonford Community College, I was intrigued to find experiments being conducted to measure the level of carbon dioxide in the air we are all breathing. Readers will of course know that high levels of carbon dioxide are associated with the greenhouse effect.

Lecturer Ray Sharp showed me round his laboratory. 'It is delicate work, measuring parts per million, but I am trying to establish what is the normal level in this area. Yesterday we took ten readings and you can see the results for yourself: 336, 334, 332, 332, 331, 331, 330, 330, 328, 326.'

When I commented that there seemed to be a lot of variation between the readings, Ray assured me that that was quite in order.

'I have taken hundreds of these measurements in the past,' he said. 'There is always a standard deviation of 2.5. That's just natural variation.'

I suggested to Ray that his students should test whether these results are significantly above the accepted value of 328 parts per million. Certainly they made me feel uneasy. Is the greenhouse effect starting here in Avonford?

Ray Sharp has been trying to establish the carbon dioxide level at Avonford. How do you interpret his figures? Do you think the correspondent has a point when she says she is worried that the greenhouse effect is already happening in Avonford?

If suitable sampling procedures have not been used, then the resulting data may be worthless, indeed positively misleading. You may wonder if this is the case with Ray's figures and about the accuracy of his analysis of the samples too. His data are used in subsequent working in this chapter, but you may well feel there is something of a question mark hanging over them. You should always be prepared to treat data with a healthy degree of caution.

Putting aside any concerns about the quality of the data, what conclusions can you draw from them?

Estimating the population mean, μ

Ray Sharp's data were as follows.

336, 334, 332, 332, 331, 331, 330, 330, 328, 326

His intention in collecting them was to estimate the mean of the parent population, the population mean.

The mean of these figures, the sample mean, is given by

$$\overline{x} = \frac{(336 + 334 + 332 + 332 + 331 + 331 + 330 + 330 + 328 + 326)}{10}$$

$$= 331.$$

What does this tell you about the population mean, μ?

It tells you that μ is about 331 but it certainly does not tell you that it is definitely and exactly 331. If Ray took another sample, its mean would probably not be 331 but you would be surprised (and suspicious) if it were very far away from it. If he took lots of samples, all of size 10, you would expect their means to be close together but certainly not all the same.

If you took 1000 such samples, each of size 10, a histogram showing their means might look like Figure 17.2.

You will notice that this distribution looks rather like the Normal distribution and so may well wonder if this is indeed the case.

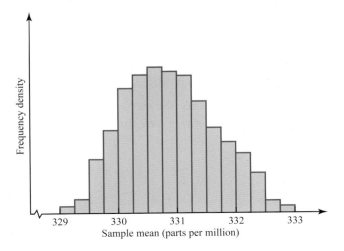

Figure 17.2

The distribution of sample means

In this section, it is assumed that the underlying population has a Normal distribution with mean μ and standard deviation σ so it can be denoted by $N(\mu, \sigma^2)$. In that case the distribution of the means of samples is indeed Normal; its mean is μ and its standard deviation is $\dfrac{\sigma}{\sqrt{n}}$. This is called the **sampling distribution of the means** and is denoted by $N\left(\mu, \dfrac{\sigma^2}{n}\right)$. This is illustrated in Figure 17.3.

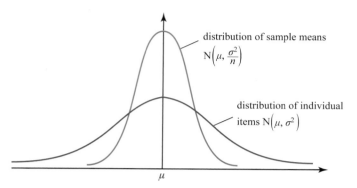

Figure 17.3

> ### Note
> Both of the distributions shown in Figure 17.3 are Normal, but the distribution of the sample means has a smaller spread. Its standard deviation is calculated by dividing the population standard deviation by $\sqrt{n}$. If you increase the sample size, n, the spread of the distribution of their means is reduced.

A hypothesis test for the mean using the Normal distribution

If your intention in collecting sample data is to test a theory, then you should set up a hypothesis test. There are many different hypothesis tests. You have already met that for the proportion in a binomial distribution. The test that follows is for the mean of a Normal distribution.

Ray Sharp was mainly interested in establishing data on carbon dioxide levels for Avonford. The correspondent, however, wanted to know whether levels were above normal and so she could have set up and conducted a test.

The following example shows you how this hypothesis test would be carried out. You will see that there are three ways in which the essential working can be expressed.

Example 17.3

Ama Williams believes that the carbon dioxide level in Avonford has risen above the usual level of 328 parts per million. A sample of 10 specimens of Avonford air are collected and the carbon dioxide level within them is determined. The results are as follows.

$$336, 334, 332, 332, 331, 331, 330, 330, 328, 326$$

Extensive previous research has shown that the standard deviation of the levels within such samples is 2.5, and that the distribution may be assumed to be Normal.

Use these data to test, at the 0.1% significance level, Ama's belief that the level of carbon dioxide at Avonford is above normal.

> This is a much smaller significance level than has been used before. This makes this test quite severe; the smaller the significance level, the more severe the test.

Solution

As usual with hypothesis tests, you use the distribution of the statistic you are measuring, in this case the Normal distribution of the sample means, to decide which values of the test statistic are sufficiently extreme as to suggest that the alternative hypothesis, not the null hypothesis, is true.

Null hypothesis, H_0:

$\mu = 328$ The level of carbon dioxide at Avonford is as expected.

Alternative hypothesis, H_1:

$\mu > 328$ The level of carbon dioxide at Avonford is above that expected.

1-tail test

Significance level: 0.1%

> The test is 1-tail because the alternative hypothesis is $\mu > 328$ and so is one sided. If the alternative hypothesis had been $\mu \neq 328$ then it would have been two sided and a 2-tail test would have been needed.

Method 1: Using probabilities

The distribution of sample means, $\overline{X}$, is $N\left(\mu, \dfrac{\sigma^2}{n}\right)$.

According to the null hypothesis, $\mu = 328$ and it is known that $\sigma = 2.5$ and $n = 10$.

So this distribution is $N\left(328, \dfrac{2.5^2}{10}\right)$, see Figure 17.5.

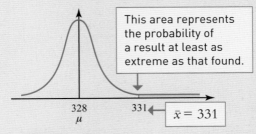

This area represents the probability of a result at least as extreme as that found.

328 μ 331 $\bar{x} = 331$

Figure 17.5

The probability of the mean, $\overline{X}$, of a randomly chosen sample being greater than the value found, i.e. 331, is given by

$$P(\overline{X} \geqslant 331) = 1 - \Phi\left(\frac{331 - 328}{\frac{2.5}{\sqrt{10}}}\right)$$

$$= 1 - \Phi(3.79)$$
$$= 1 - 0.99993$$
$$= 0.00007$$

> 0.000 07 is the p-value of Ama's test.

Since $0.000\,07 < 0.001$, the required significance level (0.1%), the null hypothesis is rejected in favour of the alternative hypothesis: that the mean carbon dioxide level is above 328, at the 0.1% significance level.

Method 2: Using critical ratios

The **critical ratio** is given by $z = \dfrac{\text{observed value} - \text{expected value}}{\text{standard deviation}}$

$$z = \left(\frac{331 - 328}{\frac{2.5}{\sqrt{10}}} \right) = 3.79$$

> This is the value of k in Method 1.

> **Note**
>
> This does not mean that the null hypothesis is definitely incorrect but that the evidence suggests, at this significance level, that this is the case.

This is now compared with the critical value for z, in this case $z = 3.09$.

Since $3.79 > 3.09$, H_0 is rejected.

Method 3: using critical regions

Since the distribution of sample means is $N\left(\mu, \dfrac{\sigma^2}{n} \right)$, critical values for a test

> k is the critical value for the standardised value, z. Make sure that you know how to get this value from your calculator.

on the sample mean are given by $\mu \pm k \times \dfrac{\sigma}{\sqrt{n}}$

In this case if H_0 is true, $\mu = 328$; $\sigma = 2.5$; $n = 10$.

The test is one-tail, for $\mu > 328$, so only the right-hand tail applies.

By using the inverse Normal function on a calculator, the value of $k = 3.09$ is obtained. For some calculators, you may need to use $1 - 0.001 = 0.999$.

For a 1-tail test at the 0.1% significance level, the critical value is thus

$328 + 3.09 \times \dfrac{2.5}{\sqrt{10}} = 330.4$, as shown in Figure 17.4.

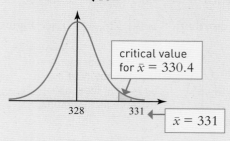

> critical value for $\bar{x} = 330.4$

328   331 $\bar{x} = 331$

Figure 17.4

However, the sample mean $\bar{x} = 331$, and $331 > 330.4$.

Therefore the sample mean lies within the critical region and so the null hypothesis is rejected in favour of the alternative hypothesis: that the mean carbon dioxide level is above 328, at the 0.1% significance level.

Notes

1   A hypothesis test should be formulated before the data are collected and not after. If sample data lead you to form a hypothesis, then you should plan a suitable test and collect further data on which to conduct it. It is not clear whether or not the test in the previous example was being carried out on the same data which were used to formulate the hypothesis.

2   If the data were not collected properly, any test carried out on them may be worthless.

ACTIVITY 17.1

The test in the example was 1-tail and the significance level was 0.1%. The critical value, k, of the standardised value of z was 3.09.

Copy and complete Table 17.3 for the values of k for other common tests of this type.

Table 17.3

	1-tail	2-tail
10%		
5%		
$2\frac{1}{2}\%$		
1%		

Known and estimated standard deviation

Notice that you can only use one of these methods of hypothesis testing if you already know the value of the standard deviation of the parent population; Ray Sharp had said that from taking hundreds of measurements he knew it to be 2.5.

It is more often the situation that you do not know the population standard deviation or variance and have to estimate it from your sample data. Provided the sample size, n, is sufficiently large – at least 50 as a rule of thumb, but the larger the better – the sample standard deviation may be used as an estimate of the population standard deviation. Theoretically, it can be shown that the sample variance, s^2, is an unbiased estimate of the population variance, σ^2.

Note

If the population standard deviation is estimated from a small sample and the population is known to have a Normal distribution, a different hypothesis test must be used. This is the t-test. It is beyond the scope of this book.

This approximation is used in the following example.

Example 17.4

Some years ago a reading test for primary school children was designed to produce a mean score of 100. Recently, a researcher put forward the theory that primary school pupils have changed in their reading skills and score differently in this test. She selects a random sample of 150 primary school pupils, all of whom have taken the test. Each score is denoted by x; the values of x are not necessarily whole numbers.

The results of the tests are summarised as follows.

$$n = 150 \qquad \sum x = 15\,483 \qquad \sum x^2 = 1\,631\,680$$

Carry out a suitable hypothesis test on the researcher's theory, at the 1% significance level. You may assume that the test scores are Normally distributed.

Solution

H_0: The parent population mean is unchanged, i.e. $\mu = 100$

H_1: The parent population mean has changed, i.e. $\mu \neq 100$.

2-tail test

Significance level 1%

From the sample summary statistics, the mean and standard deviation are as follows.

$$\bar{x} = \frac{\sum x}{n} = \frac{15\,483}{150} = 103.22$$

$$s = \sqrt{\frac{\sum x^2 - n\bar{x}^2}{n-1}} = \sqrt{\frac{1\,631\,680 - 150 \times 103.22^2}{149}}$$

$$= 15.0 \text{ (to 3 s.f.)}$$

The standardised z value corresponding to $\bar{x} = 103.22$ is calculated using $\mu = 100$ and using $s = 15.0$ as an estimate for σ.

$$z = \frac{\bar{x} - \mu_0}{\frac{\sigma}{\sqrt{n}}} = \frac{103.22 - 100}{\frac{15.0}{\sqrt{150}}} = 2.629$$

For the 1% significance level, the critical value is $z = 2.576$.

The test statistic is compared with the critical value and since $2.629 > 2.576$ the null hypothesis is rejected; there is significant evidence that scores on this reading test are now different. This is shown in Figure 17.6.

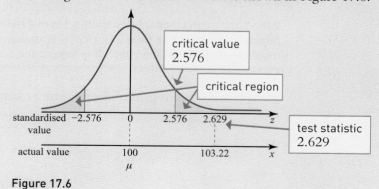

Figure 17.6

Exercise 17.1

Those questions in this exercise where the context is drawn from real life involve working with mathematical models.

① Angus is investigating whether the present day mass of a type of rare breed hen has changed since it was recorded a long time ago. He uses a 2-tail test. Under what circumstances might he use a 1-tail test?

② For each of the following, the random variable $X \sim N(\mu, \sigma^2)$, with *known* standard deviation. A random sample of size n is taken from the parent population and the sample mean, $\bar{x}$, is calculated. Carry out hypothesis tests, given H_0 and H_1, at the significance level indicated.

Table 17.4

	σ	n	$\bar{x}$	H_0	H_1	Sig. level
(i)	8	6	195	$\mu = 190$	$\mu > 190$	5%
(ii)	10	10	47.5	$\mu = 55$	$\mu < 55$	1%
(iii)	15	25	104.7	$\mu = 100$	$\mu \neq 100$	10%
(iv)	4.3	15	34.5	$\mu = 32$	$\mu > 32$	2%
(v)	40	12	345	$\mu = 370$	$\mu \neq 370$	5%

③ A machine is designed to make paperclips with mean mass 4.00 g and standard deviation 0.08 g. The distribution of the masses of the paperclips is Normal. Find the following.

(i) The probability that an individual paperclip, chosen at random, has mass greater than 4.04 g.

(ii) The probability that the mean mass of a random sample of 25 paperclips is greater than 4.04 g.

A quality control officer weighs a random sample of 25 paperclips and finds their total mass to be 101.2 g.

(iii) Conduct a hypothesis test at the 5% significance level to find out whether this provides evidence of an increase in the mean mass of the paperclips. State your null and alternative hypotheses clearly.

④ It is known that the mass of a certain type of lizard has a Normal distribution with mean 72.7 g and standard deviation 4.8 g. A zoologist finds a colony of lizards in a remote place and is not sure whether they are of the same type. In order to test this, she collects a sample of 12 lizards and weighs them, with the following results.

80.4, 67.2, 74.9, 78.8, 76.5, 75.5, 80.2, 81.9, 79.3, 70.0, 69.2, 69.1

(i) Write down, in precise form, the zoologist's null and alternative hypotheses, and state whether a 1-tail or 2-tail test is appropriate.

(ii) Carry out the test at the 5% significance level and write down your conclusion.

(iii) Would your conclusion have been the same at the 10% significance level?

⑤ Observations over a long period of time have shown that the midday temperature at a particular place during the month of June is Normally distributed with a mean value of 23.9°C with standard deviation 2.3°C.

An ecologist sets up an experiment to collect data for a hypothesis test of whether the climate is getting hotter. She selects at random twenty June days over a five-year period and records the midday temperature. Her results (in °C) are as follows.

20.1 26.2 23.3 28.9 30.4 28.4 17.3
22.7 25.1 24.2 15.4 26.3 19.3 24.0
19.9 30.3 32.1 26.7 27.6 23.1

(i) State the null and alternative hypotheses that the ecologist should use.

(ii) Carry out the test at the 10% significance level and state the conclusion.

(iii) Calculate the standard deviation of the sample data and comment on it.

⑥ A zoo has a long established colony of a particular type of rodent which is threatened with extinction in the wild. Observations over several years indicate that the life expectancy for the rodent is 470 days, with standard deviation 45 days. The staff at the zoo suspects that the life expectancy can be increased by improvements to the rodents' environment. As an experiment they allow 36 individuals to spend their whole lives in new surroundings. Their lifetimes are as follows.

491 505 523 470 468 567 512
560 468 498 471 444 511 508
508 421 465 499 486 513 500
488 487 455 523 516 486 478
470 465 487 572 451 513 483
474

(i) What do you need to assume about the parent population of lifetimes, of which the 36 is regarded as a random sample?

(ii) State the null and alternative hypotheses which these data have been collected to test.

(iii) Carry out the test at the 2% significance level and state the conclusion.

(iv) How could increased longevity help the rodent population to survive?

⑦ Some years ago the police did a large survey of the speeds of motorists along a stretch of motorway, timing cars between two bridges. They concluded that their mean speed was 80 mph with standard deviation 10 mph. Recently the police wanted to investigate whether there had been any change in motorists' mean speed. They timed the first 20 green cars between the same two bridges and calculated their speeds (in mph) to be as follows.

85	75	80	102	78	96	124
70	68	92	84	69	73	78
86	92	108	78	80	84	

(i) State an assumption you need to make about the speeds of motorists in the survey for the test to be valid.

(ii) State suitable null and alternative hypotheses and use the sample data to carry out a hypothesis test at the 5% significance level. State the conclusion.

One of the police officers involved in the investigation says that one of the cars in the sample was being driven exceptionally fast and that its speed should not be included within the sample data.

(iii) Would the removal of this outlier alter the conclusion?

⑧ The keepers of a lighthouse were required to keep records of weather conditions. Analysis of their data from many years showed the visibility at midday to have a mean value of 14 sea miles with standard deviation 5.4 sea miles. A new keeper decided he would test his theory that the air had become less clear (and so visibility reduced) by carrying out a hypothesis test on data collected for his first 36 days on duty. His figures (in sea miles) were as follows.

35	21	12	7	2	1.5
1.5	1	0.25	0.25	15	17
18	20	16	11	8	8
9	17	35	35	4	0.25
0.25	5	11	28	35	35
16	2	1	0.5	0.5	1

(i) Write down a distributional assumption for the test to be valid.

(ii) Write down suitable null and alternative hypotheses.

(iii) Carry out the test at the 2.5% significance level and state the conclusion that the lighthouse keeper would have come to.

(iv) Criticise the sampling procedure used by the keeper and suggest a better one.

⑨ The packaging on a type of electric light bulb states that the average lifetime of the bulbs is 1000 hours. A consumer association thinks that this is an overestimate and tests a random sample of 64 bulbs, recording the lifetime, in hours, of each bulb. You may assume that the distribution of the bulbs' lifetimes is Normal.

The results are summarised as follows.

$$n = 64 \qquad \sum x = 63\,910.4$$
$$\sum x^2 = 63\,824\,061$$

(i) Calculate the mean and standard deviation of the data.

(ii) State suitable null and alternative hypotheses to test whether the statement on the packaging is overestimating the lifetime of this type of bulb.

(iii) Carry out the test, at the 5% significance level, stating your conclusions carefully.

⑩ A sample of 40 observations from a Normal distribution X gave $\sum x = 24$ and $\sum x^2 = 596$. Performing a two-tail test at the 5% level, test whether the mean of the distribution is zero.

⑪ A random sample of 75 eleven-year-olds performed a simple task and the time taken, t minutes, was noted for each. You may assume that the distribution of these times is Normal.

The results are summarised as follows.

$$n = 75 \qquad \sum t = 1215 \qquad \sum t^2 = 21708$$

(i) Calculate the mean and standard deviation of the data.

(ii) State suitable null and alternative hypotheses to test whether there is evidence that the mean time taken to perform this task is greater than 15 minutes.

(iii) Carry out the test, at the 1% significance level, stating your conclusions carefully.

⑫ Bags of sugar are supposed to contain, on average, 2 kg of sugar. A quality controller suspects that they actually contain less than this amount, and so 90 bags are taken at random and the mass, x kg, of sugar in each is measured. You may assume that the distribution of these masses is Normal.

The results are summarised as follows.

$n = 90$ $\sum x = 177.9$ $\sum x^2 = 353.1916$

(i) Calculate the mean and standard deviation of the data.

(ii) State suitable null and alternative hypotheses to test whether there is any evidence that the sugar is being sold 'underweight'.

(iii) Carry out the test, at the 2% significance level, stating your conclusions carefully.

⑬ A machine produces jars of skin cream, filled to a nominal volume of 100 ml. The machine is supposed to be set to 105 ml, to ensure that most jars actually contain more than the nominal volume of 100 ml. You may assume that the distribution of the volume of skin cream in a jar is Normal.

To check that the machine is correctly set, 80 jars are chosen at random and the volume, x ml, of skin cream in each is measured.

The results are summarised as follows.

$n = 80$ $\sum x = 8376$ $\sum x^2 = 877687$

(i) Calculate the mean and standard deviation of the data.

(ii) State suitable null and alternative hypotheses for a test of whether the machine is set correctly.

(iii) Carry out the test, at the 10% significance level, stating your conclusions carefully.

Working with a large data set

⑭ Sally and Robin have slightly different conjectures about the probabilities of accidents on different days of the week.

Sally's conjecture is that an accident is equally likely to occur on any day of the week.

(i) Using Sally's conjecture, write down the probability that a randomly selected cycling accident occurred on any particular day.

Hence write down the expected number of accidents on each day of the week for a total of 92 accidents.

Name the type of distribution in this conjecture.

Table 17.5 summarises the Avonford cyclists data.

Table 17.5

Day	Frequency
Monday	19
Tuesday	21
Wednesday	14
Thursday	8
Friday	10
Saturday	12
Sunday	8

Sally carries out a test with the following hypotheses.

H_0: The probability is the same for each day of the week.

H_1: The probability is not the same for each day of the week.

The test statistic comes out to be 12.24.

Critical values for this test for different significance levels are given in Table 17.6.

Table 17.6

Significance level	10%	5%	2.5%	1%
Critical value	10.64	12.59	14.45	16.81

(ii) What should Sally conclude from the test?

Robin's conjecture, based on conversations with people from other areas, is that cycling accidents are more common on weekdays than at weekends, and so the probability of an accident occurring on a weekday is greater than $\frac{5}{7}$.

(iii) Give a common sense reason why cycling accident rates on weekdays might be different from those at the weekend.

(iv) Use a Normal approximation to the distribution $B(92, \frac{5}{7})$ to find the probability of a number of weekday accidents at least as great as the number in the Avonford data.

(v) Robin uses this probability in a hypothesis test at the 5% significance level.

State the null and alternative hypotheses for this test, and interpret the result.

2 Bivariate data: Correlation and association

Bivariate data are typically displayed on a scatter diagram and this gives an informal impression of whether there is any correlation or association between the two variables, and, if so, of its direction.

Vocabulary

The relationship is described as **correlation** when
■ a change in the value of one of the variables is likely to be accompanied by a broadly related change in the other variable
■ the relationship is linear
■ both variables are random.

Correlation is a special case of **association**. This is a more general term covering relationships that are not necessarily linear and where one of the variables need not be random.

Correlation and association are described as **positive** when both the variables increase or decrease together, and as **negative** when an increase in one is accompanied by a decrease in the other.

> **Note**
> Association does not have to be either positive or negative.

> **Discussion point**
> → Describe the relationships between the variables in the four scatter diagrams in Figure 17.7.

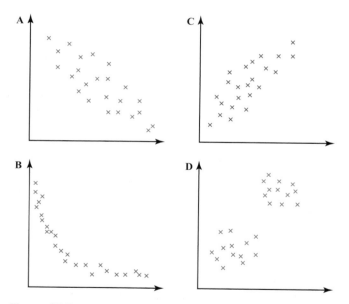

Figure 17.7

Measuring correlation

In both A and C in Figure 17.7 it looks as though there is strong correlation between the variables. The points lie reasonably close to straight lines, but how can you measure it?

The usual measure of correlation is **Pearson's product moment correlation coefficient (pmcc)**. You should be able to enter a bivariate data set into your calculator and obtain the value of the pmcc.

Note

The name Pearson is often omitted and it is common to abbreviate 'product moment correlation coefficient' to 'pmcc'.

You can calculate the value of the pmcc by hand. Seeing how this is done will help you to interpret it. The notation commonly used is as follows.

- The size of the data set is n.

- The data items are $(x_1, y_1), (x_2, y_2) \ldots (x_n, y_n)$.

- The general data item is (x_i, y_i).

- Pearson's product moment correlation coefficient is denoted by r.

Using this notation, $r = \dfrac{\sum(x_i - \bar{x})(y_i - \bar{y})}{\sqrt{\sum(x_i - \bar{x})^2 \times \sum(y_i - \bar{y})^2}}$.

To see how this formula arises, look at this scatter diagram in Figure 17.8.

> Your calculator does not know anything about the source of the data you enter and so will give you a value for the correlation coefficient whether or not the conditions for correlation apply. It is up to you to check that first.

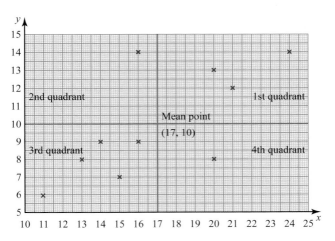

Figure 17.8

Note

The point (x, y) has a moment of $(x - 17)$ about the vertical axis and of $(y - 10)$ about the horizontal axis. So $(x - 17)(y - 10)$ is called its product moment.

There are 10 points; the mean of their x values is 17 and of their y values is 10. So the mean point, $(\bar{x}, \bar{y})$, is (17, 10) and this can be thought of as a new origin, with the red lines, $x = 17$ and $y = 10$, as new axes, dividing the graph up into 4 quadrants.

Relative to these axes the coordinates of a point that was (x, y) are now $(x - 17, y - 10)$.

If you just look at the 8 points that are now in the first and third quadrants there appears to be quite a strong positive correlation. However, there are two other points, one in the 2nd quadrant and the other in the 4th quadrant for which this is not the case. So points in the 1st and 3rd quadrants show positive correlation and points in the 2nd and 4th quadrants negative correlation.

ACTIVITY 17.2

(i)   Work out the value of the product $(x - 17)(y - 10)$ for the point (20, 8).

(ii)   Show that $\sum(x - 17)(y - 10)$, i.e. the sum of the products for all the 10 points on the scatter diagram, comes out to be 77.

> **Note**
>
> $\frac{77}{10} = 7.7$

The value of 77 you obtained in Activity 17.2 is positive and so that indicates that the overall correlation is positive but it does not make any allowance for the size of the data set or for the spread of the two variables.

- To allow for the number of items in the data set, you divide by n.

- To allow for their spread, you divide by the root mean square deviation (rmsd) for each of the variables. In this example they are
$$\sqrt{15} = 3.872 \ldots \text{ for } x \text{ and } \sqrt{8} = 2.828 \ldots \text{ for } y.$$

So the product moment correlation coefficient for the data in the scatter diagram is 0.703.

> **Note**
>
> Reminder: the rmsd for x is given by
> $$\text{rmsd} = \sqrt{\frac{\sum(x_i - \bar{x})^2}{n}},$$
> as on page 327.

> **Note**
>
> $\frac{7.7}{\sqrt{15} \times \sqrt{8}} = 0.703$

ACTIVITY 17.3

Generalise the work in the above example to obtain the formula
$$r = \frac{\sum(x_i - \bar{x})(y_i - \bar{y})}{\sqrt{\sum(x_i - \bar{x})^2 \times \sum(y_i - \bar{y})^2}}.$$

If the points on the scatter diagram all lie on a sloping straight line, there is perfect correlation.

> **Note**
>
> If you calculate a value of r and it comes out to be outside the -1 to 1 range, you have definitely made a mistake.

- If the gradient of the line is positive the value of r works out to be $+1$, corresponding to perfect positive correlation.

- If the gradient of the line is negative, $r = -1$ and there is perfect negative correlation.

These are the extreme cases. For all bivariate data $-1 \leqslant r \leqslant +1$.

> **Note**
>
> There are two other ways that the formula for the pmcc is currently written. All three are equivalent.
> $$r = \frac{\sum(x_i - \bar{x})(y_i - \bar{y})}{\sqrt{\sum(x_i - \bar{x})^2 \times \sum(y_i - \bar{y})^2}} = \frac{\sum x_i y_i - n\bar{x}\,\bar{y}}{\sqrt{\left(\sum x_i^2 - n\bar{x}^2\right)\left(\sum y_i^2 - n\bar{y}^2\right)}} = \frac{S_{xy}}{\sqrt{S_{xx}S_{yy}}}$$
> where $S_{xx} = \sum(x_i - \bar{x})^2$, $S_{yy} = \sum(y_i - \bar{y})^2$, as given on page 327.
> and $S_{xy} = \sum(x_i - \bar{x})(y_i - \bar{y})$.

Interpreting the product moment correlation coefficient

The product moment correlation coefficient provides a measure of the correlation between the two variables. There are two different ways in which it is commonly used and interpreted.

When the data cover the whole of a population the correlation coefficient tells you all that there is to be known about the level of correlation between the variables in the population and is denoted by ρ. ←

> The symbol ρ is the Greek letter rho, pronounced 'row' as in 'row a boat'. It is equivalent to rh.

> The term 'big data' is used to discribe a very large data set.

The same is true for a very large sample. This is a common situation with **big data** where the calculation is carried out by computer and might cover millions of data items. Such data sets are often not just bivariate but multivariate covering many fields. It is quite common to find unexpected correlations but at a low level, say less than 0.1, as well as those expected at a higher level, say 0.5. The level of such correlations is called the **effect size**; it is beyond the scope of this book.

A different situation occurs when a reasonably small sample is taken from a bivariate population as in Figure 17.9.

In this case the value of the sample correlation coefficient may be used

■ either to estimate the correlation coefficient, ρ, in the parent population,

■ or to test whether there is any correlation.

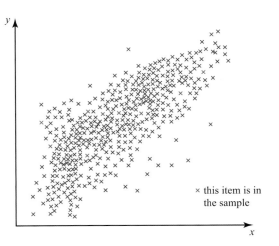

× this item is in the sample

Figure 17.9

Using the sample correlation coefficient as a test statistic

The correlation coefficient may be used as the test statistic in a test of the null hypothesis

H_0: $\rho = 0$ There is no correlation between the variables

against one of three possible alternative hypotheses.

1 H_1: $\rho \neq 0$ There is correlation between the variables (2-tail test)

2 H_1: $\rho > 0$ There is positive correlation between the variables (1-tail test)

3 H_1: $\rho < 0$ There is negative correlation between the variables (1-tail test)

The test is carried out by comparing the value of r with the appropriate critical value. This depends on the sample size, the significance level of the test and whether it is 1- or 2-tail. Critical values can be found from statistical software or tables.

The next example shows such a test.

Example 17.5

Here is a letter to a newspaper.

Dear Sir,

The trouble with young people these days is that they spend too much time watching TV. They just sit there gawping and as a result they become steadily less intelligent. I challenge you to carry out a proper test and I am sure you will find that the more TV children watch the less intelligent they are.

Yours truly,

Concerned Citizen

A reporter on the newspaper manages to collect the following data for six children.

Table 17.7

Hours of TV, x	9	11	14	7	10	9
IQ, y	142	112	100	126	109	89

The reporter then sets about conducting the 'proper test' that the letter writer requested. She uses a 5% significance level.

(i)   State the null and alternative hypotheses and describe the test.

(ii)   Determine the critical value.

(iii)   Carry out the test and comment on the result.

Note

This example is designed just to show you how to do the calculations, and no more. A sample of size 6 is so tiny that using it would be bad practice.

Solution

(i)   Null hypothesis, H_0: $\rho = 0$ There is no correlation between hours watching TV and IQ.

Alternative hypothesis H_1: $\rho < 0$ There is negative correlation between hours watching TV and IQ.

The test is 1-tailed.

(ii)   For a 1-tail test at the 5% significance level with $n = 6$, the table gives 0.7293. This means that in a test for positive correlation the critical value is 0.7293, but in a test for negative correlation (as here) the critical value is -0.7293.

	5%	$2\frac{1}{2}$%	1%	$\frac{1}{2}$%	1–tail test
	10%	5%	2%	1%	2-tail test
n					
1	–	–	–	–	
2	–	–	–	–	
3	0.9877	0.9969	0.9995	0.9999	
4	0.9000	0.9500	0.9800	0.9900	
5	0.8054	0.8783	0.9343	0.9587	
6	0.7293	0.8114	0.8822	0.9172	
7	0.6694	0.7545	0.8329	0.8745	
8	0.6215	0.7067	0.7887	0.8343	
9	0.5822	0.6664	0.7498	0.7977	
10	0.5494	0.6319	0.7155	0.7646	
11	0.5214	0.6021	0.6851	0.7348	
12	0.4973	0.5760	0.6581	0.7079	
13	0.4762	0.5529	0.6339	0.6835	
14	0.4575	0.5324	0.6120	0.6614	
15	0.4409	0.5140	0.5923	0.6411	

Figure 17.10

(iii)   The critical value is −0.7293 and the value of *r* of −0.4347 is less extreme.

So there is not enough evidence to reject the null hypothesis in favour of the alternative hypothesis.

The evidence does not support the letter writer's conjecture.

Correlation and causation

In the example above, the letter writer said '*They just sit there gawping and as a result they become steadily less intelligent*'. This was thus suggesting that watching television **causes** children to become less intelligent. Even if the result of the test had been significant, it would not have shown this. It is only one among many possible explanations. Another might be that the less intelligent children were set less homework and so had more time to watch television. No doubt you can think of other possible explanations.

In general, although there may be a high level of correlation between variables *A* and *B*, it does not mean that *A* causes *B* or that *B* causes *A*. It may well be that a third variable, *C*, causes both *A* and *B*, or it may be that there is a more complicated set of relationships.

As an example, figures for the years 1985 to 1993 showed a high correlation between the sales of personal computers and those for microwave ovens. There is of course no direct connection between the two variables. You would be quite wrong to conclude that buying a microwave oven predisposes you to buy a computer as well, or vice versa.

This is an important point: just because there is a correlation between two variables, it does not mean that one causes the other. It is summed up by the saying

'Correlation does not imply causation.'

Measuring association

You have seen that in some bivariate data sets there is association but not correlation between the variables. It may be that the relationship is non-linear or it may be that one of the variables is non-random. So sometimes you need a measure of association rather than correlation.

Such a measure can be found by a two-step process:

■   first rank each of the variables

■   then work out the product moment correlation coefficient for the ranks.

The resulting measure is sometimes called a **rank correlation coefficient** although a more accurate description would be an **association coefficient**.

Sometimes your data will already be ranked and so the first step in this process is not needed, as in the example that follows.

Punch-up at the village fete

Pandemonium broke out at the Normanton village fete last Saturday when the adjudication for the Tomato of the Year competition was announced. The two judges completely failed to agree in their rankings and so a compromise winner was chosen to the fury of everybody (except the winner).

Following the announcement there was a moment of stunned silence, followed by shouts of 'Rubbish', 'It's a fix', 'Go home' and further abuse. Then the tomatoes started to fly and before long fighting broke out.

By the time police arrived on the scene ten people were injured, including last year's winner Bert Wallis who lost three teeth in the scrap. Both judges had escaped unhurt.

Angry Bert Wallis, nursing a badly bruised jaw, said 'The competition was a nonsense. The judges were useless. Their failure to agree shows that they did not know what they were looking for'. But fete organiser Margaret Bramble said this was untrue. 'The competition was completely fair; both judges know a good tomato when they see one,' she claimed.

The judgement that caused all the trouble was as follows.

Table 17.9

Tomato	A	B	C	D	E	F	G	H
Judge 1	1	8	4	6	2	5	7	3
Judge 2	7	2	3	4	6	8	1	5
Total	8	10	7	10	8	13	8	8

Winner

You can see that both judges ranked the eight entrances 1st, 2nd, 3rd, … , 8th. The winner, C, was placed 4th by one judge and 3rd by the other. Overall the rankings of the two judges do look different so perhaps they were using different criteria on which to assess the tomatoes. A coefficient of association (or of rank correlation) might be helpful in deciding whether this was the case.

ACTIVITY 17.4
Show that the product moment correlation coefficient of the ranks in the table above is −0.690.

Although you can carry out the calculation as in Example 17.5, it is usually done a different way.

Denoting the two sets of ranks by $x_1, x_2,..., x_n$ and $y_1, y_2,..., y_n$, the coefficient of association is given by

$$r_s = 1 - \frac{6\sum d_i^2}{n(n^2 - 1)}$$

where

- r_s is called **Spearman's rank correlation coefficient**
- d_i is the difference in the ranks for a general data item (x_i, y_i); $d_i = x_i - y_i$
- n is the number of items of data.

The calculation of $\sum d_i^2$ can then be set out in a table like this.

Table 17.10

Tomato	Judge 1, x_i	Judge 2, x_2	$d_i = x_i - y_i$	d_i^2
A	1	7	−6	36
B	8	2	6	36
C	4	3	1	1
D	6	4	2	4
E	2	6	−4	16
F	5	8	−3	9
G	7	1	6	36
H	3	5	−2	4
				$\sum d_i^2 = 142$

Note

If several items are ranked equally you give them the mean of the ranks they would have had if they had been slightly different from each other.

The value of n is 8, so $r_s = 1 - \dfrac{6 \times 142}{8\left(8^2 - 1\right)} = -0.690$

You will notice that this is the same answer as you found in Activity 17.4. With a little algebra you can prove that the two methods are equivalent.

Using Spearman's rank correlation coefficient

Spearman's rank correlation coefficient is often used as a test statistic for a hypothesis test of

H_0: There is no association between the variables

against one of three possible alternative hypotheses:

either H_1: There is association between the variables (2-tail test)

or H_1: There is positive association between the variables (1-tail test)

or H_1: There is negative association between the variables (1-tail test)

The test is carried out by comparing the value of r_s with the appropriate critical value. This depends on the sample size, the significance level of the test and whether it is 1- or 2-tail. Critical values can be found from statistical software or tables.

Once you have found the value of r_s, this test follows the same procedure as that for correlation. However, the tables of critical values are not the same and so you need to be careful that you are using the right ones.

Historical note

Karl Pearson was one of the founders of modern statistics. Born in 1857, he was a man of varied interests and practised law for three years before being appointed Professor of Applied Mathematics and Mechanics at University College, London in 1884. Pearson made contributions to various branches of mathematics but is particularly remembered for his work on the application of statistics to biological problems in heredity and evolution. He died in 1936.

Figure 17.11 Karl Pearson

Charles Spearman was born in London in 1863. After serving 14 years in the army as a cavalry officer, he went to Leipzig to study psychology. On completing his doctorate there he became a lecturer, and soon afterwards a professor also at University College, London. He pioneered the application of statistical techniques within psychology and developed the technique known as factor analysis in order to analyse different aspects of human ability. He died in 1945.

Figure 17.12 Charles Spearman

Exercise 17.2

① These three scatter diagrams in Figure 17.13 cover all the countries of the world. Each point is one country. The variables are life expectancy, birth rate and mean GDP per person.

(a) For each scatter diagram describe the relationship between the two variables.

(b) Interpret each of these diagrams in terms of what is happening around the world.

(i)

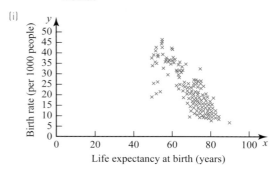

(ii)

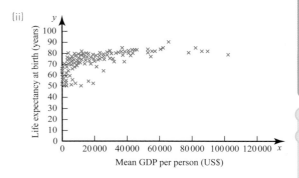

(iii)

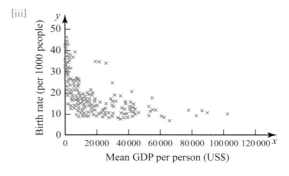

Figure 17.13

② Several sources of possible bivariate data sets are listed as (i) to (vi). The two variables are designated X and Y.

In each case

(a) Describe the relationship as Correlation, Association or Neither.

(b) Explain your decision, adding any relevant comments about the suggested data set.

(i) X The age of a girl who is no more than 20
Y Her height

(ii) X The date in May one year
Y The temperature at noon at a particular place in England

(iii) X The score on one of two dice that are rolled together
Y The score on the other one

(iv) X The distance between service stations on the same motorway
Y The time it takes to travel between them

(v) X Temperatures in Celsius, every 10° from 0° to 100°
Y The equivalent temperatures in Fahrenheit

(vi) X A person's age
Y The length of time that person has left to live

③ A small set of bivariate data points (x_i, y_i) are plotted on the scatter diagram in Figure 17.14. The two variables are both random.

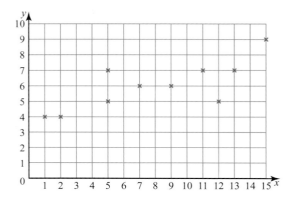

Figure 17.14

(i) Describe any possible correlation.

(ii) Find the values of $\bar{x}$ and $\bar{y}$.

(iii) Show that $\sum (x_i - \bar{x})(y_i - \bar{y}) = 51$.

④ Sachin conducts a chemistry experiment; his readings form a bivariate data set which he plots on the scatter diagram in Figure 17.15. He works out the correlation coefficient for all 16 readings. It is −0.761.

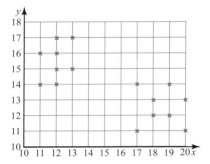

Figure 17.15

Sachin says 'The pmcc shows there is strong negative correlation between the variables'.

Is he correct? Justify your answer.

⑤ A sports reporter believes that those who are good at the high jump are also good at the long jump, and vice versa. He collects data on the best performances of nine athletes and carries out a hypothesis test, using a 5% significance level.

Table 17.10

Athlete	High jump (m)	Long jump (m)
A	2.0	8.0
B	2.1	7.6
C	1.8	6.4
D	2.1	6.8
E	1.8	5.8
F	1.9	8.0
G	1.6	5.5
H	1.8	5.5
I	1.8	6.6

(i) State suitable null and alternative hypotheses.

For these data $r = 0.715$. At the 5% significance level, the critical values for $n = 9$ are 0.582 (1-tail) and 0.666 (2-tail).

(ii) Carry out the test and comment on the result.

⑥ Charlotte is a campaigner for temperance, believing that drinking alcohol is an evil habit. Michel, a representative of a wine company, presents her with these figures which he claims show that wine drinking is good for marriages.

Table 17.11

Country	Wine consump-tion (litres per person per year)	Divorce rate (per 1000 inhabitants)
Belgium	20	2.0
Denmark	20	2.7
Germany	26	2.2
Greece	33	0.6
Italy	63	0.4
Portugal	54	0.9
Spain	41	0.6
UK	13	2.9

(i) Write Michel's claim in the form of a hypothesis test.

For these data $r = -0.854$. At the 1% significance level, the critical values for $n = 8$ are 0.789 (1-tail) and 0.834 (2-tail).

(ii) Carry out the test.

(iii) Charlotte claims that Michel is indulging in 'pseudo–statistics'. What arguments could she use to support this point of view?

⑦ The manager of a company wishes to evaluate the success of its training programme. One aspect which interests her is to see if there is any relationship between the amount of training given to employees and the length of time they stay with the company before moving on to jobs elsewhere. She does not want to waste company money training people who will shortly leave. At the same time she believes that the more training employees are given, the longer they will stay. She collects data on the average number of days training given per year to 25 employees who have recently left for other jobs, and the length of time they worked for the company.

Table 17.12

Training (days/year)	Work (days)
2.0	354
4.0	820
0.1	78
5.6	1480
9.1	980
2.6	902
0.0	134
2.6	252
7.2	867
3.4	760
1.8	125
0.0	28
5.7	1360
7.2	1520
7.5	1380
3.0	121
2.8	457
1.2	132
4.5	1365
1.0	52
7.8	1080
3.7	508
10.9	1281
3.8	945
2.9	692

(i) Plot the data on a scatter diagram.

(ii) State suitable null and alternative hypotheses.

For these data $r = 0.807$. At the 5% significance level, the critical values for $n = 25$ are 0.3365 (1-tail) and 0.3961 (2-tail).

(iii) Carry out the hypothesis test at the 5% significance level.

(iv) What conclusions would you come to if you were the manager?

⑧ Each point on the scatter diagram in Figure 17.16 shows a measurement of the height, h, of a balloon in hundreds of metres at time t minutes after it was released from ground level. The observations were not all made on the same day.

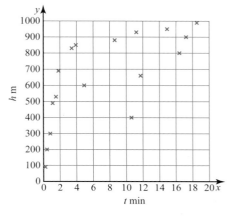

Figure 17.16

(i) Explain why it would be appropriate to calculate Spearman's rank correlation coefficient for this data set but not the product moment correlation coefficient.

(ii) Find Spearman's rank correlation coefficient. What can you conclude from its value?

⑨ During their course two trainee tennis coaches, Rachel and Leroy, were shown videos of seven people, A, B, ... , G, doing a top spin serve and were asked to rank them in order according to the quality of their style. They placed them as follows.

Table 17.13

	A	B	C	D	E	F	G
Rachel's rank	5	1	6	4	7	3	2
Leroy's rank	6	2	7	3	4	1	5

(i) Calculate the product moment correlation coefficient of the ranks from Rachel and Leroy.

(ii) Use the formula $r_s = 1 - \dfrac{6\sum d_i^2}{n(n^2 - 1)}$ to calculate Spearman's rank correlation coefficient for their ranks.

(iii) Comment on your answers to parts (i) and (ii).

⑩ The two variables in a bivariate data set are both random. Their values are denoted by x_i and y_i. You are given that

$$n = 10, \quad \sum x_i = 50, \quad \sum y_i = 120,$$

$$\sum x_i^2 = 352, \quad \sum y_i^2 = 1558,$$

$$\sum x_i y_i = 703.$$

Find the value of the product moment correlation coefficient for this set.

⑪ The two variables in the bivariate data set given in Table 17.14 are both random.

Table 17.14

x	y
1	5
4	0
6	2
4	1
10	12
3	3
3	7
3	4
4	2
2	4

(i) Find Pearson's product moment correlation coefficient.

(ii) Find Spearman's rank correlation coefficient.

(iii) Plot the data points on a scatter diagram.

(iv) Comment on the difference between your answers to parts (i) and (ii).

Is it possible to say which is the more informative measure?

Working with a large data set

⑫ The two variables represented in the scatter diagram in Figure 17.17 are the age of cyclists and their distance from home, in km, when they had their accidents. The data are taken from the Avonford cycling accidents data set. Two extreme data items are not shown on the scatter diagram and two outliers have been excluded from the data set. A spreadsheet has been used to draw the scatter diagram.

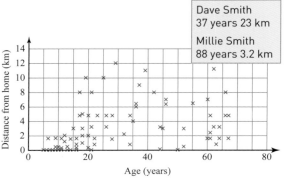

Figure 17.17

(i) Identify two different actions that have been taken in cleaning the data.

(ii) State whether each of the variables is random or non-random.

(iii) Identify one feature of the scatter diagram which suggests that some of the distances have been rounded to the nearest kilometre. Describe the rounding that has occurred for the cyclists' ages.

(iv) As the scatter diagram is drawn, which is the independent variable and which the dependent?

Explain why it is reasonable to draw the diagram in this way, rather than with the variables on the other axes.

Robin says, 'I think that older people tend to have accidents further from home'.

He uses suitable software to carry out a hypothesis test for positive association, based on the ranks of this variables. It reports the test result as a p-value of 3.7×10^{-9}.

(v) Write down the null and alternative hypotheses for the test and interpret the result in terms of the situation being investigated.

Robin also uses the spreadsheet to calculate the product moment correlation coefficient for the data. It is 0.316.

(vi) Comment on whether this figure should improve Robin's understanding of the situation.

LEARNING OUTCOMES

When you have completed this chapter, you should be able to:

➤ understand and apply the language of statistical hypothesis testing:
   - null hypothesis and alternative hypothesis
   - significance level
   - test statistic
   - 1-tail and 2-tail test
   - critical value and critical region
   - acceptance region
   - *p*-value
➤ understand that a sample is being used to make an inference about the population and appreciate that the significance level is the probability of incorrectly rejecting the null hypothesis
➤ conduct a statistical hypothesis test for the proportion in a binomial distribution and interpret the result in context
➤ conduct a statistical hypothesis test for the mean of a Normal distribution with known, given or assumed variance and interpret the results in context
➤ understand that correlation coefficients are measures of how close bivariate data points lie to a straight line on a scatter diagram
➤ use a given correlation coefficient for a sample to make an inference about correlation or association in the population for a given *p*-value or critical value.

KEY POINTS

1 **Steps for conducting a hypothesis test**
   - Establish the null and alternative hypotheses.
   - Decide on the significance level.
   - Collect suitable data using a random sampling procedure that ensures the items are independent.
   - Use the data to determine the test statistic. This is the measure that will be used to decide whether the test is significant.
   - Conduct the test doing the necessary calculations. Then:
     - either work out the *p*-value, the probability of a result at least as extreme as that obtained, and compare it with the significance level
     - or work out the critical value for the particular significance level and compare it with the test statistic
     - or work out the critical (or rejection) region for the particular significance level and determine whether the test statistic lies within it or outside it in the acceptance region. (The terms rejection and acceptance refer to the null hypothesis.)
   - Interpret the result in terms of the original claim, theory or problem.

2 **Hypothesis test for the proportion of a binomial distribution**
   - Sample data may be used to carry out a hypothesis test on the null hypothesis that the proportion, *p*, in a binomial distribution has some particular value.
   - The test statistic for a binomial test is the number of successes during the trial.
   - To find the critical region it is usual to find the cumulative probability and compare it with the significance level, and many calculators have the facility to do this.

3 **Hypothesis test for the mean of a Normal distribution**
   - For samples of size n drawn from a Normal distribution with mean μ and variance σ^2, the distribution of sample means is Normal with mean μ and variance $\dfrac{\sigma^2}{n}$, i.e. $\bar{X} \sim N\left(\mu, \dfrac{\sigma^2}{n}\right)$.

- Sample data may be used to carry out a hypothesis test on the null hypothesis that the population mean has some particular value μ_0, i.e. $H_0: \mu = \mu_0$.

- The test statistic $z = \dfrac{\overline{x} - \mu_0}{\dfrac{\sigma}{\sqrt{n}}}$ is used.

- Software packages and many calculators have the facility to calculate the cumulative Normal distribution $\Phi(z)$.

4 **Pearson's product moment correlation coefficient (pmcc)**

- Software packages and many calculators have the facility to calculate the product moment correlation coefficient (pmcc), r.

- Alternatively you can do the calculation by hand using one of three equivalent formulae

$$r = \frac{\sum(x_i - \overline{x})(y_i - \overline{y})}{\sqrt{\sum(x_i - \overline{x})^2 \times \sum(y_i - \overline{y})^2}}$$

$$= \frac{\sum x_i y_i - n\overline{x}\,\overline{y}}{\sqrt{\left(\sum x_i^2 - n\overline{x}^2\right)\left(\sum y_i^2 - n\overline{y}^2\right)}} = \frac{S_{xy}}{\sqrt{S_{xx}S_{yy}}}$$

- If the data are drawn from a bivariate Normal distribution the pmcc can be used as a test statistic for population correlation
  $H_0: \rho = 0$
  $H_1: \rho < 0$ or $\rho > 0$ (1-tail tests) or $\rho \neq 0$ (2-tail test)

  The test requires access to critical values, using tables or a suitable software package.

5 **Spearman's rank correlation coefficient**

- Spearman's rank correlation coefficent is a measure of association.

- It can be calculated using the formula

$$r_s = 1 - \frac{6\sum d_i^2}{n\left(n^2 - 1\right)}$$

- r_s can be used as a test statistic.

- The test requires access to critical values, using tables or a suitable software package.

6 **Hypothesis testing checklist**

- Was the test set up before or after the data were known?

- Was the sample involved chosen at random and are the data independent?

- Is the statistical procedure actually testing the original claim?

① Two cards are drawn from a normal pack of 52 playing cards. Which of the following calculations gives the probability that there is at least one heart?

A $\dfrac{13}{52} \times \dfrac{12}{51}$

B $1 - \dfrac{1}{4} \times \dfrac{1}{4}$

C $1 - \dfrac{3}{4} \times \dfrac{38}{51}$

D $1 - \dfrac{1}{4} \times \dfrac{12}{51}$

② The table shows the results of a survey about whether the people in a group play tennis.

	male	female
tennis	14	18
not tennis	36	42

Which of the following represents the probability that a randomly chosen female from the group plays tennis?

A $\dfrac{18}{110}$

B $\dfrac{18}{60}$

C $\dfrac{60}{110} \times \dfrac{32}{110}$

D $\dfrac{18}{32}$

③ A set of data is Normally distributed with mean 28 and standard deviation 5. What is the probability that a randomly chosen item is less than 34?

A 0.3849

B 0.1151

C 0.8849

D 0.6151

④ You are using the Normal distribution to model some discrete data with mean 100 and standard deviation 15. You wish to find the probability that a randomly chosen item is between 80 and 90, inclusive. Which of the following is the probability you require?

A $P(80 < X < 90)$

B $P(80.5 < X < 90.5)$

C $P(80.5 < X < 89.5)$

D $P(79.5 < X < 90.5)$

Questions 3 and 8 are based on large data sets published by MEI and located at www.mei.org.uk/data-sets. You may find it valuable to visit this site and familiarise yourself with the data sets.

PS ① A story carried on a news website in 2017 had the headline 'Women work 39 days a year more than men'. It went on to say that women work, on average, 50 minutes per day more than men. The extra minutes worked by women per day compared to men is shown in Figure 1 for each of 29 different countries.

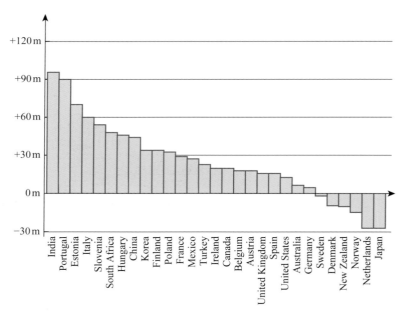

Figure 1

(i) Discuss briefly whether or not the chart supports the figure quoted of 50 minutes per day. [2 marks]

(ii) Determine whether or not the figure of 50 minutes per day is consistent with headline figure of 39 days per year. [2 marks]

② A textbook on agriculture has a chapter on statistics with the diagram shown in Figure 2.

The text below the diagram states that 'in a Normal distribution 95% of the distribution is within 1 standard deviation of the mean'.

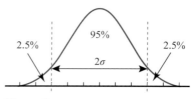

Figure 2

Correct this statement and suggest how the diagram should be modified.

[2 marks]

M PS ③ Table 1 and Figure 3 are based on a data set about slugs. They show, for each month of the year, the numbers of slugs of a particular species observed in the categories 'Hatchling', 'Immature' and 'Adult'. The observations were made by volunteers in Britain. Month 1 is January.

Table 1

Month	Hatchling	Immature	Adult
1	6	7	0
2	26	53	30
3	38	133	26
4	106	196	83
5	83	223	95
6	79	197	104
7	59	206	114
8	50	169	93
9	61	235	107
10	51	191	99
11	59	209	93
12	50	47	35

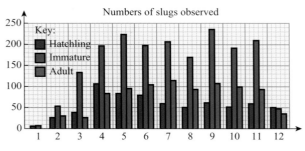

Figure 3

(i) Suggest two possible reasons for the fact that the numbers of observations are low at the beginning and the end of the year. [2 marks]

(ii) Identify, for the Adult category, a figure that appears to be inconsistent with the general pattern of observations over the year. How would you account for this inconsistency? [2 marks]

(iii) The information given with the data says that some species of slugs have an annual life cycle. Discuss briefly whether or not the data suggest an annual life cycle is a reasonable model for this species. [2 marks]

④ At my local supermarket there are two check-outs. I always choose whichever check-out looks as though it will serve me more quickly. Past experience suggests that I get this correct about 65% of the time.

(i) Find the probability that, in a random sample of 10 visits, I choose the correct check-out on more than half of the occasions. [2 marks]

The supermarket intends to open a third check-out, but I suspect that having a third check-out will make it harder to choose the quickest.

(ii) Write down the null and alternative hypotheses for a statistical test I could carry out. [2 marks]

Suppose that in a random sample of 20 visits after the third check-out has been introduced I correctly choose the quickest check-out on k occasions.

(iii) Find the set of values of k for which I should decide, using a 1% significance level, that it is now harder to choose the quickest check-out. [3 marks]

(iv) In fact, rather than taking a random sample of 20 visits, I am considering using my first 20 visits after the third check-out is opened as my sample. Discuss briefly whether or not doing that would invalidate the statistical test. [2 marks]

4

⑤ In a test of manual dexterity, children are asked to carry out three particular tasks as quickly as possible.

A model for a child's performance assumes that successes are independent with probabilities of $0.4, 0.7$ and 0.8 for the 1st, 2nd and 3rd tasks respectively.

(i) Use the model to calculate the following probabilities for a randomly chosen child:

   (a) the child is successful on exactly two tasks [2 marks]

   (b) the child is successful on at least one task. [2 marks]

(ii) Given that the child is successful on at least one task, find the probability that the child is successful on all three tasks. [3 marks]

⑥ A farmer is investigating the yield per plant of a particular variety of potato, A, that he grows. He estimates that 15% of plants yield over $4\,\text{kg}$ of potatoes, and that 20% of plants yield less than $2\,\text{kg}$.

(i) Assuming that the yields are Normally distributed, obtain estimates of the mean and standard deviation for variety A. [4 marks]

The farmer is considering growing a new variety of potato, B, hoping for a higher yield per plant. He discovers data from an agricultural research centre stating that, for variety B, the mean yield is $3.2\,\text{kg}$ with a standard deviation of $0.9\,\text{kg}$. However, the farmer suspects that he won't be able to achieve a mean as high as that because his land has low fertility. He decides to carry out a trial with variety B and to use an appropriate hypothesis test.

(ii) State the null and alternative hypotheses that the farmer should use.

[2 marks]

The farmer grows 100 plants of variety B and finds the total yield, for all 100 plants, to be $306\,\text{kg}$.

(iii) Carry out the appropriate hypothesis test using a 5% significance level.

[4 marks]

The hypothesis test makes a number of assumptions.

(iv) Identify a statistical assumption relating to the agricultural research centre's data. Discuss briefly whether it is safe to make that assumption.

[2 marks]

(v) Identify a statistical assumption relating to how the farmer grows his trial crop. Explain briefly how, if this assumption is not met, the results of the trial could be misleading. [2 marks]

 ⑦ The discrete random variable X has the following probability function.

$$P(X = r) = \frac{k}{r} \quad \text{for } r = 1, 2, 3, 4,$$

$$P(X = r) = 0 \quad \text{otherwise.}$$

(i) Prove that the constant k takes the value $\frac{12}{25}$. [2 marks]

(ii) Sketch the distribution. [1 mark]

Two independent values of X are generated. They are denoted by X_1 and X_2.

(iii) Find $P(X_1 = X_2)$. Hence find $P(X_1 < X_2)$. [4 marks]

M **T** ⑧ The data in this question are taken from a large data set about blackbirds.

Table 2 shows the means and standard deviations for wing length and weight for large samples of adult male and adult female blackbirds captured at random over a number of years.

Table 2

	Sample size	Wing length (mm)		Weight (g)	
		Mean	**SD**	**Mean**	**SD**
Males	963	134.1	3.28	111.2	11.8
Females	687	128.6	2.96	108.4	13.7

(i)   Avis, an amateur ornithologist, looks at these data and comments that it is easier to distinguish males from females by their wing length than by their weight. Explain, with reference to the data, whether or not this comment is correct. [3 marks]

(ii)   Give one reason that might account for the fact that females have a larger standard deviation of weights than males. [1 mark]

Avis wishes to investigate whether there is any correlation between wing span and weight. She constructs the following scatter diagram for the 687 adult female blackbirds.

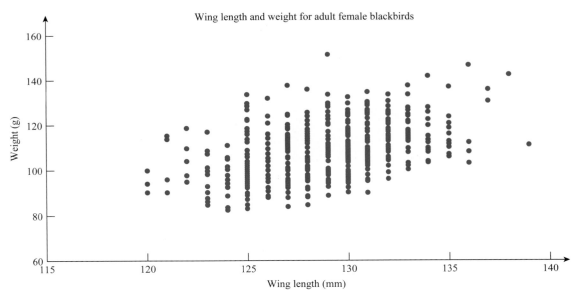

Wing length and weight for adult female blackbirds

Figure 4

(iii) Explain why the data points appear in vertical lines. [1 mark]

(iv)   What does the shape of the set of data points suggest? How is this relevant to Avis's investigation? [2 marks]

Avis uses a spreadsheet to find that the product moment correlation for this dataset is 0.47. The spreadsheet reports that the corresponding p-value is 'less than 0.000 001'.

(v)   How should Avis interpret this result? [2 marks]

(vi)   Avis now says that it wouldn't be worthwhile to repeat this analysis for adult male blackbirds. She also says that, with hindsight, it wasn't necessary to have separated out males and females when investigating the correlation. Briefly discuss these two comments. [2 marks]

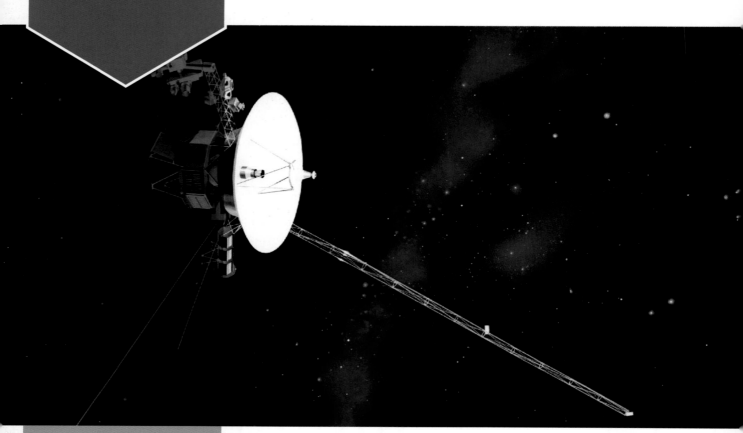

18   Kinematics

The motion of a spacecraft like Voyager on its way to the outer reaches of the solar system involves travel along curved paths. In this chapter you will see how to deal with motion in more than one dimension.

Review: Motion in one dimension

Direction of motion in one dimension

In one dimension motion is along a straight line. So there are only two possible directions and according to the context they can be described using words like right–left, forwards–backwards, up the slope–down the slope, east–west, etc. However, it is most satisfactory to define one of the directions as positive (+) and the other as negative (−).

Figure 18.1

Vocabulary for motion

- **Displacement** is the distance moved in a particular direction so is + or −. It is a vector quantity.

- **Distance** is always positive whatever the direction. It is a scalar quantity.

- **Velocity** is a vector quantity. This is the rate of change of the displacement. In one dimension, the direction is usually indicated by a + or − sign.

- **Speed** is a scalar quantity. This is the magnitude of the velocity or the rate of change of distance.

- Since velocity is the rate of change of displacement, **average velocity** is the total displacement divided by the time taken:

$$\text{average velocity} = \frac{\text{displacement}}{\text{time taken}}.$$

- Similarly, **average speed** is given by:

$$\text{average speed} = \frac{\text{total distance travelled}}{\text{total time taken}}.$$

- **Acceleration** is strictly a vector quantity and so is + or −. It is the rate of change of the velocity.

- However, you need to be careful with the word acceleration because it is often used loosely to mean the **magnitude of acceleration.**

Conventions for motion in one dimension

Certain letters are commonly used to denote these quantities in one dimension:

- s, h, r, x, y and z for displacement

- t for time

- u and v for velocity

- a for acceleration.

Units

The S.I. (Système International d'Unités) unit for **distance** is the metre (m), that for **time** is the second (s) and that for **mass** is the kilogram (kg). Other units follow from these, so speed is measured in $m\,s^{-1}$ and acceleration in $m\,s^{-2}$.

Example 18.1

James walks 150 m due north in 2 minutes. He stops for 5 minutes and then walks 600 m due south in 10 minutes.

(i) Draw a diagram of the journey.

(ii) Find

    (a) his average velocity

    (b) his average speed.

Solution

(i)

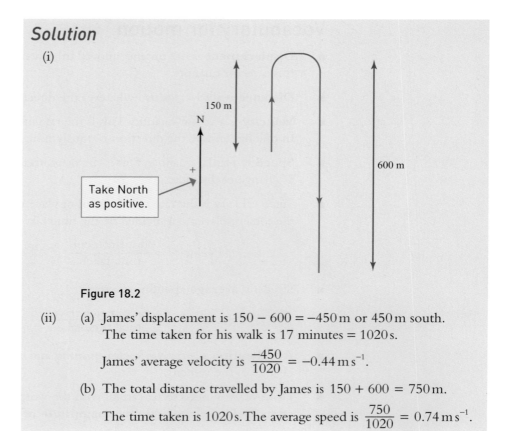

Figure 18.2

(ii)   (a) James' displacement is $150 - 600 = -450\,\text{m}$ or $450\,\text{m}$ south. The time taken for his walk is 17 minutes $= 1020\,\text{s}$.

James' average velocity is $\dfrac{-450}{1020} = -0.44\,\text{m s}^{-1}$.

(b) The total distance travelled by James is $150 + 600 = 750\,\text{m}$.

The time taken is $1020\,\text{s}$. The average speed is $\dfrac{750}{1020} = 0.74\,\text{m s}^{-1}$.

Graphical representation

James's walk is illustrated in a **displacement–time** graph (Figure 18.3).

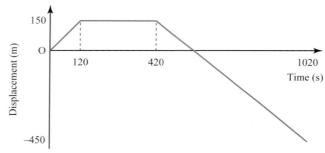

Figure 18.3

The corresponding **velocity–time** graph is shown in Figure 18.4.

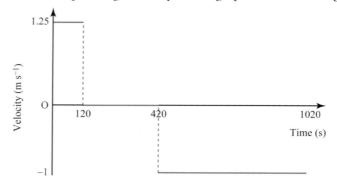

Figure 18.4

The **distance–time** graph for James's walk is shown in Figure 18.5.

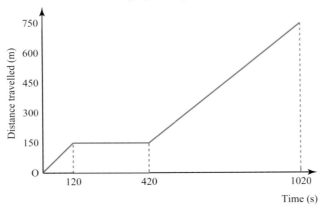

Figure 18.5

Acceleration is the rate at which the velocity changes.

Over a period of time

$$\text{average acceleration} = \frac{\text{change in velocity}}{\text{time}}.$$

Acceleration is represented by the gradient of a velocity–time graph.

The area between a velocity–time graph and the time axis

Example 18.2

Lucy cycles east for 90 s at $10\,\mathrm{m\,s^{-1}}$ and then west for 30 s at $5\,\mathrm{m\,s^{-1}}$.
Draw

(i)   the speed–time graph

(ii)   the velocity–time graph.

Solution

(i)   Speed–time graph

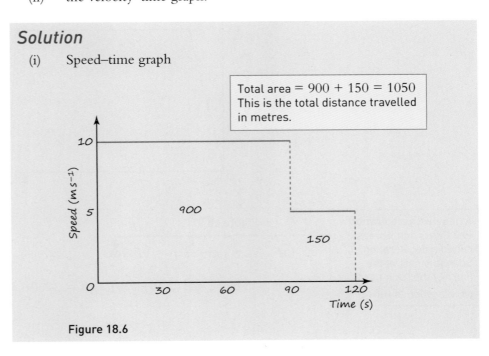

Total area $= 900 + 150 = 1050$
This is the total distance travelled in metres.

Figure 18.6

(ii) Taking east as the positive direction gives this velocity–time graph.
Velocity–time graph

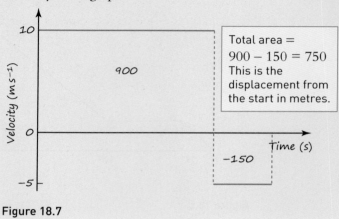

Figure 18.7

The area between a velocity–time graph and the time axis represents the change in position, that is, the displacement.

Example 18.3

A train starts from rest and accelerates at $0.9\,\mathrm{m\,s^{-2}}$ for 20 seconds. It then travels at constant speed for 100 s and finally decelerates uniformly to rest in a further 30 s.

(i) Sketch the speed–time graph.

(ii) Find the acceleration of the train in the last part of the journey.

(iii) Find the total distance travelled.

Solution

(i) The train reaches a speed of $0.9 \times 20 = 18\,\mathrm{m\,s^{-1}}$.

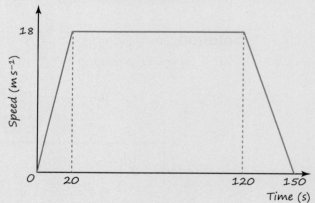

Figure 18.8

A negative acceleration like this is sometimes called a deceleration. So in this part of the journey the train has a deceleration of $0.6\,\mathrm{m\,s^{-2}}$.

(ii) $a = \dfrac{-18}{150 - 120} = -0.6\,\mathrm{m\,s^{-2}}$

(iii) distance $= \frac{1}{2}(100 + 150) \times 18 = 2\,250\,\mathrm{m}$

The constant acceleration formulae

A common situation for motion in one dimension occurs when a particle moves along a straight line with constant acceleration.

- It starts at the origin.
- Its initial velocity is u.
- Time is denoted by t.
- Its position along the line at time t is denoted by s.
- Its speed at time t is v.
- The constant acceleration is a.

There are five relationships (the constant acceleration formulae) each one involving four out of s, u, v, a and t.

- $v = u + at$
- $s = \frac{1}{2}(u + v)t$
- $s = ut + \frac{1}{2}at^2$
- $s = vt - \frac{1}{2}at^2$
- $v^2 = u^2 + 2as$

> **Note**
>
> These are sometimes called the *suvat* equations.

If the particle does not start at the origin but at position s_0 instead, then s is replaced by $(s - s_0)$ in the last four of these formulae.

These formulae can only be used when it has already been established that the acceleration is constant, as in the next example.

Example 18.4

A skier increases her speed uniformly from $8\,\text{m}\,\text{s}^{-1}$ to $20\,\text{m}\,\text{s}^{-1}$ in $10\,\text{s}$.

(i) How far does she travel in this time?

(ii) What is her acceleration?

Solution

(i) $u = 8$, $v = 20$ and $t = 10$.

> You are given u, v and t and want to find s. You need the formula that does not include a.

$$s = \frac{1}{2}(u + v)t$$

$$= \frac{1}{2}(8 + 20) \times 10 = 140$$

The skier travels $140\,\text{m}$.

(ii) $u = 8$, $v = 20$ and $t = 10$.

> You are given u, v and t and want to find a. So use $v = u + at$.

$$a = \frac{v - u}{t} = \frac{20 - 8}{10} = 1.2$$

The acceleration of the skier is $1.2\,\text{m}\,\text{s}^{-2}$.

Vertical motion under gravity

A common situation involves an object falling under gravity. This is often modelled by ignoring air resistance. In such cases, all objects fall with the same constant acceleration, g. The value of g actually varies slightly from one place on

the Earth's surface to another but, as another modelling assumption, it is assumed that all situations encountered here occur in a place where it is $9.8\,\text{m s}^{-2}$.

Example 18.5

A coin is thrown vertically upwards with a speed of $5\,\text{m s}^{-1}$ and is caught at the same height.

(i)   For how long is it in the air?

(ii)   How high does it go?

(iii)   What is its velocity when it is caught?

Solution

Start by defining the positive direction to be vertically upwards $\uparrow +$.

(i)   You are looking for the time the coin was in the air. You know that the initial velocity is $+5\,\text{m s}^{-1}$, the acceleration is $-9.8\,\text{m s}^{-2}$ and the displacement is zero. You can use the formula $s = ut + \frac{1}{2}at^2$.

$$0 = 5t - 4.9t^2 = t(5 - 4.9t)$$

$$t = 0 \text{ or } t = \frac{5}{4.9} = 1.02$$

The coin starts at $t = 0$ and returns at $t = 1.02$. It is in the air for $1.02\,\text{s}$.

(ii)   You are looking for the displacement s and are given $u = 5$, $a = -9.8$ and $v = 0$. You can use the formula $v^2 = u^2 + 2as$.

$$0 = 5^2 - 2 \times 9.8 \times s$$

$$s = \frac{25}{19.6} = 1.28$$

The coin reaches a height of $1.28\,\text{m}$.

(iii)   You are looking for v and are given $u = 5$, $a = -9.8$ and $s = 0$.

Using the formula $v^2 = u^2 + 2as$ gives you $v^2 = 25 \Rightarrow v = \pm 5$.

The velocity of the coin as it is caught is $-5\,\text{m s}^{-1}$.

Note

The value of $+5\,\text{m s}^{-1}$ occurred when the coin was thrown up in the air initially.

General motion in a straight line

Variable acceleration

It is not, of course, always the case that an object has constant acceleration. In such cases you cannot use the constant acceleration formulae and must resort to calculus instead.

The velocity of the object is the rate of change of its displacement so at any time

$$v = \frac{ds}{dt}.$$

Similarly, its acceleration is the rate of change of its velocity. So that

$$a = \frac{dv}{dt} = \frac{d^2s}{dt^2}.$$

In the reverse process, the displacement may be found from the velocity, which itself may be found from the acceleration

$$s = \int v \, dt$$

and $\quad v = \int a \, dt$.

The use of calculus is shown in the next two examples.

Example 18.6

The displacement s m of a particle at time t s is given by

$$s = \tfrac{1}{3}t^3 - t^2 + 2t - 1 .$$

Find (i) the velocity and (ii) the acceleration of the moving point when $t = 4$.

Solution

> **Note**
>
> Notice that a is not constant so you could not have used the constant acceleration formulae.

(i) $\qquad v = \dfrac{ds}{dt} = t^2 - 2t + 2$

When $t = 4 \quad v = 4^2 - 2 \times 4 + 2 = 10$

The velocity is $+10 \, \mathrm{m\,s^{-1}}$.

(ii) $\qquad a = \dfrac{dv}{dt} = 2t - 2$

When $t = 4 \quad a = 2 \times 4 - 2 = 6$

The acceleration is $+6 \, \mathrm{m\,s^{-2}}$.

Example 18.7

A particle leaves the point O with a velocity $+4 \, \mathrm{m\,s^{-1}}$ and its acceleration a after time t is given by

$$a = 2 - t.$$

> **Note**
>
> Notice that the acceleration is not constant so you cannot use the constant acceleration formulae.

Find

(i) its velocity v after time t

(ii) its displacement after time t

(iii) the velocity and displacement after 3 seconds.

Solution

(i) $\qquad v = \int a \, dt = \int (2 - t) \, dt$

$v = 2t - \tfrac{1}{2}t^2 + c$

When $t = 0, v = 4$

so substituting in the equation for v gives

$$4 = 0 - 0 + c \implies c = 4$$

$$v = 2t - \tfrac{1}{2}t^2 + 4$$

(ii) $s = \int v \, dt = \int \left(2t - \tfrac{1}{2}t^2 + 4\right) dt$

$s = t^2 - \tfrac{1}{6}t^3 + 4t + k$

When $t = 0$, $s = 0$

$0 = 0 - 0 + 0 + k \implies k = 0$

$s = t^2 - \tfrac{1}{6}t^3 + 4t$

(iii)   When $t = 3$

$v = 2 \times 3 - \tfrac{1}{2} \times 3^2 + 4 = 5\tfrac{1}{2}$

$s = 3^2 - \tfrac{1}{6} \times 3^3 + 4 \times 3 = 16\tfrac{1}{2}$

The particle has a velocity of $+5.5\,\mathrm{m\,s^{-1}}$ and is at $+16.5\,\mathrm{m}$ after 3 seconds.

Review exercise

① A car is moving at $10\,\mathrm{m\,s^{-1}}$ when it begins to accelerate at a constant rate of $0.5\,\mathrm{m\,s^{-2}}$ until it reaches its maximum speed of $25\,\mathrm{m\,s^{-1}}$ which it retains.

(i)   Draw a speed–time graph for the car's motion.

(ii) When does the car reach its maximum speed of $25\,\mathrm{m\,s^{-1}}$?

(iii) Find the distance travelled by the car after $100\,\mathrm{s}$.

(iv) Write down expressions for the speed of the car t seconds after it begins to speed up.

② A, B and C are three points on a straight road with $AB = 1200\,\mathrm{m}$ and $BC = 200\,\mathrm{m}$ and B is between A and C. Claire cycles from A to B at $10\,\mathrm{m\,s^{-1}}$, pushes her bike from B to C at an average speed of $0.5\,\mathrm{m\,s^{-1}}$ and then cycles back from C to B at an average speed of $15\,\mathrm{m\,s^{-1}}$.

(i)   Find Claire's average speed for the whole journey.

(ii) Find Claire's average velocity for the journey.

③ The velocity–time graph in Figure 18.9 shows the motion of a particle along a straight line.

(i)   The particle starts at A at $t = 0$ and moves to B in the next $20\,\mathrm{s}$. Find the distance AB.

(ii) T seconds after leaving A the particle is at C, $50\,\mathrm{m}$ from B. Find T.

(iii) Find the displacement of C from A.

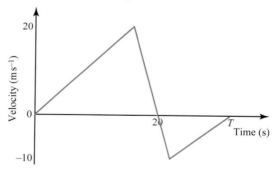

Figure 18.9

④ A car accelerates uniformly from $5\,\mathrm{m\,s^{-1}}$ to $20\,\mathrm{m\,s^{-1}}$ in a distance of $100\,\mathrm{m}$.

(i)   Find the acceleration.

(ii) Find the speed when the car has covered half the distance.

⑤ (i) Find v when $u = 5$, $a = -2$ and $t = 1$.

(ii) Find s when $v = 10$, $a = 1$ and $t = 10$.

(iii) Find a when $v = 10$, $u = 2$ and $s = 5$.

(iv) Find s when $u = 2$, $v = 5$ and $t = 10$.

(v) Find u when $s = 5$ $a = -2$ and $t = 4$.

⑥ A stone is dropped down a well. It takes 4 seconds to reach the bottom. How deep is the well?

⑦ A ball is projected vertically upward from a point which is $0.5\,\text{m}$ above ground level with a speed of $20\,\text{m s}^{-1}$.

(i) Find the time for which the ball is in the air.

(ii) Find the speed with which the ball hits the ground.

⑧ A particle starts from rest and moves in a straight line with constant acceleration until it reaches a speed of $25\,\text{m s}^{-1}$. It is then brought to rest by a constant deceleration of $2.5\,\text{m s}^{-2}$.

The particle is then $200\,\text{m}$ from its starting point.

Find the time for which the particle is moving.

⑨ A particle is moving in a straight line so that its displacement s at time t is given by

$$s = 8t - \frac{1}{6}t^3.$$

(i) Calculate the velocity and acceleration of the particle after $3\,\text{s}$.

(ii) Find the distance travelled by the particle, when it first comes to rest.

⑩ Find the velocity v and displacement s of a particle at time t, given that its acceleration a is given by

(i) $a = 6t - 8$ when
$t = 0$, $s = 4$ and $v = 6$

(ii) $a = -10$ when
$t = 1$, $s = 5$ and $v = 2$.

1 Motion in two or three dimensions

Prior knowledge

You have seen how vectors can be expressed in terms of components as well as in terms of magnitude and direction.

For motion in two or three dimensions, the equivalent vector quantities to those you used in one dimension are written using bold type, or as column vectors.

- The **position vector** of a point is denoted by **r**. It is the **displacement** from the origin and can be written as $x\mathbf{i} + y\mathbf{j}$ or alternatively as the column vector $\begin{pmatrix} x \\ y \end{pmatrix}$.

- **Distance** from the origin is the magnitude of the position vector and is the scalar quantity $\sqrt{x^2 + y^2}$.

- **Velocity** is given by $\mathbf{v} = \dfrac{\mathrm{d}\mathbf{r}}{\mathrm{d}t}$. Differentiating a vector with respect to the scalar t involves differentiating each component with respect to t, so that

$$\mathbf{v} = \frac{\mathrm{d}\mathbf{r}}{\mathrm{d}t} = \frac{\mathrm{d}x}{\mathrm{d}t}\mathbf{i} + \frac{\mathrm{d}y}{\mathrm{d}t}\mathbf{j}. \text{ Written as a column vector } \mathbf{v} = \begin{pmatrix} \dfrac{\mathrm{d}x}{\mathrm{d}t} \\ \dfrac{\mathrm{d}y}{\mathrm{d}t} \end{pmatrix}.$$

- **Speed** is the magnitude of velocity and so is $\sqrt{\left(\dfrac{\mathrm{d}x}{\mathrm{d}t}\right)^2 + \left(\dfrac{\mathrm{d}y}{\mathrm{d}t}\right)^2}$.

- **Acceleration** is the rate of change of the velocity and so is given by

$$\mathbf{a} = \frac{d\mathbf{v}}{dt} = \frac{d^2\mathbf{r}}{dt^2}$$

It can be written as

$$\mathbf{a} = \frac{d^2x}{dt^2}\mathbf{i} + \frac{d^2y}{dt^2}\mathbf{j} \text{ or as } \mathbf{a} = \begin{pmatrix} \dfrac{d^2x}{dt^2} \\ \dfrac{d^2y}{dt^2} \end{pmatrix}.$$

- **Magnitude of acceleration** is given by $\sqrt{\left(\dfrac{d^2x}{dt^2}\right)^2 + \left(\dfrac{d^2y}{dt^2}\right)^2}$.

Newton's notation

When you write derivatives in column vectors, the notation becomes very cumbersome so many people use Newton's notation when differentiating with respect to time. In this notation a dot is placed over the variable for each differentiation.

For example $\dot{x} = \dfrac{dx}{dt}$ and $\ddot{x} = \dfrac{d^2x}{dt^2}$.

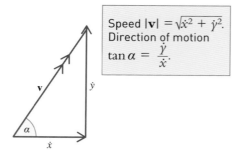

Speed $|\mathbf{v}| = \sqrt{\dot{x}^2 + \dot{y}^2}$.
Direction of motion
$\tan\alpha = \dfrac{\dot{y}}{\dot{x}}$.

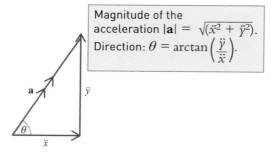

Magnitude of the
acceleration $|\mathbf{a}| = \sqrt{(\ddot{x}^2 + \ddot{y}^2)}$.
Direction: $\theta = \arctan\left(\dfrac{\ddot{y}}{\ddot{x}}\right)$.

Figure 18.10

$$\mathbf{v} = \dot{\mathbf{r}} = \dot{x}\mathbf{i} + \dot{y}\mathbf{j} \quad \text{or} \quad \begin{pmatrix} \dot{x} \\ \dot{y} \end{pmatrix}$$

$$\mathbf{a} = \dot{\mathbf{v}} = \ddot{\mathbf{r}}$$

$$\mathbf{a} = \ddot{x}\mathbf{i} + \ddot{y}\mathbf{j} \quad \text{or} \quad \begin{pmatrix} \ddot{x} \\ \ddot{y} \end{pmatrix}$$

The vector forms of the constant acceleration formulae

When the acceleration $\mathbf{a}$ is constant, then

$$\mathbf{v} = \int \mathbf{a}\, dt = \mathbf{a}t + \mathbf{c}$$

> $\mathbf{c}$ is a vector constant so it has components in the x and y directions.

where $\mathbf{c}$ is a vector constant. Substituting $t = 0$ shows that $\mathbf{c}$ is equal to the initial velocity vector $\mathbf{u}$. So

$$\mathbf{v} = \mathbf{u} + \mathbf{a}t$$

integrating again gives

$$\mathbf{r} = \int \mathbf{v}\, dt = \int (\mathbf{u} + \mathbf{a}t)\, dt$$

$$\mathbf{r} = \mathbf{u}t + \tfrac{1}{2}\mathbf{a}t^2 + \mathbf{r}_0$$

where $\mathbf{r}_0$ is the initial position vector. If the motion starts at the origin, the vector $\mathbf{r}_0$ is $\mathbf{0}$, or $\begin{pmatrix} 0 \\ 0 \end{pmatrix}$. In that case this equation is written as $\mathbf{r} = \mathbf{u}t + \tfrac{1}{2}\mathbf{a}t^2$.

As you may have noticed, the vector forms of the constant acceleration equations are similar to those for motion in one dimension.

In one dimension		Vector form
$v = u + at$	$\rightarrow$	$\mathbf{v} = \mathbf{u} + \mathbf{a}t$
$s = ut + \tfrac{1}{2}at^2$	$\rightarrow$	$\mathbf{r} = \mathbf{u}t + \tfrac{1}{2}\mathbf{a}t^2$
$s = \tfrac{1}{2}(u + v)t$	$\rightarrow$	$\mathbf{r} = \tfrac{1}{2}(\mathbf{u} + \mathbf{v})t$
$s = vt - \tfrac{1}{2}at^2$	$\rightarrow$	$\mathbf{r} = \mathbf{v}t - \tfrac{1}{2}\mathbf{a}t^2$

If the motion does not start at the origin, s is replaced by $s - s_0$ in the one-dimensional equations and in the vector equations $\mathbf{r}$ is replaced by $\mathbf{r} - \mathbf{r}_0$.

In one dimension there is a fifth constant acceleration equation, $v^2 = u^2 + 2as$. There is a vector form of this equation but it involves more advanced vector operations that are beyond the scope of this book.

ACTIVITY 18.1

Use the equations $\mathbf{v} = \mathbf{u} + \mathbf{a}t$ and $\mathbf{r} = \mathbf{u}t + \tfrac{1}{2}\mathbf{a}t^2$ to derive the other two vector equations given above: $\mathbf{r} = \tfrac{1}{2}(\mathbf{u} + \mathbf{v})t$ and $\mathbf{r} = \mathbf{v}t - \tfrac{1}{2}\mathbf{a}t^2$.

The path

When a ball is thrown into the air, its position can be represented by a vector $\mathbf{r} = \begin{pmatrix} x \\ y \end{pmatrix}$ or $= x\mathbf{i} + y\mathbf{j}$. For example, it might be given by

$$\mathbf{r} = \begin{pmatrix} 5t \\ 12t - 5t^2 \end{pmatrix} \text{ so in this case } x = 5t \text{ and } y = 12t - 5t^2.$$

You can plot the path of the ball by finding the values of x and y and hence of $\mathbf{r}$ for several values of t. The initial position of the ball is a point on the ground which is taken to be the origin.

Discussion point

→ Why is 2.4 chosen as the last value for t?

Table 18.1

t	0	0.5	1	1.5	2	2.4
$\mathbf{r}$	$\begin{pmatrix} 0 \\ 0 \end{pmatrix}$	$\begin{pmatrix} 2.5 \\ 4.75 \end{pmatrix}$	$\begin{pmatrix} 5 \\ 7 \end{pmatrix}$	$\begin{pmatrix} 7.5 \\ 6.75 \end{pmatrix}$	$\begin{pmatrix} 10 \\ 4 \end{pmatrix}$	$\begin{pmatrix} 12 \\ 0 \end{pmatrix}$

You can also plot the path of the ball by finding the Cartesian equation. This is obtained by elimination of t between the two equations

$$x = 5t \qquad\qquad\qquad\qquad ①$$
$$y = 12t - 5t^2 \qquad\qquad\qquad ②$$

From ① you can derive $t = \dfrac{x}{5}$ and then substitute for t in ② to give

$$y = 12\left(\frac{x}{5}\right) - 5\left(\frac{x}{5}\right)^2$$

$$y = 2.4x - 0.2x^2$$

Figure 18.11 shows the path of the ball and also its position $\mathbf{r}$ when $t = 2$.

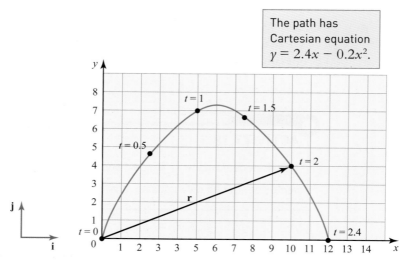

The path has Cartesian equation $y = 2.4x - 0.2x^2$.

Figure 18.11

The equation for **r** can be differentiated to give the velocity and acceleration.

In this case $\mathbf{r} = \begin{pmatrix} 5t \\ 12t - 5t^2 \end{pmatrix}$

and so $\mathbf{v} = \begin{pmatrix} 5 \\ 12 - 10t \end{pmatrix}$

and $\mathbf{a} = \begin{pmatrix} 0 \\ -10 \end{pmatrix}.$

Example 18.8

Relative to an origin on a long, straight beach, the position of a speedboat is modelled by the vector

$$\mathbf{r} = (2t + 2)\mathbf{i} + \left(12 - t^2\right)\mathbf{j}$$

where **i** and **j** are unit vectors perpendicular and parallel to the beach. Distances are in metres and the time t is in seconds.

(i) Calculate the distance of the boat from the origin, O, when the boat is 6 m from the beach.

(ii) Sketch the path of the speedboat for $0 \leqslant t \leqslant 3$.

(iii) Find expressions for the velocity and acceleration of the speedboat at time t. Is the boat ever at rest? Explain your answer.

(iv) For $t = 3$, calculate the speed of the boat and the angle its direction of motion makes to the line of the beach.

(v) Suggest why this model for the motion of the speedboat is unrealistic for large t.

Solution

Notice that the beach is in the y direction. The x direction is out to sea.

(i) $\mathbf{r} = (2t + 2)\mathbf{i} + \left(12 - t^2\right)\mathbf{j}$ so the boat is 6 m from the beach when $x = 2t + 2 = 6$ then $t = 2$ and $y = 12 - 2^2 = 8$.

The distance from O is $\sqrt{6^2 + 8^2} = 10$ m.

(ii) The table shows the position at different times and the path of the boat is shown on the graph in Figure 18.12.

Table 18.2

t	0	1	2	3
$\mathbf{r}$	$2\mathbf{i} + 12\mathbf{j}$	$4\mathbf{i} + 11\mathbf{j}$	$6\mathbf{i} + 8\mathbf{j}$	$8\mathbf{i} + 3\mathbf{j}$

(iii) $\mathbf{r} = (2t + 2)\mathbf{i} + \left(12 - t^2\right)\mathbf{j}$

$\Rightarrow \mathbf{v} = \dot{\mathbf{r}} = 2\mathbf{i} - 2t\mathbf{j}$

and $\mathbf{a} = \dot{\mathbf{v}} = -2\mathbf{j}$

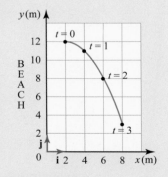

Figure 18.12

The boat is at rest if both components of velocity ($\dot{x}$ and $\dot{y}$) are zero at the same time. But $\dot{x}$ is always 2, so the velocity can never be zero.

> ! Notice how in part (iv) the direction of motion is found using the velocity and not the position. That is always the case.

(iv)   When $t = 3$ $\mathbf{v} = 2\mathbf{i} - 6\mathbf{j}$

The angle $\mathbf{v}$ makes with the beach is α as shown in Figure 18.13 where

$$\tan \alpha = \frac{2}{6}$$
$$\alpha = 18.4°$$

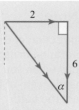

Figure 18.13

(v)   According to this model, the speed after time t is

$$|\mathbf{v}| = |2\mathbf{i} - 2t\mathbf{j}| = \sqrt{2^2 + (-2t)^2} = \sqrt{4 + 4t^2}$$

As t increases, the speed increases at an increasing rate so there must come a time when the boat is incapable of going at the predicted speed and the model cannot then apply.

Example 18.9

A particle is moving in two dimensions with constant acceleration.

The unit vectors $\begin{pmatrix} 1 \\ 0 \end{pmatrix}$ and $\begin{pmatrix} 0 \\ 1 \end{pmatrix}$ are pointing east and north respectively.

Initially the particle is at $\begin{pmatrix} -1 \\ 2 \end{pmatrix}$ m, with velocity $\begin{pmatrix} 2 \\ -3 \end{pmatrix}$ m s^{-1};

after 4 s the particle has velocity $\begin{pmatrix} 22 \\ 27 \end{pmatrix}$ m s^{-1}.

(i)   Calculate the acceleration of the particle.

(ii)   Find the velocity of the particle at time t and use it to find the time at which the particle is moving north-east.

(iii)   Calculate the position of the particle after 4 s. How far has it moved from its original position?

Solution

(i)   The acceleration is constant so you can use the constant acceleration formula $\mathbf{v} = \mathbf{u} + \mathbf{a}t$ to find $\mathbf{a}$, giving

$$\mathbf{a} = \frac{\mathbf{v} - \mathbf{u}}{t}$$

$$= \frac{\begin{pmatrix} 22 \\ 27 \end{pmatrix} - \begin{pmatrix} 2 \\ -3 \end{pmatrix}}{4}$$

$$= \frac{1}{4}\begin{pmatrix} 20 \\ 30 \end{pmatrix}$$

$$= \begin{pmatrix} 5 \\ 7.5 \end{pmatrix}$$

The acceleration of the particle is $\begin{pmatrix} 5 \\ 7.5 \end{pmatrix}$ m s^{-2}.

(ii)   Now using $\mathbf{v} = \mathbf{u} + \mathbf{a}t$ with $\mathbf{u} = \begin{pmatrix} 2 \\ -3 \end{pmatrix}$ and $\mathbf{a} = \begin{pmatrix} 5 \\ 7.5 \end{pmatrix}$ gives

$$\mathbf{v} = \begin{pmatrix} 2 \\ -3 \end{pmatrix} + \begin{pmatrix} 5 \\ 7.5 \end{pmatrix} t$$

$$= \begin{pmatrix} 2+5t \\ -3+7.5t \end{pmatrix}$$

The velocity of the particle at time t is $\begin{pmatrix} 2+5t \\ -3+7.5t \end{pmatrix}$.

The particle is moving north-east when the bearing is 045°, i.e when the components of $\mathbf{v}$ are equal.

$$2 + 5t = -3 + 7.5t$$

$$5 = 2.5t$$

$$t = 2$$

The particle is moving north-east when $t = 2$.

(iii)   Now using $\mathbf{r} = \mathbf{r}_0 + \mathbf{u}t + \frac{1}{2}\mathbf{a}t^2$, with $\mathbf{r}_0 = \begin{pmatrix} -1 \\ 2 \end{pmatrix}$, $\mathbf{u} = \begin{pmatrix} 2 \\ -3 \end{pmatrix}$

and $\mathbf{a} = \begin{pmatrix} 5 \\ 7.5 \end{pmatrix}$ gives

$$\mathbf{r} = \begin{pmatrix} -1 \\ 2 \end{pmatrix} + \begin{pmatrix} 2 \\ -3 \end{pmatrix} \times 4 + \frac{1}{2} \times \begin{pmatrix} 5 \\ 7.5 \end{pmatrix} \times 4^2$$

$$\mathbf{r} = \begin{pmatrix} -1 \\ 2 \end{pmatrix} + \begin{pmatrix} 8 \\ -12 \end{pmatrix} + \begin{pmatrix} 40 \\ 60 \end{pmatrix}$$

$$\mathbf{r} = \begin{pmatrix} 47 \\ 50 \end{pmatrix}$$

The position of the particle after 4 s is $\begin{pmatrix} 47 \\ 50 \end{pmatrix}$.

The distance from the starting position, $\mathbf{r}_0$, is

$$|\mathbf{r} - \mathbf{r}_0| = \left| \begin{pmatrix} 47 \\ 50 \end{pmatrix} - \begin{pmatrix} -1 \\ 2 \end{pmatrix} \right| = \left| \begin{pmatrix} 48 \\ 48 \end{pmatrix} \right|$$

$$= \sqrt{48^2 + 48^2}$$

$$= 48\sqrt{2} = 67.88$$

The particle is 67.9 m from its starting position.

When you are given the velocity or acceleration and wish to work backwards to the displacement, you need to integrate. The next two examples show how you can do this with vectors.

Example 18.10

An aircraft is dropping a crate of supplies onto level ground. Relative to an observer on the ground, the crate is released at the point with position vector $\begin{pmatrix} 650 \\ 576 \end{pmatrix}$ m and with initial velocity $\begin{pmatrix} -100 \\ 0 \end{pmatrix}$, where the directions are horizontal and vertical. Its acceleration is modelled by

$$\mathbf{a} = \begin{pmatrix} -t + 12 \\ \frac{1}{2}t - 10 \end{pmatrix} \text{ for } t \leqslant 12\,\text{s}$$

(i)   Find an expression for the velocity vector of the crate at time t.

(ii)   Find an expression for the position vector of the crate at time t.

(iii)   Verify that the crate hits the ground 12 s after its release and find how far from the observer this happens.

Solution

(i) $\mathbf{a} = \dfrac{d\mathbf{v}}{dt} = \begin{pmatrix} -t + 12 \\ \frac{1}{2}t - 10 \end{pmatrix}$ ①

> You can treat horizontal and vertical motion separately if you wish.

Integrating gives $\mathbf{v} = \begin{pmatrix} -\frac{1}{2}t^2 + 12t + c_1 \\ \frac{1}{4}t^2 - 10t + c_2 \end{pmatrix}$

c_1 and c_2 are constants of integration which can be found by using the initial conditions.

> ⚠ When you integrate a vector in two dimensions you need a constant of integration for each component, for example c_1 and c_2 in this example.

At $t = 0$ $\mathbf{v} = \begin{pmatrix} -100 \\ 0 \end{pmatrix} \Rightarrow \begin{matrix} 0 + 0 + c_1 = -100 \\ 0 - 0 + c_2 = 0 \end{matrix} \Rightarrow \begin{matrix} c_1 = -100 \\ c_2 = 0 \end{matrix}$

Velocity $\mathbf{v} = \begin{pmatrix} -\frac{1}{2}t^2 + 12t - 100 \\ \frac{1}{4}t^2 - 10t \end{pmatrix}$ ②

(ii) $\mathbf{v} = \dfrac{d\mathbf{r}}{dt} = \begin{pmatrix} -\frac{1}{2}t^2 + 12t - 100 \\ \frac{1}{4}t^2 - 10t \end{pmatrix}$

Integrating again gives $\mathbf{r} = \begin{pmatrix} -\frac{1}{6}t^3 + 6t^2 - 100t + k_1 \\ \frac{1}{12}t^3 - 5t^2 + k_2 \end{pmatrix}$

k_1 and k_2 are found by using the initial conditions at $t = 0$.

$\mathbf{r} = \begin{pmatrix} 650 \\ 576 \end{pmatrix} \Rightarrow \begin{matrix} k_1 = 650 \\ k_2 = 576 \end{matrix}$

Position vector $\mathbf{r} = \begin{pmatrix} -\frac{1}{6}t^3 + 6t^2 - 100t + 650 \\ \frac{1}{12}t^3 - 5t^2 + 576 \end{pmatrix}$ ③

➜

(iii) When $t = 12$ $\mathbf{r} = \begin{pmatrix} -\frac{1}{6} \times 12^3 + 6 \times 12^2 - 100 \times 12 + 650 \\ \frac{1}{12} \times 12^3 - 5 \times 12^2 + 576 \end{pmatrix}$

$$\mathbf{r} = \begin{pmatrix} 26 \\ 0 \end{pmatrix}$$

Since $y = 0$, the crate hits the ground after 12 s and it is $x = 26$ m in front of the observer.

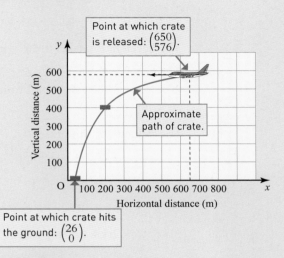

Point at which crate is released: $\begin{pmatrix} 650 \\ 576 \end{pmatrix}$.

Approximate path of crate.

Point at which crate hits the ground: $\begin{pmatrix} 26 \\ 0 \end{pmatrix}$.

Figure 18.14

Force as a function of time

When the force acting on an object is given as a function of t you can use Newton's second law to find out about its motion. You can now write this as $\mathbf{F} = m\mathbf{a}$ because force and acceleration are both vectors.

Example 18.11

A force of $(12\mathbf{i} + 3t\mathbf{j})$ N, where t is the time in seconds, acts on a particle of mass 6 kg. The directions of **i** and **j** correspond to east and north respectively.

(i) Show that the acceleration is $(2\mathbf{i} + 0.5t\mathbf{j})$ m s^{-2} at time t.

(ii) Find the acceleration and the magnitude of the acceleration when $t = 12$.

(iii) At what time is the acceleration directed north-east (i.e. a bearing of 045°)?

(iv) If the particle starts with a velocity of $(2\mathbf{i} - 3\mathbf{j})$ m s^{-1} when $t = 0$, what will its velocity be when $t = 3$?

(v) When $t = 3$, a second constant force begins to act. Given that the acceleration of the particle at that time due to both forces is 4 m s^{-2} due south, find the second force. [MEI]

Solution

(i) By Newton's second law the force = mass × acceleration

$$(12\mathbf{i} + 3t\mathbf{j}) = 6\mathbf{a}$$

$$\mathbf{a} = \tfrac{1}{6}(12\mathbf{i} + 3t\mathbf{j})$$

$$\mathbf{a} = 2\mathbf{i} + 0.5t\mathbf{j} \qquad \text{①}$$

(ii) When $t = 12$ $\qquad \mathbf{a} = 2\mathbf{i} + 6\mathbf{j}$

magnitude of $\mathbf{a}$ $\quad |\mathbf{a}| = \sqrt{2^2 + 6^2} = 6.32$

The acceleration is $2\mathbf{i} + 6\mathbf{j}$ with magnitude $6.32\,\mathrm{m\,s^{-1}}$.

(iii) The acceleration is north-east when its northerly component is equal to its easterly component. From ①, this happens when $2 = 0.5t$, i.e. when $t = 4$.

(iv) The velocity at time t is $\qquad \int \mathbf{a}\,dt = \int (2\mathbf{i} + 0.5t\mathbf{j})\,dt$

$$\Rightarrow \quad \mathbf{v} = 2t\mathbf{i} + 0.25t^2\mathbf{j} + \mathbf{c} \quad \longleftarrow \boxed{\mathbf{c} \text{ is a constant vector such as } c_1\mathbf{i} + c_2\mathbf{j}.}$$

When $t = 0$, $\mathbf{v} = 2\mathbf{i} - 3\mathbf{j}$

so $\quad 2\mathbf{i} - 3\mathbf{j} = 0\mathbf{i} + 0\mathbf{j} + \mathbf{c} \quad \Rightarrow \quad \mathbf{c} = 2\mathbf{i} - 3\mathbf{j}$

$$\mathbf{v} = 2t\mathbf{i} + 0.25t^2\mathbf{j} + 2\mathbf{i} - 3\mathbf{j}$$

$$\mathbf{v} = (2t + 2)\mathbf{i} + \left(0.25t^2 - 3\right)\mathbf{j} \qquad \text{②}$$

When $t = 3$ $\quad \mathbf{v} = 8\mathbf{i} - 0.75\mathbf{j}$

(v) Let the second force be F so the total force when $t = 3$ is $\left(12\mathbf{i} + 3 \times 3\mathbf{j}\right) + \mathbf{F}$.

The acceleration is $-4\mathbf{j}$, so by Newton's second law

$$\left(12\mathbf{i} + 9\mathbf{j}\right) + \mathbf{F} = 6 \times (-4\mathbf{j})$$

$$\mathbf{F} = -24\mathbf{j} - 12\mathbf{i} - 9\mathbf{j}$$

$$\mathbf{F} = -12\mathbf{i} - 33\mathbf{j}$$

The second force is $-12\mathbf{i} - 33\mathbf{j}$.

> **Note**
>
> You can use the same methods in three dimensions just by including a third direction, z, perpendicular to both x and y. The unit vector in the z direction is usually denoted by $\mathbf{k}$.

$\mathbf{r} = x\mathbf{i} + y\mathbf{j} + z\mathbf{k}$, $\mathbf{v} = \dot{x}\mathbf{i} + \dot{y}\mathbf{j} + \dot{z}\mathbf{k}$ and

$\mathbf{a} = \ddot{x}\mathbf{i} + \ddot{y}\mathbf{j} + \ddot{z}\mathbf{k}$

And similarly in column vectors

$$\mathbf{r} = \begin{pmatrix} x \\ y \\ z \end{pmatrix} \quad \mathbf{v} = \begin{pmatrix} \dot{x} \\ \dot{y} \\ \dot{z} \end{pmatrix} \quad \text{and} \quad \mathbf{a} = \begin{pmatrix} \ddot{x} \\ \ddot{y} \\ \ddot{z} \end{pmatrix}$$

The distance of a point with position vector $\mathbf{r}$ from the origin is

$$r = |\mathbf{r}| = \sqrt{(x^2 + y^2 + z^2)}$$

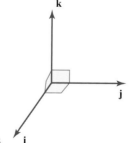

Figure 18.15

The speed is given by $v = |\mathbf{v}| = \sqrt{(\dot{x}^2 + \dot{y}^2 + \dot{z}^2)}$

The acceleration has magnitude $a = |\mathbf{a}| = \sqrt{(\ddot{x}^2 + \ddot{y}^2 + \ddot{z}^2)}$

Historical note

Figure 18.15a **Figure 18.15b**
Isaac Newton Gottfried Leibniz

Newton's work on motion required more mathematical tools than were generally used at the time. He had to invent his own ways of thinking about continuous change and in about 1666 he produced a theory of 'fluxions' in which he imagined a quantity 'flowing' from one magnitude to another. This was the beginning of calculus. He did not publish his methods, however, and when Leibniz published his version in 1684 there was an enormous amount of controversy amongst their supporters about who was first to discover the calculus. The sharing of ideas between mathematicians in Britain and the rest of Europe was hindered for a century. The contributions of both men are remembered today by their notation. Leibniz's $\frac{dx}{dt}$ is common and Newton's $\dot{x}$ is widely used in mechanics.

Exercise 18.1

Those questions in this exercise where the context is drawn from real life involve working with mathematical models.

① The first part of a race track is a bend. As the leading car travels round the bend its position, in metres, is modelled by:

$$\mathbf{r} = 2t^2\mathbf{i} + 8t\mathbf{j}$$

where t is in seconds.

(i) Find an expression for the velocity of the car.

(ii) Find the position of the car when $t = 0, 1, 2, 3$ and 4.

Use this information to sketch the path of the car.

(iii) Find the velocity of the car when $t = 0, 1, 2, 3$ and 4.

Add vectors to your sketch to represent these velocities.

(iv) Find the speed of the car as it leaves the bend at $t = 5$.

② As a boy slides down a slide his position vector in metres at time t is

$$\begin{pmatrix} x \\ y \end{pmatrix} = \begin{pmatrix} 16 - 4t \\ 20 - 5t \end{pmatrix}.$$

Find his velocity and acceleration.

③ Calculate the magnitude and direction of the acceleration of a particle that moves so that its position vector in metres is given by

$$\mathbf{r} = \left(8t - 2t^2\right)\mathbf{i} + \left(6 + 4t - t^2\right)\mathbf{j}$$

where t is the time in seconds.

④ A rocket moves with a velocity (in m s^{-1}) modelled by

$$\mathbf{v} = \tfrac{1}{10}t\mathbf{i} + \tfrac{1}{10}t^2\mathbf{j}$$

where $\mathbf{i}$ and $\mathbf{j}$ are horizontal and vertical unit vectors respectively and t is in seconds. Find

(i) an expression for its position vector relative to its starting position at time t

(ii) the displacement of the rocket after 10 s of its flight.

⑤ A particle is initially at rest at the origin. It experiences an acceleration given by

$$\mathbf{a} = 4t\mathbf{i} + (6 - 2t)\mathbf{j}$$

Find expressions for the velocity and position of the particle at time t.

⑥ A particle has position vector $\mathbf{r} = 3t^2\mathbf{i} - 4t\mathbf{j}$ at time t. Find the angle between its position vector and its direction of motion at time $t = 2$.

⑦ While a hockey ball is being hit it experiences an acceleration (in m s^{-2}) modelled by

$$\mathbf{a} = 1000[6t(t - 0.2)\mathbf{i} + t(t - 0.2)\mathbf{j}]$$

$$\text{for } 0 \leqslant t \leqslant 0.2 \text{ (in seconds)}$$

and $\mathbf{i}$ and $\mathbf{j}$ are unit vectors along and perpendicular to the side of the pitch.

The ball is initially at rest. At $t = 0.2$ it loses contact with the hockey stick.

Find its speed when $t = 0.2$.

⑧ A speedboat is initially moving at $5\,\text{m s}^{-1}$ on a bearing of 135°.

(i) Express the initial velocity as a vector in terms of **i** and **j**, which are vectors east and north respectively.

The boat then begins to accelerate with an acceleration modelled by

$$\mathbf{a} = 0.1t\mathbf{i} + 0.3t\mathbf{j} \text{ in m s}^{-2}.$$

(ii) Find the velocity of the boat 10 s after it begins to accelerate and its displacement over the 10 s period.

⑨ A girl throws a ball and, t seconds after she releases it, its position in metres relative to the point where she is standing is modelled by

$$\begin{pmatrix} x \\ y \end{pmatrix} = \begin{pmatrix} 15t \\ 2 + 16t - 5t^2 \end{pmatrix}$$

where the directions are horizontal and vertical.

(i) Find expressions for the velocity and acceleration of the ball at time t.

(ii) The vertical component of the velocity is zero when the ball is at its highest point. Find the time taken for the ball to reach this point.

(iii) When the ball hits the ground the vertical component of its position vector is zero. What is the speed of the ball when it hits the ground?

(iv) Find the equation of the trajectory of the ball.

⑩ The position (in metres) of a tennis ball t seconds after leaving a racquet is modelled by

$$\mathbf{r} = 20t\mathbf{i} + \left(2 + t - 5t^2\right)\mathbf{j}$$

where **i** and **j** are horizontal and vertical unit vectors.

(i) Find the position of the tennis ball when $t = 0, 0.2, 0.4, 0.6$ and 0.8.

Use these to sketch the path of the ball.

(ii) Find an expression for the velocity of the tennis ball.

Use this to find the velocity of the ball when $t = 0.2$.

(iii) Find the acceleration of the ball.

(iv) Find the equation of the trajectory of the ball.

⑪ An owl is initially perched on a tree. It then goes for a short flight which ends when it dives onto a mouse on the ground. The position vector (in metres) of the owl t seconds into its flight is modelled by

$$\mathbf{r} = t^2(6 - t)\mathbf{i} + \left(12.5 + 4.5t^2 - t^3\right)\mathbf{j}$$

where the foot of the tree is taken to be the origin and the unit vectors **i** and **j** are horizontal and vertical.

(i) Draw a graph showing the bird's flight.

(ii) For how long (in s) is the owl in flight?

(iii) Find the speed of the owl when it catches the mouse and the angle that its flight makes with the horizontal at that instant.

(iv) Show that the owl's acceleration is never zero during the flight.

⑫ A particle P of mass 5 units moves under the action of a force $\begin{pmatrix} 10 \\ -5 \\ 20 \end{pmatrix}$.

Initially P has velocity $\begin{pmatrix} -3 \\ 2 \\ 0 \end{pmatrix}$ and is at the point with position vector $\begin{pmatrix} 2 \\ -5 \\ 8 \end{pmatrix}$.

Find, at time $t = 4$,

(i) the speed of P

(ii) the position vector of P.

⑬ Ship A is 5 km due west of ship B and is travelling on a course 035° at a constant but unknown speed v km h^{-1}. Ship B is travelling at a constant 10 km h^{-1} on a course 300°.

(i) Write the velocity of each ship in terms of unit vectors **i** and **j** with directions east and north.

(ii) Find the position vector of each ship at time t hours, relative to the starting position of ship A.

The ships are on a collision course.

(iii) Find the speed of ship A.

(iv) How much time elapses before the collision occurs?

⑭ A particle of mass 0.5 kg is acted on by a force, in newtons,

$$\mathbf{F} = t^2\mathbf{i} + 2t\mathbf{j}.$$

The particle is initially at rest at the origin and t is measured in seconds.

(i) Find the acceleration of the particle at time t.

(ii) Find the velocity of the particle at time t.

(iii) Find the position vector of the particle at time t.

(iv) Find the speed and direction of motion of the particle at time $t = 2$.

⑮ The position vector $\mathbf{r}$ of a moving particle at time t after the start of the motion is given by

$$\mathbf{r} = (5 + 20t)\mathbf{i} + \left(95 + 10t - 5t^2\right)\mathbf{j}.$$

(i) Find the initial velocity of the particle.

At time $t = T$ the particle is moving at right angles to its initial direction of motion.

(ii) Find the value of T and the distance of the particle from its initial position at this time.

⑯ A small, delicate microchip which is initially at rest is to be moved by a robot arm so that it is placed *gently* onto a horizontal assembly bench. Two mathematical models have been proposed for the motion which will be programmed into the robot. In each model the unit of length is the centimetre and time is measured in seconds. The unit vectors $\mathbf{i}$ and $\mathbf{j}$ have directions which are horizontal and vertical respectively and the origin is the point O on the surface of the bench, as shown in Figure 18.16.

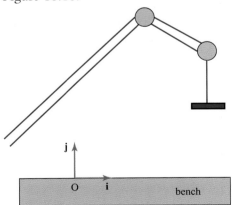

Figure 18.16

Model A for the position vector of the microchip at time t is

$$\mathbf{r}_A = 5t^2\mathbf{i} + \left(16 - 4t^2\right)\mathbf{j} \ (t \geqslant 0).$$

(i) How far above the bench is the microchip initially (i.e. when $t = 0$)?

(ii) Show that this model predicts that the microchip reaches the bench after 2 s and state the horizontal distance moved in this time.

(iii) Calculate the predicted horizontal and vertical components of velocity when $t = 0$ and $t = 2$.

Model B for the position vector at time t of the microchip is

$$\mathbf{r}_B = \left(15t^2 - 5t^3\right)\mathbf{i} + \left(16 - 24t^2 + 16t^3 - 3t^4\right)\mathbf{j}$$
$$(t \geqslant 0).$$

(iv) Show that model B predicts the same positions for the microchip at $t = 0$ and $t = 2$ as model A.

(v) Calculate the predicted horizontal and vertical components of velocity for the microchip at $t = 0$ and $t = 2$ from model B and comment, with brief reasons, on which model you think describes the more suitable motion.

⑰ The position vector of a motorcycle of mass 150 kg on a track is modelled by

$$\mathbf{r} = 4t^2\mathbf{i} + \tfrac{1}{8}t(8 - t)^2\mathbf{j} \quad 0 \leqslant t \leqslant 8$$
$$\mathbf{r} = (64t - 256)\mathbf{i} \quad 8 < t \leqslant 20$$

where t is the time in seconds after the start of the race.

The vectors $\mathbf{i}$ and $\mathbf{j}$ are in directions along and perpendicular to the direction of the track as shown in Figure 18.17. The origin is in the middle of the track. The vector $\mathbf{k}$ has direction vertically upwards.

Figure 18.17

(i) Draw a sketch to show the motorcycle's path over the first 10 seconds. The track is 20 m wide. Does the motorcycle leave it?

(ii) Find, in vector form, expressions for the velocity and acceleration of the motor cycle at time t for $0 \leqslant t \leqslant 20$.

(iii) Find in vector form an expression for the resultant horizontal force acting on the motorcycle during the first 8 s, in terms of t.

(iv) Why would you expect the driving force from the motorcycle's engine to be substantially greater than the component in the **i** direction of your answer to part (iii)?

When $t = 22$ s the motorcycle's velocity is given by $\mathbf{v} = 60\mathbf{i} + 6\mathbf{k}$.

(v) What has happened?

⑱ A hawk H and a sparrow S fly with constant velocities $\mathbf{v}_H = \mathbf{i} + 2\mathbf{j} + 3\mathbf{k}$ and $\mathbf{v}_S = -\mathbf{i} + \mathbf{j} + 2\mathbf{k}$ respectively.

At time $t = 0$ the position vector of the hawk is $\mathbf{r}_H = \mathbf{j}$ and that of the sparrow is $\mathbf{r}_S = 5\mathbf{i}$.

(i) Write down the velocity of H relative to S.

(ii) Write down the position vector of H relative to S at time t.

(iii) Find the time at which the birds are closest together and determine their least distance apart.

⑲ A particle moves in the xy plane and at time t has acceleration $\mathbf{a} = 2\mathbf{i}$. Initially the particle is at (3, 1) and is moving with velocity $\mathbf{u} = -2\mathbf{i} + \mathbf{j}$. Show that the path of the particle is a parabola and find its Cartesian equation.

⑳ At time t two points P and Q have position vectors **p** and **q** respectively, where

$$\mathbf{p} = 2\mathbf{i} + \cos\omega t\mathbf{j} + \sin\omega t\mathbf{k}$$

$$\mathbf{q} = \sin\omega t\mathbf{i} - \cos\omega t\mathbf{j} + 3\mathbf{k}$$

where ω is a constant. Find **r**, the position vector of P relative to Q, and **v** the velocity of P relative to Q.

LEARNING OUTCOMES

When you have completed this chapter, you should be able to:

➤ understand, use and derive the formulae for constant acceleration for motion in:
  ○ one dimension
  ○ two (or more) dimensions using vectors
➤ use calculus in kinematics for motion in:
  ○ one dimension
  ○ two dimensions (or more) using vectors
➤ use standard notation for motion in one and two dimensions:

  ○ in one dimension $v = \dfrac{ds}{dt}(=\dot{s}); a = \dfrac{dv}{dt}(=\dot{v}) = \dfrac{d^2s}{dt^2}(=\ddot{s});$

  $$s = \int v\,dt, v = \int a\,dt$$

  ○ in two (or more) dimensions $\mathbf{v} = \dfrac{d\mathbf{r}}{dt}(=\dot{\mathbf{r}}); \mathbf{a} = \dfrac{d\mathbf{v}}{dt}(=\dot{\mathbf{v}}) = \dfrac{d^2\mathbf{r}}{dt^2}(=\ddot{\mathbf{r}});$

  $$\mathbf{r} = \int \mathbf{v}\,dt, \mathbf{v} = \int \mathbf{a}\,dt$$

➤ use trigonometric functions to solve problems in context, including problems involving vectors, kinematics
➤ use vectors to solve problems in context, including kinematics.

KEY POINTS

1 Constant acceleration formulae

In one dimension

$v = u + at$

$s = ut + \frac{1}{2}at^2$

$s = \frac{1}{2}(u + v)t$

$s = vt - \frac{1}{2}at^2$

$v^2 = u^2 + 2as$

Vector form

$\mathbf{v} = \mathbf{u} + \mathbf{a}t$

$\mathbf{r} = \mathbf{u}t + \frac{1}{2}\mathbf{a}t^2$

$\mathbf{r} = \frac{1}{2}(\mathbf{u} + \mathbf{v})t$

$\mathbf{r} = \mathbf{v}t - \frac{1}{2}\mathbf{a}t^2$

If the motion does not start at the origin, s is replaced by $s - s_0$ and $\mathbf{r}$ is replaced by $\mathbf{r} - \mathbf{r}_0$.

2 Relationships between the variables describing general motion

Displacement $\rightarrow$ Velocity $\rightarrow$ Acceleration

$s \quad \rightarrow \quad v = \dfrac{ds}{dt} \quad \rightarrow \quad a = \dfrac{dv}{dt} = \dfrac{d^2s}{dt^2}$

$\mathbf{r} = x\mathbf{i} + y\mathbf{j} \quad \rightarrow \quad \mathbf{v} = \dfrac{d\mathbf{r}}{dt} = \dot{x}\mathbf{i} + \dot{y}\mathbf{j} = \dfrac{dx}{dt}\mathbf{i} + \dfrac{dy}{dt}\mathbf{j}$

$\rightarrow \quad \mathbf{a} = \dfrac{d\mathbf{v}}{dt} = \dfrac{d^2\mathbf{r}}{dt^2} = \ddot{x}\mathbf{i} + \ddot{y}\mathbf{j} = \dfrac{d^2x}{dt^2}\mathbf{i} + \dfrac{d^2y}{dt^2}\mathbf{j}$

$\mathbf{r} = \begin{pmatrix} x \\ y \end{pmatrix} \quad \mathbf{v} = \begin{pmatrix} \dot{x} \\ \dot{y} = \left(\dfrac{d^2x}{dt^2} \dfrac{d^2y}{dt^2} \right) \end{pmatrix} \quad \mathbf{a} = \begin{pmatrix} \ddot{x} \\ \ddot{y} = \left(\dfrac{d^2x}{dt^2} \dfrac{d^2y}{dt^2} \right) \end{pmatrix}$

Acceleration $\rightarrow$ Velocity $\rightarrow$ Displacement

$\mathbf{a} \quad \rightarrow \quad \mathbf{v} = \displaystyle\int \mathbf{a}\, dt \quad \rightarrow \quad \mathbf{r} = \displaystyle\int \mathbf{v}\, dt$

3 Vocabulary

Distance $\quad r = |\mathbf{r}| = \sqrt{x^2 + y^2}$

Speed $\quad v = |\mathbf{v}| = \sqrt{\dot{x}^2 + \dot{y}^2} = \sqrt{\left(\dfrac{dx}{dt}\right)^2 + \left(\dfrac{dy}{dt}\right)^2}$

Magnitude of acceleration $a = |\mathbf{a}| = \sqrt{\ddot{x}^2 + \ddot{y}^2} = \sqrt{\left(\dfrac{d^2x}{dt^2}\right)^2 + \left(\dfrac{d^2y}{dt^2}\right)^2}$

4 The equation of the path of a moving point

To find the Cartesian equation (y on x) of the path of a moving point with coordinates given in terms of t [$x = f(t)$; $y = g(t)$]

- express t in terms of one of the coordinates e.g. $t = f^{-1}(x)$
- then substitute for t in the expression for the other coordinate
- $y = g(f^{-1}(x))$.

Forces and motion

→ This cable car is stationary. How is this possible?

Review: Forces and motion

Modelling vocabulary

Mechanics is about modelling the real world. In order to do this, suitable simplifying assumptions are often made so that mathematics can be applied to situations and problems. This process often involves identifying factors that can be neglected without losing too much accuracy. Listed below are some commonly used modelling terms which are used to describe such assumptions:

- negligible   small enough to ignore
- inextensible   for a string with negligible stretch
- light   for an object with negligible mass
- particle   an object with negligible dimensions
- smooth   for a surface with negligible friction
- uniform   the same throughout.

Forces

A force is defined as the physical quantity that causes a change in motion. As it depends on magnitude and direction, it is a vector quantity.

Forces can start motion, stop motion, speed up or slow down objects, change the direction of their motion. In real situations, several forces usually act on an object. The sum of these forces, known as the resultant force, determines whether there is a change of motion or not.

There are several types of force that you often use.

The force of gravity

Every object on or near the earth's surface is pulled vertically downwards by the force of gravity. The size of the force on an object of mass M kg is Mg newtons where g is a constant whose value is about $9.8\,\mathrm{m\,s^{-2}}$. The force of gravity is also known as the **weight** of the object.

Tension and thrust

When a string is pulled, it exerts a **tension** force opposite to the pull. The tension acts along the string and is the same throughout the string (Figure 19.1). A rigid rod can exert a tension force in a similar way to a string when it is used to support or pull an object (Figure 19.2). It can also exert a **thrust** force when it is in compression. The thrust acts along the rod and is the same throughout the rod.

W (weight of block)

T (thrust pushing up on block)

T (thrust pushing down on floor)

R (reaction from floor)

Figure 19.1 Forces on the end of a string

Figure 19.2 Forces in a rigid rod

The tension on either side of a smooth pulley is the same (Figure 19.3).

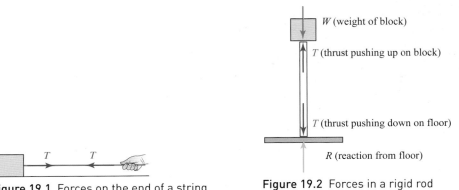

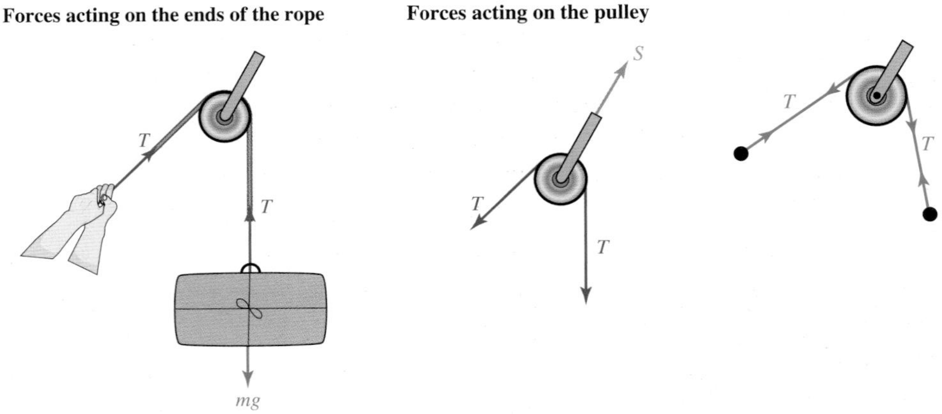

Forces acting on the ends of the rope

Forces acting on the pulley

Figure 19.3

Normal reaction

A book resting on a table is subjected to two forces: its weight and the **normal reaction** of the table. It is called normal because its line of action is normal (at right angles) to the surface of the table. Since the book is in equilibrium, the normal reaction is equal and opposite to the weight of the book.

> **Note**
>
> In this diagram the normal reaction is vertical but this is not always the case. For example, the normal reaction on an object on a slope is perpendicular to the slope.

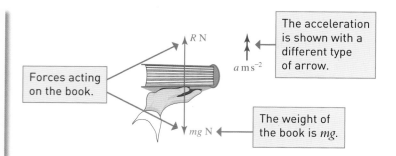

Forces acting on the book.

R N

The acceleration is shown with a different type of arrow.

$a \, \text{ms}^{-2}$

The weight of the book is mg.

mg N

Figure 19.4

Frictional force

In Figure 19.5, the book on the table is being pushed by a force P parallel to the surface. The book remains at rest because P is balanced by a **frictional force**, F, in the opposite direction to P. The magnitude of the frictional force is equal to the pushing force: $P = F$.

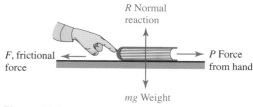

R Normal reaction

F, frictional force

P Force from hand

mg Weight

Figure 19.5

If P is increased and the book starts to move, F is still present but now $P > F$. Friction always acts in the direction opposed to the motion. Friction may prevent the motion of an object or slow it down if it is moving.

Driving force

In problems about moving objects such as cars, all the forces acting along the line of motion can usually be reduced to two or three, the **driving force**, the **resistance** to motion and possibly a **braking force**.

resistance

driving force

braking force

Figure 19.6

Example 19.1

Draw force diagrams illustrating the forces acting on a sledge which is lying on a slope which makes an angle α with the horizontal.

(i)   The sledge is at rest on the slope.

(ii)   The sledge is being pulled up the slope by a rope inclined at an angle θ to the slope.

(iii)   The sledge is prevented from sliding down the slope by a rope inclined at an angle θ to the slope.

Solution

(i)   The sledge is at rest on the slope.

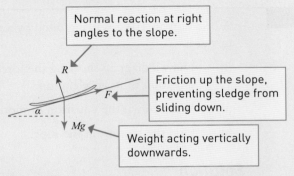

Figure 19.7

(ii)   The sledge is pulled up the slope.

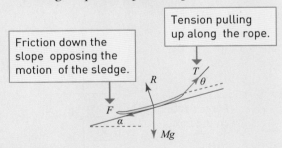

Figure 19.8

(iii)   The sledge is prevented from sliding down the slope.

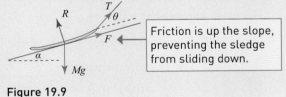

Figure 19.9

> !   This is a simplified version of what Newton actually said. It assumes the mass of the object is constant. Motion with a variable mass, e.g. a rocket taking off, is beyond the scope of this book.

Newton's laws of motion

The relationship between the forces acting on a body and its motion is summarised by Newton's laws of motion. These are fundamental to the study of mechanics.

1   Every object continues in a state of rest or uniform motion in a straight line unless it is acted on by a resultant external force.

2   The acceleration of an object is proportional to, and in the same direction as, the resultant of the forces acting on the object.

> **F** is the resultant force, m is the mass of the object, **a** is the acceleration.

 $\mathbf{F} = m\mathbf{a}$

Notice that this is a vector equation since both the magnitudes and directions of the resultant force and the acceleration are involved. If the motion is along a straight line it is often written in scalar form as $F = ma$.

3   When one object exerts a force on another there is always a reaction which is equal and opposite in direction to the acting force.

Equation of motion

The equation resulting from Newton's second law is often described as an **equation of motion**, as in the next example.

Example 19.2

A box of mass 10 kg is being lifted by a rope. The acceleration is $0.5\,\text{m s}^{-2}$ upwards.

(i)   Draw a diagram showing the forces acting on the box and the direction of its acceleration.

(ii)   Write down the equation of motion of the box.

(iii) Find the tension in the rope.

Solution

(i)   The forces acting on the box and the acceleration are shown in Figure 19.10.

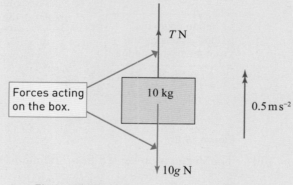

T N

Forces acting on the box.

10 kg

$0.5\,\text{m s}^{-2}$

$10g$ N

Figure 19.10

(ii)   The resultant force acting on the box is $T - 10g$ upwards.

The resulting equation

$$T - 10g = 10a = 5 \longleftarrow \boxed{\text{Since } 10a = 10 \times 0.5 = 5}$$

is called the **equation of motion**.

(iii) $T - 10 \times 9.8 = 5$

$$T = 103$$

The tension in the rope is 103 N.

Example 19.3

A car of mass 800 kg travels at constant speed along a straight horizontal road. Its engine is producing a driving force of 400 N.

(i)   What is the resistance to its motion?

Later the driving force is increased to 1000 N.

(ii)   Find the acceleration of the car, given that the resistance force remains the same.

Solution

(i)   The car is travelling at constant speed, so that the resultant force acting on the car is zero.

Figure 19.11

Let the resistance force be R N.

$$400 - R = 0$$

$$R = 400\,\text{N}$$

The resistance force is 400 N.

(ii)   The resultant force is now $1000 - 400 = 600\,\text{N}$.

Figure 19.12

The equation of motion is $600 = 800a$.

$$a = \frac{600}{800} = 0.75$$

The acceleration of the car is $0.75\,\text{ms}^{-2}$.

Motion of connected particles

Example 19.4

Two particles, A of mass 2 kg and B of mass 5 kg, are connected by a light inextensible string passing over a smooth fixed pulley. Find the acceleration of the particles and the tension in the string.

Note

A *smooth* pulley turns freely so the tension of the string is the same on both sides of it.

Solution

Equation of motion for A

$$T - 2g = 2a \qquad ①$$

Equation of motion for B

$$5g - T = 5a \qquad ②$$

$$① + ② \quad 3g = 7a$$

$$a = \frac{3g}{7} = \frac{3 \times 9.8}{7} = 4.2$$

Substitute in ①

$$T = 2a + 2g = 2 \times 4.2 + 2 \times 9.8 = 28$$

The acceleration of the particles is $4.2\,\mathrm{m\,s^{-2}}$ and the tension in the string is $28\,\mathrm{N}$.

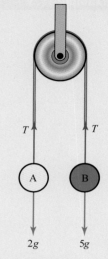

Figure 19.13

> ❗ In Example 19.4 there is just one string and it passes over a smooth pulley so the tension is the same in both parts of it. You will meet other situations, like the one illustrated in Figure 19.14, where there are two or more different strings; in such cases the tensions will not be the same even though the pulleys they pass over are smooth.

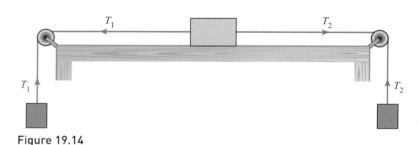

Figure 19.14

Review exercise

① Find the accelerations produced when a force of $5\,\mathrm{N}$ acts on an object

(i)   of mass $15\,\mathrm{kg}$

(ii) of mass $10\,\mathrm{g}$.

② A box of mass $25\,\mathrm{kg}$ is falling with an acceleration of $7.2\,\mathrm{m\,s^{-2}}$. What is the resistance acting on the box?

③ A car of mass $800\,\mathrm{kg}$ is travelling along a straight level road.

(i)   Calculate the acceleration of the car when a resultant force of $2400\,\mathrm{N}$ acts on it in the direction of its motion. How long does it take the car to increase its speed from $4\,\mathrm{m\,s^{-1}}$ to $12\,\mathrm{m\,s^{-1}}$?

The car has an acceleration of $1.2\,\mathrm{m\,s^{-2}}$ when there is a driving force of $2400\,\mathrm{N}$.

(ii) Find the resistance to motion of the car.

④ A load of mass $2.5\,\mathrm{kg}$ is held on the end of a string. In each of the following cases calculate the tension in the string.

(i)   The load is raised with an acceleration of $2\,\mathrm{m\,s^{-2}}$.

(ii) The load is lowered with an acceleration of $2\,\mathrm{m\,s^{-2}}$.

(iii) The load is raised with a constant speed of $2\,\mathrm{m\,s^{-1}}$.

⑤ Two boxes A and B are descending vertically supported by a parachute. Box A has mass $50\,\mathrm{kg}$. Box B has mass $25\,\mathrm{kg}$ and is suspended from box A by a light vertical wire. Both boxes are descending with acceleration $2\,\mathrm{m\,s^{-2}}$.

(i)   Draw a labelled diagram showing all the forces acting on box A in Figure 19.15 (overleaf) and another diagram showing all the forces acting on box B in Figure 19.15.

(ii) Write down separate equations of motion for box A and for box B.

(iii) Find the tensions in both wires.

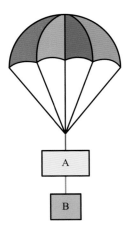

Figure 19.15

⑥ A block A of mass 5 kg is lying on a smooth horizontal table. A light inextensible string connects A to a block B of mass 4 kg which hangs freely over a smooth pulley at the edge of the table (Figure 19.16).

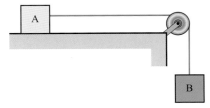

Figure 19.16

(i) Draw force diagrams to show the forces acting on each block.

(ii) Write down separate equations of motion for A and B.

(iii) Find the acceleration of the system and the tension in the string.

⑦ A particle A of mass 8 kg is connected to a particle B of mass 2 kg by a light inextensible string which passes over a smooth fixed pulley. B is connected to a third particle C of mass 5 kg by another string as shown in Figure 19.17.

(i) Draw diagrams showing all the forces acting on each one of the three particles.

(ii) Write down equations of motion for each of A, B and C.

(iii) Find the acceleration of the system and the tension in each string.

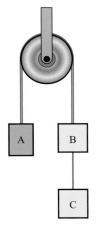

Figure 19.17

⑧ A light scale pan holding two blocks A and B, of masses 100 g and 500 g is being lifted vertically upwards with an acceleration of 0.8 m s⁻² (Figure 19.18).

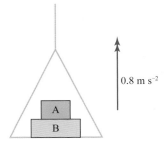

Figure 19.18

(i) Draw diagrams showing all the forces acting on each one of the masses.

(ii) Write down equations of motion for each of A and B.

(iii) Find the reaction forces between A and B and between B and the scale pan.

⑨ A car of mass 900 kg tows a caravan of mass 800 kg along a horizontal road. The engine of the car produces a driving force of 2250 N. The car is subjected to a resistance of 250 N and the caravan to a resistance of 300 N.

(i) Show in separate diagrams the horizontal forces acting on the car and the caravan.

(ii) Find the acceleration of the car and caravan.

(iii) Find the tension in the coupling between the car and the caravan.

⑩ A train consists of an engine and two trucks with masses and resistances to motion as shown in Figure 19.19. The engine provides a driving force of 18 000 N. All the couplings are light, rigid and horizontal.

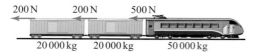

200 N   200 N   500 N

20 000 kg   20 000 kg   50 000 kg

Figure 19.19

(i) Show that the acceleration of the train is 0.19 m s⁻².

(ii) Find the force in the coupling between the two trucks.

(iii) Find the force in the coupling between the engine and the first truck.

With the driving force removed, brakes are applied, so adding an additional resistance of 18 000 N to the total of the resistances shown in Figure 19.19.

(iv) Find the new acceleration of the train.

(v) Find the force in the coupling between the two trucks.

1 Forces in equilibrium

Resolving forces

It is often helpful to replace a single force by two perpendicular forces acting in the directions in which you are most interested. This is called **resolving** the force.

Thus in Figure 19.20 the force F is acting at angle θ to the horizontal. It can be resolved into a horizontal component of $F \cos\theta$ and a vertical component of $F \sin\theta$. These two resolved components are exactly equivalent to the single force, F.

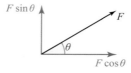

Figure 19.20

Figure 19.21 illustrates another common situation. A block is on a surface sloping at an angle of α to the horizontal. Its weight, W, acts vertically downwards but it can be resolved into components.

- $W\cos\alpha$ perpendicular to the slope
- $W\sin\alpha$ parallel to it.

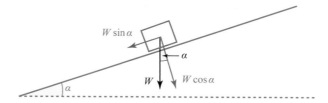

Figure 19.21

The reverse process involves finding the resultant of two perpendicular forces, using Pythagoras' theorem and simple trigonometry. The single force is the resultant of the other two. This is illustrated in Figure 19.22 overleaf.

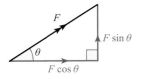

Figure 19.22

Resultant

$$= \sqrt{\left((F\sin\theta)^2 + (F\cos\theta)^2\right)}$$

$$= \sqrt{\left(F^2\sin^2\theta + F^2\cos^2\theta\right)}$$

$$= \sqrt{F^2\left(\sin^2\theta + \cos^2\theta\right)}$$

$$= F$$

It is not only forces that you can resolve but any vector quantity.

Equilibrium

When forces are in equilibrium their vector sum is zero and the sum of the resolved parts in *any* direction is zero.

Example 19.5

A brick of mass 3 kg is at rest on a rough plane inclined at an angle of 30° to the horizontal. Find the frictional force F N, and the normal reaction R N of the plane on the brick.

Solution

Figure 19.23 shows the forces acting on the brick.

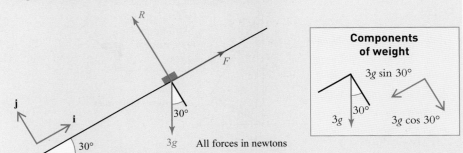

Figure 19.23

Take unit vectors **i** and **j** parallel and perpendicular to the plane as shown.

Since the brick is in equilibrium the resultant of the three forces acting on it is zero.

$3g = 29.4$

Resolving in the **i** direction: $F - 29.4\sin 30° = 0$ ①

$$F = 14.7$$

Resolving in the **j** direction: $R - 29.4\cos 30° = 0$ ②

$$R = 25.5$$

Written in vector form the equivalent is

$$\begin{pmatrix} F \\ 0 \end{pmatrix} + \begin{pmatrix} 0 \\ R \end{pmatrix} + \begin{pmatrix} -29.4\sin 30° \\ -29.4\cos 30° \end{pmatrix} = \begin{pmatrix} 0 \\ 0 \end{pmatrix}$$

Or alternatively

$$F\mathbf{i} + R\mathbf{j} - 29.4\sin 30°\mathbf{i} - 29.4\cos 30°\mathbf{j} = 0$$

Both these lead to the equations ① and ②.

The triangle of forces

When there are only three (non-parallel) forces acting and they are in equilibrium, you can draw a triangle to show the forces as shown for the brick on the plane (Figures 19.24 and 19.25).

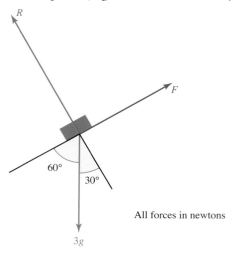

All forces in newtons

Figure 19.24

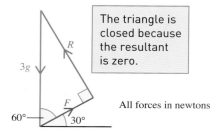

The triangle is closed because the resultant is zero.

All forces in newtons

Figure 19.25

Then $\qquad \dfrac{F}{3g} = \cos 60°$

$$F = 29.4 \cos 60° = 14.7 \text{ N}$$

And similarly $\;R = 29.4 \sin 60° = 25.5 \text{ N}$

This is an example of the theorem known as the **triangle of forces**.

 Triangle of forces

When a body is in equilibrium under the action of three non-parallel forces, then
1 the forces can be represented in magnitude and direction by the sides of a triangle
2 the lines of action of the forces pass through the same point.

When more than three forces are in equilibrium the first statement still holds but the triangle is then a polygon. The second statement is not necessarily true.

The next example illustrates two methods for solving problems involving forces in equilibrium. With experience, you will find it easy to judge which method is best for a particular problem.

Example 19.6

A sign of mass 10 kg is to be suspended by two strings arranged as shown in Figure 19.26. Find the tension in each string.

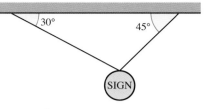

Figure 19.26

Solution

The force diagram for this situation is given in Figure 19.27.

> Notice that there are two different strings and the tensions in them, called T_1 and T_2 here, are not the same.

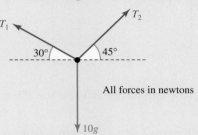

All forces in newtons

Figure 19.27

Method 1: Resolving forces

Vertically ($\uparrow$): $T_1 \sin 30° + T_2 \sin 45° - 10g = 0$

$0.5T_1 + 0.707T_2 = 98$ ①

Horizontally ($\rightarrow$): $-T_1 \cos 30° + T_2 \cos 45° = 0$

$-0.8666T_1 + 0.707T_2 = 0$ ②

Subtracting ② from ① $1.3667T_1 = 98$

$T_1 = 71.74$

Back substitution gives $T_2 = 87.87$

The tensions are 71.7 N and 87.9 N (to 1 d.p.).

Method 2: Triangle of forces

Since the three forces are in equilibrium they can be represented by the sides of a triangle taken in order.

You can estimate the tensions by measurements. This will tell you that $T_1 \approx 72$ and $T_2 \approx 88$ in newtons.

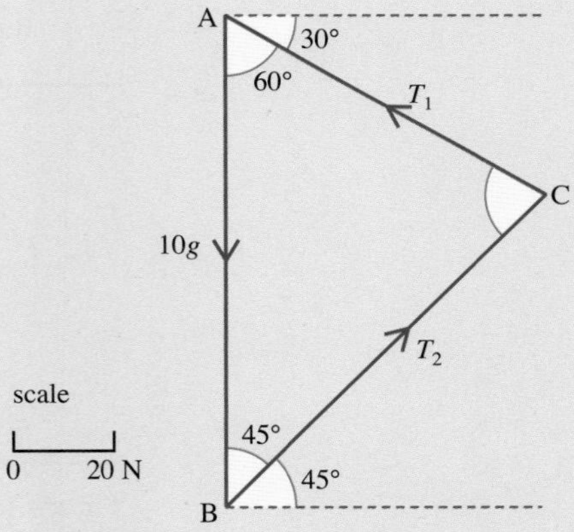

scale

0   20 N

Alternatively, you can use the sine rule to calculate T_1 and T_2 accurately.

Figure 19.28

> **Discussion point**
> → In what order would you draw the three lines in this diagram?

In the triangle ABC, $\angle CAB = 60°$ and $\angle ABC = 45°$, so $\angle BCA = 75°$.

So $\dfrac{T_1}{\sin 45°} = \dfrac{T_2}{\sin 60°} = \dfrac{98}{\sin 75°}$

giving $T_1 = \dfrac{98 \sin 45°}{\sin 75°}$ and $T_2 = \dfrac{98 \sin 60°}{\sin 75°}$

As before the tensions are found to be 71.7 N and 87.9 N.

Discussion point

Lami's theorem states that when three forces acting at a point as shown in Figure 19.29 are in equilibrium then

$$\frac{F_1}{\sin \alpha} = \frac{F_2}{\sin \beta} = \frac{F_3}{\sin \gamma}.$$

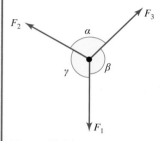

Figure 19.29

→ Sketch a triangle of forces and say how the angles in the triangle are related to α, β and γ.

→ Hence explain why Lami's theorem is true.

Example 19.7

Figure 19.30 shows three people involved in moving a packing case up to the top floor of a warehouse. Brian is pulling on a rope which passes round smooth pulleys at X and Y and is then secured to the point Z at the end of the loading beam.

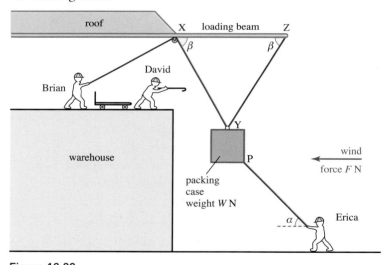

Figure 19.30

The wind is blowing directly towards the building. To counteract this, Erica is pulling on another rope, attached to the packing case at P, with just enough force and in the right direction to keep the packing case central between X and Z.

At the time of the picture the people are holding the packing case motionless.

(i)   Draw a diagram showing all the forces acting on the packing case using T_1 and T_2 for the tensions in Brian and Erica's ropes, respectively.

→

(ii)   Write down equations for the horizontal and vertical equilibrium of the packing case.

In one particular situation, $W = 200$, $F = 50$, $\alpha = 45°$ and $\beta = 75°$.

(iii)   Find the tension T_1.

(iv)   Explain why Brian has to pull harder if the wind blows more strongly.

[MEI adapted]

Solution

(i)   Figure 19.31 shows all the forces acting on the packing case and the relevant angles.

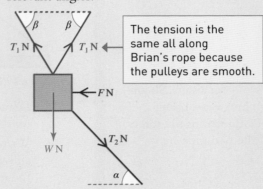

> The tension is the same all along Brian's rope because the pulleys are smooth.

Figure 19.31

(ii)   Equilibrium equations

Resolving horizontally $(\rightarrow)$

$$T_1 \cos \beta + T_2 \cos \alpha - F - T_1 \cos \beta = 0$$
$$T_2 \cos \alpha - F = 0 \qquad ①$$

Resolving vertically $(\uparrow)$

$$T_1 \sin \beta + T_1 \sin \beta - T_2 \sin \alpha - W = 0$$
$$2T_1 \sin \beta - T_2 \sin \alpha - W = 0 \qquad ②$$

> This tells you that
> $$T_2 = \frac{50}{\cos 45°}$$ but you don't need to work it out because $\cos 45° = \sin 45°$

(iii)   When $F = 50$ and $\alpha = 45°$ equation ① gives

$$T_2 \cos 45° = 50$$
$$\Rightarrow T_2 \sin 45° = 50$$

Substituting in ② gives $2T_1 \sin \beta - 50 - W = 0$

So when $W = 100$ and $\beta = 75°$ $2T_1 \sin 75° = 150$

$$T_1 = \frac{150}{2 \sin 75°} = 77.65$$

The tension in Brian's rope is 78 N (to the nearest N).

> **Note**
> ------
> Or $F = T_2 \cos \alpha$, so as F increases,
> T_2 increases $\Rightarrow$
> $T_2 \sin \alpha + W$ increases $\Rightarrow$
> $2T_1 \sin \beta$ increases.
> Hence T_1 increases.

(iv)   When the wind blows harder, F increases. Given that all the angles remain unchanged, Erica will have to pull harder so the vertical component of T_2 will increase. This means that T_1 must increase and Brian must pull harder.

Exercise 19.1

Those questions in this exercise where the context is drawn from real life involve working with mathematical models.

① A system of forces is in equilibrium. Marc drew this diagram to show the forces. What is wrong with his diagram?

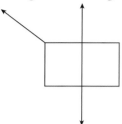

② Figure 19.32 shows a boy, Halley, holding onto a post while his two older sisters, Sheuli and Veronica, try to pull him away. Using perpendicular horizontal directions the forces, in newtons, exerted by the two girls are:

Sheuli $\begin{pmatrix} 24 \\ 18 \end{pmatrix}$

Veronica $\begin{pmatrix} 25 \\ 60 \end{pmatrix}$

Figure 19.32

(i) Calculate the magnitude and direction of the force of each of the girls.

(ii) Use a scale drawing to estimate the magnitude and direction of the resultant of the forces exerted by the two girls.

(iii) Write the resultant as a vector and so calculate (to 3 significant figures) its magnitude and direction.

Check that your answers agree with those obtained by scale drawing in part (ii).

③ Figure 19.33 shows a girder CD of mass 20 tonnes being held stationary by a crane (which is not shown). The rope from the crane (AB) is attached to a ring at B. Two ropes BC and BD, of equal length attach the girder to B; the tension in each of these ropes is T N.

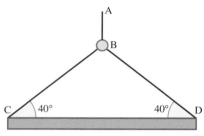

Figure 19.33

(i) Draw a diagram showing the forces acting on the girder.

(ii) Write down, in terms of T, the horizontal and vertical components of the tensions in the ropes acting at C and D.

(iii) Hence show that the tension in the rope BC is 152.5 kN (to 1 d.p.).

(iv) Draw a diagram to show the three forces acting on the ring at B.

(v) Hence calculate the tension in the rope AB.

(vi) How could you have known the answer to part (v) without any calculations?

④ A box of mass 8 kg is at rest on a horizontal floor.

(i) Find the value of the normal reaction of the floor on the box.

The box remains at rest on the floor when a force of 15 N is applied to it at an angle of 35° to the upward vertical as shown in Figure 19.34.

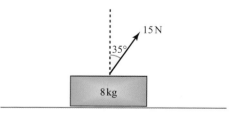

Figure 19.34

(ii) Draw a diagram showing all the forces acting on the box.

(iii) Calculate the new value of the normal reaction of the floor on the box and also the frictional force.

⑤ A block of weight 75 N is on a rough plane that is inclined at 25° to the horizontal. The block is in equilibrium with a horizontal force of 25 N acting on it as shown. Calculate the frictional force acting on the block.

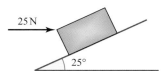

Figure 19.35

⑥ A particle is in equilibrium under the three forces shown in Figure 19.36. Find the force F and the angle θ.

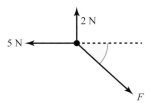

Figure 19.36

⑦ Figure 19.37 shows a simple model of a crane. The structure is at rest in a vertical plane. The rod and cables are of negligible mass and the load suspended from the joint at A is 30 N.

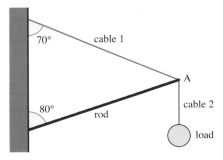

Figure 19.37

(i) Draw a diagram showing the forces acting on

    (a) the load

    (b) the joint at A.

(ii) Calculate the forces in the rod and cable 1 and state whether they are in tension or compression.

⑧ Each of three light strings has a particle attached to one of its ends. The other ends of the strings are tied together at a point A. The strings are in equilibrium with two of them passing over fixed smooth pulleys and with the particles hanging freely. The weights of the particles, and the angles between the sloping parts of the strings and the vertical, are as shown in Figure 19.38. Find the values of W_1 and W_2.

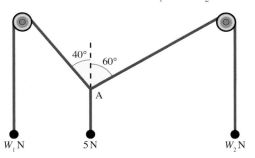

Figure 19.38

⑨ An angler catches a very large fish. When he tries to measure the mass, he finds that it is more than the 10 kg limit of his spring balance. He borrows another spring balance of exactly the same design and uses the two to measure the fish, as shown in diagram A, both balances read 8 kg.

    (i) What is the mass of the fish?

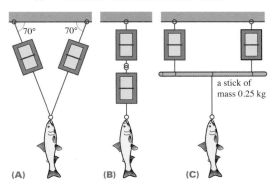

Figure 19.39

The angler believes the mass of the fish is a record and asks a witness to confirm it.

The witness agrees with the measurements but cannot follow the calculations. He asks the angler to measure the mass of the fish in two different positions, still using both balances. These are shown in diagrams B and C. Assuming the spring balances to have negligible mass, state the readings of the balances as set up in

    (ii) diagram B   (iii) diagram C.

    (iv) Which of the three methods do you think is best?

⑩ Figure 19.40 shows a device for crushing scrap cars. The light rod AB is hinged at A and raised by a cable which runs from B round a pulley at D and down to a winch at E. The vertical strut EAD is rigid and strong and AD = AB. A block of mass 1 tonne is suspended from B by the cable BC. When the block is correctly situated above the car it is released and falls onto the car.

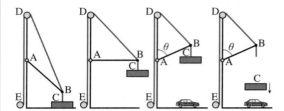

Figure 19.40

Just before the block is released the rod AB makes an angle θ with the upward vertical AD and the weight is at rest.

(i) Draw a diagram showing the forces acting at point B in this position.

(ii) Explain why the rod AB must be in thrust and not in tension.

(iii) Draw a diagram showing the vector sum of the forces at B (i.e. the polygon of forces).

(iv) Calculate each of the three forces acting at B when

(a) $\theta = 90°$ (b) $\theta = 60°$.

⑪ Four wires, all of them horizontal, are attached to the top of a telegraph pole as shown in the plan view in Figure 19.41. The pole is in equilibrium and tensions in the wires are as shown.

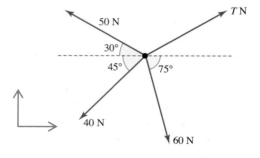

Figure 19.41

(i) Using perpendicular directions as shown in the diagram, show that the force of 60 N may be written as $\begin{pmatrix} 15.5 \\ -58 \end{pmatrix}$ N (to 3 significant figures).

(ii) Find T in both component form and magnitude and direction form.

(iii) The force T is changed to $\begin{pmatrix} 40 \\ 35 \end{pmatrix}$ N. Show that there is now a resultant force on the pole and find its magnitude and direction.

⑫ A ship is being towed by two tugs (Figure 19.42). Each tug exerts forces on the ship as indicated. There is also a drag force on the ship.

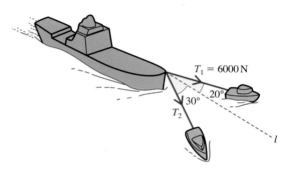

Figure 19.42

(i) Write down the components of the tensions in the towing cables along and perpendicular to the line of motion, l, of the ship.

(ii) There is no resultant force perpendicular to the line l. Find T_2.

(iii) The ship is travelling with constant velocity along the line l. Find the magnitude of the drag force acting on it.

⑬ A skier of mass 50 kg is skiing down a 15° slope.

(i) Draw a diagram showing the forces acting on the skier.

(ii) Resolve these forces into components parallel and perpendicular to the slope.

(iii) The skier is travelling at constant speed. Find the normal reaction of the slope on the skier and the resistance force on her.

(iv) The skier later returns to the top of the slope by being pulled up it at constant speed by a rope parallel to the slope. Assuming the resistance on the skier is the same as before, calculate the tension in the rope.

⑭ Figure 19.43 shows a block of mass 5 kg on a rough inclined plane. The block is attached to a 3 kg weight by a light string which passes over a smooth pulley, and is on the point of sliding up the slope.

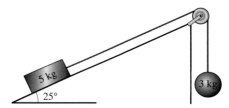

Figure 19.43

(i) Draw a diagram showing the forces acting on the block.

(ii) Resolve these forces into components parallel and perpendicular to the slope.

(iii) Find the force of resistance to the block's motion.

The 3 kg mass is replaced by one of mass m kg.

(iv) Find the value of m for which the block is on the point of sliding down the slope, assuming the resistance to motion is the same as before.

⑮ Two husky dogs are pulling a sledge. They both exert forces of 60 N but at different angles to the line of the sledge, as shown in Figure 19.44. The sledge is moving straight forwards.

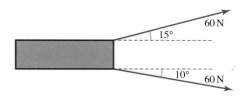

Figure 19.44

(i) Resolve the two forces into components parallel and perpendicular to the line of the sledge.

(ii) Hence find

(a) the overall forward force from the dogs

(b) the sideways force.

The resistance to motion is 20 N along the line of the sledge but up to 400 N perpendicular to it.

(iii) Find the magnitude and direction of the overall horizontal force on the sledge.

(iv) How much force is lost due to the dogs not pulling straight forwards?

⑯ One end of a string of length 1 m is fixed to a particle P of mass 1 kg and the other end is fixed to a point A. Another string is fixed to the mass and passes over a frictionless pulley at B which is 1 m horizontally from A but 2 m above it. The tension in the second string is such that the particle is held at the same horizontal level as the point A.

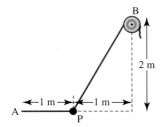

Figure 19.45

(i) Show that the tension in the horizontal string AP is 4.9 N and find the tension in the string which passes over the pulley at B. Find also the angle that this second string makes with the horizontal.

(ii) The tension in this second string is slowly increased by drawing more of it over the pulley at B. Describe the path followed by P. Will the points A, P and B, ever lie in a straight line? Give reasons for your answer.

⑰ Figure 19.46 shows a man suspended by means of a rope which is attached at one end to a peg at a fixed point A on a

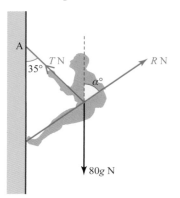

Figure 19.46

vertical wall and at the other end to a belt round his waist. The man has weight $80g$ N, the tension in the rope is T and the reaction of the wall on the man is R. The rope is inclined at $35°$ to the vertical and R is inclined at α to the vertical as shown. The man is in equilibrium.

(i) Explain why $\alpha > 0$.

(ii) By considering the horizontal and vertical equilibrium separately, obtain two equations connecting T, R and α.

(iii) Given that $\alpha = 45°$, show that T is about 563 N and find R.

(iv) What is the magnitude and the direction of the force on the peg at A?

The peg at A is replaced by a smooth pulley. The rope is passed over the pulley and tied to a hook at B directly below A. Calculate

(v) the new value of the tension in the rope section BA.

(vi) the magnitude of the force on the pulley at A.

2 Finding resultant forces

When forces are in equilibrium their resultant is zero, but of course forces may not always be in equilibrium. The next example shows you how to find the resultant of forces that are not in equilibrium. You know from Newton's second law that the acceleration of the body will be in the same direction as the resultant force; remember that force and acceleration are both vector quantities.

Example 19.8

A child on a sledge is being pulled up a smooth slope of $20°$ by a rope which makes an angle of $40°$ with the slope. The mass of the child and sledge together is 20 kg and the tension in the rope is 170 N.

(i) Draw a diagram to show the forces acting on the child and sledge together.

(ii) Find the resultant of these forces.

 Note

When the child and sledge are modelled as a particle, all the forces can be assumed to be acting at a point.

There is no friction force because the slope is smooth.

→

Solution

(i)   Here is the force diagram.

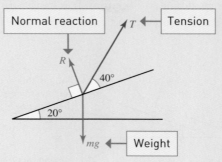

Figure 19.47

(ii)   You can find the normal reaction and the resultant force on the sledge using two methods.

Method 1: Scale drawing

Draw a scale diagram with the three forces represented by three of the sides of a quadrilateral taken in order (with the arrows following each other) as shown in Figure 19.48. The resultant is represented by the fourth side AD.

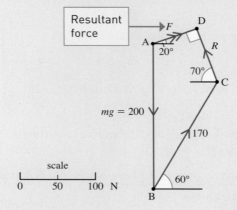

Figure 19.48

From the diagram you can estimate the normal reaction to be about 80 N and the resultant 60 N. This is a good estimate but more precise values are obtained using the second method.

> **Discussion point**
>
> → The sledge is sliding along the slope. What direction is the resultant force acting on it?

> **Discussion point**
>
> → In what order would you draw the lines in the diagram?

> **Note**
>
> Remember that in this case the resultant force must be parallel to the slope (Figure 19.48).

> ### Discussion points
>
> → What can you say about the acceleration of the sledge in the cases when
>
> (i)   the length AD is not zero?
>
> (ii)   the length AD is zero so that the starting point on the quadrilateral is the same as the finishing point?
>
> (iii) BC is so short that the point D is to the left of A as shown in Figure 19.49?

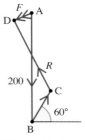

Figure 19.49

Method 2: Using components

This method involves resolving forces into components in two perpendicular directions. It is easiest to use the components of the forces parallel and perpendicular to the slope in the directions of **i** and **j** as shown.

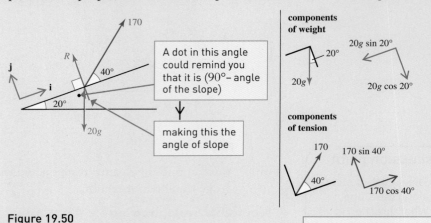

A dot in this angle could remind you that it is (90° – angle of the slope)

↓

making this the angle of slope

components of weight

components of tension

Figure 19.50

$\cos(90° - 20°) = \sin 20°$

The force R is perpendicular to the slope so it has no component in this direction.

Resolve parallel to the slope:

The resultant is $F = 170\cos 40° - 20g\sin 20°$

$= 63.191 \ldots$

Resolve perpendicular to the slope:

$R + 170\sin 40° - 20g\cos 20° = 0$

There is no resultant in this direction because the motion is parallel to the slope.

$R = 20g\cos 20° - 170\sin 40°$

$= 74.905 \cdots$

To 3 significant figures, the normal reaction is 74.9 N and the resultant is 63.2 N up the slope.

Alternatively, you could have worked in column vectors as follows.

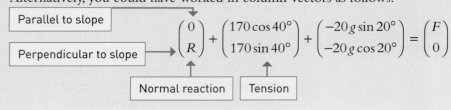

Parallel to slope

Perpendicular to slope

$$\begin{pmatrix} 0 \\ R \end{pmatrix} + \begin{pmatrix} 170\cos 40° \\ 170\sin 40° \end{pmatrix} + \begin{pmatrix} -20g\sin 20° \\ -20g\cos 20° \end{pmatrix} = \begin{pmatrix} F \\ 0 \end{pmatrix}$$

Normal reaction   Tension

Note

Try resolving horizontally and vertically. You will obtain two equations in the two unknowns F and R. It is perfectly possible to solve these equations, but quite a lot of work. It is much easier to resolve in the directions which ensure that one component of at least one of the unknown forces is zero.

Once you know the resultant force, you can work out the acceleration of the sledge using Newton's second law.

$$F = ma$$

$$63.191\cdots = 20a$$

$$a = \frac{63.191\cdots}{20} = 3.159\cdots$$

The acceleration is $3.2\,\mathrm{m\,s^{-2}}$ (correct to 1 d.p.).

Example 19.9

Two forces **P** and **Q** act at a point O. Force **P** has magnitude 20 N and acts along a bearing of 060°. Force **Q** has magnitude of 15 N and acts along a bearing of 330°.

Find the magnitude and bearing of the resultant force **P** + **Q**.

Note

Notice that **P** and **Q** are written as vectors.

Solution

Forces **P** and **Q** are shown in Figure 19.51.

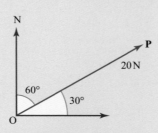

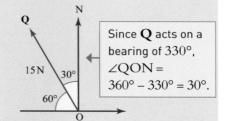

Since **Q** acts on a bearing of 330°, ∠QON = 360° − 330° = 30°.

Figure 19.51

$$\mathbf{P} = \begin{pmatrix} 20\cos 30° \\ 20\sin 30° \end{pmatrix}$$

$$= \begin{pmatrix} 17.32... \\ 10 \end{pmatrix}$$

$$\mathbf{Q} = \begin{pmatrix} -15\cos 60° \\ 15\sin 60° \end{pmatrix}$$

$$= \begin{pmatrix} -7.5 \\ 12.99... \end{pmatrix}$$

$$\mathbf{P} + \mathbf{Q} = \begin{pmatrix} 17.32... \\ 10 \end{pmatrix} + \begin{pmatrix} -7.5 \\ 12.99... \end{pmatrix}$$

$$= \begin{pmatrix} 9.82... \\ 22.99... \end{pmatrix}$$

The resultant is shown in Figure 19.52.

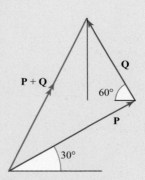

Figure 19.52

Figure 19.53 shows how you can calculate the magnitude and bearing from this resultant.

Magnitude $\quad |\mathbf{P} + \mathbf{Q}| = \sqrt{9.82...^2 + 22.99...^2}$

$$= \sqrt{625}$$

$$= 25$$

Direction $\quad \tan\theta = \dfrac{22.99...}{9.82...}$

$$= 2.34...$$

$$\theta = 66.86...°$$

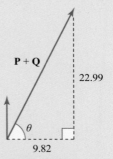

Figure 19.53

The bearing is 90° − 66.86...° = 23.13...°

The force **P** + **Q** has magnitude 25 N and bearing 023°.

Sometimes, as in the next example, it is just as easy to work with the trigonometry of the diagram as with the components of the forces.

Example 19.10

Find the magnitude and direction of the resultant of two forces **P** and **Q** inclined at an angle θ.

Solution

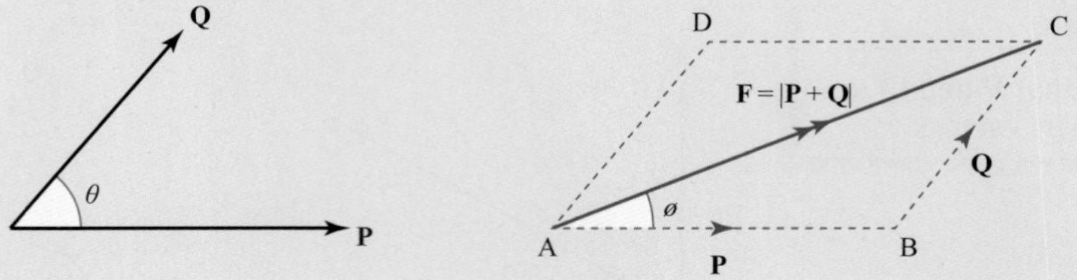

Figure 19.54

Use the cosine rule in triangle ABC. The magnitude of the resultant is F.

$$F = |\mathbf{P} + \mathbf{Q}| = \sqrt{AB^2 + BC^2 - 2AB \times BC \times \cos(\angle ABC)}$$

$$= \sqrt{P^2 + Q^2 - 2PQ \cos(180° - \theta)}$$

$$= \sqrt{P^2 + Q^2 + 2PQ \cos \theta}$$

Use the sine rule in triangle ABC. The resultant makes an angle ϕ with the **P** force.

$$\frac{\sin \angle CAB}{BC} = \frac{\sin \angle ABC}{AC}$$

$$\frac{\sin\phi}{Q} = \frac{\sin(180° - \theta)}{F}$$

$$\sin \phi = \frac{Q}{F} \sin \theta$$

$$\phi = \arcsin\left(\frac{Q}{F} \sin \theta\right)$$

The resultant of the two forces **P** and **Q** inclined at θ has magnitude $F = \sqrt{P^2 + Q^2 + 2PQ \cos \theta}$ and makes an angle $\phi = \arcsin\left(\frac{Q}{F} \sin \theta\right)$ with the **P** force.

Exercise 19.2

Those questions in this exercise where the context is drawn from real life involve working with mathematical models.

For questions 1–6, carry out the following steps. All forces are in newtons.

(i) Draw a scale diagram to show the forces and their resultant.

(ii) State whether you think the forces are in equilibrium and, if not, estimate the magnitude and direction of the resultant.

(iii) Write the forces in component form, using the directions indicated and so obtain the components of the resultant. Hence find the magnitude and direction of the resultant.

(iv) Compare your answers to parts (ii) and (iii).

⑦ Four horizontal wires are attached to a telephone post and exert the following tensions on it: 2 N in the north direction, 3 N in the east direction, 4 N in the south-west direction and 5 N in the south-east direction. Calculate the resultant of the tensions on the post and find its direction.

⑧ Forces of magnitude 12 N, 8 N and 5 N act on a particle in the directions shown in Figure 19.56.

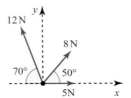

Figure 19.56

(i) Find the components of the resultant of the three forces in the **i** and **j** directions.

(ii) Find the magnitude and direction of the resultant.

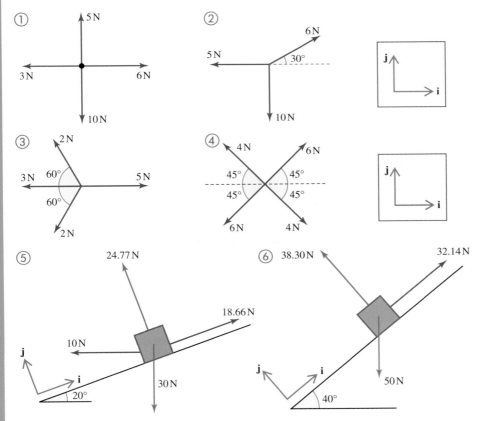

Figure 19.55

⑨ Find the resultant of the set of four forces whose magnitudes and directions are shown in Figure 19.57

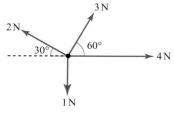

Figure 19.57

⑩ The resultant of two forces **P** and **Q** acting on a particle has magnitude P = |**P**|. The resultant of the two forces 2**P** and **Q** acting in the same directions as before also has magnitude P. Find the magnitude of **Q** and show that the direction of **Q** makes an angle of 150° with **P**.

3 Newton's second law in two dimensions

When the forces acting on an object are not in equilibrium the object will have an acceleration and you can use Newton's second law to solve problems about its motion.

The equation **F** = m**a** is a vector equation. The resultant force acting on a particle is equal in both magnitude and direction to the mass × acceleration. It can be written in component form as

$$\begin{pmatrix} F_1 \\ F_2 \end{pmatrix} = m \begin{pmatrix} a_1 \\ a_2 \end{pmatrix} \quad \text{or} \quad F_1\mathbf{i} + F_2\mathbf{j} = m(a_1\mathbf{i} + a_2\mathbf{j})$$

so that $F_1 = ma_1$ and $F_2 = ma_2$.

Discussion point

Figure 19.58

A child is on a sledge sliding down a smooth straight slope inclined at 15° to the horizontal.

➜ In what direction is the resultant force acting on the sledge?

Example 19.11

Sam is sledging. His sister gives him a push at the top of a smooth straight 15° slope and lets go when he is moving at $2\,\mathrm{ms^{-1}}$. He continues to slide for 5 seconds before using his feet to produce a braking force of 95 N parallel to the slope. This brings him to rest. Sam and his sledge have a combined mass of 30 kg.

How far does he travel?

➜

Solution

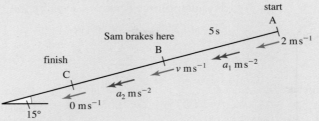

Figure 19.59

To answer this question, you need to know Sam's acceleration for the two parts of his journey. In both cases it is constant so you can use the constant acceleration formulae.

Sliding freely

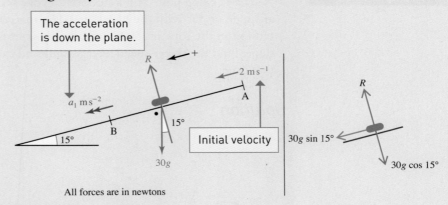

The acceleration is down the plane.

Initial velocity

All forces are in newtons

Figure 19.60

Using Newton's second law in the direction of the acceleration gives

$$30g \sin 15° = 30a_1$$
$$a_1 = 2.54$$

Resultant force down the plane = mass × acceleration.

Now you know a_1 you can find how far Sam slides (s_1 m) and his speed (v ms^{-1}) before braking.

$$s = ut + \tfrac{1}{2}at^2$$

Given $u = 2$, $t = 5$, $a = 2.54$

$$s_1 = 2 \times 5 + 0.5 \times 2.54 \times 25 = 41.75$$

$$v = u + at$$
$$v = 2 + 2.54 \times 5 = 14.7$$

Braking

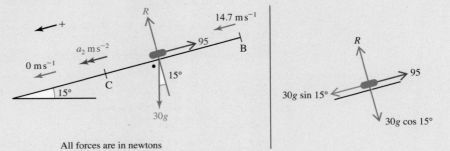

All forces are in newtons

Figure 19.61

By Newton's second law down the plane

Resultant force = mass × acceleration

$$30g \sin 15° - 95 = 30a_2$$

$$a_2 = -0.63$$

To find s_2 use $v^2 = u^2 + 2as$

$$0 = 14.7^2 - 2 \times 0.63 \times s_2 \quad\longleftarrow \boxed{\text{Given } u = 14.7,\ v = 0,\ a = -0.63}$$

$$s_2 = \frac{14.7^2}{1.26} = 171.5$$

Sam travels a distance of $41.75 + 171.5 = 213\,\text{m}$ to the nearest metre.

Discussion point

➜ Make a list of the modelling assumptions used in this example. What would be the effect of changing these?

Example 19.12

A skier is being pulled up a smooth 25° dry ski slope by a rope which makes an angle of 35° with the horizontal. The mass of the skier is 75 kg and the tension in the rope is 350 N. Initially the skier is at rest at the bottom of the slope. Find the skier's speed after 5 s and find the distance he has travelled in that time.

Solution

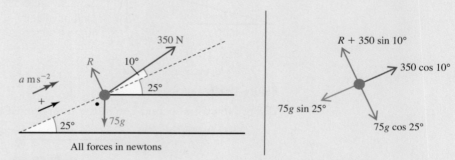

All forces in newtons

Figure 19.62

In Figure 19.62 the skier is modelled as a particle. Since the skier moves parallel to the slope consider motion in that direction.

$$\text{Resultant force} = \text{mass} \times \text{acceleration}$$

$$350 \cos 10° - 75 \times 9.8 \times \sin 25° = 75 \times a$$

$$a = \frac{34.06}{75} = 0.454 \quad (\text{to 3 d.p.})$$

This is a constant acceleration so use the constant acceleration formulae.

$$v = u + at$$

$$v = 0 + 0.454 \times 5 = 2.27 \quad\longleftarrow \boxed{u = 0,\ a = 0.454,\ t = 5}$$

The speed after 5 seconds is $2.27\,\text{m}\,\text{s}^{-1}$ (to 2 d.p.)

$$s = ut + \tfrac{1}{2}at^2$$

$$s = 0 + \tfrac{1}{2} \times 0.454 \times 25 = 5.68$$

The distance travelled is $5.68\,\text{m}$ (to 2 d.p.)

Example 19.13

A car of mass 1000 kg, including its driver, is being pushed along a horizontal road by three people, Kelly, Dean and Emma as indicated in Figure 19.63. The car is moving in the direction PQ.

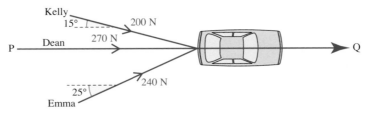

Figure 19.63

(i)   Calculate the total force exerted by the three people in the direction PQ.

(ii)   Calculate the force exerted overall by the three people in the direction perpendicular to PQ.

(iii) Explain briefly why the car does not move in the direction perpendicular to PQ.

Initially the car is stationary and 5 s later it has a speed of $2\,\text{m s}^{-1}$ in the direction PQ.

(iv)   Calculate the force of resistance to the car's movement in the direction PQ, assuming the three people continue to push as described above.

[MEI part]

Solution

(i)   Resolving in the direction PQ ($\rightarrow$) the components in newtons are:

Kelly $200\cos15° = 193$

Dean 270

Emma $240\cos25° = 218$

Total force in the direction PQ = 681 N.

(ii)   Resolving perpendicular to PQ ($\uparrow$) the components are:

Kelly $-200\sin15° = -51.8$

Dean 0

Emma $240\sin25° = 101.4$

Total force in the direction perpendicular to PQ = 49.6 N.

(iii) The car does not move perpendicular to PQ because the force in this direction is balanced by a sideways (lateral) friction force between the tyres and the road.

(iv)   The acceleration of the car is constant, so you can use the constant acceleration formula

$$v = u + at$$
$$2 = 0 + 5a \qquad \longleftarrow \boxed{u = 0,\ v = 2,\ t = 5}$$
$$a = 0.4$$

When the resistance to motion in the direction QP is R N, Figure 19.64 shows all the horizontal forces acting on the car and its acceleration.

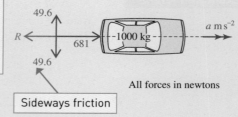

The weight of the car is in the third dimension, perpendicular to this plane and is balanced by the normal reaction of the ground.

All forces in newtons

Sideways friction

Figure 19.64

The resultant force in the direction PQ is $(681 - R)$ N. So by Newton's second law.

$$681 - R = 1000a$$

$$R = 681 - 400$$

The resistance to motion in the direction PQ is 281 N.

Exercise 19.3

Those questions in this exercise where the context is drawn from real life involve working with mathematical models.

① The forces $\mathbf{F_1} = 4\mathbf{i} - 5\mathbf{j}$ and $\mathbf{F_2} = 2\mathbf{i} + \mathbf{j}$, in newtons, act on a particle of mass 4 kg.

   (i) Find the acceleration of the particle in component form.

   (ii) Find the magnitude of the particle's acceleration.

② Two forces $\mathbf{P_1}$ and $\mathbf{P_2}$ act on a particle of mass 2 kg giving it an acceleration of $5\mathbf{i} + 5\mathbf{j}$ (in m s^{-2}).

   (i) If $\mathbf{P_1} = 6\mathbf{i} - \mathbf{j}$ (in newtons), find $\mathbf{P_2}$.

   (ii) If instead $\mathbf{P_1}$ and $\mathbf{P_2}$ both act in the same direction but $\mathbf{P_1}$ is four times as big as $\mathbf{P_2}$ find both forces.

③ Figure 19.65 shows a girl pulling a sledge at steady speed across level snow-covered ground using a rope which makes an angle of 30° to the horizontal. The mass of the sledge is 8 kg and there is a resistance force of 10 N.

Figure 19.65

   (i) Draw a diagram showing the forces acting on the sledge.

   (ii) Find the magnitude of the tension in the rope.

The girl comes to an area of ice where the resistance force on the sledge is only 2 N. She continues to pull the sledge with the same force as before and with the rope still taut at 30°.

   (iii) What acceleration must the girl have in order to do this?

   (iv) How long will it take to double her initial speed of 0.4 m s^{-1}?

④ A block of mass 12 kg is placed on a smooth plane inclined at 40° to the horizontal. It is connected by a light inextensible string, which passes over a smooth pulley at the top of the plane, to a mass of 7 kg hanging freely (Figure 19.66). Find the common acceleration and the tension in the string.

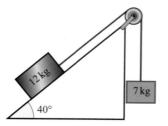

Figure 19.66

⑤ Figure 19.67 shows a situation which has arisen between two anglers, Davies and Jones, standing at the ends of adjacent jetties. Their lines have become entangled under the water with the result that they have both hooked the same fish, which has a mass 1.9 kg. Both are reeling in their lines as hard as they can in order to claim the fish.

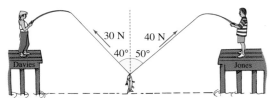

Figure 19.67

(i) Draw a diagram showing the forces acting on the fish.

(ii) Resolve the tensions in both anglers' lines into horizontal and vertical components and so find the total force acting on the fish.

(iii) Find the magnitude and direction of the acceleration of the fish.

(iv) At this point Davies' line breaks. What happens to the fish?

⑥ A crate of mass 30 kg is being pulled up a smooth slope inclined at 30° to the horizontal by a rope which is parallel to the slope. The crate has acceleration $0.75\,\text{m s}^{-2}$.

(i) Draw a diagram showing the forces acting on the crate and the direction of its acceleration.

(ii) Resolve the forces in directions parallel and perpendicular to the slope.

(iii) Find the tension in the rope.

(iv) The rope suddenly snaps. What happens to the crate?

⑦ Two toy trucks are travelling down a slope inclined at an angle of 5° to the horizontal (Figure 19.68). Truck A has a mass of 6 kg, truck B has a mass of 4 kg.

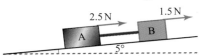

Figure 19.68

The trucks are linked by a light rigid rod which is parallel to the slope.

The resistances to motion of the trucks are 2.5 N for truck A and 1.5 N for truck B.

The initial speed of the trucks is $2\,\text{m s}^{-1}$.

Calculate the speed of the trucks after 3 seconds and also the force in the rod connecting the trucks, stating whether the rod is in tension or in thrust.

⑧ A car of mass 1000 kg is towing a trailer of mass 250 kg along a slope of 1 in 20 (i.e. at an angle θ to the horizontal and $\sin\theta = \frac{1}{20}$). The driving force of the engine is 2500 N and there are resistances to the motion of both the car, 700 N, and the trailer, 300 N.

Find the acceleration and the tension in the tow bar.

⑨ A cyclist of mass 60 kg rides a cycle of mass 7 kg. The greatest forward force that she can produce is 200 N but she is subject to air resistance and friction totalling 50 N.

(i) Draw a diagram showing the forces acting on the cyclist when she is going uphill.

(ii) What is the angle of the steepest slope that she can ascend?

The cyclist reaches a slope of 8° with a speed of $5\,\text{m s}^{-1}$ and rides as hard as she can up it.

(iii) Find her acceleration and the distance she travels in 5 s.

(iv) What is her speed now?

⑩ A builder is demolishing the chimney of a house and slides the old bricks down to the ground on a straight chute 10 m long inclined at 42° to the horizontal. Each brick has a mass of 3 kg.

(i) Draw a diagram showing the forces acting on a brick as it slides down the chute, assuming the chute to have a flat cross section and a smooth surface.

(ii) Find the acceleration of the brick.

(iii) Find the time the brick takes to reach the ground.

In fact the chute is not smooth and the brick takes 3 s to reach the ground.

(iv) Find the frictional force acting on the brick, assuming it to be constant.

⑪ A box of mass 80 kg is to be pulled along a horizontal floor by means of a light rope. The rope is pulled with a force of 100 N and the rope is inclined at 20° to the horizontal, as shown in Figure 19.69.

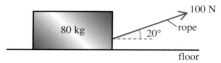

Figure 19.69

(i) Explain briefly why the box cannot be in equilibrium if the floor is smooth.

In fact the floor is not smooth and the box is in equilibrium.

(ii) Draw a diagram showing all the external forces acting on the box.

(iii) Calculate the frictional force between the box and the floor and also the normal reaction of the floor on the box, giving your answers correct to three significant figures.

The maximum value of the frictional force between the box and the floor is 120 N and the box is now pulled along the floor with the rope always inclined at 20° to the horizontal.

(iv) Calculate the force with which the rope must be pulled for the box to move at a constant speed. Give your answers correct to three significant figures.

(v) Calculate the acceleration of the box if the rope is pulled with a force of 140 N.

[MEI]

⑫ A block of mass 5 kg is at rest on a plane which is inclined at 30° to the horizontal. A light inelastic string is attached to the block, passes over a smooth pulley and supports a mass m kg which hangs freely. The part of the string between the block and the pulley is parallel to a line of greatest slope of the plane. A friction force of 15 N opposes the motion of the block. Figure 19.70 shows the block when it is slipping up the plane at a constant speed.

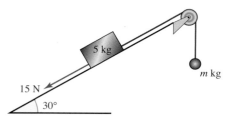

Figure 19.70

Give your answers to two significant figures.

(i) Copy the diagram and mark in all the forces acting on the block and the hanging mass, including the tension in the string.

(ii) Calculate the value of m when the block slides up the plane at a constant speed and find the tension in the string.

(iii) Calculate the acceleration of the system when $m = 6$ kg and find the tension in the string in this case. [MEI]

⑬ A sledge is found to travel with uniform speed down a slope of 1 in 50 (i.e. at an angle θ with the horizontal such that $\sin\theta = \frac{1}{50}$). If the sledge starts from the bottom of the same slope with a speed of 2 m s⁻¹, how far will it travel up the slope before coming to rest?

⑭ A train of mass 200 tonnes is travelling uniformly on level ground at 10 m s⁻¹ when it begins an ascent of 1 in 50. The driving force exerted by the engine is equal to 25 kN and the resistance force on the train is a constant 10 kN. Show that the train comes to a standstill after climbing for 413 m.

⑮ A railway truck runs down a slope of 1 in 250 and at the foot continues along level ground. Find how far it will run on the level if the speed was a constant 5 m s⁻¹ on the slope and the resistance is unchanged on the level.

LEARNING OUTCOMES

When you have completed this chapter, you should be able to:

➤ understand and use Newton's second law for motion in a straight line

➤ resolve forces in two dimensions to solve problems

➤ solve problems involving the equilibrium of a particle under coplanar forces

➤ find the resultant of several forces in one or two dimensions

➤ apply Newton's laws to a resultant force

➤ use trigonometric functions to solve problems in context, including problems involving vectors, kinematics and forces

➤ use vectors to solve problems in context, including forces and kinematics.

KEY POINTS

1 **Resolving a force**

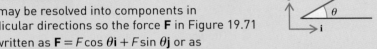

A force may be resolved into components in perpendicular directions so the force **F** in Figure 19.71 can be written as $\mathbf{F} = F\cos\theta\,\mathbf{i} + F\sin\theta\,\mathbf{j}$ or as

$$\mathbf{F} = \begin{pmatrix} F\cos\theta \\ F\sin\theta \end{pmatrix}$$

Figure 19.71

2 **The resultant force**

The forces acting on a body can be combined to form the **resultant force**. Its magnitude and direction can be found using scale drawing or by calculation using components.

3 **Scale drawing**

Draw an accurate diagram, then measure the length and direction of the resultant. This is less accurate than calculation.

4 **Calculating the resultant**

To calculate the resultant of several forces, resolve all of them in two perpendicular directions and then add the various components. Finally combine them to find the magnitude and direction of the resultant.

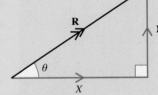

5 **Combining components**

When **R** is $X\mathbf{i} + Y\mathbf{j}$

Magnitude $|\mathbf{R}| = \sqrt{X^2 + Y^2}$

Direction $\tan\theta = \dfrac{Y}{X}$

Figure 19.73

6 **Equilibrium** When the forces are in equilibrium the resultant **R** is zero.

7 **Triangle of forces** If a body is in equilibrium under three non-parallel forces, their lines of action are concurrent and they can be represented by a triangle.

8 **Newton's second law** When the resultant **R** is not zero there is an acceleration **a** and $\mathbf{R} = m\mathbf{a}$.

9 When a particle is on a slope, it is usually helpful to resolve in directions parallel and perpendicular to the slope.

20 Moments of forces

The photo shows a swing bridge over a canal. It can be raised to allow barges and boats to pass. It is operated by hand, even though it is very heavy.

➡ How is this possible?

The bridge depends on the turning effects or **moments** of forces. To understand these you might find it helpful to start by looking at a simpler situation.

Two children sit on a simple see-saw, made of a plank balanced on a fulcrum as in Figure 20.1. Will the see-saw balance?

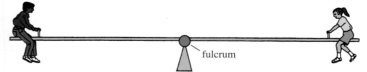
fulcrum

Figure 20.1

If both children have the same mass and sit the same distance from the fulcrum, then you expect to see the see-saw balance.

Now consider possible changes to this situation:

- If one child is heavier than the other then you expect the heavier one to go down.

- If one child moves nearer to the centre you expect that child to go up.

You can see that both the weights of the children and their distances from the fulcrum are important.

What about this case? One child has mass 35 kg and sits 1.6 m from the fulcrum and the other has mass 40 kg and sits on the opposite side 1.4 m from the fulcrum (see Figure 20.2).

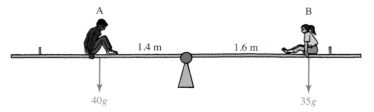
A   1.4 m   1.6 m   B

40g   35g

Figure 20.2

Taking the products of their weights and their distances from the fulcrum, gives

A: $40\,g \times 1.4 = 56\,g$

B: $35\,g \times 1.6 = 56\,g$

So you might expect the see-saw to balance and this indeed is what would happen.

1 Rigid bodies

Rigid bodies

Until now the particle model has provided a reasonable basis for the analysis of the situations you have met. In examples like the see-saw, however, where turning is important, this model is inadequate because the forces do not all act through the same point.

In such cases you need the **rigid body model**. In this model an object, or **body**, is recognised as having size and shape, but is assumed not to be deformed when forces act on it.

Suppose that you push a tray lying on a smooth table with one finger so that the force acts parallel to one edge and through the centre of mass (Figure 20.3)

Figure 20.3

The particle model is adequate here: the tray travels in a straight line in the direction of the applied force. If you push the tray equally hard with two fingers as in Figure 20.4, symmetrically either side of the centre of mass, the particle model is still adequate.

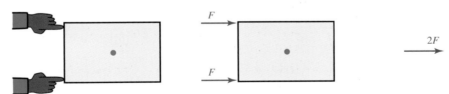

Figure 20.4

However, if the two forces are not equal or are not symmetrically placed, or as in Figure 20.5 are in different directions, the particle model cannot be used.

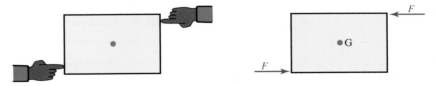

Figure 20.5

The resultant force is now zero, since the individual forces are equal in magnitude but opposite in direction. What happens to the tray? Experience tells you that it starts to rotate about G. How fast it starts to rotate depends, among other things, on the magnitude of the forces and the width of the tray. The rigid body model allows you to analyse the situation.

Moments

In the example of the see-saw you looked at the product of each force and its distance from a fixed point. This product is called the **moment** of the force about the point.

The see-saw balances because the moments of the forces on either side of the fulcrum are the same magnitude and in opposite directions. One would tend to make the see-saw turn clockwise, the other anticlockwise. By contrast, the moments about G of the forces on the tray in the last situation do not balance. They both tend to turn it anticlockwise, so rotation occurs.

Notation

The moment of a force F about a point O is defined by

$$\text{moment} = Fd$$

where d is the perpendicular distance from the point O to the line of action of the force (Figure 20.6).

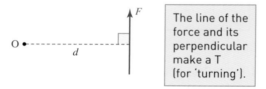

Figure 20.6

In two dimensions, the sense of a moment is described as either positive (anticlockwise) or negative (clockwise) as shown in Figure 20.7.

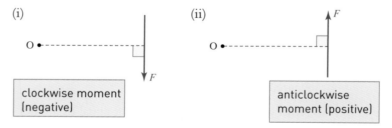

Figure 20.7

If you imagine putting a pin at O and pushing along the line of F, your page would turn clockwise for (i) and anticlockwise for (ii).

In the S.I. system the unit for moment is the newton metre (Nm), because a moment is the product of a force, the unit of which is the newton, and distance, the unit of which is the metre.

Remember that moments are always taken about a point and you must always specify what that point is. A force acting through the point will have no moment about that point because in that case $d = 0$.

> **Discussion point**
> → Figure 20.8 shows two tools for undoing wheel nuts on a car. Discuss the advantages and disadvantages of each.
>
>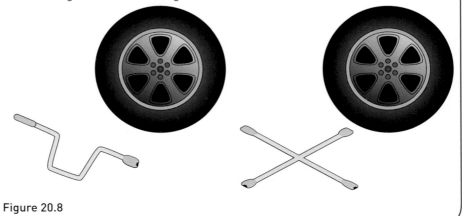
>
> **Figure 20.8**

When using the spider wrench (the tool with two 'arms'), you apply equal and opposite forces either side of the nut. These produce moments in the same direction. One advantage of this method is that there is no resultant force and hence no tendency for the nut to snap off.

Couples

Whenever two forces of the same magnitude act in opposite directions along different lines, they have a zero resultant force, but do have a turning effect. In fact the moment will be Fd about any point, where d is the perpendicular distance between the forces. This is demonstrated in Figure 20.9.

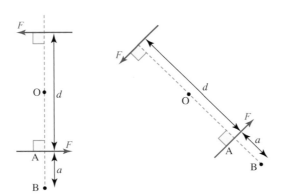

Figure 20.9

In each of these situations:

anticlockwise is positive

Moment about O:	$F\dfrac{d}{2} + F\dfrac{d}{2} = Fd$
Moment about A:	$0 + Fd = Fd$
Moment about B:	$-aF + (a + d)F = Fd$

Any set of forces like these with a zero resultant but a non-zero total moment is known as a **couple**. The effect of a couple on a rigid body is to cause rotation.

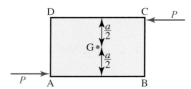

Figure 20.10

Equilibrium revisited

In Chapter 19 an object was said to be in equilibrium if the resultant force acting on it is zero. This definition is adequate provided all the forces act through the same point on the object. However, in a situation where forces act at different points, there may be an overall moment even if the resultant of the forces is zero.

Figure 20.10 shows a tray on a smooth surface being pushed equally hard at opposite corners.

The resultant force on the tray is clearly zero, but the resultant moment about its centre point, G, is

$$P\frac{a}{2} + P\frac{a}{2} = Pa.$$

The tray will start to rotate about its centre and so it is clearly not in equilibrium.

So the mathematical definition of equilibrium now needs to be tightened to include moments. For an object to remain at rest (or moving at constant velocity) when a system of forces is applied, both the resultant force and the total moment must be zero.

To check that an object is in equilibrium under the action of a system of forces, you need to check two things:

1   that the resultant force is zero;
2   that the resultant moment about any point is zero. (You only need to check one point.)

> **Note**
>
> You could have taken moments about any of the corners, A, B, C or D, or any other point in the plane of the paper and the answer would have been the same, Pa anticlockwise.

Example 20.1

Two children are playing with a door (Figure 20.11). Kerry tries to open it by pulling on the handle with a force of 50 N at right angles to the plane of the door, at a distance 0.8 m from the hinges. Peter pushes at a point 0.6 m from the hinges, also at right angles to the door and with sufficient force to just stop Kerry opening it.

(i)   What is the moment of Kerry's force about the hinges?

(ii)   With what force does Peter push?

(iii)   Describe the resultant force on the hinges.

Figure 20.11

→

Solution

Looking down from above, the line of the hinges becomes a point, H. The door opens clockwise. Anticlockwise is taken as positive.

(i)

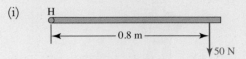

Figure 20.12

Kerry's moment about $H = -50 \times 0.8$

$$= -40 \, \text{Nm}$$

The moment of Kerry's force about the hinges is $-40 \, \text{Nm}$.

(Note that it is a clockwise moment and so negative.)

(ii)

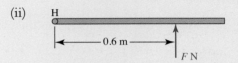

Figure 20.13

Peter's moment about $H = + F \times 0.6$

Since the door is in equilibrium the total moment on it must be zero, so

$$F \times 0.6 - 40 = 0$$

$$F = \frac{40}{0.6}$$

$$= 66.7 \text{ (to 3 s.f.)}$$

Peter pushes with a force of $66.7 \, \text{N}$.

(iii) Since the door is in equilibrium the overall resultant force on it must be zero. All the forces are at right angles to the door, as shown in Figure 20.14.

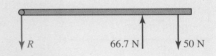

Figure 20.14

Resolving perpendicular to door:

$$R + 50 = 66.7$$

$$R = 16.7 \text{ (to 3 s.f.)}$$

The total reaction at the hinges is a force of $16.7 \, \text{N}$ in the same direction as Kerry is pulling.

Note

The reaction force at a hinge may act in any direction, according to the forces elsewhere in the system. A hinge can be visualised in cross section as shown in Figure 20.15. If the hinge is well oiled, and the friction between the inner and outer parts is negligible, the hinge cannot exert any moment. In this situation the door is said to be 'freely hinged'.

Contact may occur anywhere inside this circle.

Figure 20.15

Example 20.2

The diagram shows a man of weight 600 N standing on a footbridge that consists of a uniform wooden plank just over 2 m long of weight 200 N. Find the reaction forces exerted on each end of the plank.

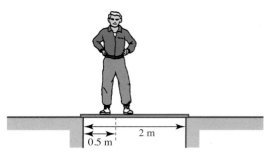

Figure 20.16

Notes

1 You cannot solve this problem without taking moments.

2 You can take moments about any point. For example, if you take moments about B you get the same answer.

3 The whole weight of the plank is being considered to act at its centre of mass.

4 When a force acts through the point about which moments are being taken, its moment about that point is zero.

Solution

Figure 20.17 shows the forces acting on the plank.

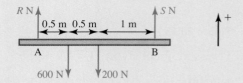

Figure 20.17

For equilibrium both the resultant force and the total moment must be zero.

Forces:

$$R + S - 800 = 0 \qquad ①$$

> All forces act vertically.

Moments:

Taking moments about the point A gives

$$R \times 0 - 600 \times 0.5 - 200 \times 1 + S \times 2 = 0 \qquad ②$$

> You could take moments about B instead, or any other point.

From equation ② $S = 250$ and so equation ① gives $R = 550$.

The reaction forces are 250 N at A and 550 N at B.

Levers

A lever can be used to lift or move a heavy object using a relatively small force. Levers depend on moments for their action.

Two common lever configurations are shown overleaf. In both cases a load W is being lifted by an applied force F, using a lever of length l. The calculations assume equilibrium.

Case 1 (2nd class levers)

The fulcrum is at one end of the lever (Figure 20.18).

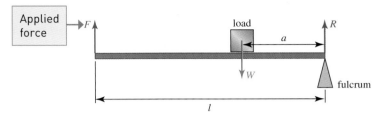

Figure 20.18

Taking moments about the fulcrum

$$F \times l - W \times a = 0$$

$$F = W \times \frac{a}{l}$$

Since a is much smaller than l, the applied force F is much smaller than the load W.

Case 2 (1st class levers)

The fulcrum is within the lever (Figure 20.19).

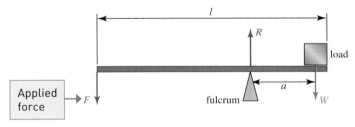

Figure 20.19

Taking moments about the fulcrum:

$$F \times (l - a) - W \times a = 0$$

$$F = W \times \frac{a}{l - a}$$

Provided that the fulcrum is nearer the end with the load, the applied force is less than the load.

These examples also indicate how to find a single force equivalent to two parallel forces. The force equivalent to F and W should be equal and opposite to R and with the same line of action.

Discussion points

→ How do you use moments to open a screw-top jar?

→ Why is it an advantage to press hard when it is stiff?

Discussion points

→ Describe the single force equivalent to P and Q in each of these cases.

→ In each case state its magnitude and line of action.

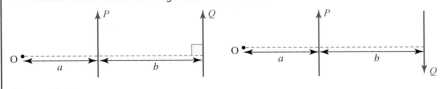

Figure 20.20

Exercise 20.1

Those questions in this exercise where the context is drawn from real life involve working with mathematical models.

① In each of the situations shown below, find the moment of the force about the point and state whether it is positive (anticlockwise) or negative (clockwise).

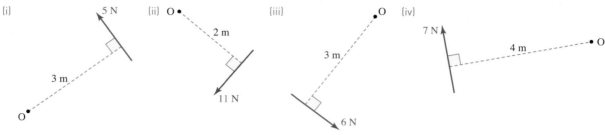

Figure 20.21

② The situations below involve several forces acting on each object. For each one, find the total moment.

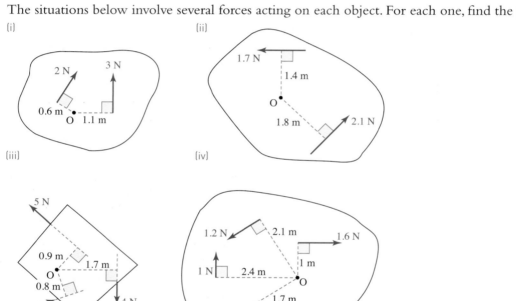

Figure 20.22

③ A uniform horizontal bar of mass 5 kg has length 30 cm and rests on two vertical supports, 10 cm and 22 cm from its left-hand end. Find the magnitude of the reaction force at each of the supports.

④ Figure 20.23 shows a motorcycle of mass 250 kg, and its rider whose mass is 80 kg. The centre of mass of the motorcycle lies on a vertical line midway between its wheels. When the rider is on the motorcycle, his centre of mass is 1 m behind the front wheel.

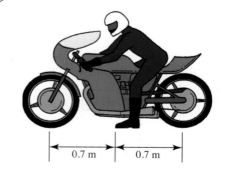

Figure 20.23

Find the vertical reaction forces acting through the front and rear wheels when

(i) the rider is not on the motorcycle

(ii) the rider is on the motorcycle.

⑤ Find the reaction forces on the hi-fi shelf shown below. The shelf itself has weight 25 N and its centre of mass is midway between A and D.

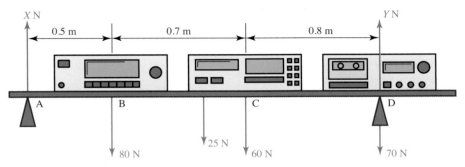

Figure 20.24

⑥ Kara and Jane are trying to find the positions of their centres of mass. They place a uniform board of mass 8 kg symmetrically on two bathroom scales whose centres are 2 m apart. When Kara lies flat on the board, Jane notes that scale A reads 37 kg and scale B reads 26 kg.

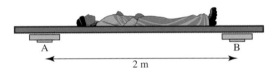

Figure 20.25

(i) Draw a diagram showing the forces acting on Kara and the board and calculate Kara's mass.

(ii) How far from the centre of scale A is her centre of mass?

⑦ The diagram shows two people, an adult and a child, sitting on a uniform bench of mass 40 kg; their positions are as shown. The mass of the child is 50 kg; that of the adult is 85 kg.

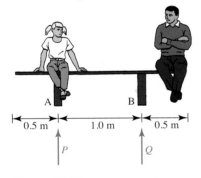

Figure 20.26

(i) Find the reaction forces, P and Q (in N), from the ground on the two supports of the bench.

(ii) The child now moves to the mid-point of the bench. What are the new values of P and Q?

(iii) Is it possible for the child to move to a position where $P = 0$? What is the significance of a zero value for P?

(iv) What happens if the child leaves the bench?

⑧ Figure 20.27 shows a diving board which some children have made. It consists of a uniform plank of mass 20 kg and length 3 m, with 1 m of its length projecting out over a pool. They have put a boulder of mass 25 kg on the end over the land; and there is a support at the water's edge.

(i) Find the forces at the two supports when nobody is using the diving board.

(ii) A child of mass 50 kg is standing on the end of the diving board over the pool. What are the forces at the two supports?

(iii) Some older children arrive and take over the diving board. One of these is a heavy boy of mass 90 kg. What is the reaction at A if the board begins to tip over?

(iv) How far can the boy walk from B before the board tips over?

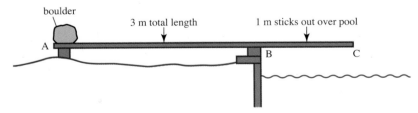

Figure 20.27

⑨ A lorry of mass 5000 kg is driven across a bridge of mass 20 tonnes. The bridge is a roadway of length 10 m which is supported at both ends.

(i) Find expressions for the reaction forces at each end of the bridge in terms of the distance x in metres travelled by the lorry from the start of the bridge.

(ii) From what point of the lorry is the distance x measured?

Two identical lorries cross the bridge at the same speed, starting at the same instant, from opposite directions.

(iii) How do the reaction forces of the supports on the bridge vary as the lorries cross the bridge?

⑩ A non-uniform rod AB of length 20 cm rests horizontally on two supports at C and D, where AC = BD = 4 cm. The greatest mass that can be hung from A without disturbing equilibrium is 8 g, and the greatest mass that can be hung from B is 10 g. Find the mass of the rod and the distance of its centre of mass from A.

⑪ A uniform plank of mass 80 kg is 12 m long and 5 m of it project over the side of a quay. What is the minimum load that must be placed on the end of the plank so that a woman of mass 45 kg can walk to the other end of the plank without tipping into the water?

⑫ A simple suspension bridge across a narrow river consists of a uniform beam, 4 m long and of mass 60 kg, supported by vertical cables attached at a distance 0.75 m from each end of the beam.

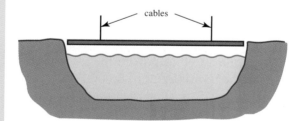

Figure 20.28

(i) Find the tension in each cable when a boy of mass 50 kg stands 1 m from the end of the bridge.

(ii) Can a couple walking hand-in-hand cross the bridge safely, without it tipping, if their combined mass is 115 kg?

(iii) What is the mass of a person standing on the end of the bridge when the tension in one cable is four times that in the other cable?

⑬ Find the magnitude, direction and line of action of the resultant of this system of forces.

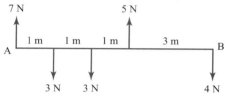

Figure 20.29

⑭ Figure 20.30 shows a stone slab AB of mass 1 tonne resting on two supports, C and D. The stone is uniform and has length 3 m. The supports are at distances 1.2 m and 0.5 m from the end.

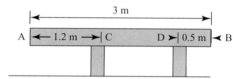

Figure 20.30

(i) Find the reaction forces at the two supports.

Local residents are worried that the arrangement is unsafe since their children play on the stone.

(ii) How many children, each of mass 50 kg would need to stand at A in order to tip the stone over?

The stone's owner decides to move the support at C to a point nearer to A. To take the weight of the slab while doing this, he sets up the lever system shown in the diagram. The distance XF is 1.25 m and FY is 0.25 m.

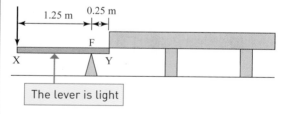

Figure 20.31

(iii) What downward force applied at X would reduce the reaction force at C to zero (and so allow the support to be moved)?

⑮ Four sailors are using a light capstan to pull in their ship's anchor cable at a steady rate. One of them is shown in the diagram. The diameter of the capstan's drum is 1 m and the spokes on which the sailors are pushing each project horizontally from the centre of the capstan. Each sailor is pushing with a force of 300 N, horizontally and at right angles to his spoke at a distance of 1.5 m from the centre of the capstan. The anchor cable is taut; it passes over a frictionless pulley and then makes an angle of 20° with the horizontal.

Note: Only one of the four sailors is shown.

Figure 20.32

(i) Find the tension in the cable.
The mass of the ship is 2000 tonnes.

(ii) Find the acceleration of the ship, assuming that no other horizontal forces act on the ship.

In fact the acceleration of the ship is 0.0015 m s⁻². Part of the difference can be explained by friction with the capstan, resulting in a resisting moment of 100 Nm, the rest by the force of resistance, R N, to the ship passing through the water.

(iii) Find the value of R.

⑯

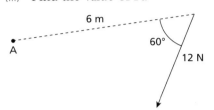

Laura says the moment of the force about A is 31.2 N (to 3 s.f.),

Meena says it is 36 N m,

Nat says it is 72 N m,

Olwyn says it is $18\sqrt{3}$ N m.

Say which one of them is correct and identify the mistakes that the others probably made.

LEARNING OUTCOMES

Now you have finished this chapter, you should be able to

➤ understand and use moments in simple static contexts

➤ know the meaning of the term couple

➤ know the conditions for equilibrium of a rigid body

➤ solve problems involving equilibrium of a rigid body.

KEY POINTS

1 The moment of a force F about a point O is given by the product Fd where d is the perpendicular distance from O to the line of action of the force.

Figure 20.33

2 The S.I. unit for moment is the newton metre (Nm).

3 Anticlockwise moments are usually called positive, clockwise negative.

4 A body is in equilibrium under the action of a system of forces if
  (i) the resultant force is zero
  (ii) the resultant moment about any point is zero.

5 A set of forces with zero resultant but non-zero moment is called a couple.

21 Projectiles

The water in a water jet follows its own path which is called its *trajectory*. You can see the same sort of trajectory if you throw a small object across a room. Its path is a parabola. Objects moving through the air like this are called projectiles.

Modelling assumptions for projectile motion

The path of a cricket ball looks parabolic, but what about a boomerang? There are modelling assumptions which must be satisfied for the motion of a projectile to be parabolic. These are

- the projectile is a particle
- it is not powered
- the air has no effect on its motion
- it has no spin.

1 Equations for projectile motion

A projectile moves in two dimensions under the action of only one force, the force of gravity, which is constant and acts vertically downwards. This means that the acceleration of the projectile is $g\,\mathrm{m\,s^{-2}}$ vertically downwards and there is no horizontal acceleration. You can treat the horizontal and vertical motion separately using the equations for constant acceleration.

To illustrate the ideas involved, think of a ball being projected with a speed of $20\,\mathrm{m\,s^{-1}}$ at $60°$ to the ground as illustrated in Figure 21.1. This could be a first model for a football, a chip shot from the rough at golf or a lofted shot at cricket.

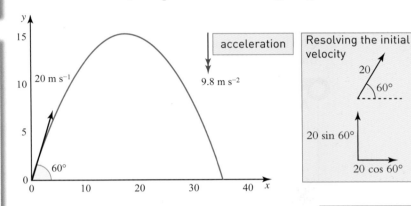

Figure 21.1

This is negative because the positive y-axis is upwards.

Using axes as shown, the components are:

	Horizontal	Vertical
Initial position	0	0
Acceleration	$a_x = 0$	$a_y = -9.8$
Initial velocity	$u_x = 20\cos 60°$	$u_y = 20\sin 60°$
	$= 10$	$= 17.32$

as vectors

$$\mathbf{a} = \begin{pmatrix} 0 \\ -9.8 \end{pmatrix}$$

$$\mathbf{u} = \begin{pmatrix} 20\cos 60° \\ 20\sin 60° \end{pmatrix}$$

Using $v = u + at$ in the two directions gives the components of the velocity.

$a_x = 0 \Rightarrow$ v_x is constant

$$\mathbf{v} = \begin{pmatrix} 10 \\ 17.32 - 9.8t \end{pmatrix}$$

Velocity	Horizontal	Vertical
	$v_x = 20\cos 60°$	$v_y = 20\sin 60° - 9.8t$
	$v_x = 10$ ①	$v_y = 17.32 - 9.8t$ ②

Using $s = ut + \frac{1}{2}at^2$ in both directions gives the components of position.

$$\mathbf{r} = \begin{pmatrix} 10t \\ 17.32t - 4.9t^2 \end{pmatrix}$$

Position	Horizontal	Vertical
	$x = (20\cos 60°)t$	$y = (20\sin 60°)t - 4.9t^2$
	$x = 10t$ ③	$y = 17.32t - 4.9t^2$ ④

These results are summarised in Table 21.1. It is assumed that the projectile has started at the origin.

Table 21.1

	Horizontal motion		Vertical motion	
a	$a_x = 0$		$a_y = -9.8$	
u	$u_x = 20 \cos 60° = 10$		$u_y = 20 \sin 60° = 17.32$	
v	$v_x = 10$	①	$v_y = 17.32 - 9.8t$	②
r	$x = 10t$	③	$y = 17.32t - 4.9t^2$	④

The four equations ①, ②, ③ and ④ for velocity and position can be used to find several things about the motion of the ball.

ACTIVITY 21.1

(i) What, if anything, can you say about the values of the variables x, y, v_x and v_y in Table 21.1 when the projectile is at

    (a) the top-most point of the path of the ball?

    (b) the point where it is just about to hit the ground?

(ii) What about the other four quantities in the table, a_x, a_y, u_x and u_y?

When you have decided the answer to these questions you have sufficient information to find the greatest height reached by the ball, the time of flight and the range (the total distance travelled horizontally before it hits the ground).

The maximum height

When the ball is at its maximum height, the *vertical* component of its velocity is zero. It still has a horizontal component of $10\,\text{m}\,\text{s}^{-1}$ which is constant. Equation ② from Table 21.1 above gives the vertical component as

$$v_y = 17.32 - 9.8t$$

At the top:
$$0 = 17.32 - 9.8t$$

$$t = \frac{17.32}{9.8}$$

$$= 1.767$$

To find the maximum height, you now need to find y at this time. Substituting for t in equation ④,

$$y = 17.32t - 4.9t^2$$

$$y = 17.32 \times 1.767 - 4.9 \times 1.767^2$$

$$= 15.3$$

The maximum height is $15.3\,\text{m}$.

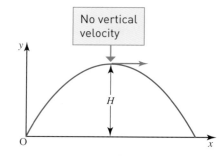

Figure 21.2

The time of flight

The flight ends when the ball returns to the ground, that is when $y = 0$.

Substituting $y = 0$ in equation ④,

$$y = 17.32t - 4.9t^2$$

$$0 = 17.32t - 4.9t^2$$

$$0 = t(17.32 - 4.9t)$$

$$t = 0 \text{ or } t = 3.53$$

Clearly $t = 0$ is the time when the ball is projected, so $t = 3.53$ is the time when it lands and the time of flight is $3.53\,\text{s}$.

The range

The range, $R\,\text{m}$, of the ball is the horizontal distance it travels before landing.

R is the value of x when $y = 0$.

R can be found by substituting $t = 3.53$ in equation ③: $x = 10t$.

The range is $\quad 10 \times 3.53 = 35.3\,\text{m}$

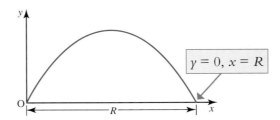

$$y = 0, \ x = R$$

Figure 21.3

Discussion point

→ Explain why, according to the standard projectile model, it is the vertical motion that determines the time of flight across a horizontal surface.

Discussion points

1 Notice in this example that the time to maximum height is half the flight time. Is this always the case?
2 Decide which of the following have flights that could be modelled as projectiles.

| a balloon | a bird | a bullet shot from a gun | a glider |
| a golf ball | a parachutist | a rocket | a tennis ball |

→ Which modelling assumptions do not apply in the particular cases?

Exercise 21.1

Those questions in this exercise where the context is drawn from real life involve working with mathematical models.

In this exercise take upwards to be positive and use $9.8\,\text{m}\,\text{s}^{-2}$ for g unless otherwise stated. All the projectiles start at the origin.

 ① In each case you are given the initial velocity of a projectile. For each part

    (a) draw a diagram showing the initial velocity and the path

(b) write down the horizontal and vertical components of the initial velocity

(c) write down equations for the velocity after time t seconds

(d) write down equations for the position after time t seconds.

(i) $10\,\text{m s}^{-1}$ at $35°$ above the horizontal

(ii) $2\,\text{m s}^{-1}$ horizontal, $5\,\text{m s}^{-1}$ vertically

(iii) $4\,\text{m s}^{-1}$ horizontally

(iv) $10\,\text{m s}^{-1}$ at $13°$ below the horizontal

(v) $U\,\text{m s}^{-1}$ at angle α to the horizontal

(vi) $u_x\,\text{m s}^{-1}$ horizontally, $u_y\,\text{m s}^{-1}$ vertically.

② In each case find
   (a) the time taken for the projectile to reach its highest point and
   (b) the maximum height.

(i) Initial velocity $5\,\text{m s}^{-1}$ horizontally and $14.7\,\text{m s}^{-1}$ vertically.

(ii) Initial velocity $10\,\text{m s}^{-1}$ at $30°$ above the horizontal.

③ In each case find
   (a) the time of flight of the projectile
   (b) the horizontal range.

(i) Initial velocity $20\,\text{m s}^{-1}$ horizontally and $19.6\,\text{m s}^{-1}$ vertically.

(ii) Initial velocity $5\,\text{m s}^{-1}$ at $60°$ above the horizontal.

④ A ball is projected from ground level with initial velocity $\mathbf{u} = \begin{pmatrix} u_x \\ u_y \end{pmatrix}$. Find

(i) the maximum height

(ii) the time of flight

(iii) the range.

2 Projectile problems

Representing projectile motion by vectors

Figure 21.4 shows a possible path for a marble which is thrown across a room from the moment it leaves the hand until just before it hits the floor.

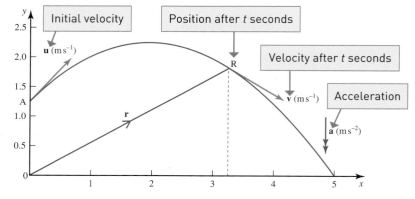

Figure 21.4

> **!** Notice that Figure 21.4 shows the trajectory of the marble. It is its path through space, not a position–time graph.

The vector $\mathbf{r} = \overrightarrow{OR}$ is the position vector of the marble after a time t seconds and the vector $\mathbf{v}$ represents the velocity in m s^{-1} at that instant of time.

You can use equations for constant acceleration in vector form to describe the motion.

Velocity $\qquad \mathbf{v} = \mathbf{u} + \mathbf{a}t$

Displacement $\qquad \mathbf{r} - \mathbf{r}_0 = \mathbf{u}t + \frac{1}{2}\mathbf{a}t^2 \qquad$ so $\qquad \mathbf{r} = \mathbf{r}_0 + \mathbf{u}t + \frac{1}{2}\mathbf{a}t^2$

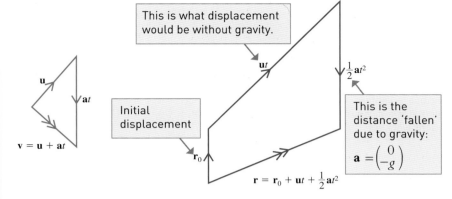

Figure 21.5

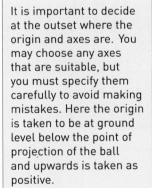

> Always check whether or not the projectile starts at the origin. The change in position is the vector $r - r_0$. This is the equivalent of $s - s_0$ in one dimension.

When working with projectile problems, you can treat each direction separately or you can write them both together as vectors. The next example shows both methods.

Example 21.1

A ball is thrown horizontally at $5\,\text{m s}^{-1}$ out of a window $4\,\text{m}$ above the ground.

(i)   How long does it take to reach the ground?

(ii)   How far from the building does it land?

(iii)   What is its speed just before it lands and at what angle to the ground is it moving?

Solution

Figure 21.6 shows the path of the ball.

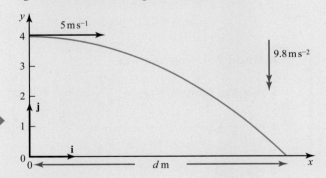

Figure 21.6

It is important to decide at the outset where the origin and axes are. You may choose any axes that are suitable, but you must specify them carefully to avoid making mistakes. Here the origin is taken to be at ground level below the point of projection of the ball and upwards is taken as positive.

s_0 is the initial position so that the displacement at time t is $s - s_0$.

Method 1: Resolving into components

Position: Using axes as shown and $s = s_0 + ut + \frac{1}{2}at^2$ in each of the two directions

Horizontally $x_0 = 0$, $u_x = 5$, $a_x = 0$

$$x = 5t \qquad \qquad \text{①}$$

Vertically $\quad y_0 = 4$, $u_y = 0$, $a_y = -9.8$

$$y = 4 - 4.9t^2 \qquad \qquad \text{②}$$

(i) The ball reaches the ground when $y = 0$. Substituting in equation ② gives

$$0 = 4 - 4.9t^2$$

$$t^2 = \frac{4}{4.9}$$

$$t = 0.904\ldots = 0.904 \text{ (to 3 s.f.)}$$

The ball hits the ground after $0.904\,$s.

> 0.904... means 0.904 and subsequent figures. This number is kept in your calculator for use in subsequent working.

(ii) When the ball lands $x = d$ so, from equation ①,

$$d = 5t = 5 \times 0.904\ldots = 4.52 \text{ (to 3 s.f.)}$$

The ball lands $4.52\,$m from the building.

(iii) *Velocity:* Using $v = u + at$ in the two directions

Horizontally $\quad v_x = 5 + 0$

Vertically $\quad\quad v_y = 0 - 9.8t$

To find the speed and direction just before it lands:

The ball lands when $t = 0.904$ so
$v_x = 5$ and $v_y = -9.8 \times 0.904\ldots = -8.886\ldots$

The components of velocity are shown in Figure 21.7. The speed of the ball is

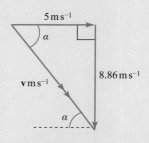

$$\sqrt{5^2 + 8.86\ldots^2} = 10.2 \text{ ms}^{-1} \text{ (to 3 s.f.)}.$$

It hits the ground moving downwards at an angle α to the horizontal where

$$\tan\alpha = \frac{8.86\ldots}{5}$$

$$\alpha = 60.6° \text{ (to 3 s.f.)}$$

Figure 21.7

Method 2: Using vectors

The initial position is $\mathbf{r}_0 = \begin{pmatrix} 0 \\ 4 \end{pmatrix}$ and the ball hits the ground when

$\mathbf{r} = \begin{pmatrix} d \\ 0 \end{pmatrix}$. The initial velocity is $\mathbf{u} = \begin{pmatrix} 5 \\ 0 \end{pmatrix}$ and the acceleration is

$\mathbf{a} = \begin{pmatrix} 0 \\ -9.8 \end{pmatrix}$.

> **Note**
>
> These vectors could just as well have been written in terms of unit vectors $\mathbf{i}$ and $\mathbf{j}$ along the x and y directions.

Using $\quad \mathbf{r} = \mathbf{r}_0 + \mathbf{u}t + \frac{1}{2}\mathbf{a}t^2$

$$\begin{pmatrix} d \\ 0 \end{pmatrix} = \begin{pmatrix} 0 \\ 4 \end{pmatrix} + \begin{pmatrix} 5 \\ 0 \end{pmatrix}t + \frac{1}{2}\begin{pmatrix} 0 \\ -9.8 \end{pmatrix}t^2$$

$$d = 5t \qquad \qquad \qquad \text{①}$$

and $\qquad 0 = 4 - 4.9t^2 \qquad \qquad \qquad \text{②}$

(i)   Equation ② gives $t = 0.904\ldots$ and substituting this into ① gives

(ii) $d = 4.52$ (to 3 s.f.).

(iii)   The speed and direction of motion are the magnitude and direction of the velocity of the ball. Using

$$\mathbf{v} = \mathbf{u} + \mathbf{a}t$$

$$\begin{pmatrix} v_x \\ v_y \end{pmatrix} = \begin{pmatrix} 5 \\ 0 \end{pmatrix} + \begin{pmatrix} 0 \\ -9.8 \end{pmatrix}t$$

So when $t = 0.904\ldots,$ $\qquad \begin{pmatrix} v_x \\ v_y \end{pmatrix} = \begin{pmatrix} 5 \\ -8.86\ldots \end{pmatrix}$

You can find the speed and angle as before.

Discussion point

Explain why the direction of a particle is determined by its velocity and not by its position vector.

Notice that in both methods the time forms a link between the motions in the two directions. You can often find the time from one equation and then substitute it in another to find out more information.

Exercise 21.2

In this exercise take upwards as positive and use $9.8\,\text{m}^{-2}$ for g in numerical questions.

① In each case

    (a)   draw a diagram showing the initial velocity and path

    (b)   write the velocity after time t s in vector form

    (c)   write the position after time t s in vector form.

    (i)   Initial position $(0, 10\,\text{m})$; initial velocity $4\,\text{m}\,\text{s}^{-1}$ horizontally.

(ii)   Initial position $(0, 7\,\text{m})$; initial velocity $10\,\text{m}\,\text{s}^{-1}$ at 35° above the horizontal.

(iii)   Initial position $(0, 20\,\text{m})$; initial velocity $10\,\text{m}\,\text{s}^{-1}$ at 13° below the horizontal.

(iv)   Initial position O; initial velocity $\begin{pmatrix} 7 \\ 24 \end{pmatrix}\text{m}\,\text{s}^{-1}$.

(v)   Initial position $(a, b)\,\text{m}$; initial velocity $\begin{pmatrix} u_0 \\ v_0 \end{pmatrix}\text{m}\,\text{s}^{-1}$.

② In each case find
  (a) the time taken for the projectile to reach its highest point
  (b) the maximum height above the origin.
  (i) Initial position (0, 15 m) velocity $5\,\text{m s}^{-1}$ horizontally and $14.7\,\text{m s}^{-1}$ vertically.
  (ii) Initial position (0, 10 m); initial velocity $\begin{pmatrix} 5 \\ 3 \end{pmatrix}\text{m s}^{-1}$.

③ Find the range for these projectiles which start from the origin.
  (i) Initial velocity $\begin{pmatrix} 2 \\ 7 \end{pmatrix}\text{m s}^{-1}$.
  (ii) Initial velocity $\begin{pmatrix} 7 \\ 2 \end{pmatrix}\text{m s}^{-1}$.
  (iii) Sketch the path of these two projectiles using the same axes.

④ A projectile starts at $\begin{pmatrix} 0 \\ h \end{pmatrix}$ and is projected with initial speed u at an angle θ with the horizontal under the acceleration $\begin{pmatrix} 0 \\ -g \end{pmatrix}$.
  Find the time taken for the projectile to hit the ground and its horizontal range.

3 Further examples

Example 21.2

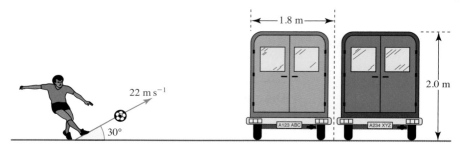

Figure 21.8

In this question use $10\,\text{m s}^{-2}$ for g and neglect air resistance.

In an attempt to raise money for a charity, participants are sponsored to kick a ball over some vans. The vans are each 2 m high and 1.8 m wide and stand on horizontal ground. One participant kicks the ball at an initial speed of $22\,\text{m s}^{-1}$ inclined at 30° to the horizontal.

(i)   What are the initial values of the vertical and horizontal components of velocity?

(ii) Show that while in flight the height y metres at time t seconds satisfies the equation $y = 11t - 5t^2$.

Calculate at what times the ball is at least 2 m above the ground.

The ball should pass over as many vans as possible.

(iii) Deduce that the ball should be placed about 3.8 m from the first van and find how many vans the ball will clear.

(iv) What is the greatest vertical distance between the ball and the top of the vans?
[MEI]

→

Solution

(i) Initial velocity

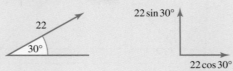

Figure 21.9

Horizontal component: $22\cos 30° = 19.05...\,\mathrm{m\,s^{-1}}$

Vertical component: $22\sin 30° = 11\,\mathrm{m\,s^{-1}}$

(ii) When the ball is above 2 m

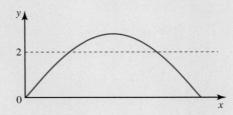

Figure 21.10

Using axes as shown and

$$s = ut + \tfrac{1}{2}at^2 \quad \text{vertically}$$

$a = -10\,\mathrm{m\,s^{-2}}$ because the positive direction is upwards.

$$\Rightarrow y = 11t - 5t^2$$

The ball is 2 m above the ground when $y = 2$, so

$$2 = 11t - 5t^2$$
$$5t^2 - 11t + 2 = 0$$
$$(5t - 1)(t - 2) = 0$$
$$t = 0.2 \quad \text{or} \quad 2$$

The ball is at least 2 m above the ground when $0.2 \leqslant t \leqslant 2$.

(iii) How many vans?

Horizontally, $s = ut + \tfrac{1}{2}at^2$ with $a = 0$

$$\Rightarrow x = 19.05...t$$

When $t = 0.2$, $x = 3.81...$ (at A)

When $t = 2$, $\quad x = 38.1...$ (at B)

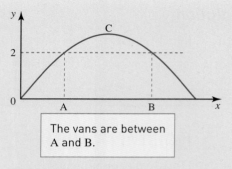

The vans are between A and B.

Figure 21.11

To clear as many vans as possible, the ball should be placed about 3.8 m in front of the first van. The distance between the first and last van cleared is AB

$$AB = 38.1\ldots -3.81\ldots = 34.29\ldots \text{ m}$$

The number of vans is

$$\frac{34.29\ldots}{1.8} = 19.05\ldots$$

The maximum possible number of vans is 19.

(iv) Maximum height

At the top (C), vertical velocity = 0, so using $v = u + at$ vertically

$$\Rightarrow \quad 0 = 11 - 10t$$

$$t = 1.1$$

Substituting in $y = 11t - 5t^2$, the maximum height is found to be

$$11 \times 1.1 - 5 \times 1.1^2 = 6.05 \text{ m}$$

The greatest vertical distance between the ball and the top of the vans is

$$6.05 - 2 = 4.05 \text{ m}.$$

Example 21.3

Sharon is diving into a swimming pool. During her dive she may be modelled as a particle. Her initial velocity is 1.8 m s^{-1} at an angle of 30° above the horizontal and initial position 3.1 m above the water. Air resistance may be neglected.

(i) Find the greatest height above the water that Sharon reaches during her dive.

(ii) Show that the time t, in seconds, that it takes Sharon to reach the water is given by $4.9t^2 - 0.9t - 3.1 = 0$ and solve this equation to find t. Explain the significance of the other root to the equation.

Just as Sharon is diving a small boy jumps into the swimming pool. He hits the water at a point in line with the diving board and 1.5 m from its end.

(iii) Is there an accident?

→

Solution

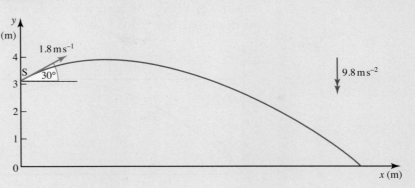

Figure 21.12

Referring to the axes shown

	Horizontal motion	Vertical motion
Initial position	0	3.1
a	0	−9.8
u	$u_x = 1.8\cos 30° = 1.56\ldots$	$u_y = 1.8\sin 30° = 0.9$
v	$v_x = 1.56\ldots$ ①	$v_y = 0.9 - 9.8t$ ②
r	$x = 1.56\ldots t$ ③	$y = 3.1 + 0.9t - 4.9t^2$ ④

(i) At the top $v_y = 0$ $\quad 0 = 0.9 - 9.8t$ $\qquad$ from ②

$$t = \frac{0.9}{9.8} = 0.0918\ldots$$

When $t = 0.0918\ldots$ $y = 3.1 + 0.9 \times 0.0918\ldots - 4.9 \times 0.0918\ldots^2$

$$= 3.14 \text{ (to 2 d.p.)}$$

Sharon's greatest height above the water is 3.14 m.

(ii) Sharon reaches the water when $y = 0$

$$0 = 3.1 + 0.9t - 4.9t^2 \qquad \text{from } ④$$

$$4.9t^2 - 0.9t - 3.1 = 0$$

$$t = \frac{0.9 \pm \sqrt{0.9^2 + 4 \times 4.9 \times 3.1}}{9.8}$$

$$t = -0.71 \text{ or } 0.89 \text{ (to 2 d.p.)}$$

Sharon hits the water after 0.89 s. The negative value of t gives the point on the parabola at water level to the left of the point where Sharon dives.

(iii) At time t the horizontal distance from the diving board is

$$x = 1.56\ldots t \qquad\qquad \text{from } ③$$

When Sharon hits the water $x = 1.56\ldots \times 0.89\ldots = 1.39$ (to 2 d.p.)

Assuming that the particle representing Sharon and the boy are located at their centres of mass, the difference of 11 cm between 1.39 m and 1.5 m is not sufficient to prevent an accident.

> **Note**
>
> When the point at which Sharon dives is taken as the origin in Example 21.3, the initial position is $(0, 0)$ and
> $y = 0.9t - 4.9t^2$. In this case Sharon hits the water when $y = -3.1$. This gives the same equation for t.

Example 21.4

A boy kicks a small ball from the floor of a gymnasium with an initial velocity of $12\,\mathrm{m\,s^{-1}}$ inclined at an angle α to the horizontal. Air resistance may be neglected. Use $10\,\mathrm{m\,s^{-2}}$ for g.

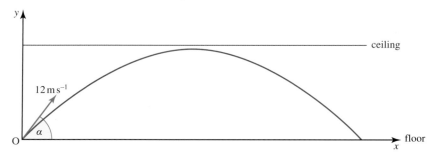

Figure 21.13

(i)   Write down expressions in terms of α for the vertical speed of the ball and the height of the ball after t seconds.

The ball just fails to touch the ceiling which is $4\,\mathrm{m}$ high. The highest point of the motion of the ball is reached after T seconds.

(ii) Use one of your expressions to show that $6\sin\alpha = 5T$ and the other to form a second equation involving $\sin\alpha$ and T.

(iii) Eliminate $\sin\alpha$ from your two equations to show that T has a value of about 0.89.

(iv) Find the range of the ball when kicked at $12\,\mathrm{m\,s^{-1}}$ from the floor of the gymnasium so that it just misses the ceiling.

[MEI adapted]

Solution

(i)   Vertical components

      speed $v_y = 12\sin\alpha - 10t$ ①

      height $y = (12\sin\alpha)t - 5t^2$ ②

(ii) Time to highest point

      At the top $v_y = 0$ and $t = T$, so equation ① gives

      $$12\sin\alpha - 10T = 0$$

      $$12\sin\alpha = 10T$$

      $$6\sin\alpha = 5T \quad ③$$

      When $t = T$, $y = 4$ so from ②

      $$4 = 12\sin\alpha T - 5T^2 \quad ④$$

(iii) Substituting for $6\sin\alpha$ from ③ into ④ gives

      $$4 = 2 \times 5T \times T - 5T^2$$

      $$4 = 5T^2$$

      $$T = \sqrt{0.8} = 0.894\ldots = 0.90 \text{ (to 2 d.p.)}$$

➡

Discussion point

Two marbles start simultaneously from the same height. One (P) is dropped and the other (Q) is projected horizontally.

➜ Which reaches the ground first?

(iv)   Range

The path is symmetrical so the time of flight is $2T$ seconds.

Horizontally $a = 0$ and $u_x = 12\cos\alpha$

$$\Rightarrow x = (12\cos\alpha)t$$

The range is $12\cos\alpha \times 2T = 21.466\ldots\cos\alpha$

From ③ $6\sin\alpha = 5T = 4.472\ldots$

$$\alpha = 48.189\ldots$$

The range is $21.466 \times \cos 48.189\ldots° = 14.310 = 14$ m (to 2 s.f.)

Exercise 21.3

Those questions in this exercise where the context is drawn from real life involve working with mathematical models.

Use $9.8\,\mathrm{m\,s^{-2}}$ for g in this exercise unless otherwise specified.

① A ball is thrown from a point at ground level with velocity $20\,\mathrm{m\,s^{-1}}$ at $30°$ to the horizontal. The ground is level and horizontal and you should ignore air resistance.

(i)   Find the horizontal and vertical components of the ball's initial velocity.

(ii) Find the horizontal and vertical components of the ball's acceleration.

(iii) Find the horizontal distance travelled by the ball before its first bounce.

(iv) Find how long the ball takes to reach maximum height.

(v)   Find the maximum height reached by the ball.

② A golf ball is hit with a velocity of $45\,\mathrm{m\,s^{-1}}$ at an elevation of $30°$, along a level fairway. Find

(i)   the greatest height reached

(ii) the time of flight

(iii) the distance travelled along the fairway.

③ Nick hits a golf ball with initial speed $50\,\mathrm{m\,s^{-1}}$ at $35°$ to the horizontal.

(i)   Find the horizontal and vertical components of the ball's initial velocity.

(ii) Specify suitable axes and calculate the position of the ball at one second intervals for the first six seconds of its flight.

(iii) Draw a graph of the path of the ball (its trajectory) and use it to estimate

(a)   the maximum height of the ball

(b)   the horizontal distance the ball travels before bouncing.

(iv) Calculate the maximum height the ball reaches and the horizontal distance it travels before bouncing. Compare your answers with the estimates you found from your graph.

(v)   State the modelling assumptions you made in answering this question.

④ Clare scoops a hockey ball off the ground, giving it an initial velocity of $19\,\mathrm{m\,s^{-1}}$ at $25°$ to the horizontal.

(i)   Find the horizontal and vertical components of the ball's initial velocity.

(ii) Find the time that elapses before the ball hits the ground.

(iii) Find the horizontal distance the ball travels before hitting the ground.

(iv) Find how long it takes for the ball to reach maximum height.

(v)   Find the maximum height reached.

A member of the opposing team is standing $20\,\mathrm{m}$ away from Clare in the direction of the ball's flight.

(vi) How high is the ball when it passes her? Can she stop the ball?

⑤ A footballer is standing $30\,\mathrm{m}$ in front of the goal. He kicks the ball towards the goal with velocity $18\,\mathrm{m\,s^{-1}}$ and angle $55°$ to the horizontal. The height of the goal's crossbar is $2.5\,\mathrm{m}$. Air resistance and spin may be neglected.

(i)   Find the horizontal and vertical components of the ball's initial velocity.

(ii) Find the time it takes for the ball to cross the goal line.

(iii) Does the ball bounce in front of the goal, go straight into the goal or go over the crossbar?

In fact the goalkeeper is standing 5 m in front of the goal and will stop the ball if its height is less than 2.8 m when it reaches him.

(iv) Does the goalkeeper stop the ball?

⑥ A plane is flying at a speed of 300 m s⁻¹ and maintaining an altitude of 10 000 m when a bolt becomes detached. Ignoring air resistance, find

(i) the time that the bolt takes to reach the ground

(ii) the horizontal distance between the point where the bolt leaves the plane and the point where it hits the ground

(iii) the speed of the bolt when it hits the ground

(iv) the angle to the horizontal at which the bolt hits the ground.

⑦ *Use g = 10 m s⁻² in this question.*

A firework is buried so that its top is at ground level and it projects sparks at a speed of 8 m s⁻¹. Air resistance may be neglected.

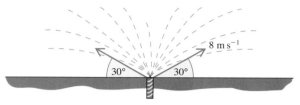

Figure 21.14

(i) Calculate the height reached by a spark projected vertically and explain why no spark can reach a height greater than this.

(ii) For a spark projected at 30° to the horizontal over horizontal ground, show that its height in metres t seconds after projection is $4t - 5t^2$ and hence calculate the distance it lands from the firework.

(iii) For what angle of projection will a spark reach a maximum height of 2 m?   [MEI]

⑧

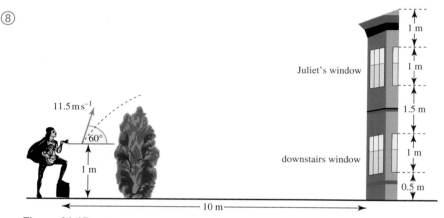

Figure 21.15

The picture shows Romeo trying to attract Juliet's attention without her nurse, who is in a downstairs room, noticing.

He stands 10 m from the house and lobs a small pebble at her bedroom window. Romeo throws the pebble from a height of 1 m with a speed of 11.5 m s⁻¹ at an angle of 60° to the horizontal.

(i) How long does the pebble take to reach the house?

(ii) Does the pebble hit Juliet's window, the wall of the house or the downstairs room window?

(iii) What is the speed of the pebble when it hits the house?   [MEI]

⑨ To kick a goal in rugby you must kick the ball over the crossbar of the goal posts (height 3.0 m), between the two uprights. Dafydd attempts a kick from a distance of 35 m. The initial velocity of the ball is 20 m s⁻¹ at 30° to the horizontal. The ball is aimed between the uprights and no spin is applied.

(i) How long does it take for the ball to reach the goal posts?

(ii) Does it go over the crossbar?

Later in the game Daffyd takes another kick from the same position and hits the crossbar.

(iii) Given that the initial velocity of the ball in this kick was also at 30° to the horizontal, find the initial speed.

⑩ Reena is learning to serve in tennis. She hits the ball from a height of 2 m. For her serve to be allowed it must pass over the net which is 12 m away from her and 0.91 m high, and it must land within 6.4 m of the net.

Make the following modelling assumptions to answer the questions.

- She hits the ball horizontally.
- Air resistance may be ignored.
- The ball may be treated as a particle.
- The ball does not spin.

(i) How long does the ball take to fall to the level of the top of the net?

(ii) How long does the ball take from being hit to first reaching the ground?

(iii) What is the lowest speed with which Reena must hit the ball to clear the net?

(iv) What is the greatest speed with which she may hit the ball if it is to land within 6.4 m of the net?

⑪ A stunt motorcycle rider attempts to jump over a gorge 50 m wide. He uses a ramp at 25° to the horizontal for his take-off and has a speed of $30\,\text{m s}^{-1}$ at this time.

(i) Assuming that air resistance is negligible, find out whether the rider crosses the gorge successfully.

The stuntman actually believes that in any jump the effect of air resistance is to reduce his distance by 40%.

(ii) Calculate his minimum safe take-off speed for this jump.

⑫ A ball is kicked from a point A on level ground and hits a wall at a point 4 m above the ground. The wall is at a distance of 32 m from A. Initially the velocity of the ball makes an angle of $\arctan\left(\frac{3}{4}\right)$ with the ground. Find the initial speed of the ball and the speed when it hits the goalpost.

⑬ In this question take g to be $10\,\text{m s}^{-2}$. A catapult projects a small pellet at speed $20\,\text{m s}^{-1}$ and can be directed at any angle to the horizontal.

(i) Find the range of the catapult when the angle of projection is
(a) 30° (b) 40° (c) 45°
(d) 50° (e) 60°.

(ii) Show algebraically that the range is the same when the angle of projection is α as it is when the angle is $90 - \alpha$.

The catapult is angled with the intention that the pellet should hit a point on the ground 36 m away.

(iii) Verify that one appropriate angle of projection would be 32.1° and write down another suitable angle.

In fact the angle of projection from the catapult is liable to error.

(iv) Find the distance by which the pellet misses the target in each of the cases in (iii) when the angle of projection is subject to an error of 0.5°. Which angle should you use for greater accuracy?

⑭ A cricketer hits the ball on the half-volley, that is when the ball is at ground level. The ball leaves the ground at 30° to the horizontal and travels towards a fielder standing on the boundary 60 m away.

(i) Find the initial speed of the ball if it hits the ground for the first time at the fielder's feet.

(ii) Find the initial speed of the ball if it is at a height of 3.2 m (well outside the fielder's reach) when it passes over the fielder's head.

In fact the fielder is able to catch the ball without moving provided that its height, h m, when it reaches him satisfies the inequality $0.25 \leq h \leq 2.1$.

(iii) Find a corresponding range of values of u, the initial speed of the ball.

⑮ A horizontal tunnel has a height of 3 m. A ball is thrown inside the tunnel with an initial speed of $18\,\text{m s}^{-1}$. What is the greatest horizontal distance that the ball can travel before it bounces for the first time?

Discussion points

→ What is the initial velocity of the projectile?

→ What is its initial position?

→ What value of g is assumed?

4 The path of a projectile

Prior knowledge

You will have seen how to obtain the Cartesian equation of the path in Chapter 18 on motion in two and three dimensions and in Chapter 11 on parametric equations.

Look at the equations

$$x = 20t$$
$$y = 6 + 30t - 5t^2$$

They represent the path of a projectile.

These equations give x and y in terms of a third variable t. They are called **parametric equations** and t is the **parameter**.

You can find the **Cartesian equation** connecting x and y directly by eliminating t as follows:

$$x = 20t \implies t = \frac{x}{20}$$

So

$$y = 6 + 30t - 5t^2$$

can be written as

$$y = 6 + 30 \times \frac{x}{20} - 5 \times \left(\frac{x}{20}\right)^2$$

$$y = 6 + 1.5x - \frac{x^2}{80} \quad \leftarrow \boxed{\text{This is the Cartesian equation.}}$$

Exercise 21.4

Those questions in this exercise where the context is drawn from real life involve working with mathematical models.

① Find the Cartesian equation of the path of each of these projectiles by eliminating the parameter t.

   (i) $x = 4t$ $y = 5t^2$

   (ii) $x = 5t$ $y = 6 + 2t - 5t^2$

   (iii) $x = 2 - t$ $y = 3t - 5t^2$

   (iv) $x = 1 + 5t$ $y = 8 + 10t - 5t^2$

   (v) $x = ut$ $y = 2ut - \frac{1}{2}gt^2$

② A particle is projected with initial velocity $50\,\text{m s}^{-1}$ at an angle of $36.9°$ to the horizontal. The point of projection is taken to be the origin, with the x-axis horizontal and the y-axis vertical in the plane of the particle's motion.

   (i)   Show that at time $t\,\text{s}$, the height of the particle in metres is given by

$$y = 30t - 5t^2$$

       and write down the corresponding expression for x.

   (ii) Eliminate t between your equations for x and y to show that

$$y = \frac{3x}{4} - \frac{x^2}{320}$$

   (iii) Plot the graph of y against x using a scale of $2\,\text{cm}$ for $10\,\text{m}$ along both axes.

   (iv) Mark on your graph the points corresponding to the position of the particle after $1, 2, 3, 4, \ldots$ seconds.

③ A golfer hits a ball with initial velocity $50\,\text{m s}^{-1}$ at an angle α to the horizontal where $\sin \alpha = 0.6$.

   (i)   Find the equation of its trajectory, assuming that air resistance may be neglected. The flight of the ball is recorded on film and its position vector, from the point where it was hit, is calculated. The results (to the nearest $0.5\,\text{m}$) are as shown in Table 21.2 overleaf.

Table 21.2

Time (s)	Position (m)
0	$\begin{pmatrix} 0 \\ 0 \end{pmatrix}$
1	$\begin{pmatrix} 39.5 \\ 24.5 \end{pmatrix}$
2	$\begin{pmatrix} 78 \\ 39 \end{pmatrix}$
3	$\begin{pmatrix} 116.5 \\ 44 \end{pmatrix}$
4	$\begin{pmatrix} 152 \\ 39 \end{pmatrix}$
5	$\begin{pmatrix} 187.5 \\ 24.5 \end{pmatrix}$
6	$\begin{pmatrix} 222 \\ 0 \end{pmatrix}$

(ii) On the same piece of graph paper draw the trajectory you found in part (i) and that found from analysing the film. Compare the two graphs and suggest a reason for any differences.

(iii) It is suggested that the horizontal component of the resistance to motion of the golf ball is almost constant. Are the figures consistent with this?

④ A particle is projected from a point O with initial velocity having components u_x and u_y along the horizontal and vertical directions, respectively.

(i) If (x, y) is a point on the trajectory of the projectile, show that

$$yu_x^2 - u_xu_yx + 4.9x^2 = 0$$

(ii) Find the speed of projection and the elevation, if the particle passes through the points with coordinates (2, 1) and (10, 1).

5 General equations

The work done in this chapter can now be repeated for the general case using algebra. Assume a particle is projected from the origin with speed u at an angle α to the horizontal and that the only force acting on the particle is the force due to gravity. The x- and y-axes are horizontal and vertical through the origin, O, in the plane of motion of the particle.

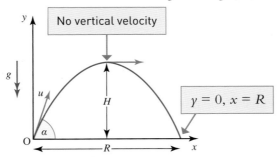

Figure 21.16

The components of velocity and position

	Horizontal motion		Vertical motion	
Initial position	0		0	
a	0		$-g$	
u	$u_x = u\cos\alpha$		$u_y = u\sin\alpha$	
v	$v_x = u\cos\alpha$	①	$v_y = u\sin\alpha - gt$	②
r	$x = ut\cos\alpha$	③	$y = ut\sin\alpha - \frac{1}{2}gt^2$	④

$ut\cos\alpha$ is preferable to $u\cos\alpha t$ because this could mean $u\cos(\alpha t)$ which is incorrect.

The maximum height, H

At its greatest height, the vertical component of the velocity is zero.

From equation ②

$$u \sin \alpha - gt = 0$$

$$t = \frac{u \sin \alpha}{g}$$

Substitute in equation ④ to obtain the height of the projectile:

$$y = u \times \frac{u \sin \alpha}{g} \times \sin \alpha - \tfrac{1}{2} g \times \frac{(u \sin \alpha)^2}{g^2}$$

$$= \frac{u^2 \sin^2 \alpha}{g} - \frac{u^2 \sin^2 \alpha}{2g}$$

The greatest height is

$$H = \frac{u^2 \sin^2 \alpha}{2g}$$

The time of flight, T

When the projectile hits the ground $y = 0$.

From equation ④ $\quad y = ut \sin \alpha - \tfrac{1}{2} g t^2$

$$0 = ut \sin \alpha - \tfrac{1}{2} g t^2$$

$$0 = t(u \sin \alpha - \tfrac{1}{2} g t)$$

> The solution $t = 0$ is at the start of the motion.

$$t = 0 \text{ or } t = \frac{2u \sin \alpha}{g}$$

The time of flight is $T = \dfrac{2u \sin \alpha}{g}$.

The range, R

The range of the projectile is the value of x when $t = \dfrac{2u \sin \alpha}{g}$

From equation ④: $x = ut \cos \alpha$

$$\Rightarrow R = u \times \frac{2u \sin \alpha}{g} \times \cos \alpha$$

$$R = \frac{2u^2 \sin \alpha \cos \alpha}{g}$$

It can be shown that $2 \sin \alpha \cos \alpha = \sin 2\alpha$, so the range can be expressed as

$$R = \frac{u^2 \sin 2\alpha}{g}$$

The range is a maximum when $\sin 2\alpha = 1$, that is when $2\alpha = 90°$ or $\alpha = 45°$. The maximum possible horizontal range for projectiles with initial speed u is

$$R_{max} = \frac{u^2}{g}.$$

The equation of the path

Discussion point

→ What are the assumptions on which this work is based?

From equation ③ $t = \dfrac{x}{u\cos\alpha} = \dfrac{x}{u}\sec\alpha$

Substitute into equation ④ to give

$$y = u \times \frac{x}{u\cos\alpha} \times \sin\alpha - \frac{1}{2}g \times \frac{x^2}{u^2}\sec^2\alpha$$

$$y = x\frac{\sin\alpha}{\cos\alpha} - \frac{gx^2}{2u^2}\sec^2\alpha$$

So the equation of the trajectory is

$$y = x\tan\alpha - \frac{gx^2}{2u^2}\sec^2\alpha$$

Discussion point

Explain how you would use the equation of the trajectory of a projectile to determine whether it passes through a given point.

Using the identity $\sec^2\alpha = 1 + \tan^2\alpha$ gives

$$y = x\tan\alpha - \frac{gx^2}{2u^2}(1 + \tan^2\alpha)$$

Exercise 21.5

Those questions in this exercise where the context is drawn from real life involve working with mathematical models.

In this exercise use the modelling assumptions that air resistance can be ignored and the ground is horizontal.

① A projectile is launched from the origin with an initial velocity $30\,\text{m}\,\text{s}^{-1}$ at an angle of $45°$ to the horizontal.

   (i)   Write down the position of the projectile after time t.

   (ii) Show that the equation of the path is the parabola $y = x - 0.0108\dot{x}^2$.

   (iii) Find y when $x = 10$.

   (iv) Find x when $y = 20$.

② Jack throws a cricket ball with velocity $10\,\text{m}\,\text{s}^{-1}$ at $14°$ to the horizontal. The ball leaves his hand $1.5\,\text{m}$ above the origin.

   (i)   Show that the equation of the path is the parabola $y = 1.5 + 0.25x - 0.052x^2$.

   (ii) Jack is aiming at a stump $0.7\,\text{m}$ high. How far from the stump is he standing if the ball just hits the top?

③ While practising his tennis serve, Matt hits the ball from a height of $2.5\,\text{m}$ with a velocity of magnitude $25\,\text{m}\,\text{s}^{-1}$ at an angle of $5°$ above the horizontal as shown in Figure 21.17.

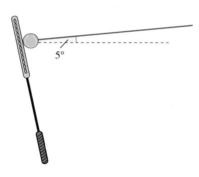

Figure 21.17

   (i)   Show that while in flight
$$y = 2.5 + 0.087x - 0.0079x^2.$$

   (ii) Find the horizontal distance from the serving point to the spot where the ball lands.

   (iii) Determine whether the ball would clear the net, which is $1\,\text{m}$ high and $12\,\text{m}$ from the serving position in the horizontal direction.

④ Ching is playing volleyball. She hits the ball with initial speed $u\,\text{m}\,\text{s}^{-1}$ from a height of $1\,\text{m}$ at an angle of $35°$ to the horizontal.

   (i)   Define a suitable origin and x and y-axes and find the equation of the trajectory of the ball in terms of x, y and u.

The rules of the game require the ball to pass over the net, which is at height 2 m, and land inside the court on the other side, which is of length 5 m. Ching hits the ball straight along the court and is 3 m from the net when she does so.

(ii) Find the minimum value of u for the ball to pass over the net.

(iii) Find the maximum value of u for the ball to land inside the court.

⑤ *Use $10\,m\,s^{-2}$ for g in this question.*

The equation of the trajectory of a projectile which is projected from a point P is given by
$$y = 1 + 0.16x - 0.008x^2$$

where y is the height of the projectile above horizontal ground and x is the horizontal displacement of the projectile from P.

The projectile hits the ground at a point Q.

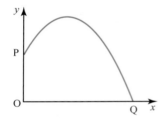

Figure 21.18

(i) Write down the height of P and find the coordinates of Q.

(ii) Find the horizontal distance x from P of the highest point of the trajectory and show that this point is 1.8 m above the ground.

(iii) Find the time taken for the projectile to fall from its highest point to the ground.

(iv) Find the horizontal component of the velocity. Deduce from this the time of flight for the projectile to travel from P to Q.

(v) Calculate the speed of the projectile when it hits the ground.

⑥ A particle is projected from a point O and passes through a point P on its trajectory when it is travelling horizontally. The coordinates of P are (16, 12). Find the angle of projection and the magnitude of the initial velocity.

⑦ *In this question take g to be $10\,m\,s^{-2}$.*
A golf ball is driven from the tee with speed $30\sqrt{2}\,m\,s^{-1}$ at an angle α to the horizontal.

(i) Show that during its flight the horizontal and vertical displacements x and y of the ball from the tee satisfy the equation
$$y = x\tan\alpha - \frac{x^2}{360}(1 + \tan^2\alpha).$$

(ii) The golf ball just clears a tree 5 m high which is 150 m horizontally from the tee. Find the two possible values of $\tan\alpha$.

(iii) Use the discriminant of the quadratic equation in $\tan\alpha$ to find the greatest distance by which the golf ball can clear the tree and find the value of $\tan\alpha$ in this case.

(iv) The ball is aimed at the hole which is on the green immediately behind the tree. The hole is 160 m from the tee. What is the greatest height the tree could be without making it impossible to hit a hole in one?

⑧ A boy is firing small stones from a catapult at a target on the top of a wall. The stones are projected from a point which is 5 m from the wall and 1 m above ground level. The target is on top of the wall which is 3 m high. The stones are projected at a speed of $7\sqrt{2}\,m\,s^{-1}$ at an angle of θ with the horizontal.

(i) The stone hits the target. Show that θ must satisfy the equation
$$5\tan^2\theta - 20\tan\theta + 13 = 0$$

(ii) Find the two values of θ.

LEARNING OUTCOMES

Now you have finished this chapter, you should be able to

➤ model vertical motion under gravity in a vertical plane using vectors

➤ model projectile motion under gravity in a vertical plane using vectors

➤ formulate the equations of motion of a projectile using vectors

➤ find the position and velocity of a projectile at any time

➤ find the range and maximum height of a projectile

➤ find the equation of the trajectory of a projectile

➤ use trigonometric functions to solve problems in context, including problems involving vectors, kinematics

➤ use vectors to solve problems in context, including forces and kinematics.

KEY POINTS

1 Modelling assumptions for projectile motion with acceleration due to gravity:
   - a projectile is a particle
   - it is not powered
   - the air has no effect on its motion.

2 Projectile motion is usually considered in terms of horizontal and vertical components.

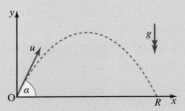

Figure 21.19

When the initial position is the origin (0, 0) and the angle of projection $= \alpha$

- Initial velocity, $\mathbf{u} = \begin{pmatrix} u\cos\alpha \\ u\sin\alpha \end{pmatrix}$

- Acceleration, $\mathbf{a} = \begin{pmatrix} 0 \\ -g \end{pmatrix}$

- At time t, velocity, $\mathbf{v} = \mathbf{u} + \mathbf{a}t \qquad \begin{pmatrix} v_x \\ v_y \end{pmatrix} = \begin{pmatrix} u\cos\alpha \\ u\sin\alpha \end{pmatrix} + \begin{pmatrix} 0 \\ -g \end{pmatrix}t$

$$v_x = u\cos\alpha$$
$$v_y = u\sin\alpha - gt$$

- Displacement, $\mathbf{r} = \mathbf{u}t + \frac{1}{2}\mathbf{a}t^2 \qquad \begin{pmatrix} x \\ y \end{pmatrix} = \begin{pmatrix} u\cos\alpha \\ u\sin\alpha \end{pmatrix}t + \frac{1}{2}\begin{pmatrix} 0 \\ -g \end{pmatrix}t^2$

$$x = ut\cos\alpha$$
$$y = ut\sin\alpha - \frac{1}{2}gt^2$$

3 At maximum height $v_y = 0$.

4 $y = 0$ when the projectile lands.

5 The time to hit the ground is twice the time to maximum height.

6 The equation of the trajectory of a projectile is

$$y = x\tan\alpha - \frac{gx^2}{2u^2\cos^2\alpha} \quad \text{or} \quad y = x\,\tan\alpha - \frac{gx^2}{2u^2}(1 + \tan^2\alpha)$$

7 When the point of projection is (x_0, y_0) rather than (0, 0)

$$\mathbf{r} = \mathbf{r}_0 + \mathbf{u}t + \frac{1}{2}\mathbf{a}t^2 \qquad \begin{pmatrix} x \\ y \end{pmatrix} = \begin{pmatrix} x_0 \\ y_0 \end{pmatrix} + \begin{pmatrix} u\cos\alpha \\ u\sin\alpha \end{pmatrix}t + \frac{1}{2}\begin{pmatrix} 0 \\ -g \end{pmatrix}t^2$$

Fireworks and aeroplanes

Robert lives underneath the flight path of aeroplanes coming in to land at an airport. They take $1\frac{1}{2}$ minutes from passing over his house to touching down.

Robert buys some fireworks from a supermarket. Will it be safe for him to let them off in his garden?

1   **Problem specification and analysis**

This problem has already been specified.

The basic analysis is straightforward. You need to know the height of the aeroplanes as they pass over Robert's garden and the maximum height that a firework is likely to reach. For it to be safe the aeroplanes need to be much higher than the fireworks. So you need to find ways of estimating these two heights.

The situation for the aeroplanes is illustrated in Figure 1 where R is Robert's garden, P is the position of an aeroplane when it is overhead and L is the point where it lands. You need to find PR.

Figure 1

There are various ways to estimate the height that a firework reaches. In Robert's method he stands a measured distance from the place where it is launched and points to the place in the sky where it reaches maximum height. A friend measures the angle his arm makes with the horizontal. Robert uses this information to estimate the firework's maximum height; he assumes that the firework travels vertically upwards.

The fireworks that go highest are called shells. However, they are not available to the general public. Of the fireworks that you can buy in a supermarket (often called 'garden fireworks'), rockets go the highest; however, such rockets are not as powerful as those used in professional displays and so will not go so high. Notice too that they are not drones which should never be flown anywhere near aeroplanes.

2 **Information collection**

The information you need to estimate PR in Figure 1 can be found on the internet or by talking to someone who knows about aviation.

Robert stood 30 metres from the spot where the rocket was launched. The angle between his arm and the horizontal was judged to be 55°.

3 **Processing and representation**

Using the information you now have, calculate estimates of the height of the aeroplanes and the maximum height of a rocket.

Illustrate your answers on a suitable scale diagram.

4 **Interpretion**

State whether or not the results indicate that it is safe for Robert to light the fireworks in his garden, explaining your answer.

How robust is your conclusion?

- Write down any modelling assumptions you have made.
- Write down any measurements involved in coming to your conclusion.
- List any other data you have used, for example any you may have obtained from the internet.

Your conclusion is based on your two estimated heights. Use the list you have just made to consider upper and lower bounds for these estimates.

As an example, your estimate of the height of the rocket probably depends on the modelling assumption that it travels vertically upwards. Suppose, instead, that its flight makes an angle of 20° with the vertical. How could this affect your estimate of its greatest height?

Taking everything into consideration, are you still convinced that your conclusion is correct?

22 A model for friction

This statement about a road accident was offered to a magistrate's court by a solicitor.

> 'Briefly the circumstances of the accident are that our client was driving his Porsche motor car. He had just left work at the end of the day. He was stationary at the junction with Victoria Road when a motorcyclist travelling north down Victoria Road lost control of his motorcycle due to excessive speed and collided with the front of our client's motor car.

> 'The motorcyclist was braking when he lost control and left a 26-metre skid mark on the road. Our advice from an expert witness is that the motorcyclist was exceeding the speed limit of 30 mph.'

It is the duty of a court to decide whether the motorcyclist was innocent or guilty.

→ Is it possible to deduce his speed from the skid mark? Draw a sketch map and make a list of the important factors that you would need to consider when modelling this situation.

1 A model for friction

Clearly the key information is provided by the skid marks. To interpret it, you need a model for how friction works; in this case between the motorcycle's tyres and the road.

As a result of experimental work, in 1785 Coulomb formulated a model for friction between two surfaces. The following laws are usually attributed to him.

1 Friction always opposes relative motion between two surfaces in contact.
2 Friction is independent of the relative speed of the surfaces.
3 The magnitude of the frictional force has a maximum which depends on the normal reaction between the surfaces and on the roughness of the surfaces in contact.
4 If there is no sliding between the surfaces

$$F \leqslant \mu R$$

where F is the force due to friction and R is the normal reaction. μ is called the **coefficient of friction**.
5 When sliding is just about to occur, friction is said to be **limiting** and $F = \mu R$.
6 When sliding occurs $F = \mu R$.

According to Coulomb's model, μ is a constant for any pair of surfaces. Typical values and ranges of values for the coefficient of friction μ are given in Table 22.1.

Table 22.1

Surfaces in contact	μ
Wood sliding on wood	0.2–0.6
Metal sliding on metal	0.15–0.3
Normal tyres on dry road	0.8
Racing tyres on dry road	1.0
Sandpaper on sandpaper	2.0
Skis on snow	0.02

How fast was the motorcyclist going?

You can proceed with the problem. As an initial model, you might make the following assumptions.

1 The road is level and horizontal.
2 The motorcycle was at rest just as it hit the car. (Obviously it was not, but this assumption allows you to estimate a minimum initial speed for the motorcycle.)
3 The motorcyclist's deceleration from full speed to rest was uniform.
4 The motorcycle and rider may be treated as a particle, subject to Coulomb's laws of friction with $\mu = 0.8$ (i.e. dry road conditions).

The calculation then proceeds as follows.

Taking the direction of travel as positive, let the motorcycle and rider have acceleration a m s^{-2} and mass m kg. You have probably realised that the acceleration will be negative. The forces (in N) and acceleration are shown in Figure 22.1.

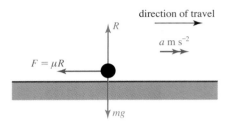

Figure 22.1

Apply Newton's second law:

Perpendicular to the road. Since there is no vertical acceleration

$$R - mg = 0 \qquad \text{①}$$

Parallel to the road. There is a constant force $-\mu R$ from friction

$$-\mu R = ma \qquad \text{②}$$

Solving for a gives $\qquad$ from ① $R = mg$

$$a = -\frac{\mu R}{m} = -\frac{\mu mg}{m} = -\mu g$$

Taking $\mu = 0.8$ and $g = 9.8$ m s^{-2} gives $a = -7.84$ m s^{-2}.

The constant acceleration formula

$$v^2 = u^2 + 2as$$

can be used to calculate the initial speed of the motorcycle. Substituting $s = 26$, $v = 0$ and $a = 7.84$ gives

$$u = \sqrt{2 \times 7.84 \times 26} = 20.2 \text{ m s}^{-1}$$

Convert this figure to miles per hour (using the fact that 1 mile = 1600 m):

$$\text{speed} = \frac{20.2 \times 3600}{1600}$$

$$= 45.5 \text{ mph}$$

So this first model suggests that the motor cycle was travelling at a speed of at least 45.5 mph before skidding began.

Modelling with friction

While there is always some frictional force between two sliding surfaces its magnitude is often very small. In such cases the frictional force is ignored and the surfaces are described as **smooth**.

Discussion points

Look carefully at the assumptions given on page 503.

→ What effect do they have on the estimate of the initial speed?

→ How good is this model and would you be confident offering the answer as evidence in court?

In situations where frictional forces cannot be ignored the surface(s) are described as **rough**. Coulomb's law is the standard model for dealing with such cases.

Frictional forces are essential in many ways. For example a ladder leaning against a wall would always slide if there were no friction between the foot of the ladder and the ground. The absence of friction in icy conditions causes difficulties for road users: pedestrians slip over, cars and motorcycles skid.

Remember that friction always opposes sliding motion.

Example 22.1

A horizontal rope is attached to a crate of mass 70 kg at rest on a flat surface. The coefficient of friction between the surface and the crate is 0.6. Find the maximum force that the rope can exert on the crate without moving it.

Solution

The forces acting on the crate are shown in Figure 22.2. Since the crate does not move, it is in equilibrium.

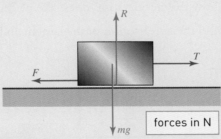

forces in N

Figure 22.2

Horizontal forces: $T = F$

Vertical forces: $R = mg$

$$= 70 \times 9.8 = 686$$

The law of friction states that

$$F \leqslant \mu R$$

for objects at rest.

So in this case

$$F \leqslant 0.6 \times 686$$

$$F \leqslant 412$$

The maximum frictional force is 412 N. As the tension in the rope and the force of friction are the only forces which have horizontal components, the crate will remain in equilibrium unless the tension in the rope is greater than 412 N.

> **Discussion point**
>
> Explain how you know whether you can use the expression μR for friction in a problem.

Example 22.2

Figure 22.3 shows a block of mass 5 kg resting on a rough table and connected by light inextensible strings passing over smooth pulleys to blocks of mass 2 kg and 7 kg. The coefficient of friction between the block and the table is 0.4.

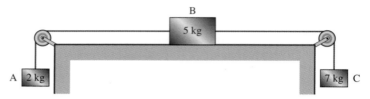

Figure 22.3

(i) Draw a diagram showing the forces acting on the three blocks and the direction of the system's acceleration.

(ii) Show that acceleration does take place.

(iii) Find the acceleration of the system and the tensions in the strings.

Solution

(i)

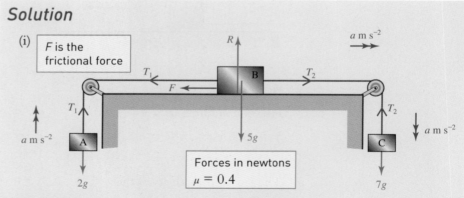

> F is the frictional force

> Forces in newtons
> $\mu = 0.4$

Figure 22.4

(ii) The direction of the acceleration is clearly from left to right. You now need to show that a must be greater than 0 for any motion to take place.

The equation of motion for the 2 kg mass is $\qquad T_1 - 2g = 2a$ ①

For the 5 kg block, the equation of motion is $T_2 - T_1 - F = 5a$ ②

For the 7 kg mass, the equation of motion is $\qquad 7g - T_2 = 7a$ ③

Adding ①, ② and ③ $\qquad\qquad\qquad\qquad\qquad 5g - F = 14a$ ④

The 5 kg block has no vertical acceleration so that $\qquad R = 5g$

The maximum possible value of F is $\qquad \mu R = 0.4 \times 5g = 2g$

In ④ a can only be zero if $F = 5g$, so that $a > 0$ and sliding occurs.

(iii)   When sliding occurs, you can replace F by $\mu R = 2g$

Then ④ gives

$$3g = 14a$$

$$a = \frac{3}{14}g = 2.1$$

Back-substituting gives $T_1 = 2(a + g) = 2 \times 11.9 = 23.8$

$$T_2 = 7(g - a) = 7 \times 7.7 = 53.9$$

The acceleration of the system is $2.1\,\text{m s}^{-2}$ and the tensions are $23.8\,\text{N}$ and $53.9\,\text{N}$.

Example 22.3

Angus is pulling a sledge of mass 12 kg at steady speed across level snow by means of a rope which makes an angle of 20° with the horizontal. The coefficient of friction between the sledge and the ground is 0.15. What is the tension in the rope?

Solution

Since the sledge is travelling at steady speed, the forces acting on it are in equilibrium. They are shown in Figure 22.5.

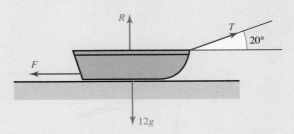

Figure 22.5

Horizontally: $T\cos 20° = F$

$$= 0.15R \longleftarrow \boxed{F = \mu R \text{ when the sledge slides.}}$$

Vertically: $T\sin 20° + R = 12g$

$$R = 12 \times 9.8 - T\sin 20°$$

Combining these gives

$$T\cos 20° = 0.15(12 \times 9.8 - T\sin 20°)$$

$$T(\cos 20° + 0.15\sin 20°) = 0.15 \times 12 \times 9.8$$

$$T = 17.8 \quad \text{(to 3 s.f.)}$$

The tension is $17.8\,\text{N}$.

! Notice that the normal reaction is reduced when the rope is pulled in an upward direction. This has the effect of reducing the friction and making the sledge easier to pull.

Example 22.4

A ski slope is designed for beginners. Its angle to the horizontal is such that skiers will either remain at rest on the point of moving or, if they are pushed off, move at constant speed. The coefficient of friction between the skis and the slope is 0.35. Find the angle that the slope makes with the horizontal.

Solution

Figure 22.6 shows the forces acting on the skier.

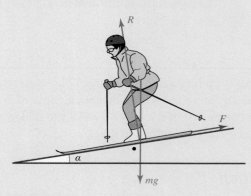

Figure 22.6

The weight mg can be resolved into components $mg\cos\alpha$ perpendicular to the slope and $mg\sin\alpha$ parallel to the slope.

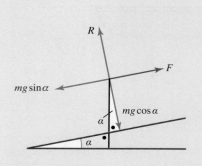

You can think of the weight mg as the resultant of two resolved components

$mg\sin\alpha$

$90° - \alpha$

mg

$mg\cos\alpha$ α

Figure 22.7

Since the skier is in equilibrium (at rest or moving with constant speed), applying Newton's second law:

Parallel to the slope: $\qquad mg\sin\alpha - F = 0$

$$\Rightarrow F = mg\sin\alpha \qquad ①$$

Perpendicular to the slope: $R - mg\cos\alpha = 0$

$$\Rightarrow R = mg\cos\alpha \qquad ②$$

In limiting equilibrium or moving at constant speed,

$$F = \mu R \quad \boxed{\text{substituting for } F \text{ and } R \text{ from } ① \text{ and } ②}$$

$$mg \sin \alpha = \mu mg \cos \alpha$$

$$\Rightarrow \mu = \frac{\sin \alpha}{\cos \alpha} = \tan \alpha$$

In this case $\mu = 0.35$, so $\tan \alpha = 0.35$ and $\alpha = 19.3°$.

Notes

1 The result is independent of the mass of the skier. This is often found in simple mechanics problems. For example, two objects of different mass fall to the ground with the same acceleration. However, when such models are refined, for example to take account of air resistance, mass is often found to have some effect on the result.

2 The angle for which the skier is about to slide down the slope is called the angle of friction. The angle of friction is often denoted by λ (lambda) and is defined by $\tan \lambda = \mu$. When the angle of the slope (α) is equal to the angle of friction (λ), it is just possible for the skier to stand on the slope without sliding. If the slope is slightly steeper, the skier will slide immediately, and if it is less steep he or she will find it difficult to slide at all without using ski poles.

Exercise 22.1

Those questions in this exercise where the context is drawn from real life involve working with mathematical models.

① A block of mass 10 kg is resting on a horizontal surface. It is being pulled by a horizontal force T (in N), and is on the point of sliding. Draw a diagram showing the forces acting and find the coefficient of friction when

(i) $T = 10$

(ii) $T = 5$.

② In each of the following situations find:

(i) the acceleration

(ii) the tension in the string(s)

(iii) the magnitude of the frictional force.

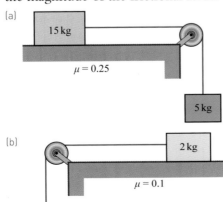

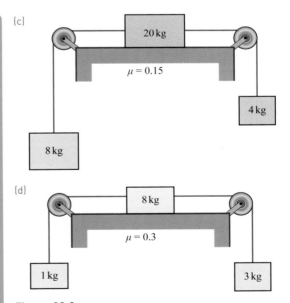

Figure 22.8

③ The brakes on a caravan of mass 700 kg have seized so that the wheels will not turn. What force must be exerted on the caravan to make it move horizontally? The coefficient of friction between the tyres and the road is 0.7.

④ A block of mass 5 kg is resting on a rough horizontal surface. The block is being pulled by a light inextensible string which makes an angle of 25° with the horizontal. The tension in the string is 15 N. Given that the block is on the point of sliding, find

(i)   the normal reaction between the block and the surface

(ii) the coefficient of friction.

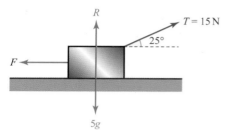

Figure 22.9

⑤ Jasmine is cycling at 12 m s⁻¹ when her bag falls off the back of her bicycle. The bag slides a distance of 9 m before coming to rest. Calculate the coefficient of friction between the bag and the road.

⑥ A boy slides a piece of ice of mass 100 g across the surface of a frozen lake. Its initial speed is 10 m s⁻¹ and it takes 49 m to come to rest.

(i)   Find the deceleration of the piece of ice.

(ii) Find the frictional force acting on the piece of ice.

(iii) Find the coefficient of friction between the piece of ice and the surface of the lake.

(iv) How far will a 200 g piece of ice travel if it too is given an initial speed of 10 m s⁻¹?

⑦ A box of mass 50 kg is being moved across a room. To help it slide a suitable mat is placed underneath the box.

(i)   Explain why the mat makes it easier to slide the box.

A force of 100 N is needed to slide the mat at a constant velocity.

(ii) What is the value of the coefficient of friction between the mat and the floor?

A child of mass 20 kg climbs onto the box.

(iii) What force is now needed to slide the mat at constant velocity?

⑧ A car of mass 1200 kg is travelling at 30 m s⁻¹ when it is forced to perform an emergency stop. Its wheels lock as soon as the brakes are applied so that they slide along the road without rotating. For the first 40 m the coefficient of friction between the wheels and the road is 0.75 but then the road surface changes and the coefficient of friction becomes 0.8.

(i)   Find the deceleration of the car immediately after the brakes are applied.

(ii) Find the speed of the car when it comes to the change of road surface.

(iii) Find the total distance the car travels before it comes to rest.

⑨ Shona, whose mass is 30 kg, is sitting on a sledge of mass 10 kg which is being pulled at constant speed along horizontal ground by her older brother, Aloke. The coefficient of friction between the sledge and the snow-covered ground is 0.15. Find the tension in the rope from Aloke's hand to the sledge when

(i)   the rope is horizontal

(ii) the rope makes an angle of 30° with the horizontal.

⑩ In each of the following situations a brick is about to slide down a rough inclined plane.

Find the unknown quantity.

(i)   The plane is inclined at 30° to the horizontal and the brick has mass 2 kg: find μ.

(ii) The brick has mass 4 kg and the coefficient of friction is 0.7: find the angle of the slope.

(iii) The plane is at 65° to the horizontal and the brick has mass 5 kg: find μ.

(iv) The brick has mass 6 kg and μ is 1.2: find the angle of the slope.

⑪ A particle of mass 2 kg is projected up an inclined plane, making an angle of 20° with the horizontal, with a speed of 6 m s⁻¹. The particle comes to rest after 4 m.

(i)   Find the deceleration of the particle.

(ii) Find the frictional force F and the normal reaction R, and hence deduce the coefficient of friction between the particle and the plane.

The particle then starts to move down the plane with acceleration a m s⁻².

(iii) Find a and the speed of the particle as it passes its starting point.

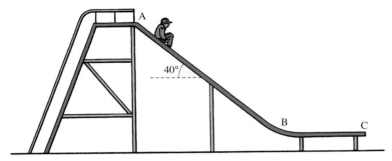

Figure 22.10

⑫ Figure 22.10 shows a boy on a simple playground slide. The coefficient of friction between a typically clothed child and the slide is 0.25 and it can be assumed that no speed is lost when changing direction at B. The section AB is 3 m long and makes an angle of 40° with the horizontal. The slide is designed so that a child, starting from rest, stops at just the right moment of arrival at C.

   (i) Draw a diagram showing the forces acting on the boy when on the sloping section AB.

   (ii) Calculate the acceleration of the boy when on the section AB.

   (iii) Calculate the speed on reaching B.

   (iv) Find the length of the horizontal section BC.

⑬ The coefficient of friction between the skis and an artificial ski slope for learners is 0.3. During a run the angle, α, which the slope makes with the horizontal varies so that initially the skier accelerates, then travels at constant speed and then slows down. What can you say about the values of α in each of these three parts of the run?

Figure 22.11

⑭ One winter day, Veronica is pulling a sledge up a hill with slope 30° to the horizontal at a steady speed. The weight of the sledge is 40 N. Veronica pulls the sledge with a rope inclined at 15° to the slope of the hill (Figure 22.12). The tension in the rope is 24 N.

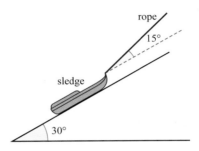

Figure 22.12

   (i) Draw a force diagram showing the forces on the sledge and find the values of the normal reaction of the ground and the frictional force on the sledge.

   (ii) Show that the coefficient of friction is slightly more than 0.1.

Veronica stops and when she pulls the rope to start again it breaks and the sledge begins to slide down the hill. The coefficient of friction is now 0.1.

   (iii) Find the new value of the frictional force and the acceleration down the slope.

                            [MEI, *adapted*]

⑮ A 5 kg block lies on a rough horizontal table. The coefficient of friction between the block and the table is $\frac{1}{5}$. The block is attached by a light inextensible string, which passes over a smooth pulley, to a mass of 2 kg hanging freely. The 5 kg block is 1.5 m from the pulley and the 2 kg mass is 1 m from the floor. The system is released from rest. Find

   (i) the acceleration of the system

   (ii) the time taken for the 2 kg mass to reach the floor

(iii) the velocity with which the 5 kg mass hits the pulley.

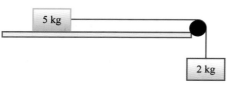

Figure 22.13

⑯ A box of mass 410 kg is placed on a rough inclined plane of slope $\arcsin\left(\frac{9}{41}\right)$ and coefficient of friction $\frac{1}{5}$. A rope is attached to the box and the direction of the rope makes an angle $\arcsin\left(\frac{3}{5}\right)$ with the upper surface of the plane. If the tension in the rope is T N, find the limiting values of T for which the box remains in equilibrium.

⑰ A box of weight 100 N is pulled at a steady speed across a rough horizontal surface by a rope which makes an angle α with the horizontal. The coefficient of friction between the box and the surface is 0.4. Assume that the box slides on its underside and does not tip up.

(i) Find the tension in the string when the value of α is
   (a) 10° (b) 20° (c) 30°.

(ii) Find an expression for the value of T for any angle α.

(iii) For what value of α is T a minimum?

⑱ If the least force which will move a body up a plane of inclination α, is 2.5 times the least force which will just prevent the body sliding down the plane, show that $\mu = \frac{3}{7}\tan\alpha$, where μ is the coefficient of friction between the body and the plane.

LEARNING OUTCOMES

Now you have finished this chapter, you should be able to:

➤ understand and use the model $f \leqslant \mu r$ for friction
➤ understand and use the coefficient of friction
➤ understand and use limiting friction
➤ solve problems involving motion of a body on a rough surface
➤ solve problems in statics
➤ use trigonometric functions to solve problems in context, including problems involving vectors, kinematics
➤ use vectors to solve problems in context, including forces and kinematics.

KEY POINTS

Coulomb's laws

1 The frictional force, **F**, between two surfaces is given by

   $F < \mu R$ when there is no sliding except in limiting equilibrium

   $F = \mu R$ in limiting equilibrium

   $F = \mu R$ when sliding occurs

   where R is the normal reaction of one surface on the other and μ is the coefficient of friction between the surfaces.

2 The frictional force always acts in the direction to oppose sliding.

3 The size of the normal reaction, and so possibly of the frictional force, is affected by any other force which has a component perpendicular to the surface on which sliding may take place.

① A box is resting on a rough horizontal surface. A light inextensible string is attached to one corner and exerts a pulling force at an angle of 45° to the horizontal. The system is in equilibrium. Which of the following is a correct diagram of the forces acting on the box?

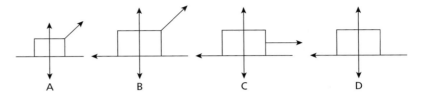

A   B   C   D

② Which of the following forces has a clockwise moment about Q of $10\,\text{N}\,\text{m}$?

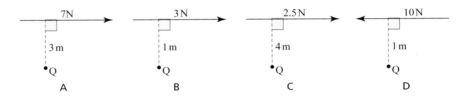

A   B   C   D

③ Two projectiles collide. Which of the following are the same for both of them when this happens?

  A   The time, horizontal distance moved and vertical distance moved

  B   The time, vertical velocity and horizontal velocity

  C   The time, vertical velocity and vertical displacement from the origin

  D   The time, horizontal displacement and vertical displacement from the origin

④ The equation of the trajectory of a projectile is $y = \dfrac{4}{3}x - \dfrac{5}{9}x^2$

Three of the following statements are true and one is false.

Which one is false?

  A   Initially its direction of flight makes an angle of arctan $\dfrac{4}{3}$ with the horizontal

  B   It passes through the point $(2.1, 0.35)$

  C   Its initial speed is $7\,\text{m}\,\text{s}^{-1}$

  D   Its range is $2.4\,\text{m}$

PRACTICE QUESTIONS: MECHANICS

① A medal of mass 0.5 kg hangs in equilibrium from two ribbons AP and BP. The ends A and B of the ribbons are fixed in place as shown in Figure 1. Calculate the tensions in the two ribbons. **[5 marks]**

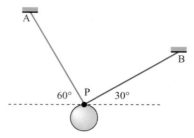

Figure 1

PS

② Juan is 1.80 m tall. He stands on accurate bathroom scales which read 72 kg.

(i) Write down the magnitude of the contact force between Juan and the scales. **[1 mark]**

He wants to know where the centre of mass of his body is. He uses a uniform plank of wood AB which is 1.80 m long and has mass 8 kg. He places it horizontally on two bathroom scales, one at each end, as shown in Figure 2.

Figure 2

He lies down on the plank with the top of his head at A and his feet at B. The scales at A read 48 kg.

(ii) Draw a diagram showing the forces acting on the plank. **[2 marks]**

(iii) Find the distance of Juan's centre of mass from his feet at B. **[3 marks]**

PS

③ A block of mass 1.2 kg is placed on a rough horizontal surface. The block is pushed by a force P which is at 20° to the horizontal as shown in Figure 3. The block remains stationary.

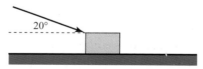

Figure 3

(i) Draw a diagram showing all the forces on the block. **[3 marks]**

The coefficient of friction between the block and the surface is $\mu = 0.4$. The force P is increased until the block begins to slide.

(ii) Find the value of P when the block on the point of sliding. **[6 marks]**

PS ④ In this question, the unit vectors $\begin{pmatrix} 1 \\ 0 \end{pmatrix}$ and $\begin{pmatrix} 0 \\ 1 \end{pmatrix}$ are in the directions of East and North respectively. Distance is measured in metres and time in seconds.

A remote controlled toy boat sails on a pond. The boat starts at the origin. Its velocity at time t seconds is given by the vector $\mathbf{v}$.

$$\mathbf{v} = \begin{pmatrix} -0.03t^2 + 0.8t - 0.4 \\ -0.4t + 8 \end{pmatrix} \text{for } 0 \leqslant t \leqslant 20.$$

(i)   Is there a time at which the boat is stationary?   [2 marks]

(ii)   Find the displacement vector $\mathbf{r}$ at time t s.   [2 marks]

(iii)   Find the bearing of the boat from the origin after 5 seconds.   [3 marks]

(iv)   Find the time at which the boat is travelling North–East. Give your answer to the nearest 0.1 s.   [3 marks]

PS ⑤ Jenny attaches a block of mass 1.5 kg to a bucket with a light inextensible string. She places the block on a smooth plane which is inclined at 40^0 to the horizontal. The string is parallel to the slope and passes over a smooth pulley. The bucket hangs down with the string vertical, as shown in Figure 4.

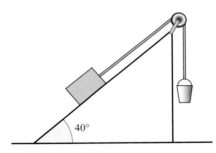

40°

Figure 4

When the bucket is empty, the system is in equilibrium.

(i)   Find the mass of the bucket.   [3 marks]

Jenny holds the block at rest on the plane. She pours water into the bucket, increasing its mass to 1.7 kg. She lets go of the block, allowing it to move up the plane.

(ii)   Find the acceleration of the block.   [3 marks]

(iii)   When the bucket has travelled 0.8 m, it reaches the ground and the string goes slack. Calculate how much further the block will travel before it comes momentarily to rest.   [5 marks]

P M T ⑥ A projectile has initial speed $u\,\mathrm{m\,s^{-1}}$ at an angle α to the horizontal. The projectile is initially on the ground which is level and horizontal. The flight time of the projectile is T s and its range is R m.

(i)   Using the standard projectile model, prove that

$$R = \frac{2u^2 \sin\alpha \cos\alpha}{g}.$$ [3 marks]

(ii)   Write down one of the modelling assumptions you have used.   [1 mark]

(iii)   Use the formula for $\sin 2\alpha$ to prove that the maximum value of

R is $\dfrac{u^2}{g}$ and that this occurs when $\alpha = 45°$.   [3 marks]

Charlie has a machine that launches table tennis balls. It states on the box that they will be launched at $10\,\mathrm{m\,s^{-1}}$. He uses columns A, B and C of the spreadsheet shown in Figure 5 to predict the time of flight and the range for various angles of projection, using the standard projectile model.

He then launches the ball at each of the angles and records the observed range in column D of the spreadsheet.

Home | Insert | Page Layout | Formulas | Data

fx

	A	B	C	D	E
1	$\alpha(°)$	T(s)	R(m)	Observed range (m)	
2	30			7.1	7.28
3	35	1.17	9.59	7.3	7.53
4	40	1.31	10.05	7.4	7.47
5	45	1.44	10.2	7.1	
6	50	1.56	10.05	6.4	6.38
7	55	1.67	9.59	5.5	5.4
8	60	1.77	8.84	4.2	4.15

Figure 5

(iv) Calculate the values for cells B2 and C2. [2 marks]

Charlie notices that the observed values are much lower than the predicted values. He adapts the model by considering a lower launch velocity u.

(v) Find, to 1 decimal place, the value of u that would give a maximum value of 7.1 m for R. [2 marks]

(vi) Without doing any further calculations, state one respect in which the observed values of the range are still different from those given by the standard projectile model. [1 mark]

Charlie tries a refined model in which the displacement of the ball is given by

$$\mathbf{r} = \begin{pmatrix} 10t\cos\alpha - 1.5t^2 \\ 10t\sin\alpha - 4.9t^2 \end{pmatrix}$$

He uses column E of the spreadsheet for the values of R predicted by this model; some of the values have already been entered.

(vii) Show that the initial speed of the ball in this model is $10\,\mathrm{m\,s^{-1}}$. [3 marks]

(viii) Explain why the values in column B can also be used in the refined model. [1 mark]

(ix) Calculate the value in the cell E5. [2 marks]

(x) State whether the revised model is an improvement of the standard projectile model, justifying your answer. [1 mark]

Data set

CYCLING ACCIDENTS DATA SET

First name	Last name	Sex M/F	Age	Distance from home	Cause	Injuries	Wearing a helmet	Nights in hospital	Time of accident	Day of accident	Month of accident	Officer reporting
Farhan	Ali	M	13	250 m	hit lamp-post	compression		1	9.15 am	Saturday	October	39014
Martin	Anderson	M	31		drunk	abrasions	n	1	23.30	Friday	February	78264
Marcus	Appleton	M	64	2 miles	car pulled out	concussion	n	2	8.25am	Monday	August	97655
Lucy	Avon	F	52	500 m	lorry turning	abrasions		0	7.50am	Thursday		39014
Thomas	Bailey	M	10	500 m	hit friend	suspected concussion	y	0	4pm	Sunday	June	78264
Andrew	Burke	M	18	1 km	car door opened	abrasions	y		3pm	Sunday	September	45211
Lee	Burnett	M	18	3 km	kerb	sprains	y	0	8am	Monday	April	78264
Clive	Burrows	M	16	about 1 mile	slipped on wet leaves	sprained wrist	y	0	6.00am	Tuesday	November	78264
Nathaniel	Carley	M	32	1.4 miles	hit by car	concussion	n	1	6.30 am	Monday	July	39813
Rory	Clark	M	44	100 m	hit by car while turning	concussion, multiple fractures		1	17.20	Friday	January	97655
Crystal	Cook	F	61	1.5 miles	didn't see car	broken arm, bruising, shock	n	1	10.45am	Wednesday	August	97655
Christopher	Court	M	60	7 km	turning lorry	dislocated elbow	y	0	8.30am	Tuesday	August	97655
Sally	Darby	F	18	2 km	knocked by bus	concussion	y	1	5pm	Tuesday	December	78264
Terry	Davidson	M	7	outside	hit tree	suspected concussion	y	1	6pm	Monday	June	78264
Matthew	de Leon	M	46	4 miles	knocked by car	sprains and shock	y	0	10am	Saturday	April	78264
Michael	Delaney	M	20	300 m	skidded into wall	abrasions	n	0	7.10am	Tuesday	February	78264
Sarah	Doyle	F	62	2 miles	hit kerb	fractured wrist	n	0	10.30	Wednesday	Oct	78264
Alexander	Duggan	M	34	0.5 miles	brakes failed	concussion	n	0	7.30 am	Monday	January	39813
Selena	Fenney	F	10	0 m	fell over	concussion	y	1	6pm	Wed	February	39813
Ceri	Flynn	F	9	300 yds	fell off	sprained wrist	y	0	3pm	Friday	August	97655
Harry	Francis	M	9	1 mile	car hit	broken leg and abrasions	y	0	10am	Monday	August	97655
Fred	Fuller	M	6	20 m	fell over	concussion	y	1	3.15	Sunday	September	39813
Toby	Geary	M	14	1.5 km	knocked off by car	broken wrist and arm	y	0	8am	Wednesday	October	78264
Stacey	Geary	F	44	2 miles	minibus turning	abrasions, shock	n	0	8.25am	Thursday	June	39813
Mary	Geraghty	F	67	1 mile	pedestrian	abrasions, shock	n	0	11.30am	Wednesday	March	39813
Luke	Grainger	M	13	0.5 km	lorry turning	broken leg, concussion		1	8pm	Saturday	Spetember	54211
Sharon	Griffin	F	7	75 m	uneven pavement	broken leg	y	1	2.35	Saturday	May	45211
Lara	Haas	F	28	3 miles		sprained wrist	n	0	1630	Tuesday	November	39813
Joseph	Hall	M	63	0.5 miles	lorry	multiple fractures, head injury	n	5	4.30 pm	Friday	November	39813
Aidan	Hart	M	66	3 miles	hit cyclist	abrasions, shock	y	0	3.30pm	Tuesday	June	39813
Penny	Hickey	F	26	3 miles	hit hedge to avoid bus	abrasions	y	0	4.45	Tuesday	February	39813
Claire	Higgins	F	19	10 km	hit by car at roundabout	fractured arm and sprained wrist	y	0	7.45am	Monday	Feb	78264
Francesca	Hill	F	22	0.5 miles	car	broken arm	n	0	5.50pm	Tuesday	September	78264
Fortune	Hilton	F	29	12 km	car changing lanes	broken arm and leg	y	2	6.50pm	Wednesday	January	97655
Rhys	Hooper	M	22	115 km	multiple vehicles	multiple fractures	y	3	8.30am	Monday	March	39813
Adam	Housman	M	39	11 km	hit by car	broken arm	y	0	5.50pm	Monday	January	97655
Eden	Howell	M	14	50 m	brakes failed	abrasions	y	none	6.30 am	Tues	March	97655
David	Huker	M	8	50 m	hit cyclist	concussion	n	1	10	Saturday	May	97655
Edward	Hunton	M	7	50 m	fell off	severe abrasions	y	0	2.30pm	Friday	August	97655
Avril	Johnson		9	at home	hit wall	dislocated knee	y	1	1700	Thursday	August	39813
George	Jones	F	37	9 km	lorry collision	punctured lung	y	8	7.15am	Wednesday	February	78264
Roger	Kenny	M	62	7 miles	car changed lanes	multiple fractures	y	1	1800	Thursday	May	39813
Henry	Kerridge	M		3 miles	hit by car	abrasions	y	0	8.30am	Tuesday	August	97655
Debbie	Lane	F	5	75 km	hit kerb	broken arm	y	0	9.00am	Sunday	April	78264
Scott	Learman	M	61	3 miles	avoiding children	sprained wrist and broken arm	n	0	4.15pm	Monday	November	39813
Gary	Lighter	M	59		car braking	concussion	n	1	1730	Friday	November	39813
Seren	Maher	F	16	20 m	hit kerb	dislocated elbow	n	0	7am	Wednesday	November	78264
Shelley	Mann	F	21	50 m	slipped on wet leaves	deep cuts to legs, strained wrist	y	0	9pm	Friday	January	97655

First name	Last name	Sex M/F	Age	Distance from home	Cause	Injuries	Wearing a helmet	Nights in hospital	Time of accident	Day of accident	Month of accident	Officer reporting
Harry	Markson	M	15	3 km	collision with cyclist	unreadable	y	0	8am	Monday	March	39813
Jeremy	Marlow	M	55	6.5 km	lorry at roundabout	multiple fractures	y	2		Monday	June	97655
Michael	Marston	M	23	3 miles	car braked suddenly	head injuries	y	1	8.20am	Friday	November	97655
Joanne	Mason	F	26	2 miles	lorry turning left	concussion and broken arm		1	10am	Saturday	May	39014
Jennifer	Massey	F	10	100 m	fell off	suspected concussion		0	8 am	Saturday	Sept	39014
Justin	Matthews	M	6.5	1 mile	hit wall	broken arm	y	0	noon	Tuesday	August	97655
Sam	Maynard	M	28	1.5 km	pothole	unreadable	n	0	5.20 pm	Thursday	July	97655
Richard	McLennan	M	22	1 mile	car	bruising	y	0	8.10am	Wednesday	April	78264
Owen	Mitchell	M	61	1 mile	pothole	suspected concussion	n	1	9.30am	Tuesday	June	39813
Lisa	Montgomery	F	16	300 m	fell over	abrasions	n	0	3pm	Saturday	June	39813
James	Moore	M	20	2 km	trying to fix chain	partially severed finger	n	0	7am	Monday	June	97655
Glyn	Morgan	M	36	6 km	lorry	concussion	y	0	8.50am	Wednesday	September	39813
Luke	Murphy	M	20	3 miles	messing around	broken arm	y	0	10pm	Saturday	April	78264
Lewis	Ofan	M	36	4 miles	hit car	concussion, sprained wrists	y	0	10.30	Sunday	August	45211
Matthew	Ogunwe	M	25	10 km	car turning	broken leg	y	0	8.30am	Thursday	May	97655
Dylan	Omerod	M	42	8 km	bus pulled out	concussion, multiple fractures	y	2	8.45am		October	78264
Patrick	O'Toole	M	21	5 miles	slipped on oil patch	broken arm	y	0	8.20am	Monday	June	39813
Zoe	Painter	F	21	2 miles	skidded	abrasions		0	10am	Wednesday	July	78264
Eric	Passant	M	67	3 miles	slipped on leaves	abrasions, shock	n	1	9.50am	Friday	November	97655
Anthony	Patrick	M	17	3 miles	dragged under lorry	severe head injuries	y	5	4.30pm	Monday	April	97655
Cath	Pickin	F	16	half mile	puncture by nail	abrasions on leg	y	0	7.30am	Tuesday		79655
Kobi	Pitts	M	34	3 miles	bus	broken arm and sprained wrist	y	0	11am	Tuesday	August	97655
Simon	Porter	M	66	5 miles	hit cyclist	broken arm	y	0	3.30pm	Tuesday	June	39813
Kate	Price	F	17	5 miles	car collision	concussion	y	0	8am	Friday	June	39813
Dee	Pugh	F	11	about 1 mile	car hit	dislocated shoulder	y	1	3.30pm	Monday	August	97655
Simon	Rice	M	11	100m	hit friend	suspected concussion	y	0	4pm	Sunday	June	78264
Bob	Roberts	M	138	3 miles	lorry collision	broken arm	n	0	8.25am	Monday	January	79264
Benjamin	Ronan	M	60	1 mile	collision with cyclist	abrasions	n	0	8am	Monday	March	39813
Michael	Root	M	13	30 m	wing mirror	deep cuts	y	0	5pm	Monday	August	39813
Jonathan	Sanders	M	9	200 m	hit car	concussion	y	1	2.30pm	Tuesday	December	39813
Simon	Sefton	M	15	3 km	car hit	concussion	n	1		Tuesday	May	97655
Arvinder	Sethi	M	12	1200 m	lorry	multiple fractures	n	40	4.40pm	Thursday	May	39813
Dave	Smith	M	37	23 km	van pulled out	concussion, multiple fractures		2	3.30pm	Saturday	June	39014
John	Smith	M	45	3 km	skid	concussion	y	0	7.45	Tuesday	May	78264
Millie	Smith	F	88	2 miles	hit by car	bruising, shock	n	1	2pm	Tuesday	April	78264
Jacob	Squires	M	46	7 km	skidded on wet road	severe abrasions	y	0	8.50am	Sunday	September	45211
Jodie	Stanton	F	18	5 km	hit brick in road	abrasions to left leg	y	0	5.15pm	Thursday	January	97655
Sam	Thomas	F	9		hit tree	broken wrist	y	0	15.45	Wednesday	December	39813
Manny	Umberton	M	9	2 m	hit kerb	dislocated thumb	y	0	9.30am	Sunday	April	78264
Ian	Wade	M	12	2 km	car	concussion with complications	y	3	10 am	Tuesday	March	39813
Natalie	Walken	F	50	50 m	brakes failed, hit kerb	broken wrist, bruising	n	0	1300	Wednesday	March	39813
Agatha	Walker	F	64	1 mile	knocked by car	fractured wrist	n	0	10am	Wednesday	October	78264
Kerry	Wilde	F	52	3 km	car didn't look	abrasions to left arm	n	0	10	Saturday	May	97655
Jordan	Williams	M	35	4 km	hit fence	broken fingers	y	0	9.15am	Tuesday	May	97655
Marion	Wren	F	8.5	300 m	cyclist collision	bruising		0	10	Saturday	May	39014

Answers

Chapter 1

Discussion point (page 3)

Rob can argue that each term that is added halves the area remaining of the square that measures 2 by 2 so has area of 4. Thus the total areal, and hence the sum, cannot be greater than 4.

To make the argument watertight Rob could argue that the sequence is a geometrical progression. You will see how that would work in Chapter 3.

Discussion point (page 3)

Since odd numbers cannot be written as the sum of two odd numbers, the only way of writing an odd number as the sum of two primes is when one of the two primes is 2. There are many odd numbers that cannot be written as the sum of 2 and a prime number, e.g. 11.

Discussion point (page 3)

There are many prime numbers which are not 5.

Discussion point (page 3)

For example, 10 is even but 5 is not.

Exercise 1.1 (page 4)

1 True
2 (i) $\Rightarrow$
   (ii) $\Leftrightarrow$
   (iii) $\Rightarrow$
   (iv) $\Leftarrow$
3 (i) No, because BC and AD may not be parallel
   (ii) ABCD is a parallelogram $\Rightarrow$ AB is parallel to CD. The converse is true.

4 (ii) The total distance if R is (5,5) is $5 + \sqrt{13}$, but if R is (4,5) the total distance is $6\sqrt{2}$ which is smaller.
   (iii) The best position for R is (4,5). The distance is equivalent to the case where Q is (8,9) (see diagram below), and (4,5) is on the straight line $y = x + 1$ between (2,3) and (8,1).

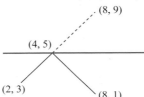

5 There are 4 different arrangements, one of which is shown below.

	7	
3	1	4
5	8	6
	2	

7 (i) $14 + 12^2 + 7^3$ and $15 + 19^2 + 5^3$

Discussion point (page 7)

No, reversing 90 does not give a two-digit number.

Exercise 1.2 (page 10)

1 1, 5, 6, 0
2 False, when $n = 5, 2^n + 3 = 35$, not prime.
3 True – direct proof
4 True – direct proof
5 True – proof by exhaustion
6 False – e.g. a hexagon (6 sides) has 9 diagonals
7 True – direct proof
8 True – proof by contradiction

9 (i) True – direct proof
   (ii) True – direct proof
10 (i) True – direct proof
    (ii) True – direct proof
11 False – e.g. $n = 10$
12 True – proof by contradiction
13 (i) True – direct proof
    (ii) False – e.g. $1^2 + 2^2 + 3^2 + 4^2 = 30$
14 True – direct proof
18 Numbers of the form 2^n
19 (i) If N^2 is the sum of the squares of two integers, then N is the sum of the squares of two different integers.
    (ii) False – e.g. $N^2 = 225 = 12^2 + 9^2$, but $N = 15$ is not the sum of the squares of two integers.

Chapter 2

Opening activity (page 12)

174.5 m

Exercise 2.1 (page 16)

1 $\dfrac{11\pi}{6}$
2 (i) $\dfrac{\pi}{4}$
   (ii) $\dfrac{\pi}{2}$
   (iii) $\dfrac{2\pi}{3}$
   (iv) $\dfrac{5\pi}{12}$
   (v) $\dfrac{5\pi}{3}$
   (vi) 0.4 radians
   (vii) $\dfrac{5\pi}{2}$
   (viii) 3.65 rad
   (ix) $\dfrac{5\pi}{6}$
   (x) $\dfrac{\pi}{25}$

3 (i) 18°
    (ii) 135°
    (iii) 114.6°
    (iv) 80°
    (v) 540°
    (vi) 240°
    (vii) 28.6°
    (viii) 450°
    (ix) 420°
    (x) 77.1°

4 (i) (a) 1 cm
       (b) $\sqrt{3}$ cm
    (ii) (a) $\frac{\pi}{3}$
       (b) $\frac{\pi}{6}$
    (iii) (a) $\frac{\sqrt{3}}{2}$
       (b) $\frac{1}{2}$
       (c) $\sqrt{3}$
       (d) $\frac{1}{2}$
       (e) $\frac{\sqrt{3}}{2}$
       (f) $\frac{\sqrt{3}}{3}$

5

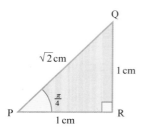

$PQ = \sqrt{1^2 + 1^2} = \sqrt{2}$, the triangle is a right-angled so $\angle QPR + \angle PQR = \frac{\pi}{2}$ and the triangle is isosceles so $\angle QPR = \angle PQR = \frac{\pi}{4}$

(i) $\sin\frac{\pi}{4} = \frac{1}{\sqrt{2}} = \frac{\sqrt{2}}{2}$

(ii) $\cos\frac{\pi}{4} = \frac{1}{\sqrt{2}} = \frac{\sqrt{2}}{2}$

(iii) $\tan\frac{\pi}{4} = \frac{1}{1} = 1$

6 $\sin\frac{9\pi}{4} = \cos\left(-\frac{\pi}{4}\right) = \frac{\sqrt{2}}{2}$

$\tan\left(-\frac{5\pi}{3}\right) = \tan\frac{4\pi}{3} = \sqrt{3}$

$\cos\frac{11\pi}{6} = \sin\frac{2\pi}{3} = \frac{\sqrt{3}}{2}$

$\cos\pi = \tan\frac{3\pi}{4} = -1$

$\sin\frac{5\pi}{6} = \cos\frac{5\pi}{3} = \frac{1}{2}$

7 $x = 0.6$ and $x = 2.5$ to 1 d.p.

8 $x = 0.8$ and $x = 3.9$ to 1 d.p.

9 (i) $k = \pi$
    (ii) $k = \frac{\pi}{2}$
    (iii) $k = \pi$
    (iv) $k = \frac{\pi}{2}$

10 (i) $x = \frac{2\pi}{7}$
    (ii) $y = \frac{8\pi}{5}$
    (iii) $k = \frac{1}{3}$

Activity 2.1 (page 18)

	Radians	Degrees
Angle	θ^c	$\alpha° \left(\alpha = \theta \times \frac{180}{\pi}\right)$
Arc length	$r\theta$	$r\alpha \times \frac{\pi}{180}$
Area of sector	$\frac{1}{2}r^2\theta$	$\frac{1}{2}r^2\alpha \times \frac{\pi}{180}$

Exercise 2.2 (page 20)

1 4 cm
2 (i) (a) 3.14 cm
       (b) 9.14 cm
       (c) 4.71 cm²
    (ii) (a) 22.0 cm
       (b) 30.0 cm
       (c) 44.0 cm²

3

r (cm)	θ (rad)	s (cm)	A (cm²)
4	$\frac{1}{2}$	2	4
$1\frac{1}{2}$	$\frac{\pi}{3}$	$\frac{\pi}{2}$	$\frac{3\pi}{8}$
5	0.8	4	10
1.875	0.8	1.5	1.41
3.46	$\frac{2\pi}{3}$	7.26	4π

4 (i) 140 yards
    (ii) 5585 square yards
5 (i) 15π cm²
    (ii) 9 cm²
    (iii) (15π − 9) cm²

8 1.05 cm³
9 (i) 5.16 m
    (ii) 3.11 m²
10 (i) $\frac{9\pi}{8} = 3.53$ m² to 2 d.p.
    (ii) $\frac{40\pi}{3} = 41.89$ m² to 2 d.p.
    (iii) 32.14 m²
11 (i) 16.9 cm²
    (ii) 19.7 cm²
12 (i) 1.98 mm²
    (ii) 43.0 mm

Discussion point (page 22)

Because the formula *area of a sector* $= \frac{1}{2}r^2\theta$ assumes θ is in radians.

Discussion point (page 23)

By the shape and symmetry of the graphs, in each case the maximum percentage error will occur for $\theta = 0.1$ radians.

$y = \sin\theta$: $\theta = 0.1$ rad
true value = 0.099 833
approximate value = 0.1
% error = 0.167%
$y = \tan\theta$: $\theta = 0.1$ rad
true value = 0.100 335
approximate value = 0.1
% error = 0.334%
$y = \cos\theta$: $\theta = 0.1$ rad
true value = 0.995 004
approximate value = 0.995
% error = 0.000 419%

Exercise 2.3 (page 24)

1 (i) θ^2
    (ii) $\frac{\theta^2}{2}$
    (iii) $\left(1 - 2\theta^2\right)$
    (iv) 2θ
2 (i) 2
    (ii) $1 - \frac{5\theta^2}{2}$
    (iii) $\frac{1}{2}$
    (iv) -3θ
3 (i) 5θ
    (ii) 5

4 (i) $\dfrac{\theta^2}{2}$

(ii) $\dfrac{1}{8}$

5 (i) $8\theta^2$

(ii) $4\theta^2$

(iii) 2

6 0.24 radians

7 0.102 radians; the other roots are not small angles

8 (ii) $\angle\text{BAE} = 90° - \angle\text{OAB}$

Review: Algebra (1)

Exercise R.1 (page 29)

1 (i) $6 - \sqrt{3}$

(ii) $-1 + 5\sqrt{2}$

(iii) $17 - 12\sqrt{2}$

(iv) 1

2 (i) $3\sqrt{3}$

(ii) $\sqrt{5} - 1$

(iii) $\dfrac{16 + 10\sqrt{2}}{7}$

(iv) $\dfrac{5\left(3\sqrt{5} + 4\right)}{29}$

3 (i) $\dfrac{1}{32}$

(ii) 3

(iii) 125

(iv) $\dfrac{3}{4}$

4 (i) x^7

(ii) x

(iii) x^{-6}

(iv) x^4

(v) x^4

5 (i) $6x^5y^5$

(ii) x^2y^4

(iii) $3x^4y^2$

(iv) $(1 + 3x)(1 + x)^2$

Exercise R.2 (page 33)

1 (i) $\log 9$

(ii) $\log 48$

(iii) $\log \frac{2}{3}$

2 (i) $2\log x$

(ii) $6\log x$

(iii) $\frac{3}{2}\log x$

3 (i) $x = -1.15$ (3 s.f.)

(ii) $x = 0.222$ (3 s.f.)

(iii) $x = 1.20$ (3 s.f.)

(iv) $x = 2$

4 (i) $x = 99\,999$

(ii) $x = \sqrt{5}$

(iii) $x = e^4 - 3$

(iv) $x = \sqrt{5}$

5 (i) $x = c^a + b$

(ii) $x = \dfrac{\ln s - q}{p}$

6 (i) 200

(ii)

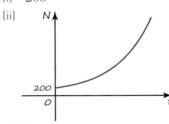

(iii) 4017 bees

(v) The model is not suitable in the long term since not enough room, no allowance made for bees dying, etc.

7 (ii) $p = 1.4, q = 0.8$

8 (ii) $k = 5, n = 0.5$

Chapter 3

Discussion point (page 36)

(i) (a) Avonford Savings

(b) 80 000, 160 000, 320 000, …

(c) Exponential geometric sequence

(d) The sequence could go on but the family will not live forever.

(ii) (a) Pizza outlet opening hours

(b) 10, 10, 10, 10, 12 …

(c) There is a cycle that repeats every week.

(d) The sequence will continue as long as they keep the same opening hours.

(iii) (a) The clock

(b) $0, -3.5, -5, -3.5, \dots$

(c) The pattern of 8 numbers keeps repeating.

(d) The sequence continues for as long as the clock is working.

(iv) (a) The stairs

(b) 120, 140, 160, …

(c) Increases by 20 each time

(d) The steps can't go on forever.

Exercise 3.1 (page 41)

1 Two out of: both start with 1, are periodic, have period 2, oscillate

2 (i) 19, 22, 25, 28, … Arithmetic, 1st term $a = 7$, common difference $d = 3$

(ii) 4, 3, 2, 1, … Arithmetic, 1st term $a = 8$, common difference $d = -1$

(iii) 3.3, 3.1, 2.9, 2.7, … Arithmetic, 1st term $a = 4.1$, common difference $d = -0.2$

(iv) 48, 96, 192, 384, … Geometric, 1st term $a = 3$, common ratio $r = 2$

(v) 4, 2, 1, 0.5, … Geometric, 1st term $a = 64$, common ratio $r = 0.5$

(vi) $16, -32, 64, -128, \dots$ Geometric, 1st term $a = 1$, common ratio $r = -2$

(vii) 2, 2, 2, 5, … Periodic with period 4

(viii) 1, 3, 5, 3, … Periodic with period 4, oscillating about 3

3 (i) 3, 5, 7, 9, …

(ii) 6, 12, 24, 48, …

(iii) 4, 8, 14, 24, …

(iv) $1, \frac{1}{2}, \frac{1}{3}, \frac{1}{4}, \dots$

(v) 4, 6, 4, 6, …

4 (i) 12, 15, 18, 21, …

(ii) $-5, 5, -5, 5, \dots$

(iii) 72, 36, 18, 9, … .

(iv) 1, 4, 9, 16, … .

(v) 4, 6, 4, 6, …

5 (i) 58
  (ii) 90
  (iii) 25
  (iv) 8
  (v) 40

6 (i) $\sum_{1}^{10} k$

  (ii) $\sum_{1}^{10}(20 + k)$

  (iii) $\sum_{1}^{10}(200 + 10k)$

  (iv) $\sum_{1}^{10}(200 + 11k)$

  (v) $\sum_{1}^{10}(200 - 10k)$

7 (i) 15
  (ii) 400
  (iii) −20
  (iv) 100
  (v) 220

8 (i) 0, 1, 1, 2, 3, … After the first term this is the same as the sequence itself.
  (ii) 13, 21, 34, …
  (iii) 1.0, 2.0, 1,5, 1.667, 1.6, 1.625, 1.615, 1.619, … These values are oscillating and appear to be converging.

9 (i) 2, 6, 2, 6, 2, 6, …
  (ii) Oscillating and periodic with period 2
  (iii) (a) 2, 8, −4, 20, −28, 68, … Oscillating and diverging
       (b) $3\frac{1}{2}, 4\frac{1}{4}, 3\frac{7}{8}, 4\frac{1}{16}, 3\frac{31}{32}, 4\frac{1}{64}, \dots$
       Oscillating sequence converging towards 4

10 (i) $t_2 = 0, t_3 = -2, t_4 = 2, t_{100} = 2$
   (ii) $2^2 - 2 = 2; t_1 = -1$
   (iii) (a) Diverges
        (b) Converges to −1
        (c) The terms oscillate between $\frac{\sqrt{5} - 1}{2}$ and $\frac{-\sqrt{5} - 1}{2}$

11 (ii) First series: 118 terms, 294 terms, 10 793 terms, 23 445 terms
   Second series: 11 terms, 145 terms, 10 308 terms, 22 388 terms

Exercise 3.2 (page 46)

1 (i) Yes, 3, 32
  (ii) No
  (iii) Yes, −3, −22
  (iv) Yes, 4, 39
  (v) No
  (vi) Yes, 1.5, 15.5

2 (i) 34
  (ii) 29 terms

3 38 terms

4 (i) 3
  (ii) 470

5 (i) 4
  (ii) −3, 1, 5, 9
  (iii) 375

6 (i) 3
  (ii) 15 150

7 (i) 120
  (ii) 22

8 (i) 5049
  (ii) 5100
  (iii) 51 terms in each sequence with the terms in the second one each 1 greater than those in the first.

9 30

10 (i) $a = d = 2$
   (ii) 330

11 267

12 (i) 16
   (ii) 2.5 cm

13 (i) $a = 1, d = 4$
   (ii) 270

14 (i) $S_n = \frac{d}{2}n^2 + \left(a - \frac{d}{2}\right)n$
   (ii) Subtracting the sum of the first $(n - 1)$ terms from the sum of the first n terms will leave only the nth term.
   nth term $= p(2n - 1) + q$
   (iii) First term $= (p + q)$; common difference $= 2p$

Discussion point (page 50)

If $r = +1$ then all of the terms are equal to the first term, a, and the sum of n terms is na so the sum is *divergent*.

If $r = -1$ then the terms alternate between $+a$ and $-a$ and the sum will alternate between $+a$ and 0.

Discussion point (page 52)

S is an infinite series with a common ratio of (-2). Since this is outside the interval $(-1, 1)$ the sum to infinity does not have a unique value.

Discussion point (page 52)

Disease, availability of food, water and shelter.

Exercise 3.3 (page 53)

1 (i) Yes, 2, 160
  (ii) Yes, −1, −1
  (iii) No
  (iv) No
  (v) Yes, $\frac{1}{2}, \frac{3}{16}$
  (vi) Yes, −2, −64

2 (i) $48, (-2), 3(-2)^{n-1}$
  (ii) $\frac{1}{8}, \frac{1}{2}, 2 \times \left(\frac{1}{2}\right)^{n-1} = \left(\frac{1}{2}\right)^{n-2}$
  (iii) ab^4, b, ab^{n-1}

3 (i) $5\frac{1}{3}$
  (ii) 8
  (iii) Diverges
  (iv) $4\frac{1}{2}$
  (v) Diverges

4 The value of r is not less than 1 in size, or the partial sums do not get closer and closer to a limit, but oscillate instead.

5 (i) 11
  (ii) 256
  (iii) 4094

6 (i) (a) 2; 12 terms
     (b) 20 475
  (ii) (a) 3; 11 terms
      (b) 177 146
  (iii) (a) $\left(-\frac{1}{2}\right)$; 16 terms
       (b) 5.333 (4 s.f.)

7   19 years

8   (i)   2

      (ii)   3

      (iii)   3 069

      (iv)   19 terms

9   (i) $\frac{2}{3}$

      (ii)   7th

      (iii)   7 terms

      (iv)   27

10   (i)   83 terms

      (ii)   73 terms

11   (i) $\dfrac{x(1 - x^n)}{(1 - x)}$

      (ii) $\dfrac{(1 - (-y)^n)}{(1 + y)}$

      (iii) $\dfrac{(1 - (-2a)^n)}{(1 + 2a)}$

13   (i)   20, 10, 5, 2.5, 1.25

      (ii)   5 operations

      (iii)   0, 10, 15, 17.5, 18.75

      (iv)   The sequence of amounts of water is geometric (first term 20, common ratio 0.5) but the sequence of antifreeze is not (a geometric sequence cannot have zero as its first term since that would imply that all terms are zero).

14   (i)   After 67 swings

      (ii)   241° (to the nearest degree)

      (iii)   600°

15   (i) $\dfrac{1}{\cos\theta}$

      (ii) $\dfrac{31\sqrt{3}}{2}$

      (iii) $2\sqrt{3}$

      (iv) $\dfrac{1}{\cos\theta \sin^2\theta}$

      (v)   No

16   (i)   27 cm, 36 cm, 48 cm, 64 cm, $85\frac{1}{3}$ cm

      (ii)   The length increases without bound $\left(\text{G.P. with } r = \frac{4}{3}\right)$.

      (iii)   No – the figure is always inside the circle which circumscribes the original triangle.

Review: Algebra(2)

Review exercise R.1 (page 60)

1   (i) $x = 2, 6$

      (ii) $a = -5, -6$

      (iii) $x = \pm 4$

      (iv) $p = -0.5, -2$

      (v) $c = 1, -1\frac{3}{4}$

      (vi) $x = 4, \frac{2}{3}$

2   (i) $x = 2.37, -3.37$

      (ii) $x = 0.78, 3.22$

      (iii) $x = -3.70, 2.70$

3   (i) $3(x - 2)^2 - 5$

      (ii) $2\left(x + \frac{3}{2}\right)^2 + \frac{1}{2}$

      (iii) $21 - (x - 4)^2$

4   (i) $x = -2, (-2, -12)$

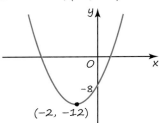

      (ii) $x = -2, (-2, -11)$

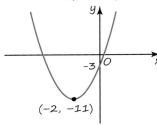

      (iii) $x = 1, (1, 5)$

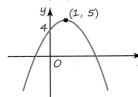

5   (i) $x = 4, y = 3$

      (ii) $x = 2, y = 3$

      (iii) $x = -1, y = -2$

6   (i) $x = 4, y = 2$ or $x = -4, y = -2$

      (ii) $x = 4, y = 1$ or $x = 1, y = 4$

      (iii) $x = 1, y = 5$ or $x = 11, y = 25$

7   (i) $\{x: x < 4\} \cap \{x: x > -2\}$

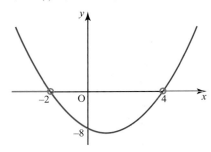

      (ii) $\{x: x \leqslant -1\} \cap \{x: x \geqslant -7\}$

      (iii) $\{x: x \geqslant 0\} \cup \{x: x \leqslant -4\}$

8   (i) $x < -2$ or $x > 4$

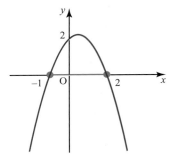

      (ii) $-1 \leqslant x \leqslant 2$

      (iii) $\frac{2}{3} < x < \frac{3}{2}$

9   (i)   Points of intersection: $(-2, 5)$ and $\left(\frac{5}{2}, 14\right)$

      (ii)

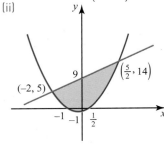

10   (i) $x = \pm 2$ or $x = \pm 1$

      (ii) $x = \pm\frac{1}{2}$ or $x = \pm\sqrt{3}$

      (iii) $x = \frac{9}{4}$ or $x = 25$

11 (i) $x = 0$ or $x = 1$
  (ii) $x = -1.58$
  (iii) $x = 0.208$
      -0.565

Review exercise R.2 (page 62)

1 (i) $x^3 + 3x^2 - x + 1$
  (ii) $5x^4 + x^3 + 4x^2 - 2x - 4$
  (iii) $x^3 + x^2 + 3x - 6$
  (iv) $3x^3 + 7x^2 - 3x + 1$

2 (i) $2x^3 + 3x^2 - 5x + 12$
  (ii) $x^4 - 2x^2 + 1$
  (iii) $x^2 + 3x + 2$
  (iv) $x^2 - 3x - 4$

3 (i)

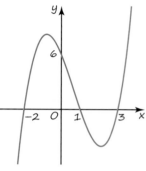

  (ii)

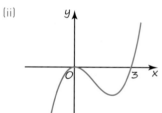

  (iii)

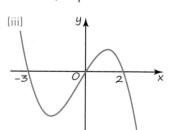

4 (i) $x = -2$ (repeated), $x = \frac{1}{2}$
  (ii)

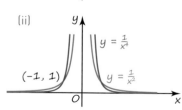

5 $y = 2x^3 + 4x^2 - 22x - 24$

6 (i) $x = 1, -\frac{1}{2}, \frac{3}{2}$
  (ii) $x = -2, 3, -\frac{1}{3}$

7 $x^3 - 2x^2 + 4x - 8$

Chapter 4

Review exercise (page 69)

1 (i)

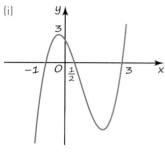

  (ii)

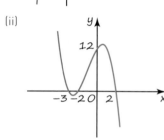

  (iii) (-1) repeated,
      (-3) repeated

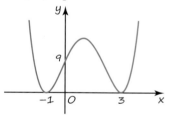

2 (i)

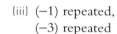

  (ii)

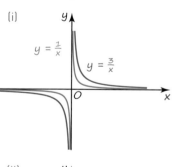

3 (i)

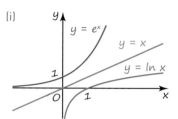

  (ii)

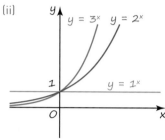

4 (i)

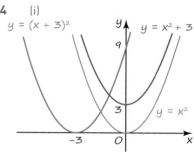

  (ii)

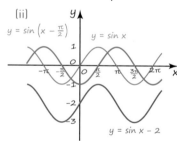

5 (i) $x^2(x + 3)$;
      translation $\begin{pmatrix} -2 \\ 0 \end{pmatrix}$

  (ii) $3(x - 2)^2(x + 1)$;
      stretch scale factor 3 in the y direction

  (iii) $(x - 2)^2(x + 1) - 2$;
      translation $\begin{pmatrix} 0 \\ -2 \end{pmatrix}$

  (iv) $-(x - 2)^2(x + 1)$;
      reflection in the x-axis

6 (i) 3.11 cm
  (ii) £30.36
  (iii) £45.36

(iv)

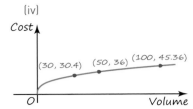

7 (i) £146
(ii)

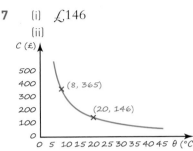

(iii) Not good for extreme
temperatures, since it
implies an infinite bill
for a temperature of 0°C.

8 (i)

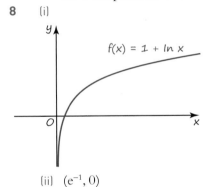

(ii) $(e^{-1}, 0)$

9 (i)

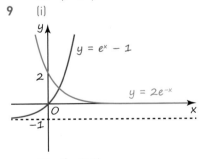

(ii) $(\ln 2, 1)$

10 (i) Can be obtained from
by a stretch scale factor
$\frac{1}{2}$ parallel to the x-axis.

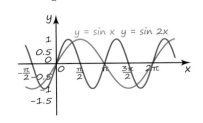

(ii) Can be obtained from
by a translation through

$$\begin{pmatrix} -\dfrac{\pi}{2} \\ 0 \end{pmatrix}$$

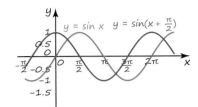

Activity 4.1 (page 75)

(i) Different
(ii) Same
(iii) Same
(iv) Different
(v) Same

Discussion point (page 77)

Translation 2 units to the right,
reflection in the x-axis, translation 5 units vertically upwards

Exercise 4.1 (page 78)

1 (i) One-to-one, yes
(ii) Many-to-one, yes
(iii) Many-to-many, no
(iv) One-to-many, no
(v) Many-to-many, no
(vi) One-to-one, yes

2 (i) (a) Examples: one → 3;
word → 4
(b) Many-to-one
(ii) (a) Examples: 1→ 4;
2.1 → 8.4
(b) One-to-one
(iii) (a) Examples: 1 → 1;
6 → 4
(b) Many-to-one
(iv) (a) Examples: 1 → −3;
−4 → −13
(b) One-to-one
(v) (a) Examples: 4 → 2;
9 → 3
(b) One-to-one

3 (i) (a) One-to-one
(b) Domain $\mathbb{R}^+$,
Range $\mathbb{R}^+$

(ii) (a) Many-to-many
(b) Domain $\mathbb{R}^+$,
Range $\mathbb{R}^+$
(iii) (a) One-to-one
(b) Domain $\mathbb{R}^+$,
Range $\mathbb{R}^+$
(iv) (b) Many-to-one
(c) Domain $\mathbb{R}$,
Range $\mathbb{R}^+$

4 (i) (a) −5
(b) 9
(c) −11
(ii) (a) 3
(b) 5
(c) 10
(iii) (a) 32
(b) 82.4
(c) 14
(d) −40

5 (i) $f(x) \leqslant 2$
(ii) $y \in \{2, 3, 6, 11, 18\}$
(iii) $\mathbb{R}$
(iv) $y \in \mathbb{R}^+$
(v) $\dfrac{1}{\sqrt{2}} \leqslant y \leqslant 1$

6 (i) $\{0.5, 1, 2, 4\}$
(ii) $0 < f(x) \leqslant 1$
(iii) $f(x) \geqslant 3$

7 For f, every value of x,
(including $x = 3$) gives a
unique output, whereas g(2)
can equal either 4 or 6.

8 (i) Translation $\begin{pmatrix} 2 \\ 0 \end{pmatrix}$; $x = 2$

(ii) Stretch parallel to the
y-axis of scale factor 3
and translation $\begin{pmatrix} 2 \\ 0 \end{pmatrix}$ in
either

order; $x = 2$

(iii) Rewrite as
$3\left[(x - 2)^2 - \dfrac{14}{3}\right]$;

translation $\begin{pmatrix} 2 \\ -\dfrac{14}{3} \end{pmatrix}$;

then stretch parallel to
the y-axis of scale factor
3; $x = 1$

9 (i) Translation $\begin{pmatrix} 0 \\ -2 \end{pmatrix}$

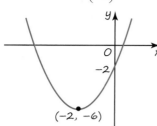
$(-2, -6)$

(ii) Translation $\begin{pmatrix} 2 \\ 0 \end{pmatrix}$

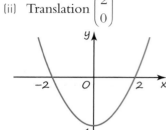

(iii) Stretch parallel to the y-axis, scale factor 2, then translation $\begin{pmatrix} 0 \\ 3 \end{pmatrix}$

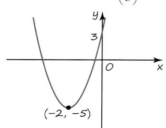
$(-2, -5)$

(iv) Stretch parallel to the x-axis, scale factor $\frac{1}{2}$, then translation $\begin{pmatrix} 0 \\ 3 \end{pmatrix}$

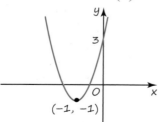
$(-1, -1)$

(v) Translation $\begin{pmatrix} 2 \\ 0 \end{pmatrix}$ then stretch parallel to the y-axis scale factor 3

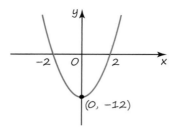
$(0, -12)$

10 (i)
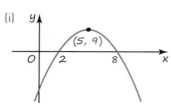
$(5, 9)$

(ii)
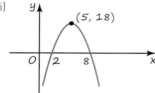
$(5, 18)$

(iii)
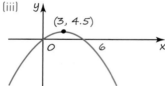
$(3, 4.5)$

(iv)
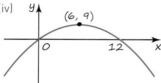
$(6, 9)$

11 Ellipse is $\dfrac{x^2}{9} + \dfrac{y^2}{4} = 1$

12 (i)
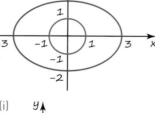
$(0.5, 1)$

(ii)
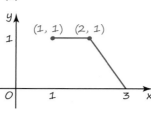
$(1, 1)$ $(2, 1)$

(iii)
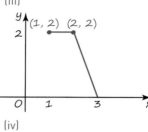
$(1, 2)$ $(2, 2)$

(iv)
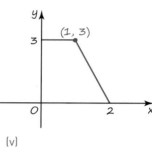
$(1, 3)$

(v)
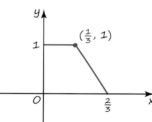
$\left(\frac{1}{3}, 1\right)$

(vi)
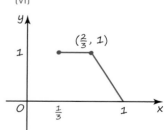
$\left(\frac{2}{3}, 1\right)$

13 (i) Stretch scale factor 3 in the y direction.

(ii) Stretch scale factor $\frac{1}{3}$ in the x direction and translation $\begin{pmatrix} 0 \\ -1 \end{pmatrix}$ in either order.

(iii) Translation $\begin{pmatrix} -30° \\ 0 \end{pmatrix}$
followed by a stretch scale factor $\frac{1}{3}$ in the x direction.

14 (i) (a)

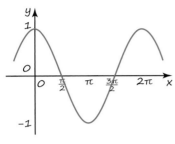

(b) $y = \cos x$

(ii) (a)

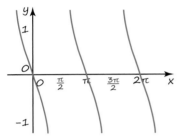

(b) $y = -\tan x$

(iii) (a)

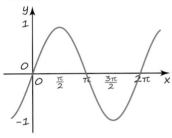

(b) $y = \sin x$

(iv) (a)

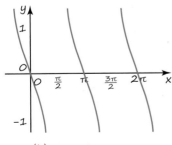

(b) $y = -\tan x$

(v) (a)

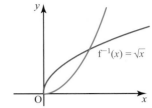

(b) $y = -\sin x$

15 (i) $a = 3, b = 5$

(ii)

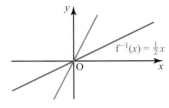

translation $\begin{pmatrix} 3 \\ 5 \end{pmatrix}$

(iii) $y = 6x - x^2 - 14$

16 Reflection in the x-axis and translation $\begin{pmatrix} 1 \\ 0 \end{pmatrix}$ in either order; $x = 1$.

17 $f(x) = (x - 1)^2 (x - 3) - 4$; $g(x) = 4 - (x - 1)^2 (x - 3)$.

18 $g(x) = \ln(9 - 2x)$.

Discussion point (page 83)

(i) (a) Function with an inverse function.

(b) $f: C \to \frac{9}{5}C + 32$;

$f^{-1}: F \to \frac{5}{9}(F - 32)$

(ii) (a) Function but no inverse function since one grade corresponds to several marks.

(iii) (a) Function with an inverse function.

(b) 1 light year $\approx 6 \times 10^{12}$ miles or almost 10^{16} metres.

$f: x \to 10^{16} x$ (approx.);

$f^{-1}: x \to 10^{-16} x$ (approx.)

(iv) (a) Function but no inverse function since fares are banded.

Activity 4.2 (page 85)

(i)

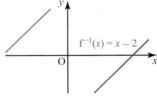

(ii)

(iii)

(iv)

$y = f(x)$ and $y = f^{-1}(x)$ appear to be reflections of each other in the line $y = x$

Discussion point (page 88)

(i) For all values of x, $\sqrt{x^2}$ will give a positive answer.

(ii) The answer will be $-19°$ since the arcsin function gives the principal value.

Exercise 4.2 (page 88)

1 x^2

2 (i) 4

(ii) 5

(iii) 9

(iv) 25

3 (i) $8x^3$
(ii) $2x^3$
(iii) $(x + 2)^3$
(iv) $x^3 + 2$
(v) $4x$
(vi) $x + 4$

4 (i) $f^{-1}(x) = \dfrac{x - 7}{2}$
(ii) $f^{-1}(x) = 4 - x$

5
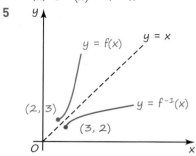

6 (i) fg
(ii) g^2
(iii) fg^2
(iv) gf

7 (i) $\dfrac{3\pi}{2}$
(ii) $\dfrac{\pi}{2}$

8 (i) $8(x + 2)^3$
(ii) $2(x^3 + 2)$
(iii) $[(x + 2)^3 + 2]^3$

9 (i) $f^{-1}(x) = \dfrac{2x - 4}{x}$
(ii) $f^{-1}(x) = \sqrt{x + 3}; x \geqslant -3$

10 (i) $f(x)$ not defined for $x = 4$;
$h(x)$ not defined for $x > 2$
(ii) $f^{-1}(x) = \dfrac{4x + 3}{x}$;
$h^{-1}(x) = 2 - x^2; x \geqslant 0$
(iii) g is a many-to-one function
(iv) $\mathbb{R}^+$
(v) No, the domain of fg excludes $x = 2$ and $x = -2$, whereas the domain of gf excludes $x = 4$

11 (i) x
(ii) $\dfrac{1}{x}$
(iii) $\dfrac{1}{x}$
(iv) $\dfrac{1}{x}$

12 (i) $a = 2, b = -5$
(ii) Translation $\begin{pmatrix} -2 \\ -5 \end{pmatrix}$

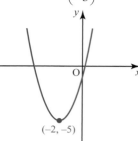

(iii) $y \geqslant -5$
(iv) $c = -2$
(v)
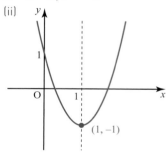

13 (i) $a = 2, b = -1$
(ii)
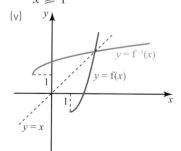

Line of symmetry is $x = 1$, minimum point $(1, -1)$
(iii) $f(x) \geqslant -1$
(iv) f is not a 1-1 function; $x \geqslant 1$
(v)
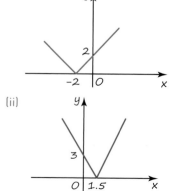

(vi) $\left(\dfrac{5 + \sqrt{17}}{4}, \dfrac{5 + \sqrt{17}}{4} \right)$

14 (i) Range of $f(x)$ is $\mathbb{R}^+$.
Range of $g(x)$ is $\mathbb{R}$.
$f(x)$ is not 1-1

(ii) $g^{-1}(x) = \dfrac{1}{2}(x + 1)$
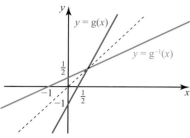

(iii) $gf(x) = 2x^2 - 1$;
$fg(x) = (2x - 1)^2$;
$x = 1$ (repeated).

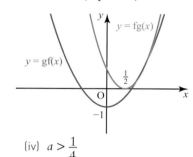

(iv) $a > \dfrac{1}{4}$

Discussion point (page 90)
$g(3) = 3, g(-3) = 3$
$|3 + 3| = 6, |3 - 3| = 0$,
$|3| + |3| = 6, |3| + |-3| = 6$

Discussion point (page 91)
$y = |x|$ lies below the line $y = 2$ for $-2 < x < 2$

Exercise 4.3 (page 94)

1 $|x| < 3$
2 (i)

(ii)

(iii)

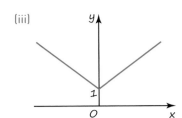

3 (i) $-8 < x < 2$
   (ii) $0 \leqslant x \leqslant 4$
   (iii) $x < -1$ or $x > 11$
   (iv) $x \leqslant -3$ or $x \geqslant 1$
4 (i) $|x - 1| < 2$
   (ii) $|x - 5| < 3$
   (iii) $|x - 1| < 3$
5 (i) $-2 < x < 5$
   (ii) $-\frac{2}{3} \leqslant x \leqslant 2$
   (iii) $x > 1$ or $x < -4$
   (iv) $x \leqslant -3\frac{1}{3}$ or $x \geqslant 2$
6 (i) $|x - 2.5| < 3.5$
   (ii) $|x - 10| < 0.1$
   (iii) $|x - 4| < 3.5$
7 (i)

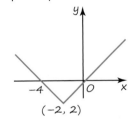

$(-2, 2)$

   (ii)

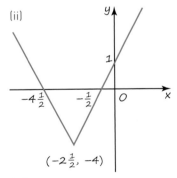

$(-2\frac{1}{2}, -4)$

   (iii)

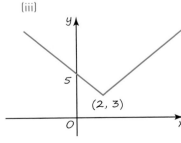

$(2, 3)$

8 (i)

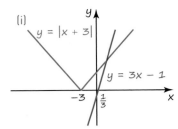

   (ii) $x > 2$
9 (i)

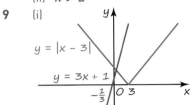

   (ii) $x > \frac{1}{2}$
10 (i) $\Leftrightarrow$
    (ii) $\Leftrightarrow$
    (iii) $\Leftarrow$
11 $x < 1$
12

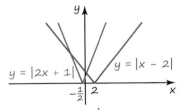

$x < -3$ or $x > \frac{1}{3}$
13 $-0.4 \leqslant x \leqslant 4$

Chapter 5

Opening activity (page 96)

(i) The rate of change of the volume with respect to the height and the rate of change of the height with respect to time.

(ii) The rate of change of the volume with respect to time.

Review exercise (page 101)

1 (i) 4
   (ii) 9
   (iii) 2
2 5890 thousand bacteria per second
3 Tangent is $4x + y - 9 = 0$
   Normal is $x - 4y - 15 = 0$

4 (i) $\dfrac{dy}{dx} = 12x^3 - x$,
        $\dfrac{d^2 y}{dx^2} = 36x^2 - 1$
   (ii) $\dfrac{dy}{dx} = 8x^3 - \dfrac{1}{3}$,
        $\dfrac{d^2 y}{dx^2} = 24x^2$
   (iii) $\dfrac{dy}{dx} = \dfrac{1}{\sqrt{x}} - 2x - \dfrac{1}{x^2}$,
        $\dfrac{d^2 y}{dx^2} = -\dfrac{1}{2x\sqrt{x}} - 2 + \dfrac{2}{x^3}$
6

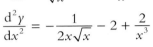
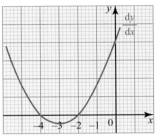

7 (i) $x < -2$
   (ii) $x < -4, x > 4$
   (iii) $0 < x < \dfrac{2}{3}$
8 (i) $\dfrac{dy}{dx} = 5x\sqrt{x} - \dfrac{2}{\sqrt{x}}$,
        $\dfrac{d^2 y}{dx^2} = \dfrac{15\sqrt{x}}{2} + \dfrac{1}{x\sqrt{x}}$
   (ii) $\dfrac{dy}{dx} = 15x^2 - \dfrac{3}{2}$,
        $\dfrac{d^2 y}{dx^2} = 30x$
9 (i) $\dfrac{dy}{dx} = 3 - 3x^2$,
        $\dfrac{d^2 y}{dx^2} = -6x$
   (ii) Minimum at $(-1, -2)$ and maximum at $(1, 2)$
   (iii)

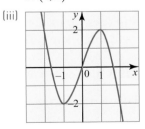

10 $8x + 7y - 95 = 0$
11 Minimum at $(9, -21)$

12 (i) $a = 1, b = 3$

(ii) No real solutions to $x^2 + 2x + 5 = 0$

13 (i) $\dfrac{dy}{dx} = 12x^3 - 12x^5$,

$\dfrac{d^2y}{dx^2} = 36x^2 - 60x^4$

(ii) Maximum point at $(-1, 1)$ and $(1, 1)$, minimum point at $(0, 0)$.

(iii)

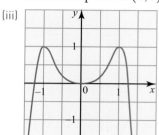

14 $c = -\dfrac{19}{2}, c = \dfrac{34}{3}$

15 $V = 486, x = 3$

Discussion point (page 102)

Gradient is zero but point is not a turning point.

Activity 5.1 (page 105)

(i) (a)

(b)

(c)

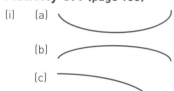

(ii) (a) concave upwards
(b) concave downwards

Discussion point (page 105)

The gradient is constant, i.e. a straight line.

Discussion point (page 106)

No, the second differential is sometimes zero at maximum points and minimum points.

Exercise 5.1 (page 110)

1 The stationary point of inflection has $\dfrac{dy}{dx} = 0$ as well as $\dfrac{d^2y}{dx^2} = 0$

2 (i) $x < 3$
(ii) $x < -7$
(iii) $x > \dfrac{1}{2}$

3 (i) $x > -\dfrac{1}{3}$
(ii) $x > 4$
(iii) $x < \dfrac{2}{3}$

4 (i) Stationary point of inflection is $(1, -4)$

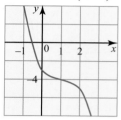

(ii) Stationary point of inflection is $\left(-\dfrac{1}{2}, 1\right)$

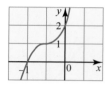

(iii) Stationary point of inflection is $(-2, 8)$

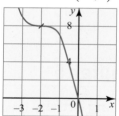

5 (i) $x < -1, x > 3$
(ii) $-1 < x < 2$

6 (i) $\dfrac{dy}{dx} = 3x^2 - 2x$,

$\dfrac{d^2y}{dx^2} = 6x - 2$

(ii) maximum at $(0, 0)$,

minimum at $\left(\dfrac{2}{3}, -\dfrac{4}{27}\right)$

(iii) $\left(\dfrac{1}{3}, -\dfrac{2}{27}\right)$

(iv) $-\dfrac{1}{3}$

(v)

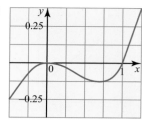

7 (i) Point of inflection at $(0, 0)$ and maximum at $(3, 27)$

(ii)

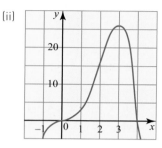

(iii) $(2, 16)$

8 (i) Minimum at $(0, -3)$ and point of inflection at $(1, -2)$

(ii)

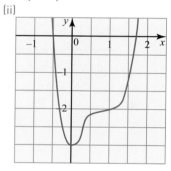

(iii) $\left(\dfrac{1}{3}, -\dfrac{70}{27}\right)$

9 Minimum points at $(2, 16)$ and $(-1, -38)$. Maximum points at $(-2, -16)$ and $(1, 38)$. Points of inflection at $(0, 0), \left(\dfrac{\sqrt{10}}{2}, \dfrac{65\sqrt{10}}{8}\right)$ and $\left(-\dfrac{\sqrt{10}}{2}, -\dfrac{65\sqrt{10}}{8}\right)$

Discussion point (page 110)

(i) Can be done easily by expanding the brackets.

(ii) Can also be done by expanding the brackets, but this is time-consuming and tedious.

(iii) and (iv) Cannot be done by
expanding the brackets,
and an alternative
method is needed.

Activity 5.2 (page 111)

Table 5.1

$y = f(x)$	u	$y = f(u)$
E.g. $y = \tan\left(\dfrac{1}{x^3}\right)$	$u = \dfrac{1}{x^3}$	$y = \tan u$
(i) $y = \left(x^3 + 3\right)^4$	$u = x^3 + 3$	$y = u^4$
(ii) $y = \sin\left(x^2 - 4\right)$	$u = x^2 - 4$	$y = \sin u$
(iii) $y = e^{2x+5}$	$u = 2x + 5$	$y = e^u$
(iv) $y = \ln\left(x^2 - 2x\right)$	$u = x^2 - 2x$	$y = \ln u$

Exercise 5.2 (page 114)

1   $f(x) = 2(1 + x)^9$, because it
    is the composition of two
    functions and the other
    function has two functions
    multiplied together.

2   (i) B; an alternative is to
         expand the brackets and
         differentiate directly
    (ii) C; differentiate directly
    (iii) A

3   (i) $\dfrac{dy}{dx} = -3(3x + 2)^{-2}$

    (ii) $\dfrac{dy}{dx} = 28x\left(2x^2 + 6\right)^6$

    (iii) $\dfrac{dy}{dx} = \dfrac{3}{\sqrt{6x - 2}}$

4   (i) $y = x^6 - 6x^4 + 12x^2 - 8$

    $\Rightarrow \dfrac{dy}{dx} = 6x^5 - 24x^3 + 24x$

    (ii) $\dfrac{dy}{dx} = 6x\left(x^2 - 2\right)^2$

    (iii) Expanding the answer to
         (ii) gives the answer to (i)

5   (i) 48
    (ii) $\dfrac{3}{2}$

    (iii) -3

6   (i) $\dfrac{dy}{dx} = 9(3x - 5)^2$
    (ii) $y - 9x + 17 = 0$

7   (i) $\dfrac{dx}{dt} = \dfrac{2\left(\sqrt{t} - 1\right)^3}{\sqrt{t}}$

    (ii) $\dfrac{dz}{dy} = \dfrac{1 - \dfrac{1}{y^2}}{2\sqrt{\dfrac{1}{y} + y}}$

    $= \dfrac{y^2 - 1}{2y^2\sqrt{\dfrac{1}{y} + y}}$

    (iii) $\dfrac{dp}{dr} = \dfrac{-3\sqrt{r}}{\left(2r\sqrt{r} - 6\right)^2}$

8   (i) $\dfrac{dy}{dx} = 8(2x - 1)^3$

    (ii) minimum at $\left(\dfrac{1}{2}, 0\right)$

    (iii)
    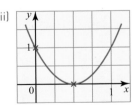

9   (i) $\dfrac{dy}{dx} = 4(2x - 1)$
         $\times \left(x^2 - x - 2\right)^3$

    (ii) Minimum at $(-1, 0)$,
         maximum at $\left(\dfrac{1}{2}, \dfrac{6561}{256}\right)$,
         minimum at $(2, 0)$

(iii)

10 (i) $\dfrac{dy}{dx} = 3\left(3x^2 - 2x\right)$
         $\times \left(x^3 - x^2 + 2\right)^2$

    (iii) $a = \dfrac{2}{3}$
    (iv) 12, $12x - y - 4 = 0$

11 (i)   72 minutes
    (ii) 67 minutes
    (iii) 1333 metres,
          66 minutes

Activity 5.3 (page 116)

(i) $\dfrac{dy}{dx} = 3x^2$

(ii) $x = y^{\frac{1}{3}}$

(iii) $\dfrac{dx}{dy} = \dfrac{1}{3}y^{-\frac{2}{3}}$

(iv) $\dfrac{dx}{dy} = \dfrac{1}{\dfrac{dy}{dx}}$

(v)   Yes

Discussion point (page 116)

$\dfrac{dy}{dx} = 2$, $\dfrac{dx}{dy} = \dfrac{1}{2}$

Exercise 5.3 (page 118)

1   $\dfrac{dx}{dy} = \dfrac{3}{2}$

2   (i) $\dfrac{1}{6}$

    (ii) $-\dfrac{1}{2}$

    (iii) -12

3   (i) $-\dfrac{1}{3}$

    (ii) $-\dfrac{3}{2}$

    (iii) $\dfrac{3}{2}$

4   (i) $\dfrac{dA}{dx} = 10x$

    (ii) $\dfrac{dA}{dt} = 5x$

    (iii) 30

5 $16\,\text{cm}^2\,\text{s}^{-1}$

6 $0.01\pi\,\text{m}^2$ per day

7 $-0.015\,\text{N}\,\text{s}^{-1}$

8 $\dfrac{1}{5\pi}\,\text{m}\,\text{s}^{-1}$

9 $\dfrac{1}{48}\,\text{m}\,\text{h}^{-1}$

10 $1.2\,\text{cm}^3\,\text{s}^{-1}$

11 $20.8\,\text{mm}^2\,\text{s}^{-1}$

Activity 5.4 (page 119)

(i) $y = x^9 \Rightarrow \dfrac{\mathrm{d}y}{\mathrm{d}x} = 9x^8$

(ii) $u = x^3 \Rightarrow \dfrac{\mathrm{d}u}{\mathrm{d}x} = 3x^2$

 $v = x^6 \Rightarrow \dfrac{\mathrm{d}v}{\mathrm{d}x} = 6x^5$

(iii) No, $9x^8 \neq 18x^7$

Activity 5.5 (page 120)

No, $\dfrac{\mathrm{d}y}{\mathrm{d}x} = 4x^3$, but

$\dfrac{\frac{\mathrm{d}u}{\mathrm{d}x}}{\frac{\mathrm{d}v}{\mathrm{d}x}} = \dfrac{7x^6}{3x^2} = \dfrac{7}{3}x^4$

Discussion point (page 121)

Yes

Exercise 5.4 (page 122)

1

Table 5.2

Function	Product or quotient rule?	u	v
e.g. $y = \dfrac{(x+1)^6}{\sqrt[3]{x-1}}$	Quotient	$u = (x+1)^6$	$v = \sqrt[3]{x-1}$
(i) $y = \sqrt{(x^2-1)}\,(x^3+3)^2$	Product	$u = (x^2-1)^{\frac{1}{2}}$	$v = (x^3+3)^2$
(ii) $y = (2x^3-3)^5$	Neither	–	–
(iii) $y = \dfrac{\sqrt{x+1}}{x^2}$	Quotient	$u = (x+1)^{\frac{1}{2}}$	$v = x^2$
(iv) $y = x^2\sqrt{x-2}$	Product	$u = x^2$	$v = (x-2)^{\frac{1}{2}}$

2 (i) $\dfrac{\mathrm{d}y}{\mathrm{d}x} = 2x(2x-1)^3(6x-1)$
$\times (4x^2+2x-1)$

(ii) $\dfrac{\mathrm{d}y}{\mathrm{d}x} = \dfrac{3x-1}{2\sqrt{x-1}}$

(iii) $\dfrac{\mathrm{d}y}{\mathrm{d}x} = \dfrac{(7x-3)(x-3)^2}{2\sqrt{x}}$

3 (i) $\dfrac{\mathrm{d}y}{\mathrm{d}x} = \dfrac{3x^2-2x}{(3x-1)^2}$

(ii) $\dfrac{\mathrm{d}y}{\mathrm{d}x} = \dfrac{-3(1-2x)^2}{x^4}$

(iii) $\dfrac{\mathrm{d}y}{\mathrm{d}x} = \dfrac{\sqrt{x}+2}{2(\sqrt{x}+1)^2}$

4 (i) $\dfrac{1}{4}$

(ii) -8

(iii) $-\dfrac{13}{108}$

5 (i) $\dfrac{\mathrm{d}y}{\mathrm{d}x} = 3x(x-2)$

(ii) Maximum at $(0,4)$ and minimum at $(2,0)$

(iii)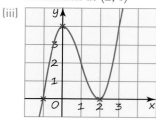

6 (i) $\dfrac{\mathrm{d}y}{\mathrm{d}x} = \dfrac{-9}{(2x-3)^2}$

(ii) $y + 9x - 6 = 0$

(iii) $y = x$

(iv) $(3,3)$

7 (i) $\dfrac{\mathrm{d}y}{\mathrm{d}x} = \dfrac{2(x+2)(x+1)}{(2x+3)^2}$

(ii) $\dfrac{\mathrm{d}^2y}{\mathrm{d}x^2} = \dfrac{2}{(2x+3)^3}$

(iii) Turning points at $(-1,-2)$ and $(-2,-3)$

8 Minimum points at $(0,0)$ and $(3,0)$, maximum at $(1,16)$

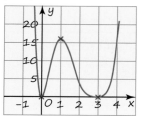

9 (i) $\dfrac{\mathrm{d}y}{\mathrm{d}x} = \dfrac{\sqrt{x}-2}{(\sqrt{x}-1)^2}$

(ii) $\dfrac{1}{4}$

(iii) $(4,8)$

(iv) Q is $\left(\dfrac{37}{4}, 8\right)$ and R is $(4,29)$. Area of triangle is $\dfrac{441}{8}$

10 (i) $\left(\dfrac{3}{2}, \dfrac{9}{4}\sqrt{2}\right)$

(ii) Gradient is 3 at the origin

(iii) $y = 0$, gradient is infinite

(iv)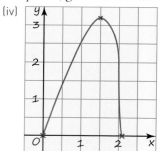

Multiple choice questions

1 C $4n+6$ is a multiple of 4

2 B Its sum converges to a limit

3 D $|x-1| \leqslant 2$

4 C $y = u^6, u = 1 - \dfrac{1}{x}$

Practice questions 1

(page 127)

1 (i) $2r + r\theta = 3$ [1]

$\theta = \dfrac{3 - 2r}{r}$ [1]

Area $= \dfrac{1}{2}r^2\left(\dfrac{3 - 2r}{r}\right)$ [1]

$A = \dfrac{3}{2}r - r^2$

(ii) $\dfrac{dA}{dr} = \dfrac{3}{2} - 2r$ [1]

$\dfrac{dA}{dr} = 0$ when $r = \dfrac{3}{4}$ [1]

$\dfrac{d^2 A}{dr^2} = -2$ so the area is

a maximum [1]

$\theta = \dfrac{3 - 1.5}{0.75} = 2$ [1]

2 (i) $\cos\left(-\dfrac{\pi}{3}\right) = 0.5$ [1]

$\cos\left(-\dfrac{\pi}{3} - \dfrac{\pi}{6}\right) + 0.5$

$= 0 + 0.5 = 0.5$

$\cos\left(-\dfrac{\pi}{2}\right) = 0$ [1]

$\cos\left(-\dfrac{\pi}{2} - \dfrac{\pi}{6}\right) + 0.5$

$= -0.5 + 0.5 = 0$

(ii) $\left(\dfrac{5\pi}{3}, 0.5\right)$ OR

$\left(\dfrac{3\pi}{2}, 0\right)$ OR either of

these with a positive

multiple of 2π added to

the x coordinate. [1, 1]

(iii) Translation [1]

$\begin{pmatrix} \dfrac{\pi}{6} \\ 0.5 \end{pmatrix}$ [1]

3 (i) $\dfrac{dy}{dx} = 8x\left(x^2 - 1\right)^3$ [1]

$\dfrac{d^2 y}{dx^2} = 8\left(x^2 - 1\right)^3 +$

$24x(x^2 - 1)^2 . 2x$ [1, 1]

$\dfrac{d^2 y}{dx^2} = 8\left(x^2 - 1\right)^2\left(x^2 - 1 + 6x^2\right)$

$= 8\left(x^2 - 1\right)^2\left(7x^2 - 1\right)$ [1]

(ii) $7x^2 - 1 < 0$ [1]

$-\dfrac{1}{\sqrt{7}} < x < \dfrac{1}{\sqrt{7}}$ [1]

4 (i) $ff(x) = \dfrac{\dfrac{x + 3}{x - 1} + 3}{\dfrac{x + 3}{x - 1} - 1}$ [1]

$\dfrac{x + 3 + 3(x - 1)}{x + 3 - (x - 1)}$ [1]

$\dfrac{4x}{4} = x$ [1]

$f^{-1}(x) = \dfrac{x + 3}{x - 1}$ [1]

(ii) $y = x$ [1]

(iii) $\dfrac{x + 3}{x - 1} = \dfrac{x - 1 + 4}{x - 1}$ [1]

$f(x) = 1 + \dfrac{4}{x - 1}$ [1]

(iv) Translation $\begin{pmatrix} 1 \\ 0 \end{pmatrix}$ [1]

Stretch parallel to y-axis,

scale factor 4 [1]

Translation $\begin{pmatrix} 0 \\ 1 \end{pmatrix}$ [1]

5 (i) Max $= a + b$ [1]

Min $= a - b$ [1]

$a + b = 16.75$ [1]

$a - b = 7.583\ldots$ [1]

$a = 12.166\ldots$

$b = 4.5833\ldots$ [1]

(ii) 21 June is day

number 172 [1]

Model says

$12.1666 - 4.58333\cos 172$

$= 16.705$ hours [1]

Close to actual

value of 16.75 hours [1]

(iii) (a) Stretch to make

period 365 days

$\left(\text{scale factor } \dfrac{365}{360}\right)$ [1]

(b) Translation to make

maximum at 21

June (oe) [1]

6 Let the terms be

$a - d$, a, $a + d$ [1]

Suppose they form a

geometric sequence

and $d \neq 0$ [1]

$\dfrac{a}{a - d} = \dfrac{a + d}{a}$ [1]

$a^2 = a^2 - d^2$ [1]

So $d^2 = 0$ and hence

$d = 0$ but this is a

contradiction so the

terms cannot form a

geometric sequence. [1]

7 Nine terms are

$1, r, r^2, \ldots, r^8$ [1]

$1 + 2 + \cdots + 8 = 36$ [1]

$r^{36} = 262144$ [1]

$r = \pm\sqrt{2}$ (oe) [1]

Sum $= \dfrac{\left(\sqrt{2}\right)^9 - 1}{\sqrt{2} - 1}$

$= 15\sqrt{2} + 31$

≈ 52.2132 [1]

OR $\dfrac{\left(-\sqrt{2}\right)^9 - 1}{-\sqrt{2} - 1}$

$= -15\sqrt{2} + 31 \approx 9.7868$ [1]

8 (i) $\dfrac{dy}{dx} = 4x^3 - 3x$ [1]

$\dfrac{d^2 y}{dx^2} = 12x^2 - 3$ [1]

At A and B

$12x^2 - 3 = 0$ [1]

$x = \pm 0.5$ [1]

$(-0.5, 2.6875)$ and

$(0.5, 2.6875)$ [1]

(ii) Gradient at A is 1. [1]

Gradient at B is -1.

$1 \times -1 = -1$ so AC is

perpendicular to BC. [1]

By symmetry, C is the

point $\left(0, \dfrac{51}{16}\right)$. [1]

$AC = BC$ [1]

Angle CAD = Angle

CBD (tangent perpen-

dicular normal) [1]

All angles are 90° and

all sides are equal so it is

a square.

Review: The sine and cosine rules

Discussion point (page 131)

Yes, cosine rule leads to same

answer of 201.5°.

Review exercise R.1 (page 132)

1 (i) (a) 63.1°

(b) 42.8 cm²

(ii) (a) 47.7°

(b) 13.0 cm²

2 (i) 4.72 cm
  (ii) 10.4 cm
  (iii) 9.1 cm
  (iv) 10.72 cm
3 114 m
4 120°
5 71.3° or 108.7°
6 AD = 57.3 m, CD = 70.2 m, height = 46.9 m
7 3.28 km
8 14.8 cm
9 (i) 87°
  (ii) 5.29 km
10 $6\sqrt{3}$ cm²
11 $(14 + 2\sqrt{13})$ cm

Chapter 6

Discussion point (page 138)

Since $\tan\theta = \dfrac{y}{x}$ and $\sin\theta = y$

and $\cos\theta = x$ then $\tan\theta = \dfrac{\sin\theta}{\cos\theta}$, $\cos\theta \neq 0$.

Using Pythagoras' theorem:
$x^2 + y^2 = 1$
So $\sin^2\theta + \cos^2\theta \equiv 1$.

Exercise 6.1 (page 140)

1 horizontal one, cos θ: vertical one sin θ
2 sec θ
3 (i) (a) $\dfrac{\sqrt{2}}{2}$
    (b) $-\dfrac{\sqrt{2}}{2}$
    (c) -1
    (d) $-\sqrt{2}$
  (ii) (a) $-\dfrac{1}{2}$
    (b) $-\dfrac{2\sqrt{3}}{3}$
    (c) $-\dfrac{\sqrt{3}}{3}$
    (d) 2
  (iii) (a) $\sqrt{3}$
    (b) $\dfrac{1}{2}$
    (c) $\dfrac{\sqrt{3}}{2}$
    (d) $-\dfrac{\sqrt{3}}{3}$

  (iv) (a) $\dfrac{3}{4}$
    (b) $\dfrac{1}{4}$
    (c) 3
    (d) $\dfrac{4}{3}$

4 (i) $B = 60°, C = 30°$
  (ii) $\sqrt{3}$
5 (i) $L = 45°, N = 45°$
  (ii) $\sqrt{2}, \sqrt{2}, 1$
6 (ii) 14.0°
7 (i) Domain: $x \in \mathbb{R}$, $x \neq \pm 90°, \pm 270°...$ range: $f(x) \geq 1, f(x) \leq -1$
  (ii) Domain: $x \in \mathbb{R}$, $x \neq n180°$ where n is an integer; range: $f(x) \geq 1$, $f(x) \leq -1$
  (iii) Domain: $x \in \mathbb{R}$, $x \neq n180°$ where n is an integer; range: $f(x) \in \mathbb{R}$
8 (i) $\dfrac{1}{2}$
  (ii) $\dfrac{1}{\sqrt{3}} = \dfrac{\sqrt{3}}{3}$
9 (i) $-\dfrac{\sqrt{7}}{4}$
  (ii) $-\dfrac{4}{3}$
11 (i) (a) $f(x) \geq 1$ or $f(x) \leq -1$
    (b) 120°
    (c)

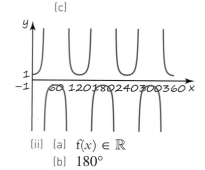

  (ii) (a) $f(x) \in \mathbb{R}$
    (b) 180°
    (c)

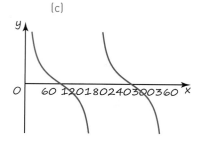

  (iii) (a) $f(x) \leq -4$, $f(x) \geq -2$
    (b) 360°
    (c)

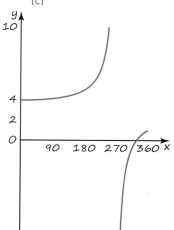

  (iv) (a) $f(x) \leq 2, f(x) \geq 4$
    (b) 1080°
    (c)

12 (i) One-way stretch, scale factor -2, parallel to y-axis
  (ii) Reflection in x-axis and then one-way stretch, scale factor $\dfrac{1}{2}$, parallel to x-axis
  (iii) Translation by $\begin{pmatrix} 0 \\ 2 \end{pmatrix}$ and then reflection in y-axis
  (iv) One-way stretch, scale factor 2, parallel to x-axis and one-way stretch, scale factor 2, parallel to y-axis
  (v) Translation by $\begin{pmatrix} 0 \\ -30 \end{pmatrix}$ and then reflection in x-axis
  (vi) Reflection in x-axis and then translation by $\begin{pmatrix} 0 \\ 2 \end{pmatrix}$

13 (i)
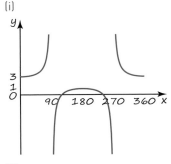

(ii)
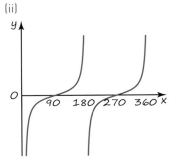

(iii)
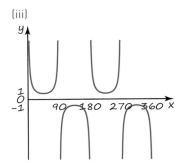

(iv)
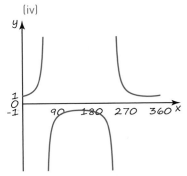

14 (i) $a = 1$, $b = -1$ and $c = 2$
(ii) $203°$, $255°$ ($\pm 2°$)
(iii) 0 as $f(2x)$ is a one way stretch scale factor $\frac{1}{2}$, parallel to the x-axis.

Exercise 6.2 (page 146)

1 The graph of cosine has only positive values between $-90°$ and $90°$ so its domain has to be different to cover all the possible values.

2 (i) $30°$, $330°$
(ii) $60°$, $120°$
(iii) $60°$, $240°$
(iv) $150°$, $210°$
(v) $240°$, $300°$
(vi) $120°$, $300°$

3 (i) 2
(ii) 2
(iii) 1
(iv) 4
(v) 4
(vi) 4

4 (i) $-1 \leqslant x \leqslant 1$; $0 \leqslant f(x) \leqslant \pi$
(ii) $-1 \leqslant x \leqslant 1$; $-\frac{\pi}{2} \leqslant f(x) \leqslant \frac{\pi}{2}$
(iii) All real numbers; $-\frac{\pi}{2} < f(x) < \frac{\pi}{2}$

5 (i) $90°$
(ii) $60°$, $300°$
(iii) $14.0°$, $194.0°$
(iv) $109.5°$, $250.5°$
(v) $135°$, $315°$
(vi) $210°$, $330°$

7 (i) $0°$, $180°$, $360°$
(ii) $45°$, $225°$
(iii) $60°$, $300°$
(iv) $54.7°$, $125.3°$, $234.7°$, $305.3°$
(v) $18.4°$, $71.6°$, $198.4°$, $251.6°$
(vi) $45°$, $135°$, $225°$, $315°$

Exercise 6.3 (page 147)

1 (i) 2
(ii) $\dfrac{2}{\sqrt{3}}$
(iii) $\dfrac{1}{\sqrt{3}}$
(iv) $\dfrac{2}{\sqrt{3}}$
(v) 2
(vi) $-\dfrac{1}{\sqrt{3}}$
(vii) $-\sqrt{2}$
(viii) $-\sqrt{2}$
(ix) -1

2 (i) $\dfrac{\pi}{6}$, $\dfrac{11\pi}{6}$
(ii) $\dfrac{\pi}{4}$, $\dfrac{5\pi}{4}$
(iii) $\dfrac{\pi}{4}$, $\dfrac{3\pi}{4}$
(iv) $\dfrac{7\pi}{6}$, $\dfrac{11\pi}{6}$
(v) $\dfrac{3\pi}{4}$, $\dfrac{5\pi}{4}$
(vi) $\dfrac{\pi}{3}$, $\dfrac{4\pi}{3}$

3 (i) 0.201 rads, 2.940 rads
(ii) -0.738 rads, 0.738 rads
(ii) -1.893 rads, 1.249 rads
(iv) -2.889 rads, -0.253 rads
(v) -1.982 rads, 1.982 rads
(vi) -0.464 rads, 2.678 rads

4 (i) 0.32, 1.25, 3.46, 4.39
(ii) 2.27, 6.02
(iii) 0.19, 1.76, 3.33, 4.90
(iv) $\dfrac{3\pi}{8}$, $\dfrac{5\pi}{8}$, $\dfrac{11\pi}{8}$, $\dfrac{13\pi}{8}$
(v) $\dfrac{\pi}{3}$
(vi) $\dfrac{\pi}{3}$, $\dfrac{2\pi}{3}$

5 (i) $\dfrac{\pi}{4}$, $\dfrac{3\pi}{4}$
(ii) $\dfrac{\pi}{4}$, $\dfrac{3\pi}{4}$
(iii) $\dfrac{\pi}{4}$, $\dfrac{3\pi}{4}$
(iv) $\dfrac{\pi}{3}$, $\dfrac{5\pi}{6}$
(v) $\dfrac{\pi}{3}$, $\dfrac{2\pi}{3}$
(vi) $\dfrac{\pi}{3}$, π
(vii) 0, 0.730, 2.412, π
(viii) 1.23, $\dfrac{\pi}{3}$

6 0, $\dfrac{\pi}{4}$, π, $\dfrac{5\pi}{4}$, 2π

7 2.24, 4.05

8 (ii) $\dfrac{\pi}{6}$, $\dfrac{5\pi}{6}$, $\dfrac{7\pi}{6}$, $\dfrac{11\pi}{6}$

Chapter 7

Activity 7.1 (page 151)

$k = \frac{1}{2}$, $x = 0.18$

Review exercise (page 152)

1 (i)

```
                1
              1   1
            1   2   1
          1   3   3   1
        1   4   6   4   1
      1   5  10  10   5   1
    1   6  15  20  15   6   1
```

(ii) Totals are 1, 2, 4, 8, 16, 32, 64. These can be written as $2^0, 2^1, 2^2, 2^3, 2^4, 2^5, 2^6$.

(iii) The total would be $2^{12} = 4096$.

2 (i) $1 + 3x + 3x^2 + x^3$
   (ii) $1 + 6x + 12x^2 + 8x^3$
   (iii) $1 - 6x + 12x^2 - 8x^3$

3 (i) (a) $1 + 8x + 24x^2 + 32x^3 + 16x^4$
   (b) $1 - 4x + 6x^2 - 4x^3 + x^4$
   (ii) $1 + 4x - 2x^2$

4 (i) $12x^2$
   (ii) $24x^2$
   (iii) $720x^3$

5 $1 - \frac{7}{2}x + \frac{21}{4}x^2 - \frac{35}{8}x^3 + \frac{35}{16}x^4$

Activity 7.3 (page 153)

(i) (a) $1 + \frac{1}{2}x - \frac{1}{8}x^2 + \ldots$
   (b) $1 - 2x + 3x^2 + \ldots$

(ii) (a) 1.04875
   (b) 0.83

(iii) $\sqrt{1.1} = 1.048808\ldots$ and $\frac{1}{1.1^2} = 0.8264\ldots$, so the approximations are good.

(iv) (a) The expansion gives -6.5 which is not a good approximation for $\sqrt{11}$.
   (b) The expansion gives 281 which is not a good approximation for $\frac{1}{11^2}$.

Discussion point (page 155)

It is because of the x^2 in the bracket.

Exercise 7.1 (page 157)

1 (i) $1 - 3x + 6x^2 - 10x^3$; valid for $-1 < x < 1$
   (ii) $1 - 6x + 24x^2 - 80x^3$; valid for $-\frac{1}{2} < x < \frac{1}{2}$
   (iii) $1 + 6x + 24x^2 + 80x^3$; valid for $-\frac{1}{2} < x < \frac{1}{2}$

2 (i) $1 + \frac{1}{2}x - \frac{1}{8}x^2$; valid for $-1 < x < 1$
   (ii) $1 - \frac{1}{2}x + \frac{3}{8}x^2$; valid for $-1 < x < 1$

(iii) $1 + \frac{1}{4}x - \frac{3}{32}x^2$; valid for $-1 < x < 1$
(iv) $1 - \frac{1}{4}x + \frac{5}{32}x^2$; valid for $-1 < x < 1$

3 (i) $1 - \frac{2}{3}x + \frac{1}{3}x^2 - \frac{4}{27}x^3$; valid for $-3 < x < 3$
   (ii) $1 - \frac{4}{3}x + \frac{4}{3}x^2 - \frac{32}{27}x^3$; valid for $-1.5 < x < 1.5$
   (iii) $1 + \frac{4}{3}x + \frac{4}{3}x^2 + \frac{32}{27}x^3$; valid for $-1.5 < x < 1.5$

4 (i) (a) $1 - 2x + 3x^2$
   (b) $|x| < 1$
   (c) 0.43%
   (ii) (a) $1 - 2x + 4x^2$
   (b) $|x| < \frac{1}{2}$
   (c) 0.8%
   (iii) (a) $1 - \frac{x^2}{2} - \frac{x^4}{8}$
   (b) $|x| < 1$
   (c) $0.000\,006\,3\%$

5 (i) $1 - 3x + 3x^2 - x^3$
   (ii) $1 - 4x + 10x^2$; valid when $|x| < 1$
   (iii) $a = -7$, $b = 256$

6 (ii) $1 - \frac{x}{8} + \frac{3x^2}{128}$; valid when $|x| < 4$
   (iii) $1 - \frac{9x}{8} + \frac{19x}{128}$

7 (i) $1 - 3x + 7x^2$
   (ii) $|x| < \frac{1}{2}$

8 $\frac{3}{4} + \frac{11}{16}x - \frac{5}{64}x^2$, $|x| < 4$

9 (i) $a = 2$, $b = -1$, $c = \frac{3}{16}$
   (ii) $|x| < 2$

10 $2(1 - x)$; $|x| < 1$

11 (i) $1 + 2x + 5x^2$
   (ii) $|x| < \frac{1}{3}$

12 (i) $-4 - 10x - 16x^2$
   (ii) $|x| < \frac{1}{2}$

13 (i) $3\sqrt{\frac{7}{9}}$ (there are other options)
   (ii) 2.64609 (5 d.p.)
   (iii) % error 0.013% so very good

14 (i) A possible rearrangement is $\sqrt{x\left(1 - \frac{1}{x}\right)} = \sqrt{x}\left(1 - \frac{1}{x}\right)^{\frac{1}{2}}$

Since $x > 1$
$\Rightarrow 0 < \frac{1}{x} < 1$; the binomial expansion could be used but the resulting expansion would not be a series of positive powers of x.

(ii) 3.968

Exercise 7.2 (page 163)

1 Translation of $\begin{pmatrix} 3 \\ 0 \end{pmatrix}$

2 (i) $(x + 2)$
   (ii) $\frac{1}{(x + 2)}$
   (iii) $\frac{(x - 1)}{(x + 2)}$
   (iv) $\frac{(x + 2)}{(x - 1)}$

3 (i) $\frac{x + 3}{x - 6}$
   (ii) $\frac{2x - 5}{2x + 5}$
   (iii) $\frac{a - b}{2a - b}$

4 (i) $\frac{9}{20x}$
   (ii) $\frac{2a}{(a + 1)(a - 1)}$
   (iii) $\frac{7x + 3}{12}$

5 (i) $-\frac{2}{(x + 2)(x - 2)}$
   (ii) $-\frac{(3b + 1)}{(b + 1)^2}$
   (iii) $\frac{3a - 4}{(a + 2)(2a - 3)}$

6 (i) $2(2x + 1)$
   (ii) $x(2x + 1)$
   (iii) $a(2x + 1)$

7 (i) $\frac{3x^2 - 4}{x(x - 2)(x + 2)}$
   (ii) $\frac{9 + 6a - a^2}{a(3 - a)(3 + a)}$
   (iii) $\frac{2b^2 - 2b + 1}{b(b - 1)^2}$

8 (i) $y = 5 - \dfrac{8}{x+1}$

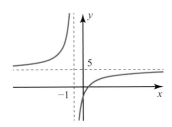

(ii) $y = 4 + \dfrac{10}{x-2}$

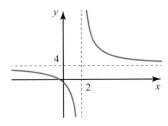

(iii) $y = 3 - \dfrac{6}{2x+1}$

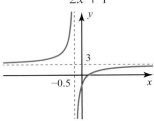

9 (i) $2x + 1 + \dfrac{4}{x+1}$

(ii) $3x - 5 + \dfrac{4}{x+2}$

(iii) $x - 5 + \dfrac{2}{x-1}$

10 $3x^2 - 4x + 3 - \dfrac{10}{x+2}$

11 Write as $y = -3 + \dfrac{2}{x}$;
Stretch scale factor 2 in y direction and translation $\begin{pmatrix} 0 \\ -3 \end{pmatrix}$.

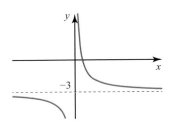

Activity 7.4 (page 163)

(i) $\dfrac{13x - 3}{(3x+1)(2x-3)}$

(ii) $1 - \dfrac{20}{3}x + \dfrac{158}{9}x^2$

(iii) Probably expanding the two separate fractions.

Discussion point (page 165)

The identity is true for all values of x. Once a particular value of x is substituted you have an equation. Equating constant terms is equivalent to putting $x = 0$.

Exercise 7.3 (page 167)

1 $\dfrac{5x + 13}{(x-1)(x+5)}$

2 (i) $\dfrac{1}{x-1} - \dfrac{1}{x}$

(ii) $\dfrac{1}{x} - \dfrac{1}{x+1}$

(iii) $\dfrac{1}{x-1} - \dfrac{1}{x+1}$

3 (i) $\dfrac{1}{x-1} - \dfrac{1}{x+2}$

(ii) $\dfrac{1}{x-2} - \dfrac{1}{x+3}$

(iii) $\dfrac{1}{x-3} - \dfrac{1}{x+4}$

4 (i) $\dfrac{1}{x-2} - \dfrac{2}{2x-1}$

(ii) $\dfrac{1}{x-3} - \dfrac{3}{3x-1}$

(iii) $\dfrac{1}{x-4} - \dfrac{4}{4x-1}$

5 (i) $\dfrac{1}{x-2} - \dfrac{1}{x}$

(ii) $\dfrac{1}{x-3} - \dfrac{1}{x-1}$

(iii) $\dfrac{1}{x-4} - \dfrac{1}{x-2}$

6 (i) $\dfrac{1}{(x-1)^2} - \dfrac{1}{x-1} + \dfrac{1}{x+2}$

(ii) $\dfrac{9}{1-3x} - \dfrac{3}{1-x} - \dfrac{2}{(1-x)^2}$

7 (i) $\dfrac{12}{1+2x} - \dfrac{2}{1+x} - \dfrac{12}{1+3x}$

(ii) $\dfrac{6}{2+x} - \dfrac{2}{1+x} - \dfrac{4}{3+x}$

8 (i) $\dfrac{5}{x-1} - \dfrac{3}{x} - \dfrac{2}{x+1}$

(ii) $\dfrac{8}{x} + \dfrac{3}{2-x} - \dfrac{5}{2+x}$

9 (i) $\dfrac{9}{(1-3x)} - \dfrac{3}{(1-x)} - \dfrac{2}{(1-x)^2}$

(ii) $4 + 20x + 72x^2$

(iii) $|x| < \dfrac{1}{3}$

10 (i) $\dfrac{7x-1}{x^2+1} - \dfrac{7}{x+1}$

(ii) $-8 + 14x - 6x^2$ valid for $|x| < 1$

11 (i) $\dfrac{2}{(x-1)} + \dfrac{1}{(x-1)^2} + \dfrac{3}{(x-1)^3}$

(ii) $-4 - 9x - 17x^2$ valid for $|x| < 1$

Chapter 8

Opening activity (page 169)

Answers will vary.

For example, $a \approx 1.6$ to 1.8, $b \approx 3 \Rightarrow y = 1.5\sin 3x$

Activity 8.1 (page 170)

$y = \sin(\theta + 60°)$ is obtained from $y = \sin\theta$ by a translation $\begin{pmatrix} -60° \\ 0 \end{pmatrix}$

$y = \cos(\theta - 60°)$ is obtained from $y = \sin\theta$ by a translation $\begin{pmatrix} 60° \\ 0 \end{pmatrix}$

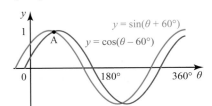

It appears that the θ coordinate of A is midway between the two maxima $(30°, 1)$ and $(60°, 1)$.

Checking: $\theta = 45° \rightarrow$
$\sin(\theta + 60°) = 0.966$

$\cos(\theta - 60°) = 0.966$.

If 60° is replaced by 35°, using the trace function on a graphic calculator would enable the solutions to be found.

Activity 8.2 (page 171)

(i) $\sin(\theta + \phi) =$
$\quad\quad \sin\theta\cos\phi + \cos\theta\sin\phi$
$\Rightarrow \sin[(90° - \theta) + \phi] =$
$\quad\quad \sin(90° - \theta)\cos\phi +$
$\quad\quad \cos(90° - \theta)\sin\phi$
$\Rightarrow \sin[90° - (\theta - \phi)] =$
$\quad\quad \cos\theta\cos\phi + \sin\theta\sin\phi$
$\Rightarrow \cos(\theta - \phi) =$
$\quad\quad \cos\theta\cos\phi + \sin\theta\sin\phi$

(ii) $\Rightarrow \cos[\theta - (-\phi)] =$
$\quad\quad \cos\theta\cos(-\phi) + \sin\theta\sin(-\phi)$
$\Rightarrow \cos(\theta + \phi) =$
$\quad\quad \cos\theta\cos\phi - \sin\theta\sin\phi$

(iii) $\tan(\theta + \phi)$
$= \dfrac{\sin(\theta + \phi)}{\cos(\theta + \phi)}$

$= \dfrac{\sin\theta\cos\phi + \cos\theta\sin\phi}{\cos\theta\cos\phi - \sin\theta\sin\phi}$

$= \dfrac{\left[\dfrac{\sin\theta\cos\phi}{\cos\theta\cos\phi} + \dfrac{\cos\theta\sin\phi}{\cos\theta\cos\phi}\right]}{\left[\left(\dfrac{\cos\theta\cos\phi}{\cos\theta\cos\phi}\right) - \left(\dfrac{\sin\theta\sin\phi}{\cos\theta\cos\phi}\right)\right]}$

$= \dfrac{\tan\theta + \tan\phi}{1 - \tan\theta\tan\phi}$

(iv)
$\tan[\theta + (-\phi)] = \dfrac{\tan\theta + \tan(-\phi)}{1 - \tan\theta\tan(-\phi)}$

$\tan(\theta - \phi) = \dfrac{\tan\theta - \tan\phi}{1 + \tan\theta\tan\phi}$

2 Yes, the formulae are valid for all values of θ and ϕ in both degrees and radians.

Exercise 8.1 (page 173)

1 (i) $\dfrac{1}{\sqrt{2}}(\sin\theta + \cos\theta)$

(ii) $\dfrac{1}{\sqrt{2}}(\cos\theta - \sin\theta)$

(iii) $\dfrac{1}{\sqrt{2}}(\cos\theta + \sin\theta)$

(iv) $\dfrac{1}{\sqrt{2}}(\cos\theta - \sin\theta)$

(v) $\dfrac{\tan\theta + 1}{1 - \tan\theta}$

(vi) $\dfrac{1 - \tan\theta}{1 + \tan\theta}$

2 $\sin 2\theta\cos\theta - \cos 2\theta\sin\theta$
$\quad = \sin\theta$
$\cos\theta\cos 3\theta - \sin\theta\sin 3\theta$
$\quad = \cos 4\theta$
$\cos\theta\cos\theta + \sin\theta\sin\theta = 1$
$\cos\theta\cos\theta - \sin\theta\sin\theta = \cos 2\theta$
$\cos 4\theta\cos 2\theta + \sin 4\theta\sin 2\theta$
$\quad = \cos 2\theta$
$\cos 2\theta\sin\theta + \sin 2\theta\cos\theta$
$\quad = \sin 3\theta$

3 (i) 0

(ii) $\dfrac{\sqrt{3}}{2}$

(iii) $\dfrac{1}{2}$

(iv) -1

4 (i) (a) $\dfrac{1 + \sqrt{3}}{2\sqrt{2}}$

(b) $\dfrac{\sqrt{3} - 1}{2\sqrt{2}}$

(c) $2 + \sqrt{3}$

(ii) (a) $\dfrac{\sqrt{3} - 1}{2\sqrt{2}}$

(b) $\dfrac{1 + \sqrt{3}}{2\sqrt{2}}$

(c) $2 - \sqrt{3}$

5 (i) $\dfrac{\sqrt{3} - 1}{2\sqrt{2}}$

(ii) $\dfrac{1 - \sqrt{3}}{2\sqrt{2}}$

(iii) $-2 - \sqrt{3}$

6 (i) 15°

(ii) 157.5°

(iii) 0° or 180°

(iv) 111.7°

(v) 165°

7 (i) $\dfrac{\pi}{8}$

(ii) $\dfrac{\pi}{2}$

(iii) 2.79 radians

8 (ii) $\dfrac{\pi}{2}$

10 (i) $1 - \theta\sqrt{3} - \dfrac{\theta^2}{2}$

(ii) $\theta\sin\alpha + \theta^2\cos\alpha$

(iii) $\dfrac{1 - \theta}{1 + \theta}$

11 (i) (a) $\dfrac{\theta^2\sqrt{3}}{2}$

(b) $2\theta^2$

(ii) $\dfrac{\sqrt{3}}{4}$

12 (i) $-0.5\,\text{m}$

(ii) $k = \dfrac{\pi}{6}$

(iii) 2π seconds, 14π seconds

(iv) $1\,\text{m}$; 8π seconds

(v) No; $\sin\left(\dfrac{t}{12} - \dfrac{\pi}{6}\right) = -1$
at $t = 20\pi$ and $20\pi > 60$

(vi)

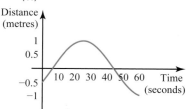

13 (i) $\sin x\cos h + \cos x\sin h$

(ii) $\sin x + h\cos x - \dfrac{h^2}{2}\sin x$

(iii) $\cos x - \dfrac{h}{2}\sin x$

(iv) $\cos x$

(v) Derivative of $\sin x$

14 (i) $\dfrac{1 + \theta}{1 - \theta}$

(ii) $1 + 2\theta + 2\theta^2$

Discussion point (page 174)

For $\sin 2\theta$ and $\cos 2\theta$, substituting $\theta = 45°$ is helpful.
You know that $\sin 45° = \cos 45°$
$= \dfrac{1}{\sqrt{2}}$ and that $\sin 90° = 1$ and
$\cos 90° = 0$.
For $\tan 2\theta$ you cannot use $\theta = 45°$.
Take $\theta = 30°$ instead;
$\tan 30° = \dfrac{1}{\sqrt{3}}$ and $\tan 60° = \sqrt{3}$.
No, checking like this is not the same as proof.

Exercise 8.2 (page 176)

1 (i) $\cos 80°$

(ii) $\dfrac{1}{2}\sin 80°$

(iii) $\tan 40°$

2 $\cot\theta$

3 (i) (a) $\frac{4}{5}$

     (b) $\frac{3}{4}$

     (c) $\frac{24}{25}$

     (d) $\frac{7}{25}$

     (e) $\frac{24}{7}$

  (ii) (a) $-\frac{4}{5}$

     (b) $-\frac{3}{4}$

     (c) $-\frac{24}{25}$

     (d) $\frac{7}{25}$

     (e) $-\frac{24}{7}$

4 (i) $14.5°, 90°, 165.5°, 270°$

  (ii) $0°, 35.3°, 144.7°, 180°,$
$215.3°, 324.7°, 360°$

 (iii) $90°, 210°, 330°$

 (iv) $30°, 150°, 210°, 330°$

  (v) $0°, 138.6°, 221.4°, 360°$

5 (i) $-\pi, 0, \pi$

  (ii) $-\pi, 0, \pi$

 (iii) $-\dfrac{2\pi}{3}, 0, \dfrac{2\pi}{3}$

 (iv) $-\dfrac{3\pi}{4}, -\dfrac{\pi}{2}, \dfrac{\pi}{4}, \dfrac{\pi}{2}$

  (v) $-\dfrac{11\pi}{12}, -\dfrac{3\pi}{4}, -\dfrac{7\pi}{12}, -\dfrac{\pi}{4},$
$\dfrac{\pi}{12}, \dfrac{\pi}{4}, \dfrac{5\pi}{12}, \dfrac{3\pi}{4}$

6 $3\sin\theta - 4\sin^3\theta, \theta = 0,$
$\dfrac{\pi}{4}, \dfrac{3\pi}{4}\pi, \dfrac{5\pi}{4}, \dfrac{7\pi}{4}, 2\pi$

7 $51°, 309°$

9 (ii) $63.4°$

Activity 8.3 (page 177)

$y = a\sin\theta + b\cos\theta$ is a translation
and vertical stretch of a sine or
cosine graph.
(i) and (ii) Answers depend on
students' choice of values of a
and b.

Discussion point (page 178)

A translation by the vector
$\begin{pmatrix} -36.9 \\ 0 \end{pmatrix}$ and a one-way
stretch, scale factor 5, parallel to
the y-axis.

Activity 8.4 (page 178)

(i) The expansion of $r\cos(\theta - \beta)$
is given by $r\cos(\theta - \beta) =$
$r(\cos\theta\cos\beta + \sin\theta\sin\beta)$.

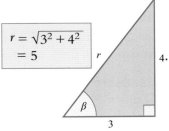

$\boxed{\begin{array}{l} r = \sqrt{3^2 + 4^2} \\ = 5 \end{array}}$

To compare this with
$3\cos\theta + 4\sin\theta$
$r = \sqrt{3^2 + 4^2} = 5$ $\cos\beta =$
$\frac{3}{5}$ $\sin\beta = \frac{4}{5} \Rightarrow \beta = 53.1°$.
So you can write
$3\cos\theta + 4\sin\theta$ in the
form $r\cos(\theta - \beta) =$
$5\cos(\theta - 53.1°)$.

(ii) $y = \cos x$ maps to $y = \sin x$
by a translation by the
vector $\begin{pmatrix} 90 \\ 0 \end{pmatrix}$

So translating $y = \cos x$ by
the vector $\begin{pmatrix} 53.1 \\ 0 \end{pmatrix}$ and
translating $y = \sin x$ by the
vector $\begin{pmatrix} -36.9 \\ 0 \end{pmatrix}$ result in
the same curve.

Exercise 8.3 (page 181)

1 (a) $r = 3, \alpha = 41.8°$

  (b) $r = \sqrt{5}, \alpha = 50.8°$

2 (i) $\dfrac{3}{2}\cos\theta + \dfrac{\sqrt{3}}{2}\sin\theta$

  (ii) $\dfrac{3}{2}\cos\theta - \dfrac{\sqrt{3}}{2}\sin\theta$

 (iii) $\dfrac{3}{2}\sin\theta - \dfrac{\sqrt{3}}{2}\cos\theta$

 (iv) $\dfrac{3}{2}\sin\theta + \dfrac{\sqrt{3}}{2}\cos\theta$

3 (i) (a) max: 2; min: −2

    (b) $x = \dfrac{\pi}{3}$

  (ii) (a) max: $\sqrt{2}$; min: $-\sqrt{2}$

    (b) $x = \dfrac{5\pi}{3}$

 (iii) (a) max: $\dfrac{1}{\sqrt{2}}$; min: $\dfrac{-1}{\sqrt{2}}$

    (b) $x = \dfrac{\pi}{6}$

 (iv) (a) max: $\dfrac{1}{2 - \sqrt{2}}$;

      min: $\dfrac{1}{2 + \sqrt{2}}$

    (b) $x = \dfrac{7\pi}{6}$

4 (i) (a) $\sqrt{2}\cos(\theta - 45°)$

    (b) $5\cos(\theta - 53.1°)$

  (ii) (a) $\sqrt{2}\sin(\theta - 45°)$

    (b) $5\sin(\theta - 53.1°)$

5 (i) (a) $\sqrt{2}\cos\left(\theta + \dfrac{\pi}{4}\right)$

    (b) $\sqrt{2}\sin\left(\theta - \dfrac{\pi}{4}\right)$

  (ii) (a) $2\cos\left(\theta + \dfrac{\pi}{6}\right)$

    (b) $2\sin\left(\theta - \dfrac{\pi}{6}\right)$

6 (i) $2\cos(\theta - (-60°))$

  (ii) $4\cos(\theta - (-45°))$

 (iii) $2\cos(\theta - 30°)$

 (iv) $13\cos(\theta - 22.6°)$

  (v) $2\cos(\theta - 150°)$

 (vi) $2\cos(\theta - 135°)$

7 (i) $13\cos(\theta + 67.4°)$

  (ii) Max 13, min −13

 (iii)

y
13
5
112.6°
O
360° θ
−13

 (iv) $4.7°, 220.5°$

8 (i) $2\sqrt{3}\sin\left(\theta - \dfrac{\pi}{6}\right)$

  (ii) Max $2\sqrt{3}, \theta = \dfrac{2}{3}\pi$;

    min $-2\sqrt{3}, \theta = \dfrac{5}{3}\pi$

 (iii)

y
$2\sqrt{3}$
$\sqrt{3}$
O
2π θ
$-\sqrt{3}$
$-2\sqrt{3}$

(iv) $\dfrac{\pi}{3}, \pi$

9 (i) $\sqrt{3}\cos(\theta - 54.7°)$

   (ii) Max $\sqrt{3}, \theta = 54.7°$;
     min $-\sqrt{3}, \theta = 234.7°$

(iii)

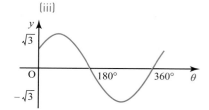

(iv) Max $\left(\dfrac{1}{3 - \sqrt{3}}\right)$,
$\theta = 234.7°$; min
$\left(\dfrac{1}{3 + \sqrt{3}}\right), \theta = 54.7°$

10 (i) $\sqrt{13}\sin(2\theta + 56.3°)$

   (ii) Max $\sqrt{13}, \theta = 16.8°$;
     min $-\sqrt{13}, \theta = 106.8°$

(iii)

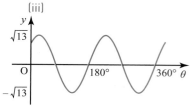

(iv) $53.8°, 159.9°, 233.8°,$
    $339.9°$

Summary exercise (page 182)

1 (i) $\sin 6\theta$
  (ii) $\cos 6\theta$
  (iii) 1
  (iv) $\cos\theta$
  (v) $\sin\theta$
  (vi) $\dfrac{3}{2}\sin 2\theta$
  (vii) $\cos\theta$
  (viii) -1

2 (i) $1 - \sin 2x$
  (ii) $\dfrac{1}{2}(5\cos 2x - 1)$

4 (i) $4.4°, 95.6°$
  (ii) $199.5°, 340.5°$
  (iii) $-\dfrac{\pi}{6}, \dfrac{\pi}{2}$
  (iv) $-15.9°, 164.1°$
  (v) $\dfrac{\pi}{6}, \dfrac{\pi}{2}, \dfrac{5\pi}{6}$
  (vi) $20.8°, 122.3°$
  (vii) $76.0°, 135°$

Chapter 9

Discussion point (page 185)
$f'(x)e^{f(x)}$

Discussion point (page 186)
The graphs are parallel. They are translations in the y direction.

Discussion point (page 186)
$\dfrac{f'(x)}{f(x)}$

Exercise 9.1 (page 187)

1 (i) $\dfrac{dy}{dx} = 3e^{3x+1}$

  (ii) $\dfrac{dy}{dx} = (2x + 3)e^{x^2+3x+1}$

2 (i) $\dfrac{dy}{dx} = 7^x \ln 7$

  (ii) $\dfrac{dy}{dx} = (5\ln 7)7^{5x}$

3 (i) $\dfrac{dy}{dx} = \dfrac{1}{x}$

  (ii) $\dfrac{dy}{dx} = \dfrac{2x}{x^2 + 1}$

  (iii) $\dfrac{dy}{dx} = \dfrac{1}{2x}$

4 (i) $\dfrac{dy}{dx} = e^{4x}(4x + 1)$

  (ii) $\dfrac{dy}{dx} = \dfrac{1 - 2\ln x}{x^3}$

  (iii) $\dfrac{dy}{dx} = \dfrac{1 - x}{e^x}$

5 (i) $\dfrac{4}{e}$
  (ii) 24
  (iii) $\dfrac{3}{8}$

6 (i) $\dfrac{dy}{dx} = e^x(x + 1)$ and
    $\dfrac{d^2y}{dx^2} = e^x(x + 2)$

  (ii) $\left(-1, -\dfrac{1}{e}\right)$

  (iii) $\left(-2, -\dfrac{2}{e^2}\right)$

7 (i) $(2, \ln 7)$
  (ii) $2x - 7y - 4 + 7\ln 7 = 0$

(iii)

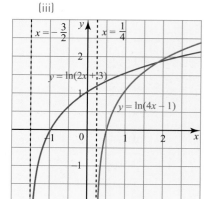

8 (i) Rotational symmetry of order 2 about origin, because $f(-x) = -f(x)$

  (ii) $f'(x) = \ln(x^2) + 2$ and
    $f''(x) = \dfrac{2}{x}$

  (iii) $\left(\dfrac{1}{e}, -\dfrac{2}{e}\right), \left(-\dfrac{1}{e}, \dfrac{2}{e}\right)$

9 (i) $a = 1$
  (ii) $x + 9y - 1 = 0$

10 (i) $\dfrac{dy}{dx} = 2x - 3 + \dfrac{1}{x}, \dfrac{d^2y}{dx^2}$
    $= 2 - \dfrac{1}{x^2}$

  (ii) Minimum at $(1, 0)$, maximum at
    $\left(\dfrac{1}{2}, \dfrac{3}{4} - \ln 2\right)$

11 Minimum at $(1, e)$ and asymptotes $x = 0, y = 0$

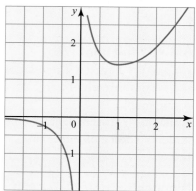

12 The proofs depend on the results that $\dfrac{dy}{dz} = \dfrac{-z}{\sqrt{2\pi}}e^{-\frac{1}{2}z^2}$
and $\dfrac{d^2y}{dz^2} = \dfrac{(z^2 - 1)}{\sqrt{2\pi}}e^{-\frac{1}{2}z^2}$

Discussion point (page 190)

No

Activity 9.2 (page 190)

$$y = \frac{\sin x}{\cos x} \Rightarrow \frac{dy}{dx}$$

$$= \frac{\cos x . \cos x - \sin x(-\sin x)}{\cos^2 x}$$

$$= \frac{\cos^2 x + \sin^2 x}{\cos^2 x} = \frac{1}{\cos^2 x}$$

$$= \sec^2 x$$

Activity 9.3 (page 191)

$$y = \sin(f(x)) \Rightarrow \frac{dy}{dx}$$

$$= f'(x)\cos(f(x))$$

$$y = \cos(f(x)) \Rightarrow \frac{dy}{dx}$$

$$= -f'(x)\sin(f(x))$$

Exercise 9.2 (page 192)

1 (i) $\dfrac{dy}{dx} = \cos x - 2\sin x$

(ii) $\dfrac{dy}{dx} = 5\cos x + 3\sin x$

(iii) $\dfrac{dy}{dx} = 3\sec^2 x - 2$

2 (i) $\dfrac{dy}{dx} = 3\sec^2 3x$

(ii) $\dfrac{dy}{dx} = \frac{1}{2}\cos\left(\frac{x}{2}\right)$

(iii) $\dfrac{dy}{dx} = 6\sin(-6x)$

3 (i) $\dfrac{dy}{dx} = 2x\cos(x^2)$

(ii) $\dfrac{dy}{dx} = e^{\tan x}\sec^2 x$

(iii) $\dfrac{dy}{dx} = -\tan x$

4 (i) $\dfrac{dy}{dx} = \tan x + x\sec^2 x$

(ii) $\dfrac{dy}{dx} = e^x(\cos x - \sin x)$

(iii) $\dfrac{dy}{dx} = \cos^2 x - \sin^2 x$

5 (i) $\dfrac{dy}{dx} = \frac{x\cos x - \sin x}{x^2}$

(ii) $\dfrac{dy}{dx} = \frac{e^x(\cos x + \sin x)}{\cos^2 x}$

(iii) $\dfrac{dy}{dx} = \frac{\sin x - x\cos x - 1}{\sin^2 x}$

6 (i) $\dfrac{dy}{dx} = \frac{3\cos 3x}{2\sqrt{\sin 3x}}$

(ii) $\dfrac{dy}{dx} = -2e^{\cos 2x}\sin 2x$

(iii) $\dfrac{dy}{dx} = \frac{\sin x \cos x - x\ln x}{x\sin^2 x}$

8 (i) $\dfrac{dy}{dx} = \cos x - x\sin x$

(ii) -1

(iii) $y + x = 0$

(iv) $x - y - 2\pi = 0$

9 (i) $\dfrac{dy}{dx} = \cos x - 2\sin 2x$,

$$\frac{d^2 y}{dx^2} = -\sin x - 4\cos 2x$$

(ii) $\left(\dfrac{\pi}{2}, 0\right), (0.253, 1.125)$

$(2.889, 1.125)$

(iii) $\left(\dfrac{\pi}{2}, 0\right)$ is a minimum,

$(0.253, 1.125)$ is a maximum, $(2.889, 1.125)$ is a maximum

(iv) $(0.883, 0.579)$, $(2.259, 0.579)$

11 (i)

$$\frac{dy}{dx} = \lim_{h \to 0}\left(\frac{\cos(x + h) - \cos x}{h}\right)$$

(ii) $\cos(x + h)$

$$= \cos x \cos h - \sin x \sin h$$

(iii) $\dfrac{dy}{dx}$

$$= \lim_{\delta x \to 0}\left(\frac{(\cos x)\left(1 - \frac{1}{2}h^2\right) - (\sin x)h - \cos x}{h}\right)$$

$$= \lim_{h \to 0}\left(\frac{\cos x - \frac{1}{2}h^2\cos x - h\sin x - \cos x}{h}\right)$$

$$= \lim_{h \to 0}\left(-\frac{1}{2}h\cos x - \sin x\right)$$

(iv) $\lim_{h \to 0}\left(-\frac{1}{2}h\cos x - \sin x\right)$

$$= -\sin x$$

12 (i) $\left(\dfrac{\pi}{6}, \dfrac{\pi}{2} - \sqrt{3}\right)$

(ii) $\sqrt{3} + \dfrac{\pi}{6}$

Activity 9.4 (page 194)

(i) $7x^6$

(ii) $10y^4\dfrac{dy}{dx}$

(iii) $-2z\dfrac{dz}{dx}$

(iv) $3\dfrac{dz}{dy}$

(v) $8x^3\dfrac{dx}{dz}$

Exercise 9.3 (page 195)

1 (i) $\dfrac{1}{2\sqrt{y}}\dfrac{dy}{dx}$

(ii) $-\dfrac{8}{y^3}\dfrac{dy}{dx}$

(iii) $3z^2\dfrac{dz}{dx} + 9y^8\dfrac{dy}{dx}$

2 (i) $2ye^{y^2}\dfrac{dy}{dx}$

(ii) $3\dfrac{dy}{dx}\cos(3y - 2)$

(iii) $\dfrac{2y}{y^2 + 2}\dfrac{dy}{dx}$

3 (i) $e^y\left(1 + x\dfrac{dy}{dx}\right)$

(ii) $-\sin x \sin y\dfrac{dy}{dx}$
      $+ \cos x \cos y$

(iii) $\dfrac{3x^2\sin y - x^3\cos y\dfrac{dy}{dx}}{\sin^2 y}$

4 (i) $\dfrac{dy}{dx} = \dfrac{-(1 + y)}{1 + x}$

(ii) $\dfrac{dy}{dx} = \dfrac{-y^2}{2(e^{2y} + xy)}$

(iii) $\dfrac{dy}{dx} = \dfrac{-\sin x}{\sin y}$

5 $\dfrac{1}{5}$

6 (i) 0

(ii) $y = -1$

7 Tangent is $x + y - 2 = 0$ and normal is $x - y + 2 = 0$

10 $(1, -2)$ and $(-1, 2)$

11 (i) $\dfrac{dy}{dx} = \dfrac{y + 4}{6 - x}$

(ii) $x - 2y - 11 = 0$

(iii) $\left(2, -\dfrac{9}{2}\right)$

12 (i) $\left(\frac{\pi}{2}, 0\right), \left(\frac{\pi}{2}, \pi\right)$

   (ii) $\left(0, \frac{\pi}{2}\right), \left(\pi, \frac{\pi}{2}\right)$

   (iii) $x = \frac{\pi}{2}, y = \frac{\pi}{2}$

   (iv)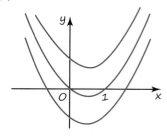

Chapter 10

Review exercise (page 201)

1 (i) $\frac{3}{2}x^8 + c$

   (ii) $\frac{4}{3}x^3 + x + c$

   (iii) $x^2 - \frac{x^5}{5} + \frac{x^9}{3} + c$

2 (i) $\frac{15}{2}$

   (ii) $\frac{75}{4}$

   (iii) -26

3 (i) $\frac{4x\sqrt{x}}{3} + c$

   (ii) $-\frac{3}{x} + c$

   (iii) $\frac{6x^3\sqrt{x}}{7} + c$

4 (i) $-\frac{103}{12}$

   (ii) $\frac{14}{9}$

5 (i) $y = x^2 - x + c$

   (ii)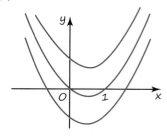

   (iii) $y = x^2 - x + 1$

6 $y = 2 - \frac{2}{x^2} - 2\sqrt{x}$

7 (i) $\frac{y^3}{3} + 2y - \frac{1}{y} + c$

   (ii) $\frac{t^4}{2} - 2\sqrt{t} + c$

8 $f(x) = x^3 - 2x^2 + x - 4$

9 $\frac{9}{2}$

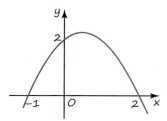

10 $\frac{1}{24}$

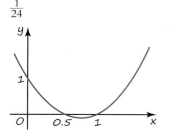

11 (i) $x^3 + 3x^2 - x - 3 = (x - 1)(x + 1)(x + 3)$

   (ii)

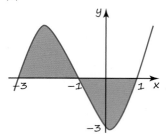

   (iii) 8

12 (i) $y + 2x - 3 = 0$

   (ii) $\left(\frac{3}{2}, 0\right)$

   (iii)

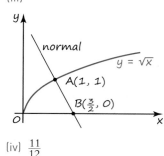

   (iv) $\frac{11}{12}$

Discussion point (page 203)

The estimate may be good enough for some contexts. It can be improved by using more, narrower rectangles.

Discussion point (page 205)

Each of these rectangles has a height which is the height of the top curve minus the height of the bottom curve. The enclosed region is the limit of the sum of the areas of these rectangles.

Discussion point (page 208)

The area under the red curve is partly above and partly below the x-axis. If you find the total area between the red curve and the x-axis, and then subtract this from the area under the blue curve, this will give the wrong answer. You would need to subtract the areas under the red curve between $x = 0$ and $x = 1$, and between $x = 3$ and $x = 4$, and add the area between $x = 1$ and $x = 3$.

Exercise 10.1 (page 208)

1 $\frac{4}{3}$

2 (ii) 6 square units

3 (i) $(3, 9), (-3, 9)$

   (ii)

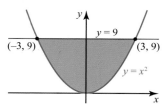

   (iii) 36 square units

4 (i) $(1, 5), (5, 5)$

   (ii)

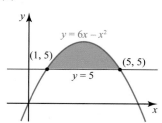

   (iii) $\frac{32}{3}$ square units

5 (i) $(0, 0), (3, 3)$

(ii)

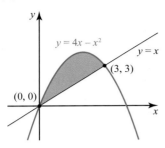

(iii) 4.5 square units

6 (i) $\frac{4}{3}$ square units, $\frac{4}{3}$ square units, $\frac{4}{3}$ square units

(ii) $\frac{32}{3}$ square units

(iii) $\int_{-2}^{2}(3-(x^2-1))\,dx$
$= \int_{-2}^{2}(4-x^2)\,dx$

(iv) $\frac{32}{3}$ square units

7 $\frac{14}{3}$ square units

8 (i) $(-1, 5)$, $(3, -3)$
(ii)

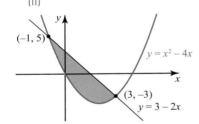

(iii) $\frac{32}{3}$ square units

9 $\frac{64}{3}$

10 (i) $(1, 4)$, $(-1, 4)$
(ii)

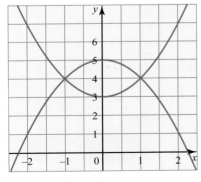

(iii) $\frac{8}{3}$ square units

11 $\frac{4}{3}$ square units

12 72 square units

13 8 square units

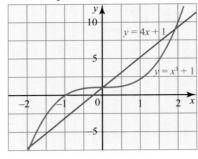

14 (i)

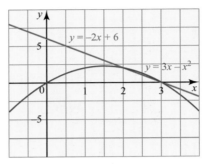

(ii) $\sqrt{5}$

(iii) $\frac{1}{6}$

15 (i) 20 square units
(ii) It is impossible to tell as the strips are partly below and partly above the curve.
(iii) 20 square units
(iv) 19.5 square units
(v) $18\frac{2}{3}$ square units. The more rectangles used, the closer the estimate gets to the true answer.

16 (i) 21 square units
(ii) 21.25 square units
(iii) $21\frac{1}{3}$ square units

17 652:77

Discussion point (page 212)
$\frac{1}{20} = \frac{1}{5 \times 4}$
5 is the derivative of the bracket and 4 is the new power.

Activity 10.1 (page 212)
(i) (a) $\frac{1}{14}(2x + 3)^7 + c$

(b) $-\frac{1}{12}(5 - 3x)^4 + c$

(c) $\frac{1}{3}(1 + 2x)^{\frac{3}{2}} + c$

(ii) $\frac{1}{a(n + 1)}(ax + b)^{n+1} + c$

Activity 10.2 (page 214)
(i) $\frac{2(x^3 + 1)^{\frac{3}{2}}}{9} + c$

Exercise 10.2 (page 215)
1 (i) $\frac{1}{5}(x - 7)^5 + c$

(ii) $\frac{1}{10}(2x - 7)^5 + c$

(iii) $\frac{1}{25}(5x - 7)^5 + c$

2 (i) $\frac{1}{10}(x - 2)^{10} + c$

(ii) $\frac{1}{8}(2x + 3)^4 + c$

(iii) $-\frac{1}{36}(1 - 6x)^6 + c$

3 (i) 0.2
(ii) 20
(iii) 20

4 $\frac{812}{625}$

5 (i) $\frac{2}{3}(x + 3)^{\frac{3}{2}} + c$

(ii) $-\frac{1}{3(3x + 1)} + c$

(iii) $\sqrt{2x - 1} + c$

6 $y = \sqrt{3} + \sqrt{2x + 3}$

7 $\frac{4}{3}$ square units

8 $\frac{175}{12}$

9 (i) $\frac{1}{14}(x^2 - 1)^7 + c$

(ii) $\frac{2}{9}(x^3 + 1)^{\frac{3}{2}} + c$

(iii) $\frac{1}{4(1 - x^4)} + c$

10 $\frac{1}{6}$

11 (i) $(2x + 1)^4 =$
$16x^4 + 32x^3 + 24x^2 + 8x + 1$
$\int (2x + 1)^4 dx =$
$\frac{16}{5}x^5 + 8x^4 + 8x^3 + 4x^2 + x + c$

(ii) $\frac{1}{10}(2x + 1)^5 + k$

(iv) In (ii) the constant term is $\frac{1}{10} + k$ which is an arbitrary constant.

12 (i) A negative number does not have a square root.

    (ii) $\left|\sqrt{x^2 - 1}\right| < |x|$

    (iii) $2\sqrt{2} - \sqrt{3}$

Activity 10.3 (page 216)

1 (i) $\cos x$

    (ii) $-\sin x$

    (iii) $\sec^2 x$

2 (i) $\sin x + c$

    (ii) $-\cos x + c$

    (iii) $\tan x + c$

3 (i) $4\cos 4x$

    (ii) $-5\sin 5x$

    (iii) $5\sec^2 5x$

4 (i) $\frac{1}{4}\sin 4x + c$

    (ii) $-\frac{1}{5}\cos 5x + c$

    (iii) $\frac{1}{5}\tan 5x + c$

Activity 10.5 (page 218)

1 (i) e^x

    (ii) $3e^{3x}$

2 (i) $e^x + c$

    (ii) $\frac{1}{3}e^{3x} + c$

Activity 10.6 (page 219)

    (i) $e^{f(x)} + c$

    (ii) $\dfrac{(f(x))^{n+1}}{n + 1} + c$

Exercise 10.3 (page 220)

1 $-\frac{1}{2}\cos 2x + c$

2 (i) $\frac{1}{2}e^{2x} + c$

    (ii) $\frac{1}{3}e^{3x+5} + c$

    (iii) $-\frac{1}{2}e^{3-2x} + c$

3 (i) $-\frac{1}{6}\cos 6x + c$

    (ii) $\frac{1}{2}\sin(2x + 1) + c$

    (iii) $\frac{1}{3}\cos(1 - 3x) + c$

4 (i) $\sin x + \cos x + c$

    (ii) $e^x - e^{-x} + c$

    (iii) $-3\cos x - \frac{1}{2}e^{2x} + c$

5 (i) 1

    (ii) $\frac{\pi}{2} - 1$

    (iii) $2(e - 1)$

6 (i) $\frac{1}{2}e^{x^2+3} + c$

    (ii) $-\frac{1}{5}\cos^5 x + c$

7 (i) $\left(\dfrac{1}{2}, \dfrac{1}{2\sqrt{e}}\right),$

    $\left(-\dfrac{1}{2}, -\dfrac{1}{2\sqrt{e}}\right)$

    (ii) 0.432

8 (i) $-e^{\cos x} + c$

    (ii) $\frac{1}{8}\left(e^x - 5\right)^8 + c$

9 (i) $(0,0), \left(\sqrt{\ln 3}, 3\sqrt{\ln 3}\right),$

    $\left(-\sqrt{\ln 3}, -3\sqrt{\ln 3}\right)$

    (ii) $3\ln 3 - 2$

10 (i) $\cos^2 x = \frac{1}{2}(1 + \cos 2x)$

    (ii) $\frac{1}{2}x + \frac{1}{4}\sin 2x + c$

    (iii) $\frac{1}{2}x - \frac{1}{4}\sin 2x + c$

11 (ii) $-\cos x + \frac{1}{3}\cos^3 x + c$

Discussion point (page 220)

All three differentiate to $\dfrac{1}{x}$.

For $x > 0, \int \frac{1}{x}\,dx = \ln kx + c$ where k is any positive constant.

This is possible because of the arbitrary constant: $\ln kx = \ln k + \ln x$, and $\ln k$ is a constant.

$\ln x$ is only defined for positive values of x.

Discussion point (page 222)

k must be positive for $\ln k$ to be defined. Yes, for every value of c we can write $k = e^c$, and therefore $k > 0$ and $c = \ln k$.

Activity 10.7 (page 224)

$\dfrac{x + 3}{(x + 1)^2} = \dfrac{x + 1 + 2}{(x + 1)^2}$

$= \dfrac{x + 1}{(x + 1)^2} + \dfrac{2}{(x + 1)^2}$

$= \dfrac{1}{x + 1} + \dfrac{2}{(x + 1)^2}$

Exercise 10.4 (page 225)

1 (i) $2\ln x + c$

    (ii) $\frac{1}{2}\ln x + c$

    (iii) $\frac{1}{2}\ln(2x + 7) + c$

2 (i) $\ln|x - 4| + c$

    (ii) $-\ln|2 - x| + c$

    (iii) $\frac{1}{2}\ln|2x - 7| + c$

3 (i) $\ln 2$

    (ii) $\ln\left(\frac{2}{3}\right)$

    (iii) $\ln\left(\frac{9}{4}\right)$

4 $\ln|x^2 + 1| + c$

5 $\ln\left(\frac{2}{3}\right)$

6 (i) $\ln|x^3 - 4| + c$

    (ii) $-\frac{1}{2}\ln|2 - x^2| + c$

7 (i) $\frac{1}{3}\ln 6$

    (ii) $-\frac{1}{2}\ln 2$

8 (i) $\ln\left|\dfrac{3x - 2}{1 - x}\right| + c$

    (ii) $\ln\left|\dfrac{x}{1 - x}\right| - \dfrac{1}{x} + c$

    (iii) $\ln\left|\dfrac{(x - 1)^2}{\sqrt{2x + 1}}\right| + c$

9 (i) $\frac{1}{2}\ln\left(\frac{3}{2}\right)$

    (ii) $\frac{2}{3} + \ln\left(\frac{21}{11}\right)$

    (iii) $\ln\left(\frac{5}{2}\right) - \frac{2}{5}$

10

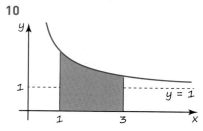

Asymptotes are $x = 0, y = 1$

Area of region is $2 + \frac{3}{2}\ln 3$

11 (i) $y = 0$, $x = \frac{1}{2}$

(ii)

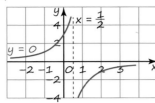

(iii) $\frac{3}{2}\ln\left(\frac{5}{3}\right)$

12 (i) $\frac{1}{1+x} + \frac{2}{1-2x}$

(ii) 0.31845

13 (i) $\left(0, \frac{1}{2}\right)$

(ii) As $x \to \infty$, $y \to 1$ (from below), as $x \to -\infty$, $y \to 0$ (from above)

(iii) 1.31 square units

14 (i) $\frac{1}{2}\ln\left(\frac{x-1}{x+1}\right) + c$

(ii) $\frac{1}{2}\ln\left|x^2 - 1\right| + c$

(iii) $x + \frac{1}{2}\ln\left(\frac{x-1}{x+1}\right) + c$

15 (i) $\left(\frac{2\pi}{3}, \frac{\sqrt{3}}{3}\right)$

(ii) Area of triangle $= \frac{\sqrt{3}}{6}\pi$

(iii) Total area $\approx \frac{\sqrt{3}}{3}\pi$

(iv) $2\ln 3$. This is a little more than the answer from (iii)

16 (i) $-\frac{1}{x} + c$

(ii) There is a discontinuity at $x = 0$

(iii) No, because the function is undefined at $x = 0$

(iv) You cannot evaluate a definite integral if the function has a discontinuity within the interval of integration.

Activity 10.9 (page 228)

(i) (a) $u = 1 + x$

(b) $\ln|1 + x| + \frac{1}{1+x} + c$

(ii) (a) $u = 2 + x$

(b) $\frac{2}{15}(2 + x)^{\frac{3}{2}}(3x - 4) + c$

Exercise 10.5 (page 228)

1 (i) $x - 2\ln|x + 1| - \frac{1}{x+1} + c$

(ii) $\frac{1}{15}(2x + 1)^{\frac{3}{2}}(3x - 1) + c$

2 (i) $\frac{7}{8}$

(ii) $100\frac{4}{5}$

3 (i) (a) $u = 1 - x$

(b) $\frac{-2(2 + x)\sqrt{1 - x}}{3} + c$

(ii) (a) $u = x + 4$

(b) $\frac{2(3x - 8)(x + 4)^{\frac{3}{2}}}{3} + c$

(iii) (a) $u = 2x + 1$

(b) $\frac{1}{4}\ln|2x + 1| - \frac{x}{2} + c$

4 (i) $y = 2x - \ln|2x + 1| + c$

(ii) $y = 2x - 1 - \ln|2x + 1|$

5 (i) $a = 1$

(ii) Minimum at $\left(-1, -\frac{1}{4}\right)$

(iii) $\ln 2 - \frac{1}{2}$

6 $\frac{(1 + \ln x)^2}{6}(2\ln x - 1) + c$

7 (i) $k = -3$

(ii) Minimum at $(-4.5, 3.93)$

(iii) 1.72

8 $y = \frac{1}{35}(2x + 1)^{\frac{5}{2}}(5x - 1) + 2$

9 0.0271

10 (ii) $\frac{\pi}{8}$

11 (ii) $\frac{\pi}{6}$

Activity 10.10 (page 229)

(i) $\int x\sin x \, dx = -x\cos x + \sin x + c$

(ii) $\int 2xe^{2x} \, dx = xe^{2x} - \frac{1}{2}e^{2x} + c$

Activity 10.11 (page 230)

(i) $\int x\sin x \, dx = \frac{x^2}{2}\sin x - \int \frac{x^2}{2}\cos x \, dx$

(ii) Parts chosen the wrong way round

Exercise 10.6 (page 233)

1 $\frac{1}{2}x\sin 2x + \frac{1}{4}\cos 2x + c$

2 $\frac{5}{e} - e$

3 (i) $\frac{1}{2}xe^{2x} - \frac{1}{4}e^{2x} + c$

(ii) $\frac{x^2}{2}\ln 3x - \frac{x^2}{4} + c$

4 (i) $1 - \frac{2}{e}$

(ii) $2 + \pi$

5 $\frac{1}{15}(5x + 2)(x - 2)^5 + c$

6 (i) $(0, 2), (2, 0)$

(ii) $e^{-2} + 1$

7 (i)

(graph)

(ii) π

8 $\ln 3 - \frac{2}{3}$

9 (i) 1

(ii) $e - 1$

(iii) e

This is the area of the rectangle formed by the red and blue regions.

10 (i) $-\frac{1}{2}x^2\cos 2x + \frac{1}{2}x\sin 2x + \frac{1}{4}\cos 2x + c$

(ii) $-e^{-x}\left(x^2 + 2x + 2\right) + c$

11 (i) $(1, 0)$

(ii) Minimum at $\left(\frac{1}{\sqrt{e}}, -\frac{1}{2e}\right)$

(iii)

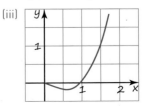

(iv) $\frac{8}{3}\ln 2 - \frac{7}{9}$

12 (ii) $\frac{1}{2}\left(e^x \sin x - e^x \cos x\right) + c$

Summary exercise (page 234)

1. $-\frac{1}{3}xe^{-3x} - \frac{1}{9}e^{-3x} + c$
2. $\frac{1}{3}x^3 - \frac{5}{2}x^2 + 6x + c$
3. $\frac{1}{22}(2x + 5)^{11} + c$
4. $\frac{1}{2}e^{x^2} + c$
5. $\frac{1}{4}(1 + \sin x)^4 + c$
6. $-\frac{1}{12(3x-4)^4} + c$
7. $\frac{1}{3}\sin 3x + \frac{1}{9}\cos 3x + c$
8. $\frac{1}{4}\ln\left|\frac{x-2}{x+2}\right| + c$
9. $x + 2\ln|x - 2| + c$
10. $\frac{1}{3}(x + 3)^3 + c$
11. $\frac{1}{7}x^7 - 16x + c$
12. $x\ln 3x - x + c$
13. $\frac{1}{2}\tan^2 x + c$
14. $\ln|x^3 + x| + c$
15. $e^{\sin x} + c$
16. $\frac{1}{2}\sin^2 x + c$
17. $\tan x + c$
18. $\ln\left|\frac{(x-3)^3}{(x-5)^2}\right| + c$
19. $\frac{2}{27}(9x + 4)^{\frac{3}{2}} + c$
20. $x + \tan x + c$
21. $\frac{1}{5}x^5 + \frac{1}{4}x^4 + \frac{1}{3}x^3 + \frac{1}{2}x^2 + x + c$
22. $2x + c$
23. $\frac{1}{2}x^2 e^{2x} - \frac{1}{2}xe^{2x} + \frac{1}{4}e^{2x} + c$
24. $2\sin x + c$
25. $\frac{1}{4}x^4 + \frac{2}{3}x^3 + \frac{1}{2}x^2 + c$
26. $\frac{1}{4}x^4 \ln x - \frac{1}{16}x^4 + c$
27. $\frac{1}{2}\ln|x^2 + 2\ln x| + c$
28. $\frac{1}{2}x^2 - x + \ln|x + 1| + c$
29. $x - \ln|\cos x| + c$
30. $\frac{1}{8}e^{2x} - \frac{1}{8}e^{-2x} + \frac{1}{2}x + c$

Multiple choice questions

1. B $-2 < x < 2$
2. D $2\cos^2\theta + \cos\theta - 1 = 0$

3. B $2y\dfrac{dy}{dx}$
4. C $u = x - 5$

Practice questions 2

(page 237)

1. $\int_2^4 \frac{1}{x}\,dx = \left[\ln x\right]_2^4$ [1]

    $= \ln 4 - \ln 2 = \ln 2$

 $\int_5^a \frac{1}{x}\,dx = \ln a - \ln 5$ [1]

 $\ln 2 = \ln\frac{a}{5}$ (oe)   [1]

 $a = 10$ [1]

2. (i) Any corect justification, e.g. $y = 5\sin x$ is the curve which passes through the origin since $5\sin 0 = 0$.   [1]

 (ii) Valid method to determine horizontal translation such as

 $4\sin x - 3\cos x = 0$ [1]

 $\tan x = \frac{3}{4}$ [1]

 Therefore the translation is described by the vector   [1]

 $\begin{pmatrix} -\arctan\frac{3}{4} \\ 0 \end{pmatrix}$ [1]

3. (i) $(1 - x)^{-2}$

 $= 1 + (-2)(-x) + \frac{(-2)(-3)}{2}(-x)^2 + \ldots$ [1]

 $= 1 + 2x + 3x^2 + 4x^3$

 $+ 5x^4 + 6x^5 + \ldots$ [2]

 ([1] if at least three correct terms)

 (ii) $\left(\frac{9}{10}\right)^{-2}$ [1]

 $= 1 + 0.2 + 0.03 + 0.004 +$ $0.0005 + 0.00006 + \ldots$ [1]

4. Area of triangle $AOB = \frac{1}{2}.r.r\sin 2\alpha$ [1]

 Area of triangle AOB $= 2 \times$ area of triangle OMA   (oe)   [1]

 $= 2 \times \frac{1}{2}$OM.AM

 $= r\cos\alpha.r\sin\alpha$ [2]

 $\frac{1}{2}.r.r\sin 2\alpha =$ $r\cos\alpha.r\sin\alpha$

 $\sin 2\alpha = 2\cos\alpha.\sin\alpha$ [1]

5. (i) $\dfrac{1}{\cos^2\theta} + \dfrac{1}{\sin^2\theta} \equiv \dfrac{\sin^2\theta + \cos^2\theta}{\cos^2\theta\sin^2\theta}$ [1]

 $\equiv \dfrac{1}{\cos^2\theta\sin^2\theta}$

 $= \dfrac{4}{(2\sin\theta\cos\theta)^2}$ [1]

 $= \dfrac{4}{\sin^2 2\theta}$ [1]

 (ii) This is equivalent to $\dfrac{4}{\sin^2 2\theta} = 4$ and so

 $\sin 2\theta = \pm 1$ [1]

 Roots $2\theta =$

 $\dfrac{\pi}{2}, \dfrac{3\pi}{2}, \dfrac{5\pi}{2}, \dfrac{7\pi}{2}, \ldots$ [1]

 at least two correct and so

 $\theta = \dfrac{\pi}{4}, \dfrac{3\pi}{4}, \dfrac{5\pi}{4}, \dfrac{7\pi}{4}, \ldots$ [1]

 all correct

6. (i) $x^2\dfrac{dy}{dx} + 2xy + 2\dfrac{dy}{dx} - 4 = 0$ [1]

 for product rule;   [1]

 for at least one correct appearance of $\dfrac{dy}{dx}$

 $\dfrac{dy}{dx}(x^2 + 2) = 4 - 2xy$ [1]

 $\dfrac{dy}{dx} = \dfrac{4 - 2xy}{2 + x^2}$

 (ii) $\dfrac{dy}{dx} = 0 \iff$

 $4 - 2xy = 0$ [1]

 Substitute $y = \dfrac{2}{x}$ or equivalent in $x^2 y + 2y - 4x = 0$ [1]

 $\dfrac{4}{x} - 2x = 0 \implies x = \pm\sqrt{2}$

 ($x = \sqrt{2}$ OK)   [1]

 Maximum point has coordinates $\left(\sqrt{2}, \sqrt{2}\right)$ [1]

7 (i)
$$\frac{x-2}{(x+1)(x-3)} \equiv \frac{A}{x+1} + \frac{B}{x-3}$$

$x - 2 \equiv A(x-3) + B(x+1)$ [1]

$x = 3:\quad 1 = 4B \Rightarrow B = \frac{1}{4}$ [1]

$x = -1: -3 = -4A \Rightarrow A = \frac{3}{4}$ [1]

$$\frac{x-2}{(x+1)(x-3)} \equiv \frac{\frac{3}{4}}{x+1} + \frac{\frac{1}{4}}{x-3}$$

(ii) $\int_0^2 \frac{\frac{3}{4}}{x+1} + \frac{\frac{1}{4}}{x-3}\, dx$ [1]

$= \left[\frac{3}{4}\ln|x+1| + \frac{1}{4}\ln|x-3|\right]_0^2$

[1] for ln, [1] for $|x - 3|$

$= \left(\frac{3}{4}\ln 3 + \frac{1}{4}\ln 1\right) - \left(\frac{3}{4}\ln 1 + \frac{1}{4}\ln 3\right)$

$= \left(\frac{3}{4}\ln 3\right) - \left(\frac{1}{4}\ln 3\right)$ [1]

$= \frac{1}{2}\ln 3$ [1]

8 (i) P clearly lies on the line $y = x$.
P lies on the curve since when
$x = \frac{\pi}{2}, y = \frac{\pi}{2}.\sin\frac{\pi}{2} = \frac{\pi}{2}$ [1]

Evidence of the need to show that at P the gradient of the curve is 1. [1]

$y = x\sin x \Rightarrow$

$\frac{dy}{dx} = \sin x + x\cos x$ [1]

When $x = \frac{\pi}{2}$,

$\frac{dy}{dx} = \sin\frac{\pi}{2} + \frac{\pi}{2}\cos\frac{\pi}{2}$

$= 1 + 0 = 1$ [1]

(ii)
$\int x\sin x\, dx = -x\cos x + \int\cos x\, dx$
[1] by parts

$= -x\cos x + \sin x + c$
[1] accuracy

$\frac{1}{2}.\frac{\pi}{2}.\frac{\pi}{2} - \int_0^{\frac{\pi}{2}} x\sin x\, dx$

or $\int_0^{\frac{\pi}{2}} x - x\sin x\, dx$ [1]

$= \frac{\pi^2}{8} - \left[-x\cos x + \sin x\right]_0^{\frac{\pi}{2}}$ [1]

$= \frac{\pi^2}{8} - 1$ [1]

9 (i)

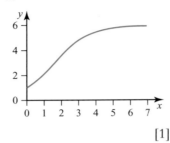

[1]
curve increasing towards asymptote

(ii) $\frac{dP}{dt} =$ [1] with attempt

$= \frac{(5 + e^t)6e^t - 6e^t.e^t}{(5 + e^t)^2}$

[1]

$= \frac{30e^t}{(5 + e^t)^2}$ [1]

$\frac{d^2P}{dt^2} = 0$ [1] attempt

$\frac{(5+e^t)^2 30e^t - 30e^t.2(5+e^t)e^t}{(5+e^t)^4}$

[1]

$= 30e^t\frac{(5 + e^t) - 2e^t}{(5 + e^t)^3}$ [1]

$5 = e^t \Rightarrow t = \ln 5$ [1]

(iii) Substitute $P = 7$ to obtain $7 = \frac{6e^t}{5 + e^t}$ [1]

Simplification to $e^t = -35$ [1]

The equation has no solution so P cannot reach a value of 7. [1]

Review: Coordinate geometry

Discussion point (page 242)

Area of triangle

$= \frac{1}{2}$ CM × MD

$= \frac{1}{2} \times \sqrt{3^2 + 6^2} \times \sqrt{12^2 + 6^2}$

$= \frac{1}{2} \times \sqrt{45} \times \sqrt{180}$

$= \frac{1}{2} \times \sqrt{8100}$

$= 45$

Exercise R.1 (page 242)

1 (i) (a) -1
    (b) $(4, 3)$
    (c) $2\sqrt{2}$
 (ii) (a) -1
    (b) $(3, 4)$
    (c) $2\sqrt{2}$
 (iii) (a) $-\frac{3}{4}$
    (b) $(-1, -1)$
    (c) 10
 (iv) (a) 1
    (b) $(-3, 4)$
    (c) $2\sqrt{2}$

2 (i) 3
 (ii) -2
 (iii) $-\frac{2}{3}$
 (iv) $\frac{5}{2}$

3 (i) $y = 2x + 3$
 (ii) $2y = x - 3$
 (iii) $2y = -x - 3$
 (iv) $2y = x + 3$
 (v) $2y = 3 - x$

4 (i) $(3, 5)$
 (ii) $(-2, 3)$
 (iii) $(-1, 4)$
 (iv) $\left(\frac{9}{8}, -\frac{1}{4}\right)$

5 $k = 9$

6 (i) Gradient of AB is $\frac{\pi}{2}$ and gradient of BC $= -\frac{4}{3}$.
$-\frac{4}{3} \times \frac{3}{4} = -1$, hence AB and BC are perpendicular and ABC is right-angled.
 (ii) 12.5 square units

7 (i) Gradient of AB is $-\frac{1}{2}$, gradient of CD is $-\frac{1}{2}$, gradient of BC is -3 and gradient of AD is $\frac{1}{3}$. Hence one pair of parallel sides $\Rightarrow$ ABCD is a trapezium.

AD = BC = $\sqrt{10}$ ⇒
ABCD is an isoceles
trapezium.

(ii) $\left(\frac{5}{3}, \frac{7}{3}\right)$

8 6.25
9 (i) $k = 2$
   (ii) $3x - 2y - 27 = 0$
10 $2y + 3x = 0$
11 (4, 0), (0, 1.6)
12 6.5 square units
13 (2, −1), (−0.8, −6.6)
14 $\frac{13}{3}$ square units

Exercise R.2 (page 245)

1 (i) (3, 1); 5
   (ii) (3, −1); $\sqrt{5}$
   (iii) (−3, −1); 5
   (iv) (1, −3); $\sqrt{5}$
2 $(x - 5)^2 + (y + 2)^2 = 16$
3 $x^2 + 6x + y^2 - 8y - 56 = 0$
4 $(x - 1)^2 + (y - 6)^2 = 10$
5 (i) $(x - 4)^2 + (y + 1)^2 = 10$
   (ii) centre (4, −1); radius $\sqrt{10}$
6 (5, −5), (8, −2)
7 $\frac{6}{5}\sqrt{5}$
8 $\left(\sqrt{2}, -2\left(\sqrt{2} - 1\right)\right)$ and
   $\left(-\sqrt{2}, 2\left(1 + \sqrt{2}\right)\right)$
9 $k = 13$ or $k = -7$
10 (i) Use the circle theorem:
   'angle in a semicircle is
   a right angle' to show
   PQR is a right-angled
   triangle with a right
   angle at R. Either show
   $PR^2 + QR^2 = PQ^2$
   $(2 + 82 = 164)$ or show
   $m_{PR} \times m_{RQ} = -1$
   $\left(\frac{1}{9} \times -9 = -1\right)$
   (ii) $(x - 3)^2 + (y + 5)^2 = 41$
   (iii) $\left(0, -5 \pm 4\sqrt{2}\right)$,
   (−1, 0), (7, 0)
11 (i) Centre lies on the
   perpendicular bisector of
   AB which is $2y = x + 4$
   (ii) $(x - 2)^2 + (y - 3)^2 = 10$

12 $12\sqrt{5}$
13 (2.4, 0.2); 5

Chapter 11

Discussion point (page 251)

At points where the rate of
change of gradient is greatest.

Exercise 11.1 (page 253)

1

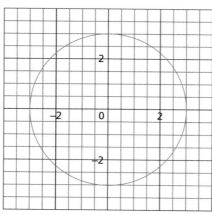

2 (−2, 1)
3 (i) (a) (4, 1)
      (b) (7, 6)
      (c) (−2, 3)
   (ii) $t = -3$
4 (i) (1, 3)
   (ii) (27, 27)
   (iii) (8, 12), (−8, 12)
5 (i) (a)

t	x	y
−2	−4	4
−1.5	−3	2.25
−1	−2	1
−0.5	−1	0.25
0	0	0
0.5	1	0.25
1	2	1
1.5	3	2.25
2	4	4

(b)

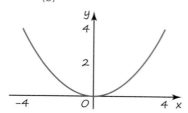

(ii) (a)

t	x	y
−2	4	−8
−1.5	2.25	−3.375
−1	1	−1
−0.5	0.25	−0.125
0	0	0
0.5	0.25	0.125
1	1	1
1.5	2.25	3.375
2	4	8

(b)

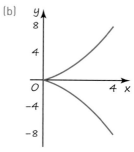

(iii) (a)

t	x	y
−2	4	6
−1.5	2.25	3.75
−1	1	2
−0.5	0.25	0.75
0	0	0
0.5	0.25	−0.25
1	1	0
1.5	2.25	0.75
2	4	2

(b)

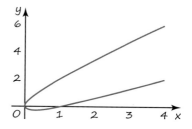

6 (i) (a)

θ	x	y
0°	1	0
30°	0.5	0.25
60°	−0.5	0.75
90°	−1	1
120°	−0.5	0.75
150°	0.5	0.25
180°	1	0
210°	0.5	0.25
240°	−0.5	0.75
270°	−1	1
300°	−0.5	0.75
330°	0.5	0.25
360°	1	0

(b)

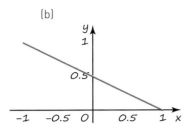

(c) Defined for
$-1 \leqslant x \leqslant 1$ and
$0 \leqslant y \leqslant 1$

(ii) (a)

θ	x	y
0°	0	1
30°	0.25	2
60°	0.75	2.73
90°	1	3
120°	0.75	2.73
150°	0.25	2
180°	0	1
210°	0.25	0
240°	0.75	−0.73
270°	1	−1
300°	0.75	−0.73
330°	0.25	0
360°	0	1

(b)

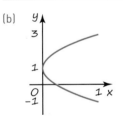

(c) Defined for
$0 \leqslant x \leqslant 1$ and
$-1 \leqslant y \leqslant 3$

(iii) (a)

θ	x	y
0°	0	3
30°	0.5	2.6
60°	1.5	1.5
90°	2	0
120°	1.5	−1.5
150°	0.5	−2.6
180°	0	−3
210°	0.5	−2.6
240°	1.5	−1.5
270°	2	0
300°	1.5	1.5
330°	0.5	2.6
360°	0	3

(b)

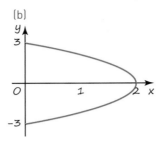

(c) Defined for $0 \leqslant x \leqslant 2$
and $-3 \leqslant y \leqslant 3$

7 (i)

t	x	y
−2	4	16
−1.5	2.25	5.0625
−1	1	1
−0.5	0.25	0.0625
0	0	0
0.5	0.25	0.0625
1	1	1
1.5	2.25	5.0625
2	4	16

(ii)

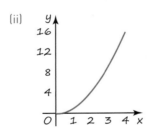

(iii) Because it should also
state 'for $x \geqslant 0$'

8 (i)

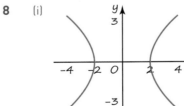

(ii) Undefined for $-2 < x < 2$

9 (i) (a)

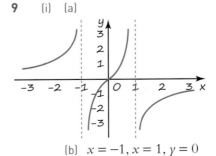

(b) $x = -1, x = 1, y = 0$

(ii) (a)

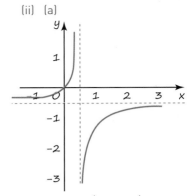

(b) $x = \frac{1}{2}, y = -\frac{1}{2}$

10 (i)

t	x	y
−2	0.14	−0.91
−1.5	0.22	−1.00
−1	0.37	−0.84
−0.5	0.61	−0.48
0	1	0
0.5	1.65	0.48
1	2.72	0.84
1.5	4.48	1.00
2	7.39	0.91

(ii) $x > 0$

(iii)

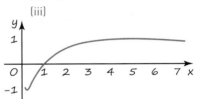

(iv) The graph oscillates
infinitely many times
from −1 to +1 for $t < -2$,
i.e. where $0 < x < 0.14$.

For $t > 2$ the graph
oscillates infinitely many
times from −1 to +1,
but successive distances
between a maximum
and a minimum become
increasingly large.

11 (i)

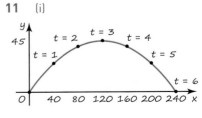

(ii) 240 m

(iii)

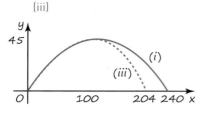

(iv) 36 m

12 (i)

θ	x	y
0	0	0
$\frac{\pi}{3}$	$0.2a$	$0.5a$
$\frac{2\pi}{3}$	$1.2a$	$1.5a$
π	$3.1a$	$2a$
$\frac{4\pi}{3}$	$5.1a$	$1.5a$
$\frac{5\pi}{3}$	$6.1a$	$0.5a$
2π	$6.3a$	0
$\frac{7\pi}{3}$	$6.5a$	$0.5a$
$\frac{8\pi}{3}$	$7.5a$	$1.5a$
3π	$9.4a$	$2a$
$\frac{10\pi}{3}$	$11.3a$	$1.5a$
$\frac{11\pi}{3}$	$12.4a$	$0.5a$
4π	$12.6a$	0
$\frac{13\pi}{3}$	$12.7a$	$0.5a$
$\frac{14\pi}{3}$	$13.8a$	$1.5a$
5π	$15.7a$	$2a$
$\frac{16\pi}{3}$	$17.6a$	$1.5a$
$\frac{17\pi}{3}$	$18.7a$	$0.5a$
6π	$18.8a$	0

(ii)

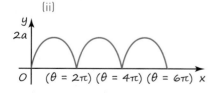

(iii) Periodic

13 (i)

θ	x	y
0	a	0
$\frac{\pi}{3}$	$0.13a$	$0.65a$
$\frac{\pi}{2}$	0	a
$\frac{2\pi}{3}$	$-0.13a$	$0.65a$
π	$-a$	0
$\frac{4\pi}{3}$	$-0.13a$	$-0.65a$
$\frac{3\pi}{2}$	0	$-a$
$\frac{5\pi}{3}$	$0.13a$	$-0.65a$
2π	a	0

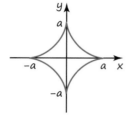

(ii)

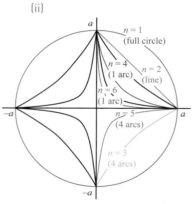

When $n = 0$ the curve becomes the single point (a, a).

(iii) (a) The larger the value of n the closer the curve is to the axes. If the power is even, the curve is only in the first quadrant.

(b) If the power is odd, the curve is in all four quadrants.

Discussion point (page 257)

A one-way stretch scale factor 3 parallel to the y-axis followed by a one-way stretch scale factor 4 parallel to the x-axis (transformations can be carried out in either order).

Discussion point (page 257)

(i)

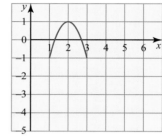

(ii)

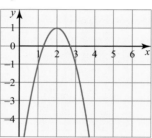

They are different because in (i) $1 \leqslant x \leqslant 3$ as $\sin \theta$ lies between -1 and 1 and $-1 \leqslant y \leqslant 1$ as $\cos 2\theta$ lies between -1 and 1 and in (ii) there are no restrictions on the domain $(x \in \mathbb{R})$ and the range is $y \leqslant 1$

Activity 11.1 (page 258)

(i)

$x = 2 + 5\sin t$	$x = 2 + 5\cos 2t$
$y = 3 + 5\cos t$	$y = 3 + 5\sin 2t$
$x = 2 - 5\cos t$	$x = 2 + 5\cos t$
$y = 3 - 5\sin t$	$y = 3 + 5\sin t$

$x = 3 + 5\sin t$	$x = 3 + 5\cos \frac{1}{2}t$
$y = 2 + 5\cos t$	$y = 2 + 5\sin \frac{1}{2}t$

$$x = 2 + 5\cos t \quad x = 2 + 5\sin t$$
$$y = 3 + 5\cos t \quad y = 3 + 5\sin t$$

$$x = 2 + 5\sin\tfrac{1}{2}t$$
$$y = 3 + 5\sin\tfrac{1}{2}t$$

$$x = 5\cos\tfrac{1}{2}t \quad x = 5\cos 2t$$
$$y = 5\sin\tfrac{1}{2}t \quad y = 5\sin 2t$$

$$x = 5\cos t$$
$$y = 5\sin t$$

(ii) Changing the interval for the parameter produces different sections of the circle.

For example, for $x = 2 + 5\sin t$, $y = 3 + 5\cos t$ the following curves are produced:

(a) $0 \leqslant t \leqslant \dfrac{\pi}{2}$

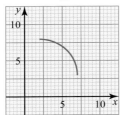

(b) $0 \leqslant t \leqslant \pi$

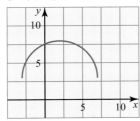

(c) $0 \leqslant t \leqslant 2\pi$

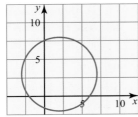

Discussion point (page 258)

Changing t to $2t$ traces out the circle more quickly, so a full circle is drawn in the interval $0 \leqslant t \leqslant \pi$ and it is described twice in the interval $0 \leqslant t \leqslant 2\pi$.

Discussion point (page 259)

(i)

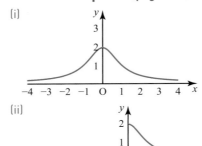

(ii)

Since $e^t > 0$, the second graph only shows half of the full curve.

Exercise 11.2 (page 259)

1   $xy = 9$ and $\begin{aligned} x &= 3t \\ y &= \dfrac{3}{t} \end{aligned}$

   $y^2 = 12x$ and $\begin{aligned} x &= 3t^2 \\ y &= 6t \end{aligned}$

   $y = (1 - x)^2$ and $\begin{aligned} x &= 1 - t \\ y &= t^2 \end{aligned}$

   $xy = 3$ and $\begin{aligned} x &= \dfrac{1}{t} \\ y &= 3t \end{aligned}$

   $4y = 4 - x^2$ and $\begin{aligned} x &= 2t \\ y &= 1 - t^2 \end{aligned}$

2   (i) $y = \dfrac{x^2}{4}$

   (ii) $y^2 = x^3$

   (iii) $y = x \pm \sqrt{x}$

3   (i) $x = 2t$, $y = 5 - 4t$ for $0 \leqslant t \leqslant 2$

   (ii) $x = u - 1$, $y = 7 - 2u$ for $1 \leqslant u \leqslant 5$

4   (i) $y = -1 + 4\sin t$

   (ii) $y = -1 + 4\cos t$

5   (i)   A segment of $y = \dfrac{1 - x}{2}$ where $-1 \leqslant x \leqslant 1$ and $0 \leqslant y \leqslant 1$

   (ii)   Part of $(y - 1)^2 = 4x$ where $0 \leqslant x \leqslant 1$ and $-1 \leqslant y \leqslant 3$

   (iii)   Part of $y^2 = \dfrac{9}{2}(2 - x)$ where $0 \leqslant x \leqslant 2$ and $-3 \leqslant y \leqslant 3$

6   (i) $x^2 + y^2 = 25$

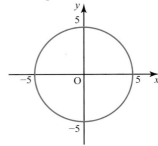

   (ii) $x^2 + y^2 = 9$

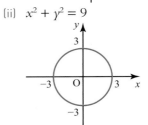

   (iii) $(x - 4)^2 + (y - 1)^2 = 9$

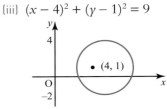

   (iv) $(x + 1)^2 + (y - 3)^2 = 4$

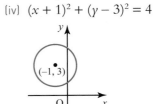

7   (i)

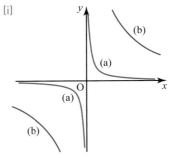

   (ii)   (a) $xy = 1$

   (b) $xy = 16$

(iii) The curve in (b) is an enlargement of the curve in (a), centre the origin, scale factor 4.

8 (i) $7 = 2 \times 3 + 1$

(ii) (a) $\begin{pmatrix} 3 \\ 7 \end{pmatrix}$

(b) $\begin{pmatrix} 1 \\ 2 \end{pmatrix}$ or any vector of

the form $\begin{pmatrix} a \\ 2a \end{pmatrix}, a \neq 0$

(iii) $\begin{pmatrix} 3 \\ 7 \end{pmatrix} + t\begin{pmatrix} 1 \\ 2 \end{pmatrix} = \begin{pmatrix} 3 + t \\ 7 + 2t \end{pmatrix}$

(iv) $x = 3 + t,\ y = 7 + 2t$

9 (i)

(ii) $y = -2$
(iii) $x = (y + 2)^2$

10 (i) $x^2 - y^2 = 4$

(ii) $y = \dfrac{2x}{1 - x^2}$

11 (i) C(4, 6); P(9, 6)
(ii) $x = 4 + 5\cos t$, $y = 6 + 5\sin t$
(iii) (a) $X\left(\dfrac{13}{2}, 6 + \dfrac{5}{2}\sqrt{3}\right)$, $XY = 5\sqrt{2}$
(b) No

Exercise 11.3 (page 263)

1 $y - 2x + 2 = 0$

2 (i) (a) $\dfrac{\mathrm{d}x}{\mathrm{d}t} = 6t$

(b) $\dfrac{\mathrm{d}y}{\mathrm{d}t} = 6t^2$

(c) $\dfrac{\mathrm{d}y}{\mathrm{d}x} = t$

(ii) (a) $\dfrac{\mathrm{d}x}{\mathrm{d}t} = 4$

(b) $\dfrac{\mathrm{d}y}{\mathrm{d}t} = 4t^3$

(c) $\dfrac{\mathrm{d}y}{\mathrm{d}x} = t^3$

(iii) (a) $\dfrac{\mathrm{d}x}{\mathrm{d}t} = 1 - \dfrac{1}{t^2}$

(b) $\dfrac{\mathrm{d}y}{\mathrm{d}t} = 1 + \dfrac{1}{t^2}$

(c) $\dfrac{\mathrm{d}y}{\mathrm{d}x} = \dfrac{t^2 + 1}{t^2 - 1}$

3 (i) (a) $\dfrac{\mathrm{d}x}{\mathrm{d}t} = 2(t + 1)$

(b) $\dfrac{\mathrm{d}y}{\mathrm{d}t} = 2(t - 1)$

(c) $\dfrac{\mathrm{d}y}{\mathrm{d}x} = \dfrac{t - 1}{t + 1}$

(ii) (a) (16, 4)

(b) $\dfrac{\mathrm{d}y}{\mathrm{d}x} = \dfrac{1}{2}$

(c) $2y = x - 8$

4 (i) $\dfrac{1 + \cos\theta}{1 + \sin\theta}$

(ii) $-\dfrac{2}{3}\cot\theta$

(iii) $-\tan\theta$

5 (i) (a) $\dfrac{(1 + t)^2}{(1 - t)^2}$

(b) 9

(ii) (a) $y = \dfrac{x}{1 - 2x}$

(b) $\dfrac{1}{(1 - 2x)^2}$

(iii) At $t = 2$, $x = \dfrac{2}{3}$ and
$\dfrac{\mathrm{d}y}{\mathrm{d}x} = \dfrac{1}{(1 - 2x)^2} = 9$

6 (i) (10, 58.75)
(ii) 70 m
(iii) 5.74 s; 114.8 m
(iv)

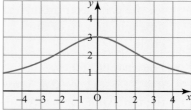

7 (i) $\dfrac{\mathrm{d}y}{\mathrm{d}x} = \dfrac{-t(t^3 - 2)}{1 - 2t^3}$

(ii) $\left(2^{\frac{1}{3}}, 2^{\frac{2}{3}}\right)$

(iii) $\left(\dfrac{3}{2}, \dfrac{3}{2}\right)$; $\dfrac{3\sqrt{2}}{2}$; P is the farthest point on the curve from the origin

8 (i) $\dfrac{1}{2e^t}$

(ii) (a) $y = \dfrac{1}{2}x$

(b) $2ey = x + e^2 - 1$

9 (i) (a) $(1, -3)$ minimum
(b)

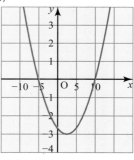

(ii) (a) $(0, 3)$ maximum

10 (i) 6
(ii) $y = 6x - \sqrt{3}$
(iii) $3x + 18y - 19\sqrt{3} = 0$

11 (i) $\left(\dfrac{1}{4}, 0\right)$

(ii) 2
(iii) $y = 2x - \dfrac{1}{2}$

(iv) $\left(0, -\dfrac{1}{2}\right)$

12 (i) $x - ty + at^2 = 0$
(ii) $tx + y = at^3 + 2at$
(iii) $(at^2 + 2a, 0),\ (0, at^3 + 2at)$

13 (ii) $(0, -1)$; minimum

14 (i) $-\dfrac{b}{at^2}$

(ii) $at^2y + bx = 2abt$

(iii) X$(2at, 0)$, Y$\left(0, \dfrac{2b}{t}\right)$

(iv) Area $= 2ab$

15 (ii) $(1, 0)$ and $(-1, 0)$
(iii) The tangent is vertical at the point $(0, 1)$.

(iv)

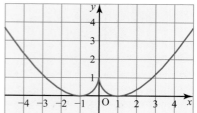

16 (i) $-\dfrac{3\cos t}{4\sin t}$

    (ii) $3x\cos t + 4y\sin t = 12$

    (iii) $t = 0.6435 + n\pi$

17 (i) $x\cos\theta + y\sin\theta =$
$3\sin\theta + 3\cos\theta + 2$

    (iii) $2.85, 5.01$ radians

    (iv)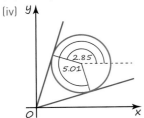

18 (i) $-\dfrac{\cos\theta}{\sin\theta}$

    (ii) $y\cos\theta - x\sin\theta =$
$5\cos\theta - 2\sin\theta$

Chapter 12

Opening activity (page 266)

Yes, by sailing at an angle to the wind. When a boat sails in the same direction as the wind it can never go faster than wind speed.

The keel stops the boat moving sideways, so only the forwards component of the wind comes into play. However, when a boat sails at an angle to the wind it can increase the relative wind velocity across the sails and so sail faster than the wind.

Discussion point (page 270)

For example, velocity, force, acceleration, weight, displacement.

Exercise 12.1 (page 272)

1 $\mathbf{a} = -2\mathbf{i} - 3\mathbf{j}$
     $\mathbf{b} = 2\mathbf{i}$
     $\mathbf{c} = \mathbf{i} + 3\mathbf{j}$
     $\mathbf{d} = -2\mathbf{i} + 2\mathbf{j}$
     $\mathbf{e} = -2\mathbf{j}$
     $\mathbf{f} = 3\mathbf{i} + 2\mathbf{j}$

2 (i) $\mathbf{b} = \mathbf{a}, \mathbf{c} = -\mathbf{a}, \mathbf{d} = 2\mathbf{a},$
$\mathbf{e} = -3\mathbf{a}$; $\mathbf{f}$ cannot be
given in terms of $\mathbf{a}$

    (ii) (a)
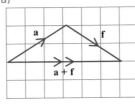

    (b)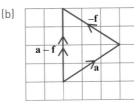

    (c)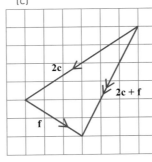

    (d)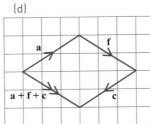

3 (i) $(13, 67.4°)$
    (ii) $(13, 22.6°)$
    (iii) $(13, -67.4°)$
    (iv) $(13, 112.6°)$
    (v) $(13, 202.6°)$

4 (i) 5
    (ii) $\sqrt{21}$
    (iii) $\sqrt{5}$
    (iv) $\sqrt{14}$

5 (i) $\begin{pmatrix} 1 \\ -3 \\ 2 \end{pmatrix}$

    (ii) $\begin{pmatrix} -2 \\ -2 \\ 4 \end{pmatrix}$

    (iii) $\begin{pmatrix} -2 \\ -3 \\ 3 \end{pmatrix}$

    (iv) $\begin{pmatrix} 16 \\ 1 \\ 1 \end{pmatrix}$

    (v) $\begin{pmatrix} -7 \\ -5 \\ 8 \end{pmatrix}$

    (vi) $\begin{pmatrix} 2 \\ 0 \\ -6 \end{pmatrix}$

6 (i) $\sqrt{\left(\tfrac{1}{3}\right)^2 + \left(-\tfrac{2}{3}\right)^2 + \left(\tfrac{2}{3}\right)^2}$
$= \sqrt{\tfrac{1}{9} + \tfrac{4}{9} + \tfrac{4}{9}} = 1$

    (ii) (a) $\tfrac{2}{7}\mathbf{i} - \tfrac{6}{7}\mathbf{j} + \tfrac{3}{7}\mathbf{k}$

    (b) $\tfrac{\sqrt{3}}{3}\mathbf{i} + \tfrac{\sqrt{3}}{3}\mathbf{j} + \tfrac{\sqrt{3}}{3}\mathbf{k}$

7 $\mathbf{a}, \mathbf{d}, \mathbf{e}, \mathbf{f},$ and $\mathbf{g}$ are parallel
    $\mathbf{c}$ and $\mathbf{h}$ are parallel

8 $x = 2, y = 3$

9 $a = -4.5, b = 10.5$

10 $k = 4, 10.8\,\text{N}$

11 (i) $\overrightarrow{AC} = -3\mathbf{i} + \mathbf{j} + 2\mathbf{k},$
$\overrightarrow{AB} = 5\mathbf{i} - 4\mathbf{j} + 3\mathbf{k},$
$\overrightarrow{BC} = -8\mathbf{i} + 5\mathbf{j} - \mathbf{k}$

    (ii) $\sqrt{98} \neq \sqrt{6} + \sqrt{46} - \sqrt{8}$

12 (i) $(6\mathbf{i} + \mathbf{j})\,\text{N}$
    (ii) $(4\mathbf{i} + 5\mathbf{j})\,\text{N}$
    (iii) $2\sqrt{5}\,\text{N}; 116.6°$ to the $\mathbf{i}$
direction
    (iv) $\sqrt{10}\,\text{N}; 71.6°$ to the $\mathbf{i}$
direction

13 $a = -6; b = -5$

14 (i) $a = -3; b = -2$
    (ii) $a = -6; b = 6$
    (iii) $a = 0; b = 1$

15 (i) $12.8\,\text{km}$
    (ii) $20\,\text{km}\,\text{h}^{-1}, 5\,\text{km}\,\text{h}^{-1}$
    (iii) After 40 minutes the
boats meet.

Discussion point (page 274)

Three different answers:

$$\overrightarrow{OD} = \begin{pmatrix} 10 \\ -1 \\ -3 \end{pmatrix}, \overrightarrow{OD} = \begin{pmatrix} -14 \\ 7 \\ -1 \end{pmatrix}$$

or $\overrightarrow{OD} = \begin{pmatrix} 16 \\ 11 \\ 21 \end{pmatrix}$

Exercise 12.2 (page 275)

1 (i)
$$\overrightarrow{AB} = -\overrightarrow{CD},$$
$$\overrightarrow{BC} = -\overrightarrow{DA} \Rightarrow \overrightarrow{AB} + \overrightarrow{CD}$$
$$+ \overrightarrow{BC} + \overrightarrow{DA}$$
$$= \overrightarrow{AB} - \overrightarrow{AB} + \overrightarrow{BC} - \overrightarrow{BC} = 0$$
　(ii) Yes, closed loop.

2 (i) (a) $2\mathbf{b}$
　　　(b) $\mathbf{c} - \mathbf{a}$
　　　(c) $-\mathbf{a} - \mathbf{b}$
　　　(d) $\mathbf{c}$ or $(\mathbf{b} - \mathbf{a})$
　(ii) $\overrightarrow{AB} = \mathbf{c} = \mathbf{b} - \mathbf{a}$
　　　$\Rightarrow \mathbf{a} + \mathbf{c} - \mathbf{b} = \mathbf{0}$

3 (i) (a) $\mathbf{a}$
　　　(b) $\mathbf{c} + \mathbf{a}$
　　　(c) $\mathbf{c} - \mathbf{a}$
　　　(d) $\mathbf{a} - \mathbf{c}$
　　　(e) $-\mathbf{a} - \mathbf{c}$
　　　(f) $\frac{1}{2}\mathbf{c}$
　　　(g) $\mathbf{a} + \frac{1}{2}\mathbf{c}$
　　　(h) $\frac{1}{2}\mathbf{c} - \mathbf{a}$
　(ii) $\overrightarrow{OP} = \frac{1}{2}(\mathbf{c} + \mathbf{a})$,
　　　$\overrightarrow{OQ} = \frac{1}{2}(\mathbf{a} + \mathbf{c})$;
　　　diagonals of a
　　　parallelogram bisect
　　　each other.

4 (i) (a) $\mathbf{q}$
　　　(b) $-\mathbf{p}$
　　　(c) $\mathbf{q} + \mathbf{p}$
　　　(d) $\mathbf{r} + \mathbf{p}$
　　　(e) $\mathbf{p} + \mathbf{q} + \mathbf{r}$
　(ii) $\mathbf{p} + \frac{3}{5}\mathbf{r}$
　(iii) Assume the diagonals
　　　bisect at N, then
　　　$\overrightarrow{ON} = \frac{1}{2}\overrightarrow{OC}$
　　　$= \frac{1}{2}(\mathbf{p} + \mathbf{q} + \mathbf{r})$

Also,
$$\overrightarrow{ON} = \mathbf{p} + \mathbf{q} + \frac{1}{2}\overrightarrow{BG}$$
$$= \mathbf{p} + \mathbf{q} + \frac{1}{2}(\mathbf{r} - \mathbf{p} - \mathbf{q})$$
$$= \frac{1}{2}\mathbf{p} + \frac{1}{2}\mathbf{q} + \frac{1}{2}\mathbf{r} = \frac{1}{2}\overrightarrow{OC}$$
as required.

5 Trapezium
6 11.74

7 (i) $\frac{3}{2}(\mathbf{b} - \mathbf{a})$
　(ii) $\overrightarrow{AB} = (\mathbf{b} - \mathbf{a})$,
　　　$\overrightarrow{PQ} = \frac{3}{2}(\mathbf{b} - \mathbf{a})$ and
　　　$\overrightarrow{XY} = \frac{2}{3}(\mathbf{b} - \mathbf{a})$ since
　　　the vectors are all scalar
　　　multiples of $\mathbf{b} - \mathbf{a}$ they
　　　are all parallel.
　(iii) 4:9

8 $\overrightarrow{AB} = \overrightarrow{DC} = \begin{pmatrix} 2 \\ 4 \\ 8 \end{pmatrix}$,
so AB and CD are parallel
and equal in length.

$\overrightarrow{AD} = \overrightarrow{BC} = \begin{pmatrix} 3 \\ -3 \\ -6 \end{pmatrix}$,
so AD and BC are also
parallel and equal in length.
Hence ABCD is a parallel-
ogram as it has two pairs of
equal, parallel sides.

9 $\overrightarrow{OD} = \frac{1}{2}(\mathbf{p} + \mathbf{q})$,
$\overrightarrow{OE} = \frac{1}{2}\mathbf{p} + \mathbf{q}$,
$\overrightarrow{DE} = \overrightarrow{OE} - \overrightarrow{OD} =$
$\frac{1}{2}\mathbf{p} + \mathbf{q} - \frac{1}{2}(\mathbf{p} + \mathbf{q}) = \frac{1}{2}\mathbf{q}$

10 (i) $\overrightarrow{OD} = \begin{pmatrix} 1 \\ 4 \\ 10 \end{pmatrix}$
　(ii) $|\overrightarrow{AB}|^2 + |\overrightarrow{AD}|^2 = |\overrightarrow{BD}|^2$
　　　$\Rightarrow 24 + 35 = 59$
　　　$\Rightarrow$ triangle ABD is
　　　right-angled, hence
　　　ABCD is a rectangle.
　(iii) $\overrightarrow{OX} = \begin{pmatrix} 0.5 \\ 5.5 \\ 6.5 \end{pmatrix}$

11 $\overrightarrow{OC} = \frac{1}{3}(\mathbf{a} + 2\mathbf{b})$ and
$\overrightarrow{OD} = \frac{1}{2}(\mathbf{a} + 2\mathbf{b})$, hence the
vectors are scalar multiples of
each other and so are parallel.
Both vectors pass through O,
so OCD is a straight line; 2:3

12 (i) $\overrightarrow{OM} = \begin{pmatrix} -0.25 \\ 3 \\ -1.25 \end{pmatrix}$
　(ii) $\overrightarrow{ON} = \begin{pmatrix} 0.8 \\ 4.4 \\ 1.2 \end{pmatrix}$

13 $a = 4$ or -2

14 $\overrightarrow{OC} = \overrightarrow{OA} + \frac{s}{s + t}\overrightarrow{AB}$
$\overrightarrow{OA} = \mathbf{a}$ and $\overrightarrow{AB} = \mathbf{b} - \mathbf{a}$
$\overrightarrow{OC} = \mathbf{a} + \frac{s}{s + t}(\mathbf{b} - \mathbf{a})$
$= \mathbf{a} + \frac{s}{s + t}\mathbf{b} - \frac{s}{s + t}\mathbf{a}$
$= \frac{s + t}{s + t}\mathbf{a} - \frac{s}{s + t}\mathbf{a} + \frac{s}{s + t}\mathbf{b}$
$= \frac{t}{s + t}\mathbf{a} + \frac{s}{s + t}\mathbf{b}$

15 No, a quadrilateral is a plane
figure and these points do
not lie on the same plane.

Chapter 13

Discussion point (page 279)
(i) First
(ii) Third
(iii) First
(iv) Second

Exercise 13.1 (page 282)

1 (i) $y = x^3 + c$

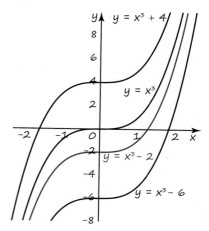

　(ii) $y = x^3 - 2$

2 $\dfrac{\mathrm{d}v}{\mathrm{d}t}$ is the rate of change of the velocity with respect to time (the acceleration). The differential equation tells you that the acceleration is increasing and so is the velocity.

3 $\dfrac{\mathrm{d}N}{\mathrm{d}t} = kN$

4 The odd one out is $\dot{y}$. It is the rate of change of y with respect to time; all the others are the rate of change of y with respect to x.

5 (i) $y = x^2 - 4x + c$

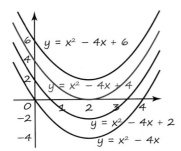

$y = x^2 - 4x + 6$
$y = x^2 - 4x + 4$
$y = x^2 - 4x + 2$
$y = x^2 - 4x$

6 $\dfrac{\mathrm{d}P}{\mathrm{d}t} = k\sqrt{P}$

7 $\dfrac{\mathrm{d}v}{\mathrm{d}t} = \dfrac{4}{\sqrt{v}}$

8 $\dfrac{\mathrm{d}T}{\mathrm{d}x} = -\dfrac{x}{4}$

9 (i) $y = 2\mathrm{e}^{-2x} + c$

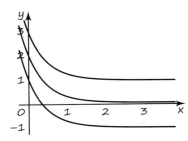

(ii) $y = 2\mathrm{e}^{-2x} - \dfrac{1}{2}$

10 $\dfrac{\mathrm{d}T}{\mathrm{d}t} = -\dfrac{(T - 15)}{160}$

11 (i) $\dfrac{\mathrm{d}r}{\mathrm{d}t} = -k\sqrt{t}$ with $k = 1.5$

(ii) $r = 8 - t^{\frac{3}{2}}$

(iii) 4 weeks

12 (i) $\dfrac{\mathrm{d}v}{\mathrm{d}t} = \dfrac{2}{t + 2}$

13 (i) $\dfrac{\mathrm{d}M}{\mathrm{d}t} = 0.05t(8 - t)$

(ii) 4.27 kg

16 (i) It is growing at a constant rate of 0.5 metres per year. After 10 years it is 5 metres tall.

(ii) $\dfrac{\mathrm{d}h}{\mathrm{d}t} = \dfrac{k}{t}$ with $k = 5$

(iv) 74 years to the nearest year

(v)

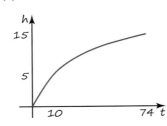

Exercise 13.2 (page 287)

1 $y = \mathrm{e}^{\frac{1}{3}} x^3$

2 (i) $y = \sqrt[3]{\dfrac{3x^2}{2} + c}$

(ii) $y = \dfrac{2}{c - x^2}$

3 (i) $y = \ln\left(\dfrac{x^2}{2} + 1\right)$

(ii) $y = \sec x$

4 11.02 a.m.

5 $\dfrac{\mathrm{d}s}{\mathrm{d}t} = \dfrac{2}{s}$, $s = \sqrt{4t + c}$

6 (i) $\theta = 20 - A\mathrm{e}^{-2t}$

(ii) $\theta = 20 - 15\mathrm{e}^{-2t}$

(iii)

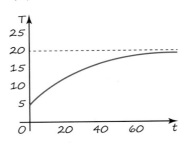

(iv) 60 seconds

(v) Temperature becomes the same as the room, 20°C

7 $y = \pm\sqrt{\dfrac{5x^2 + 3}{2}}$

8 (i) $\dfrac{4}{2x - 3} - \dfrac{1}{x - 1}$

(ii) $\dfrac{\mathrm{d}y}{\mathrm{d}x} = \dfrac{A(2x - 3)^2}{x - 1}$

9 (i) $h = 5 + \dfrac{(20 - 3t)^2}{4}$

(ii) The water stops flowing

(iii) 163 seconds

10 (i) $\dfrac{\mathrm{d}r}{\mathrm{d}t} = \dfrac{k}{r^2}$

(ii) $k = 5000$, 141m (3 s.f.)

(iii) $\dfrac{\mathrm{d}r}{\mathrm{d}t} = \dfrac{c}{r^2(2 + t)}$,

$c = 10000$

(iv) 104 m (3 s.f.)

11 (i) $\dfrac{\mathrm{d}v}{\mathrm{d}t} = \dfrac{3}{v}$

(ii) $v = \sqrt{6t + 4}$

(iii) 10 seconds

12 (i) $P = 600\mathrm{e}^{kt}$

(iii) $P = 600\mathrm{e}^{(0.005t - 0.4\sin(0.02t))}$, very good fit with the data

(iv) 549

13 (i) $q = Ap^{-\eta}$

(ii) (a) $\eta = 0.776$ (to 3 s.f.)

(b) $q = 817p^{-0.776}$

(c)

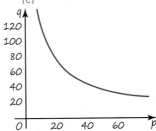

14 (i) $\dfrac{\mathrm{d}P}{\mathrm{d}t} = \lambda P$

(ii) $P = P_0\mathrm{e}^{2t}$

(iii) 499 minutes

(iv) $P = P_0\mathrm{e}^{\sin 2t}$

(v) 551 minutes

(vi) In the first model, the population increases exponentially with time. In the second model, the population oscillates with time, going above then below the initial population of P_0

15 (ii) $\dfrac{1}{c}$

Chapter 14

Activity 14.1 (page 292)
(i)   Yes: $x = 1$, $x = 3$
(ii) Yes: $x = -5 \pm \sqrt{17}$
(iii) No: $x \approx -1.6$, $x \approx 0.6$, $x \approx 1.3$
(iv) Yes: $x = 0$, $x = -1$, $x = 1$
(v)   No: $x \approx 0.35$, $x \approx 2.15$

Discussion point (page 292)
The first root is between 0.3 and 0.4, and the second is between 2.1 and 2.2.

The first root looks as if it is closer to 0.4, so you can be fairly confident to state that the root is 0.4 to 1 d.p., but you would have to zoom in further to show that the root is to the right of 0.35, to be sure. It is not clear whether the second root is nearer to 2.1 or 2.2 – you would have to zoom in to see on which side of 2.15 the root lies.

Discussion points (page 294)
Decimal search is reasonably efficient but a computer programme might need to carry out several calculations before finding a change of sign. If you are doing it yourself on a calculator, you can speed it up: if you know that a root lies in the interval [1, 2] you could try 1.5 first, to find out which half of the interval it is in, rather than working through 1.1, 1.2, etc.

Discussion points (page 295)
Interval bisection is more efficient when programming a computer because it follows the rule of halving the interval each time, so only one point needs to be tested at each stage. However, the numbers to be tested quickly

become long, so this method is more difficult to carry out by hand.

Discussion point (page 295)
9 iterations for 1 d.p. and 10 iterations for 2 d.p.

Activity 14.2 (page 298)
The iterations either diverge, or they converge to the root already found in the interval [0, 1].

Activity 14.3 (page 299)
(i) $\left[0.45, 0.55\right]$
(ii) $\left[-0.3695, -0.3685\right]$
(iii) $\left[21.63415, 21.63425\right]$

Discussion point (page 301)
Often it is not so much a failure of the method as a failure to use an appropriate starting point or an appropriate re-arrangement of the equation.

Discussion point (page 301)
The iteration diverges because of the gradient of the curve; it is steeper than the line $y = x$.

Exercise 14.1 (page 302)
1   $f(x) = e^x - 5x$
    $f(2) = -2.61\ldots < 0$
    $f(3) = 5.08\ldots > 0$
    Sign change and continuity, therefore root in the interval.

2   Equation 1 – Diagram (b).
    The iterations are converging to the root α.

    Equation 2 – Diagram (c).
    The iterations are converging to the root α.

    Equation 3 – Diagram (a).
    The iterations are diverging.

    Equation 4 – Diagram (d).
    The iterations are diverging.

4   (i) $f(x) = x^3 - x - 2$
          $f(1) = -2 < 0$
          $f(2) = 4 > 0$
          Sign change and continuity, therefore root in the interval
    (iii) $x_1 = 1$,
          $x_2 = 1.442\,249\ldots$,
          $x_3 = 1.509\,897\ldots$,
          $x_4 = 1.519\,724\ldots$,
          $x_5 = 1.521\,141\ldots$
    (iv) $f(1.5205) = -0.005 < 0$
          $f(1.5215) = 0.0007\ldots > 0$
          Sign change and continuity, therefore root in the interval [1.5205, 1.5215].
          Therefore root is 1.521 accurate to 3 d.p.

5   (i) $f(x) = e^x - x - 2$
          $f(1) = -0.28\ldots < 0$
          $f(2) = 3.38\ldots > 0$
          Sign change and continuity, therefore root in the interval
    (iii) $x_1 = 2$,
          $x_2 = 1.386\,294\ldots$,
          $x_3 = 1.219\,736\ldots$
    (iv) and (v)

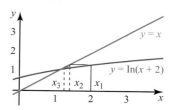

6   (i) $f(x) = 6 - x - \ln x$
          $f(4) = 0.61\ldots > 0$
          $f(5) = -0.60\ldots < 0$
          Sign change and continuity, therefore root in the interval
    (ii) $f(4.5) = -0.004\ldots < 0$
          Sign change between $x = 4$ and $x = 4.5$ therefore root lies in this interval. So root is closer to $x = 4$.

7 (i) Two roots

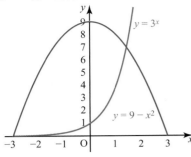

(iv) $x = 1.6549$ (4 d.p.)

8 (i) and (iii)

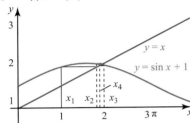

(ii) $x_1 = 1$,
$x_2 = 1.84147098\ldots$,
$x_3 = 1.96359072\ldots$,
$x_4 = 1.92384305\ldots$

(iv) You can only give the root as $x = 2$ to the nearest whole number.

(v) $x = 1.93$ (2 d.p.)

10 (i) Curves cross only once

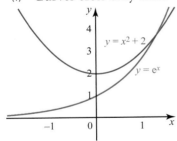

(ii) 1.32

11 (ii) 3.1623

12 0.747

Discussion point (page 306)

$f'(x_1) = 0$ so the method cannot proceed. The tangent never crosses the x-axis

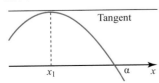

1 (i) $f'(x) = 3x^2 - 1$

(ii) $x_{n+1} = x_n - \dfrac{x_n^3 - x_n - 3}{3x_n^2 - 1}$

(iii) $1.727\ldots$

(iv)

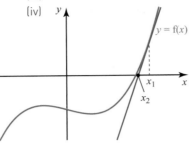

2 (i) $f(x) = x^3 - 6x^2 + 12x - 11$
$f(3) = -2 < 0$
$f(4) = 5 > 0$
Sign change and continuity, therefore the root is in the interval.

(ii) $x_2 = 3.444$,
$x_3 = 3.442$ (3 d.p.)

(iii) $f(3.435) < 0$ and $f(3.445) > 0$, so the root lies in the interval $[3.435, 3.445]$ and so the root is 3.44 to 3 s.f.

3 (i) $f(x) = e^{2x} - 15x - 2$
$f(1.5) = -4.41\ldots < 0$
$f(1.7) = 2.46\ldots > 0$
Sign change and continuity, therefore the root is in the interval.

(ii) $x_2 = 1.64$ (3 s.f.)

(iii) $f(x) = e^{2x} - 15x - 2$
$f(1.635) = -0.21\ldots < 0$
$f(1.645) = 0.16\ldots > 0$
Sign change and continuity, therefore the root is in the interval $[1.635, 1.645]$ and is therefore 1.64 to 3 s.f.

4 (i) $f'(x) = 6x^2 + 5$
$f'(x)$ is positive for all values of x, so the graph of $y = f(x)$ has no turning points and therefore crosses the x-axis once only.

(ii) $f(-1) = -5 < 0$
$f(0) = 2 > 0$
Sign change and continuity, therefore the root is in the interval

(iii) $x_2 = -0.3846153\ldots$,
$x_3 = -0.3783533\ldots$

(iv) $f(-0.3785) < 0$,
$f(-0.3775) > 0$, so root lies in the interval $[-0.3785, -0.3775]$ and is therefore -0.378 correct to 3 d.p.

5 (i) and (iii) Maximum at $(-1, 5)$, minimum at $(1, 1)$

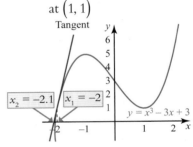

(ii) $-2.111111\ldots$

(iv) $f(-2.115) < 0$,
$f(-2.105) < 0$, so the root does not lie in the interval $[-2.115, -2.105]$ and therefore the root is not -2.11 correct to 2 d.p.
$f(-2.15) < 0$,
$f(-2.05) > 0$, so the root does lie in the interval $[-2.15, -2.05]$ and therefore the root is -2.1 correct to 1 d.p.

6 (i) $f(x) = x^4 - 7x^3 + 1$
$f(0) = 1 > 0$
$f(1) = -5 < 0$
Sign change and continuity, therefore the root is in the interval

(ii) 0.54 (2 d.p.)

(iii) Tangent to the curve at $x = 0$ is horizontal, so never meets the x-axis. Second approximation is undefined

7

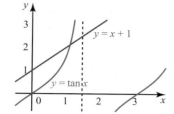

1.13 (3 s.f.)
$f(x) = \tan x - x - 1$
$f(1.125) = -0.032... < 0$
$f(1.135) = 0.012... > 0$
Sign change and continuity, therefore the root is in the interval $[1.125, 1.135]$ and is therefore 1.13 correct to 3 s.f.

8

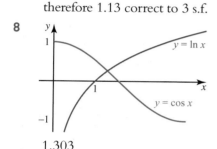

1.303
$f(x) = \ln x - \cos x$
$f(1.3025) = -0.0008... < 0$
$f(1.3035) = 0.0009... > 0$

Sign change and continuity, therefore the root is in the interval $[1.3025, 1.3035]$ and is therefore 1.303 correct to 3 d.p.

9 (i) One point of intersection

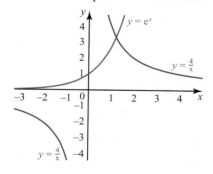

(ii)

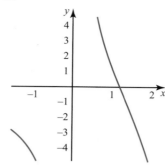

(iii) 1.202 (3 d.p.)
(iv) Still converges rapidly to root

10 Equation is
$3x^{12} - 750x + 747 = 0$
$x = 1.49513$ (5 d.p.)

Discussion point (page 309)

The shape is a semi-circle of radius 2.5 so calculating the area is easy. As you increase the number of strips, the accuracy of the approximation increases.

Discussion point (page 312)

If there is a turning point then the rectangle using the smaller y value may not lie entirely below the curve, and the top of the rectangle using the bigger y value may not be entirely above the curve, making your values unreliable as bounds.

Rectangle using smaller y value:

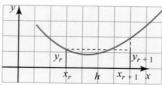

Rectangle using bigger y value:

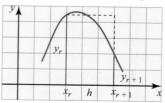

As you increase the number of strips, the accuracy of the approximation increases.

Exercise 14.3 (page 314)

1 54.75
2 (i) 0.8194
    (ii) 0.8317
3 0.512, overestimate because curve is concave upwards.

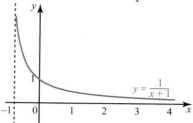

4 1.22, underestimate because curve is concave downwards.

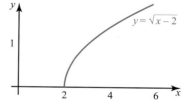

5 (i) 458 m
    (ii) Trapezium rule approximates curve with straight lines, values are not given to a very high degree of accuracy.

6 (i)

x	y
0	1
0.2	0.9615
0.4	0.8621
0.6	0.7353
0.8	0.6098
1	0.5

    (ii) Upper bound is 0.8337
        Lower bound is 0.7337
    (iii) 0.8 to 1 d.p.
    (iv) Increase the number of rectangular strips
7 (i) 0.229 383 0
    (ii) $2.3 \ln 2.3 - 2 \ln 2 - 0.3$, 0.0059%
8 Trapezium rule with 8 strips gives 0.3749, sum of 8 rectangles below the curve gives 0.3667, so the area is 0.37 to 2 d.p.
9 (i) Approx 15 strips, the area is 0.683.
    (ii) Approx 15 strips, the area is 0.683.

Multiple choice questions

1 B $y = 2(x-1)^2$

2 C $-\mathbf{i} - \frac{5}{2}\mathbf{j} - \frac{1}{2}\mathbf{k}$

3 C $x = \sqrt{\frac{4x-2}{x}}$

4 A $[0.9005, 0.9015]$

Practice questions 3

(page 319)

1 $\dfrac{dm}{dt} = \dfrac{k}{\sqrt{m}}$ [1]

When $t = 0$, $m = 9$, and

$\dfrac{dm}{dt} = 10$, so $10 = \dfrac{k}{3}$

$\Rightarrow k = 30$, so $\dfrac{dm}{dt} = \dfrac{30}{\sqrt{m}}$.   [2]

2 $\mathbf{F}_1 + \mathbf{F}_2 + \mathbf{F}_3 =$
$(2a\mathbf{i} + 3b\mathbf{j}) + (-b\mathbf{i} + a\mathbf{j}) +$
$(10\mathbf{i} - 2\mathbf{j}) =$
$(2a - b + 10)\mathbf{i} + (3b + a - 2)\mathbf{j}$
$= 0$ [1]
$\Rightarrow 2a - b = -10$, $a + 3b = 2$
$\Rightarrow b = 2$, $a = -4$ [3]

3 (i) $\overrightarrow{AB} = \begin{pmatrix} -2 \\ 1 \\ -1 \end{pmatrix}$;

$\overrightarrow{AC} = \begin{pmatrix} 2 \\ -2 \\ -6 \end{pmatrix}$;

$\overrightarrow{BC} = \begin{pmatrix} 4 \\ -3 \\ -5 \end{pmatrix}$ [2]

(ii)
$AB = \sqrt{(-2)^2 + 1^2 + (-1)^2} = \sqrt{6}$;

$AC = \sqrt{(2)^2 + (-2)^2 + (-6)^2} = \sqrt{44}$;

$BC = \sqrt{4^2 + (-3)^2 + (-5)^2} = \sqrt{50}$,

so $AB^2 + AC^2 = 6 + 44 = 50$
$= BC^2$, and by Pythagoras'
theorem triangle ABC is
right angled.   [3]

4 (i) $\overrightarrow{OD} = \mathbf{v} + \frac{1}{3}\mathbf{u}$;

$\overrightarrow{OE} = \frac{3}{4}\left(\mathbf{v} + \frac{1}{3}\mathbf{u}\right)$
$= \frac{3}{4}\mathbf{v} + \frac{1}{4}\mathbf{u}$;

$\overrightarrow{CE} = -\mathbf{v} + \frac{3}{4}\mathbf{v} + \frac{1}{4}\mathbf{u}$
$= \frac{1}{4}\mathbf{u} - \frac{1}{4}\mathbf{v}$ [3]

(ii) $\overrightarrow{CA} = \mathbf{u} - \mathbf{v}$, so

$\overrightarrow{CE} = \frac{1}{4}\overrightarrow{CA}$ and E must
therefore lie on AC
(as CE is in the same
direction as CA).   [2]

5 (i) $f'(x) = 8\cos 2x - 3x^2$ [1]

$x_{n+1} = x_n - \dfrac{4\sin 2x_n - x_n^3}{8\cos 2x_n - 3x_n^2}$ [1]

$x_0 = 1$, 1.41667,
1.29723, 1.28864,
1.28859, so
$x = 1.29$ (3 s.f.)   [2]

(ii) $f(1.285) = 0.042\ldots$,
$f(1.295) = -0.0755\ldots$
and change of sign
means root is correct
to 3 s.f.   [1]

6 (i) Interval is $[-1.5, -1,4]$ as
$f(-1.5) = -1.59375 < 0$
and $f(-1.4) = 0.42176 > 0$.   [2]

(ii) Using list function or
otherwise:
$f(-1.43) = -0.11971$,
$f(-1.42) = 0.066466$,
so root lies in
$[-1.43, -1.42]$.   [2]

(iii) $f(-1.425) = -0.0259$,
or interval to 3 d.p. is
$[-1.424, -1.423]$
so root is closer to
-1.42, to 2 d.p.   [2]

7 (i) $x = \frac{1}{2}(x^3 - 5)$, so $a = 0.5$
and $b = -2.5$ [1]

(ii) $x_1 = -2$, $x_2 = -6.5$,
$x_3 = -139.8125$.   [1]
Divergent   [1]

(iii) $x = \sqrt[3]{(2x + 5)}$,
so $c = 2$ and $d = 5$.   [1]

(iv) $x_1 = 1.912931183$,
$x_2 = 2.066580768$,
$x_3 = 2.090292423$.   [2]

(v) $f(2.085) = -0.106\ldots$,
$f(2.095) = 0.005\ldots$,
change of sign so root
lies in interval
$[2.085, 2.095]$ and
root is 2.09 to 2 d.p.   [2]

8 (i) A is when $t = -2$,
$x = -3$, $y = 0$; B is
when $t = 2$, $x = 5$,
$y = 8$.   [4]

(ii) $\dfrac{dy}{dx} = \dfrac{\frac{dy}{dt}}{\frac{dx}{dt}} = \dfrac{2 + 2t}{2}$.   [2]

$\dfrac{dy}{dx} = 0$ when

$2 + 2t = 0 \Rightarrow t = -1$ [1]
when $t = -1$, $x = -1$,
$y = -1$, so turning
point is $(-1, -1)$.   [2]

(iii) $t = \dfrac{x-1}{2}$

$\Rightarrow y = 2 \times \dfrac{x-1}{2} + \left(\dfrac{x-1}{2}\right)^2$
$= x - 1 + \frac{1}{4}x^2 - \frac{1}{2}x + \frac{1}{4}$
$= \frac{1}{4}x^2 + \frac{1}{2}x - \frac{3}{4}$ [3]

9 (i)
$\displaystyle\int \dfrac{3\,dv}{v(v-3)} = -\int kt\,dt$ [1]

$\dfrac{3}{v(v-3)} = \dfrac{A}{v} + \dfrac{B}{v-3}$

$\Rightarrow 3 = A(v-3) + Bv$
$v = 0 \Rightarrow 3 = -3A$,
so $A = -1$
$v = 3 \Rightarrow 3 = 3B$,
so $B = 1$

so $\dfrac{3}{v(v-3)} = \dfrac{1}{v-3} - \dfrac{1}{v}$ [3]

$\Rightarrow \displaystyle\int \dfrac{3\,dv}{v(v-3)} = \int\left[\dfrac{1}{v-3} - \dfrac{1}{v}\right]dv$

$= \ln(v-3) - \ln v$ [1]

so $\ln(v-3) - \ln v = -kt + c$

$\Rightarrow \dfrac{v-3}{v} = e^{-kt+c} = Ae^{-kt}$ [2]

when $t = 0$, $v = 10 \Rightarrow A = 0.7$ [1]

$\Rightarrow 1 - \frac{3}{v} = 0.7e^{-kt}$

$\Rightarrow \frac{3}{v} = 1 - 0.7e^{-kt}$

$\Rightarrow v = \dfrac{3}{1 - 0.7e^{-kt}}$ as required.   [2]

(ii) As $t \to \infty$, $e^{-kt} \to 0$, so
$v \to 3$, and terminal
velocity is $3\,\text{m s}^{-1}$ [2]

Review: Working with data

Discussion point (page 325)

The same They are illustrating the same data

Different Labels

Frequency chart	The vertical scale is labelled 'Frequency'
Histogram	The vertical scale is labelled 'Frequency density'

Scales on the vertical axes

Frequency chart	The vertical scale is the frequency
Histogram	The vertical scale is such that the areas of the bars represent frequency.

Class intervals

Frequency chart	The class intervals are all the same, 10 seconds
Histogram	The class intervals are not all the same; some are 10 seconds but one is 50 seconds.

Discussion point (page 327)

$S_{xx} = 8$, $s^2 = \frac{8}{9} = 0.888...$,

$s = \sqrt{0.888...} = 0.943 \ (3 \text{ d.p.})$

Exercise R.1 (page 328)

1 R.8
2 (i) The mean per capita GDP is much higher in Western Europe than it is in Eastern Europe.
   (ii) (a) 41 000 US$
     (b) 8000 US$
   (iii) (a) Both 30 000 to 40 000 US$ and 40 000 to 50 000 US$
     (b) 20 000 to 30 000 US$
     (c) 20 000 to 30 000 US$
   (iv) Modal group
   (v) Western Europe 39 000 US$ Eastern Europe 16 000 US$
   (vi) (a) 15 941 US$
     (b) They do not take the sizes of the populations of the various countries into account. Weighted means would be better.
   (vii) 19th or 20th

3 (i) 1.623 m and 0.152 m
   (ii) 31.7° C and 0.61° C
4 (i) Positively skewed
   (ii) Median = 0, $Q_1 = 0$, $Q_3 = 1$
   (iii) Since the lower quartile and median are both zero you can't draw the box properly and there is no whisker on the left.
   (iv) The large frequency for 0 gold medals means that this group dominates displays such as vertical line charts and stem and leaf diagrams, making it difficult to see the details.
5 A FALSE. 309 medals were awarded and between them the top 4 countries won 118 of them, so less than half.
   B TRUE. There were 309 medals and 207 countries, giving a mean of $\frac{309}{207} = 1.492... = 1.5$ to 1 d.p.
   C TRUE. 148 + 21 = 169 countries got 0 or 1 gold

medals. This is 81.6% of the 207 countries.
   D UNCERTAIN. The information is not provided to draw this conclusion. It was in fact false with several large countries, e.g. India, Pakistan and Nigeria, getting no gold medals but that information was not given.
   E TRUE. $1.49 + 2 \times 4.79 = 11.07$ so countries with 12 or more medals are identified as outliers.
   F FALSE. It is true that the mode is 0 but the midrange is $\frac{0 + 46}{2} = 23$ and not 0. So the combined statement is false.
6 (ii) (a) 5%
     (b) 12%
     (c) About 11%
   (iii) The distribution is negatively skewed. You would expect this as babies can arrive very early but if a baby is getting long overdue it is common for the birth to be induced.
   (iv) The label 'births' means their frequency. The unit on the horizontal scale is 1 day so 'per day' is the same as 'per unit' or 'density'.
7 (i) 4.79
   (ii)

$$\sum (x - \bar{x})^2 = \sum \left(x^2 - 2x\bar{x} + \bar{x}^2 \right)$$
$$= \sum x^2 - 2\sum (x\bar{x}) + \sum \bar{x}^2$$
$$= \sum x^2 - 2\bar{x}\sum x + n\bar{x}^2$$
$$= \sum x^2 - 2\bar{x} \times (n\bar{x}) + n\bar{x}^2$$
$$= \sum x^2 - 2n\bar{x}^2 + n\bar{x}^2$$
$$= \sum x^2 - n\bar{x}^2$$

(iii) If you are working by hand and the mean, $\bar{x}$, has several decimal places, working out the individual deviations can be tedious. So in such cases it is easier to use the $\sum x^2 - n\bar{x}^2$ form.

8   Cleaning involves looking through a data set to check for the following:

Outliers: Example – Bob Roberts's age is given as 138

Errors: Example – Debbie Lane's distance from home is 75 km

Missing entries: Example – the time of Simon Sefton's accident is missing

Format: Example – use of 12 hour and 24 hour clock for times of accidents

9   Averages could be calculated for age, distance from home; nights in hospital and possibly time of accident. It would be best to look at the distributions before deciding which average. The other fields are categorical (including the number of the police officer reporting the accident), and so no average is appropriate. However for some of these the modal class may be helpful, for example, the day of the week.

10 Robin would want his case studies to be as representative as possible so he should look for different groups, or strata within the data. Obvious fields are: age, cause and injuries. Next he should decide which of these he thinks is the most important, e.g. age, and classify the data for it. Then he should write down some possible stratified samples. He should then classify the data in his other chosen fields and see in which of the possible samples those strata are as well represented as possible, given the small size of the sample he is selecting. Note that a random sample may not turn out to be all that representative when, like this, the size is small. So, in this case, a stratified sample is preferable.

11 With the exception of people's names, all the fields could be relevant to this comparison, some more obviously so than others. So the first thing is to decide which fields you are going to use. Then, starting with the most important field, divide the data for each field into two subsets: younger cyclists and older cyclists. Compare their total frequencies and their distributions. This comparison may show there are no differences, which is interesting because it suggests that both groups may be equally at risk. On the other hand, the comparison may show up differences and that too is interesting. It suggests respects in which one group is more at risk than the other. This in turn could lead to intervention policies to reduce the risk.

Chapter 15

Opening activity (page 334)

(i)   A civil case between two parties is decided on the 'balance of probabilities'.

(ii) In a criminal case there is an assumption that the accused is innocent and must be proved guilty 'beyond reasonable doubt', i.e. with a very high probability.

Review exercise (page 337)

1   (i) $\frac{2}{6}$

    (ii) $\frac{3}{6}$

    (iii) $\frac{1}{6}$

    (iv) 1

    (v)   0

2   (i)   True

    (ii) False. There are two ways of selecting those two cards: first $7\blacklozenge$, then $8\blacktriangledown$ or first $8\blacktriangledown$ then $7\blacklozenge$. Each has a probability of $\frac{1}{52} \times \frac{1}{51} = \frac{1}{2652}$ so the overall probability is $2 \times \frac{1}{2652} = \frac{2}{2652}$.

    (iii) True

    (iv) False. There are 4 fours and 13 hearts but the $4\blacktriangledown$ is repeated. So the total number is $4 + 13 - 1 = 16$ and the probability is $\frac{16}{52}$.

    (v)   False. The probability that the first card is a spade is $\frac{13}{52} = \frac{1}{4}$ but for the second card there would be only 12 spades left out of 51 cards and $\frac{12}{51} \neq \frac{1}{4}$.

    (vi) True.

3 0.9

4 0.42

5 0.28

6 (i) $x = 3, y = 14, z = 11$

(ii) $\dfrac{17}{30}, \dfrac{25}{30}, \dfrac{28}{30}, \dfrac{14}{30}$

(iii) $\dfrac{28}{30} + \dfrac{14}{30} = \dfrac{42}{30}$
$= \dfrac{17}{30} + \dfrac{25}{30}$

(iv) Since $P(B \cap F) \neq 0$ both B and F can occur. There are nights when both badgers and foxes visit the garden.

7 0.57

8 0.6

9 (i)

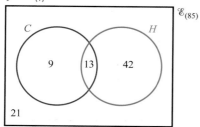

(ii) (a) $\dfrac{64}{85} = 0.753$

(b) $\dfrac{13}{85} = 0.153$

(iii) $\dfrac{64}{85} = \dfrac{22}{85} + \dfrac{55}{85} - \dfrac{13}{85}$

(iv) $\dfrac{21}{85} = 1 - \dfrac{64}{85}$

(v) 30% of those not wearing helmets suffered from concussion compared with 24% of those wearing helmets.

You would expect lower concussion rates among those wearing helmets and this is what the data indicate. However the actual percentages, 30% and 24%, are too close for it to be safe to draw any conclusions on statistical grounds, particularly given the small number of cyclists involved.

Discussion point (page 342)

It seems unrealistic to think that any witness would select someone in an identity parade entirely at random, let alone all four doing so. It is also unlikely that any court would convict someone on the evidence of only one out of four witnesses.

However, this example raises an important statistical idea about how we make judgements and decisions.

Judgements are usually made under conditions of uncertainty and involve us in having to weigh up the plausibility of one explanation against that of another. Statistical judgements are usually made on such a basis. We choose one explanation if we judge the alternative explanation to be sufficiently unlikely, that is, if the probability of its being true is sufficiently small. Exactly how small this probability has to be will depend on the individual circumstances and is called the significance level.

Somebody on trial is assumed innocent until shown to be guilty beyond reasonable doubt; reasonable doubt must mean a very small probability that the person is innocent.

Exercise 15.1 (page 343)

1 0.4

2

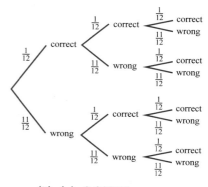

(i) 0.2401

(ii) 0.5002

(iii) 0.4998

3

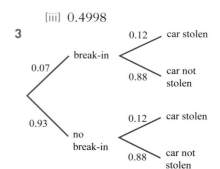

(i) 0.0084

(ii) 0.1732

(iii) 0.1816

4 (i)

(ii) (a) 0.00058

(b) 0.77

(c) 0.020

(iii) (a) 0.0052

(b) 0.52

(c) 0.094

5 0.93

6 (i) 0.2436

(ii) 0.7564

(iii) 0.2308

(iv) 0.4308

7 (i) 0.0741

(ii) 0.5787

(iii) 0.5556

8 For a sequence of events you multiply the probabilities. However, $\dfrac{1}{6} \times \dfrac{1}{6} \times \dfrac{1}{6} \times \dfrac{1}{6} \times \dfrac{1}{6} \times \dfrac{1}{6}$ gives the probability of six 6s in six throws. To find the probability of at least one 6 you need $1 - P(\text{No 6s})$ and that is $1 - \dfrac{5}{6} \times \dfrac{5}{6} \times \dfrac{5}{6} \times \dfrac{5}{6} \times \dfrac{5}{6} \times \dfrac{5}{6}$

= 0.665.

9 0.5833

10 0.31

11 (i) 0.1667
 (ii) 0.1389
 (iii) 0.1157
 (iv) 0.5787

12 (i)

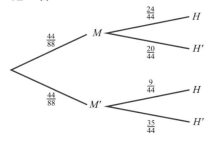

 (ii) (a) $\frac{24}{88}$
 (b) $\frac{55}{88}$

 (iii)

	H	H'	Total
M	24	20	44
M'	9	35	44
Total	33	55	88

 (iv) The probability of being kept in hospital after an accident involving a motor vehicle, $\frac{24}{44}$, is greater than that after an accident not involving a motor vehicle, $\frac{9}{44}$.

Discussion point (page 347)

$P(T \mid S) = \frac{109}{169} = 0.645$

$P(T \mid S') = \frac{43}{87} = 0.494$

So $P(T \mid S) \neq P(T \mid S')$

Exercise 15.2 (page 350)

1 (i) 0.6
 (ii) 0.556
 (iii) 0.625
 (iv) 0.047
 (v) 0.321
 (vi) 0.075
 (vii) 0.028
 (viii) 0.0022
 (ix) 0.00095
 (x) 0.48

2 (i) (a) 0.031
  (b) 0.078
  (c) 0.043
  (d) 0.061
  (e) 0.00086
  (f) 0.11
  (g) 0.014

 (ii) Those sentences for motoring offences would probably have shorter sentences than others so are likely to represent less than 1.5% of the prison population at any time.

3 (i) $\frac{35}{100}$
 (ii) $\frac{42}{100}$
 (iii) $\frac{15}{65}$

4 (i) $\frac{1}{6}$
 (ii) $\frac{5}{12}$
 (iii) $\frac{2}{5}$

5 (i) 30
 (ii) $\frac{7}{40}$
 (iii) $\frac{7}{10}$
 (iv) $\frac{7}{15}$

6 (i)

	Hunter dies	Hunter lives	Total
Quark dies	$\frac{1}{12}$	$\frac{5}{12}$	$\frac{1}{2}$
Quark lives	$\frac{2}{12}$	$\frac{1}{3}$	$\frac{1}{2}$
	$\frac{1}{4}$	$\frac{3}{4}$	1

 (ii) $\frac{1}{12}$
 (iii) $\frac{5}{12}$

 (iv) $\frac{5}{6}$

7 (i)

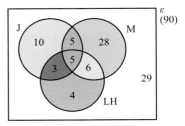

Key:
J = juniors
M = males
LH = left-handed players

 (ii) (a) $\frac{1}{4}$
  (b) $\frac{1}{6}$
  (c) $\frac{28}{45}$
  (d) $\frac{4}{5}$
  (e) $\frac{19}{24}$
  (f) $\frac{10}{39}$

8 (i)

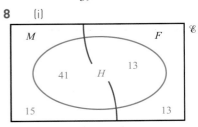

 (ii) (a) $\frac{26}{82}$
  (b) $\frac{54}{82}$
  (c) $\frac{13}{54}$
  (d) $\frac{13}{26} = \frac{1}{2}$

 (iii) Robin might say, 'Our figures suggest that males are more likely to be wearing a helmet than females. Using them we estimate that the probability that someone is wearing a helmet given that the person is male, $\frac{41}{56} = 0.73$, and the probability that someone is wearing a helmet given that the person is female, $\frac{13}{26} = 0.5$.'

Chapter 16

Opening activity (page 353)

For accurate estimates you must leave your thought experiment behind and go out and collect some real data.

Review exercise (page 354)

1 (i) $X \sim B(100, 0.25)$
   (ii) 25 (given by np)
   (iii) 0.049
   (iv) 0.149
2 (i) $B\left(7, \frac{1}{6}\right)$ where p is the probability of scoring a 6
   (ii) $_7C_7 \left(\frac{1}{6}\right)^7 \left(\frac{5}{6}\right)^0 =$ 0.000 003 57
   (iii) 280 000
3 (i) (a) 0.282
       (b) 0.526
   (ii) 75
   (iii) 0.049
   (iv) They may well be related and so their characteristics (i.e. the colour of their spots) may not be independent.
4 (i) (a) 20
       (b) 0.3
       (c) 0.7
   (ii) It is $1 - 0.95^2 = 0.048$, so 4.8%
   (iii) $a = 3$
5 $P(X \leqslant 11) = 0.697$
6 (i) 0.392
   (ii) 0.556
   (iii) 3
   (iv) It must be assumed that the probability of the train being late is the same on each day, and that whether or not the train is late in one day does not influence whether or not it is late on another day. The first assumption may not be unreasonable for a particular time of the year (but it might be

argued that lateness is more likely on certain days of the week, for various reasons). The second is more dubious; if a train is late on one day it may be due to a cause which is also present the next day, such as bad weather, engineering works, track problems, etc.
7 (i) (a) 0.016 59
       (b) 0.003 69
       (c) 0.000 39
       (d) 0.020 67
   (ii) If 70% of this age group are wearing helmets, then the probability of 20 or more out of a sample of 22 wearing helmets is given by the answer to part (i)(d); it is 0.020 67. This is indeed small so it is very likely that the true figure is greater than 70%. However, it is not certain, as the organiser claims.
   (iii) 70.4% for the 15 to 29 group and 44% for those over 29. These lower figures suggest that the campaign is influencing the behaviour of young people, particularly children, but is not very effective with older people. However, they are based on the small numbers of cyclists involved in accidents so can only be estimates of the overall figures in Avonford, where the campaign has been held.

Discussion point (page 358)

The scores on a fair six-sided dice. The outcomes on a spinner with equal sectors, labelled differently.

Exercise 16.1 (page 360)

1 0.22
2 (i)

r	$P(X = r)$
2	$\frac{1}{36}$
3	$\frac{2}{36}$
4	$\frac{3}{36}$
5	$\frac{4}{36}$
6	$\frac{5}{36}$
7	$\frac{6}{36}$
8	$\frac{5}{36}$
9	$\frac{4}{36}$
10	$\frac{3}{36}$
11	$\frac{2}{36}$
12	$\frac{1}{36}$

   (ii) The distribution is symmetrical.

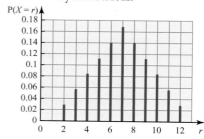

   (iii) (a) $\frac{5}{18}$
         (b) $\frac{1}{2}$
         (c) $\frac{2}{3}$
3 (i) $\frac{1}{20}$
   (ii) Uniform distribution
   (iii) 0.5
4 (i)

r	0	1	2	3	4	5
$P(Y = r)$	$\frac{3}{18}$	$\frac{5}{18}$	$\frac{4}{18}$	$\frac{3}{18}$	$\frac{2}{18}$	$\frac{1}{18}$

   (ii) The distribution has positive skew.

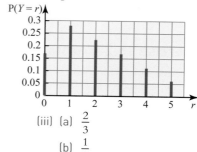

   (iii) (a) $\frac{2}{3}$
         (b) $\frac{1}{2}$
5 (i) $k = 0.4$

r	2	4	6	8
$P(X = r)$	0.1	0.2	0.3	0.4

   (ii) (a) 0.3
         (b) 0.35

6 (i) $k = \dfrac{20}{49}$

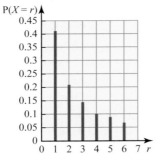

(ii) 0.248 (to 3 s.f.)

7 (i)

r	0	1	2	3
$P(X = r)$	$\frac{1}{8}$	$\frac{3}{8}$	$\frac{3}{8}$	$\frac{1}{8}$

(ii) The distribution is symmetrical.

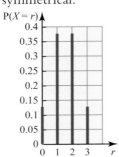

(iii) $\dfrac{1}{2}$

(iv) No. Could get 2H, 2T

8 (i)

r	1	2	3	4	6	8	9	12	16
$P(X = r)$	$\frac{1}{16}$	$\frac{2}{16}$	$\frac{2}{16}$	$\frac{3}{16}$	$\frac{2}{16}$	$\frac{2}{16}$	$\frac{1}{16}$	$\frac{2}{16}$	$\frac{1}{16}$

(ii) $\dfrac{1}{4}$

9 (i) $a = 0.42$

(ii) $k = \dfrac{1}{35}$

(iii) Since the probability distributions look quite different, the model is not a good one.

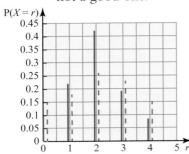

10 (i) $P(X = 1) = \dfrac{1}{216}$

(ii) $P(X \leqslant 2) = \dfrac{8}{216}$

(iii) $P(X \leqslant 3) = \dfrac{27}{216}$,
$P(X = 3) = \dfrac{19}{216}$;
$P(X \leqslant 4) = \dfrac{64}{216}$,
$P(X = 4) = \dfrac{37}{216}$;
$P(X \leqslant 5) = \dfrac{125}{216}$,
$P(X = 5) = \dfrac{61}{216}$;
$P(X \leqslant 6) = 1$,
$P(X = 6) = \dfrac{91}{216}$

(iv) The distribution has negative skew.

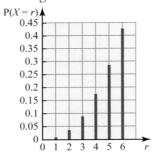

11 (i)

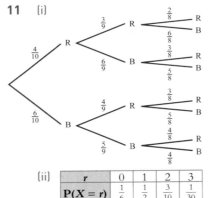

(ii)

r	0	1	2	3
$P(X = r)$	$\frac{1}{6}$	$\frac{1}{2}$	$\frac{3}{10}$	$\frac{1}{30}$

12 (i)

Nights	0	1	2	3	4	5	>5	No data
Frequency	58	23	5	2	0	2	2	1

(ii)
$$\text{Sum} = (1 - p)\left(1 + p + p^2 + \ldots\right)$$
$$= (1 - p) \times \frac{1}{(1 - p)} = 1$$

(iii)

Nights	0	1	2	3	4	5	>5
Frequency	59.8	20.9	7.3	2.6	0.9	0.3	0.2

(iv) They are a good match, but the real data have more high values, including the two outliers, 8 and 40.

(v) p is the probability that someone who is at hospital in the daytime will be detained overnight, and $(1 - p)$ is the probability that the person will be discharged.

So if there are N people at hospital on one day, $N(1 - p)$ of them will be discharged and Np of them will be kept in overnight. So at the start of the next day Np of the people will be in hospital. Of these $Np(1 - p)$ will be discharged and Np^2 will be kept for a further night, and so on. The initial value of N is the number of people in the sample, in this case 92.

Discussion point (page 368)
In both cases the vertical scale is frequency density and so the area under the graph represents frequency.

Discussion point (page 373)
Some people may have refused to answer and those planning to vote Labour may have been more likely to do that. The sampling may have been concentrated at a time of day when non-Labour voters were more likely to be there. Also, although it's very unlikely, it is still possible to get that result from a random sample.

Exercise 16.2 (page 374)
1 (i) 0.631
(ii) 0.252
(iii) 0.117
2 (i) 0.8413
(ii) 0.0228
(iii) 0.1359
3 N (5.349, 4.825)

4
  (i) 0.0668
  (ii) 0.6915
  (iii) 0.2417

5
  (i) 0.0668
  (ii) 0.1587
  (iii) 0.7745

6
  (i) 78.9%
  (ii) 0.5

7
  (i) 106
  (ii) 75 and 125
  (iii) 39 or 40

8
  (i) 0.246
  (ii) 0.0796
  (iii) 0.0179
  (iv) The Normal distribution is used for continuous data and so covers non-integer values that have to be rounded to find the nearest integer values of the binomial distribution.

9
  (i) 0.0038
  (ii) 0.5
  (iii) 0.495

10
  (i) 31, 1.98
  (ii) 2.0, 13.5, 34.4, 34.4, 13.5, 2.0
  (iii) More data would need to be taken to say reliably that the weights are Normally distributed.

11
  (i) 0.309
  (ii) 0.383
  (iii) 7881 daffodils

12
  (i) (a) 0.315
      (b) 0.307
  (ii) Assuming the answer to part (i)(b) is correct, there is a 7.6% error. Worse.
  (iii) 0.5245

13
  (i) The areas of the bars give the numbers of people they represent. $14 + 25 + 16 + 10 + 6 + 19 + 1 = 91$.
  (ii) It is not symmetrical (positively skewed). It is bimodal.

  (iii)

Age, a years	Frequency density
$0 \leqslant a < 10$	0.836
$10 \leqslant a < 20$	1.363
$20 \leqslant a < 30$	1.743
$30 \leqslant a < 40$	1.743
$40 \leqslant a < 50$	1.363
$50 \leqslant a < 70$	0.618
$70 \leqslant a < 90$	0.098

  (iv) It would represent $8.36 + 13.63 + 17.43 + 17.43 + 13.63 + 12.37 + 1.96 = 84.81$. The left hand tail of the Normal distribution (i.e. the part for negative ages) has been excluded.
  (v) It would give a partial explanation. However, a Normal population with mode in the 60s would inevitably be quite small. Taking it away from the main population would still leave a positive skew.

Chapter 17

Opening activity (page 378)
Medicines are only licensed after undergoing extensive trials supported by rigorous statistical tests.

Review exercise (page 382)
1
  (i) 2; 1.195
  (ii) P(2 defectives in 10) = 0.302; in 50 samples of 10, the expected number of samples with two defectives is 15.1 which agrees well with the observed 15
  (iii) H_0: P(mug defective) = 0.2; H_1: P(mug defective) < 0.2; $n = 20$

P(0 or 1 defective mug) = 0.0692; Accept H_0 since 0.0692 > 5%

It is not reasonable to assume that the proportion of defective mugs has been reduced.
  (iv) Opposite conclusion since 0.0692 > 10%

2
  (i) Let p = P(business operates no smoking policy) H_0: $p = 0.7$, H_1: $p < 0.7$
  (ii) $k \leqslant 10$
  (iii) $k \leqslant 1$
  (iv) For 19 businesses, P(H_0 rejected) = 0.1855

For 4 businesses, P(H_0 rejected) = 0.1265

The 19 business test is preferable because it gives a greater probability of rejecting H_0 when it should be rejected.

3
  (i) 0.430
  (ii) 0.9619
  (iii) 0.0046
  (iv) H_0: $p = 0.9$, H_1: $p < 0.9$
  (v) $n = 17$; P($X \leqslant 13$) = 0.0826 > 5%; not sufficient evidence to reject H_0.
  (vi) Critical region is $X \leqslant 12$, since P($X \leqslant 12$) = 0.0221

4
  (i) (a) 0.0278
      (b) 0.0384
  (ii) Let p = P(blackbird is male) H_0: $p = 0.5$, H_1: $p > 0.5$
  (iii) Result is significant at the 5% significance level. Critical region is $X \geqslant 12$.
  (iv) You would be more reluctant to accept H_1. Although H_0 is still $p = 0.5$, the sampling method is likely to give a non-random sample.

5 (i) (a) 0.0991

(b) 0.1391

(ii) Let $p = $ P(seed germinates)

$H_0: p = 0.8$,
$H_1: p > 0.8$, since a higher germination rate is suspected.

(iii) Critical region is $X \geqslant 17$, since

$P(X \geqslant 17) = 0.0991 < 10\%$ but

$P(X \geqslant 16) = 0.2713 > 10\%$.

(iv) (a) When $p = 0.8$ he reaches the wrong conclusion if he rejects H_0, i.e. if $X \geqslant 17$, with probability 0.0991.

(b) When $p = 0.82$ he reaches the wrong conclusion if he fails to reject H_0, i.e. if $X \leqslant 16$, with probability $1 - 0.1391 = 0.8609$.

6 (i) 0.0417

(ii) 0.0592

(iii) 0.0833

(iv) 0.1184

(v) Let $p = $ P(man selected)
$H_0: p = 0.5$, $H_1: p \neq 0.5$
$P(X \leqslant 4 \text{ or } X \geqslant 11) = 0.1184 > 5\%$
There is not sufficient evidence to reject H_0, so it is reasonable to suppose that the process is satisfactory.

(vi) $4 \leqslant w \leqslant 11$

7 (i) Let p be the proportion of those in this age group having accidents who are boys.

$H_0: p = 0.5$ Boys and girls of this age group are equally likely to have cycling accidents

$H_1: p > 0.5$ Boys are more likely than girls to have accidents.

1-tail test, using binomial B(20, 0.5)

Significance level 10%

The p-value for the observed number of boys is $P(X \geqslant 13)$, where $X \sim$ B(20, 0.5) and this is 0.1316.

Since $0.1316 > 0.10$, the result is not significant. H_0 is accepted.

The data do not provide sufficient evidence to support Sally's conjecture.

(ii) Overall 66% of the accidents are males. Among the under-13s the figure is 65%. This suggests that this age group are no different in this respect from people of all ages.

(iii) The data do not provide evidence as to the cause of the disparity. Further data, directly related to the suggestions, would need to be collected to establish to what extent they are valid.

Activity 17.1 (page 388)

	1-tail	2-tail
10%	1.282	1.645
5%	1.645	1.960
$2\frac{1}{2}$%	1.960	2.240
1%	2.320	2.576

Exercise 17.1 (page 391)

1 Either he wants to test whether the mass of the hens of that type has increased, or he wants to test whether the mass of the hens of that type has decreased.

2 (i) $Z = 1.53$, not significant

(ii) $Z = -2.37$, significant

(iii) $Z = 1.57$, not significant

(iv) $Z = 2.25$, significant

(v) $Z = -2.17$, significant

3 (i) 0.3085

(ii) 0.0062

(iii) $H_0: \mu = 4.00\,\text{g}$,
$H_1: \mu > 4.00\,\text{g}$
$z = 3$, significant

4 (i) $H_0: \mu = 72.7\,\text{g}$,
$H_1: \mu \neq 72.7\,\text{g}$;
Two-tail test.

(ii) $z = 1.84$, not significant

(iii) No, significant

5 (i) $H_0: \mu = 23.9°$,
$H_1: \mu > 23.9°$

(ii) $z = 1.29$, significant

(iii) 4.54; this is much greater than 2.3 so the ecologist should be asking whether the temperature has become more variable.

6 (i) You must assume it has a Normal distribution.

(ii) $H_0: \mu = 470$ days,
$H_1: \mu > 470$ days

(iii) $z = 3.02$, significant

(iv) More time to produce offspring

7 (i) You must assume that the speeds are Normally distributed.

(ii) $H_0: \mu = 80$ mph,
$H_1: \mu \neq 80$ mph
$z = 2.28$, significant

(iii) Yes: $z = 1.33$, not significant

8 (i) You must assume that the visibilities are Normally distributed.

(ii) $H_0: \mu = 14$ sea miles,
$H_1: \mu < 14$ sea miles

(iii) $z = -2.284$, significant

(iv) Choosing 36 consecutive days to collect data is not a good ideas because weather patterns will ensure that the data are not independent. A better sampling procedure would be to choose every tenth day. In this way the effects of

9 (i) 998.6, 7.055
   (ii) $H_0: \mu = 1000$,
        $H_1: \mu < 1000$
   (iii) $z = -1.59$, not
         significant

10 $H_0: \mu = 0, H_1: \mu \neq 0$;
   $z = 0.98$,
   not significant

11 (i) 16.2, 5.231
   (ii) $H_0: \mu = 15, H_1: \mu > 15$
   (iii) $z = 1.986$, not
         significant

12 (i) 1.977, 0.132
   (ii) $H_0: \mu = 2, H_1: \mu < 2$
   (iii) $z = -1.68$, not
         significant

13 (i) 104.7, 3.019
   (ii) $H_0: \mu = 105$,
        $H_1: \mu \neq 105$
   (iii) $z = -0.89$, not
         significant

14 (i) Probability = $\frac{1}{7}$,
       Expectation = 13.143,
       the distribution is
       uniform.
   (ii) At the 10%
        significance level the
        evidence suggests the
        distribution is not
        uniform, but at the
        5%, 2.5% and 1% levels
        the evidence is not
        strong enough for the
        hypothesis that the
        distribution is uniform
        to be rejected.

        So it is marginal
        whether or not the
        distribution is uniform.
   (iii) Patterns of cycling are
         different. On weekdays
         some people cycle to
         work, and students
         cycle to school or
         college. At the weekend
         people cycle more for
         leisure.

(iv) The number in the
     Avonford data set is 92.
     Standard deviation =
     $\sqrt{92 \times \frac{5}{7} \times \frac{2}{7}} = 4.333$,
     mean =
     $92 \times \frac{5}{7} = 65.714$

     Using continuity
     correction
     $z = \dfrac{71.5 - 65.714}{4.333} = 1.3352$
     $\Phi(1.335) = 0.9090$
     Probability = $1 - 0.9090$
     $= 0.0910$

(v) $H_0: p = \frac{5}{7}$
    $H_1: p > \frac{5}{7}$
    1-tail test at 5%
    significance level
    Since $0.0910 > 0.05$,
    the null hypothesis is
    accepted.
    The evidence does not
    suggest that accidents
    are more common
    on weekdays than at
    weekends.

Discussion point (page 394)

A   Negative correlation,
    providing both variables
    are random
B   Positive correlation,
    providing both variables
    are random
C   Negative association
D   Two distinct groups
    showing neither correlation
    nor association

Activity 17.3 (page 396)

The calculation in the example
followed these steps.

Work out $\sum(x_i - 17)(y_i - 10)$.
Then divide it by the number of
points. Then divide by the rmsd
for x and the rmsd for y.

In this 17 is the mean value of
x so can be replaced by $\bar{x}$.

Similarly 10 can be replaced by
$\bar{y}$. The number of points is n.

The root mean squared
deviation of x is given by
$$\text{rmsd} = \sqrt{\dfrac{\sum(x_i - \bar{x})^2}{n}}$$
and similarly for y.

So the calculation becomes
$$\dfrac{\sum(x_i - \bar{x})(y_i - \bar{y})}{n}$$
$$\div \left[\sqrt{\dfrac{\sum(x_i - \bar{x})^2}{n}} \times \sqrt{\dfrac{\sum(y_i - \bar{y})^2}{n}} \right]$$

The n cancels out and so this
becomes
$$r = \dfrac{\sum(x_i - \bar{x})(y_i - \bar{y})}{\sqrt{\sum(x_i - \bar{x})^2 \times \sum(y_i - \bar{y})^2}}.$$

Activity 17.4 (page 400)

This can be done by calculator
or by using the method in
Example 17.5.

Exercise 17.2 (page 403)

1 (i) (a) Negative correlation
      (b) Countries with high
          life expectancy tend
          to have low birth
          rates.
  (ii) (a) Positive association
       (b) Countries with high
           GDP per capita
           tend to have high
           life expectancy.
  (iii) (a) Negative association
        (b) Countries with high
            GDP per capita
            tend to have low
            birth rates.

2 (i) (a) Positive association
      (b) Both variables are
          random but the
          relationship is not
          linear.

(ii) (a) Positive association
    (b) This is a time series and so correlation is not appropriate because the time (in this case X) is not random.
(iii) (a) Neither
    (b) The two variables are independent so there is neither correlation nor association.
(iv) (a) Positive correlation
    (b) Both variables are random and the relationship is linear.
(v) (a) Neither
    (b) X is controlled and so not random and the value of Y is determined by it so is not random either.
(vi) (a) Negative correlation
    (b) Both variables are random but the relationship is linear.

3 (i) Positive correlation
  (ii) $\bar{x} = 8$, $\bar{y} = 6$

4 Sachin is not correct. The data clearly belong to two distinct groups and so it is not appropriate to calculate a correlation coefficient. In fact there is no significant correlation in either of the two groups; if you work out the values of r, they are 0.316 for $x < 15$ and -0.1 for $x > 15$.

5 (i) $H_0: \rho = 0$, $H_1: \rho > 0$
  (ii) Accept H_1. The evidence suggest that there is a positive correlation between performance in the high jump and the long jump.

6 (i) $H_0: \rho = 0$, $H_1: \rho < 0$
  (ii) Accept H_1
  (iii) Correlation does not imply causation.

There may be other explanations for the high divorce rates in some countries and the low rates in others. Perhaps Charlotte should collect some data herself to highlight the dangers of drinking alcohol, e.g. wine consumption/liver disease.

7 (ii) $H_0: \rho = 0$, $H_1: \rho > 0$
  (iii) Accept H_1
  (iv) Giving more training to employees does tend to keep staff with the company.

8 (ii) $r_s = 0.79$, strong positive association

9 (i) 0.54
  (ii) 0.54
  (iii) They are the same because they use alternative ways to calculate this measure of association.

10 0.94

11 (i) 0.50
  (ii) -0.242
  (iv) Apart from one outlier, $(10, 12)$, the data show negative correlation. Without knowing more about the outlier it is not possible to evaluate the two measures.

12 (i) Outliers have been considered. Missing data items have been excluded.
  (ii) Both are random.
  (iii) There are places where several points lie on straight horizontal lines corresponding to whole numbers of kilometres. Ages are rounded down to the nearest integer.
  (iv) The independent variable is Age and the dependent variable is Distance from home.

It is reasonable to think that people cycle further from home because they are older.
It is not reasonable to think that people are older because they cycle further from home.
(v) H_0: There is no association between the distance from home at which people have accidents and their ages.
H_1: There is a positive association between the distance from home at which people have accidents and their ages.
The p-value provides extremely strong evidence in favour of the alternative hypothesis. There is virtually no chance of seeing data like these unless there is an association in the underlying population.
(vi) The scatter diagram suggests that the data are not drawn from a bivariate Normal distribution; the departure from Normality looks clear and considerable. Consequently, no useful interpretation can be made of the value of the product moment correlation coefficient. Knowing the value of the pmcc can add nothing new to Robin's understanding of the situation.

Multiple choice questions

1 C $1 - \frac{3}{4} \times \frac{38}{51}$

2 B $\frac{18}{60}$

3 C 0.8849

4 D $P(79.5 < X < 90.5)$

Practice questions: Statistics (page 410)

1 (i) Only 5 countries show a value over 50 minutes, and 24 countries show a value below. Of the two countries with the biggest populations, India is considerably above 50 minutes and China is just below. It isn't possible to be certain, but the figure may be plausible, particularly if it is rounded. [2]

   (ii) 50 minutes per day is about 304 hours per year. This is roughly 39 days of 7.8 hours, so 39 working days. [2]

2 It should say 'in a Normal distribution (about) 95% of the distribution is within 2 standard deviations of the mean' (or about $\frac{2}{3}$ within 1 standard deviation). [1]
   Replace 2σ with 4σ (or change the percentage) in the diagram. [1]

3 (i) There may be fewer slugs to be seen in the winter months. There may be fewer observers looking for slugs in the winter months. [2]

   (ii) 93 (in month 8) is an anomalous dip. It could be just randomness in the sampling. Or, noting that the other observations in month 8 are low, it could

be that there are fewer observers in August: perhaps they are on holiday. [2]

   (iii) The rise and fall in the figures is consistent with an annual life cycle. However the low figures are also consistent with other hypotheses – for example that slugs hibernate in the winter – so it isn't possible to say. [2]

4 (i) Use $X \sim B(10, 0.65)$ to find $P(X > 5) = 0.751$. [2]

   (ii) $H_0: p = 0.65$ where p is the probability of choosing the quickest check-out when there are three available.
   $H_1: p < 0.65$ [2]

   (iii) $P(X < 8) = 0.0060$, $P(X < 9) = 0.020$, so $k = 0, 1, 2, \ldots, 7$. [3]

   (iv) The first 20 visits might be sufficiently random if, for example, they were at different times of day. However, patterns of behaviour, by staff and customers, might be different immediately after opening a third check-out. Also, it is possible that the person asking the question is still learning in the first 20 trials. [2]

5 (i) (a) $0.4 \times 0.7 \times 0.2$
              $+ 0.4 \times 0.3 \times 0.8$
              $+ 0.6 \times 0.7 \times 0.8$
              $= 0.488$ [2]
          (b) $1 - 0.6 \times 0.3 \times 0.2$
              $= 0.964$ [2]

   (ii) P(successful on all 3)
   $= 0.4 \times 0.7 \times 0.8 = 0.224$
   P(successful on all 3 | successful on at least 1)
   $= \dfrac{\text{P(successful on all 3} \cap \text{successful on at least 1)}}{\text{P(successful on at least 1)}}$

$= \frac{0.224}{0.964} = 0.232$ [3]

6 (i) $4 = \mu + 1.0364\sigma$ and
   $2 = \mu - 0.8416\sigma$ [2]
   Solve: $\mu = 2.90$
   and $\sigma = 1.06$ [2]

   (ii) $H_0: \mu = 3.2$, where μ is the population mean yield for variety B
   $H_1: \mu < 3.2$ [2]

   (iii) Sample mean is 3.06 kg. Critical ratio is
   $\dfrac{3.06 - 3.2}{0.9 \div \sqrt{100}} = -1.556$ [2]
   Compare with
   $z = -1.645$ [1]
   Accept H_0; insufficient reason to suppose his yield will be below 3.2. [1]

   (iv) Assumption: the SD is still 0.9. This is the best information available, but the SD for the farmer's crop could be different. [2]

   (v) Assumption: the farmer's crop is a random sample (of typical conditions on his land). If, for example, all the trial plants were grown in the same location this might be untypical of the land as whole. In that case the mean yield is not a reliable estimate. [2]

7 (i) $k\left(\frac{1}{1} + \frac{1}{2} + \frac{1}{3} + \frac{1}{4}\right) = 1$,
   so $k = \frac{60}{125} = \frac{12}{25}$. [2]

   (ii) Vertical line chart. [1]

   (iii) $P(X_1 = X_2) =$
   $k^2\left(\frac{1}{1} + \frac{1}{4} + \frac{1}{9} + \frac{1}{16}\right) = \frac{41}{125}$ [2]

   $P(X_1 < X_2) = \frac{1}{2}\left(1 - \frac{41}{125}\right)$
   $= \frac{42}{125}$ [2]

8 (i) Mean wing lengths differ by about 2 SDs; mean weights differ by about $\frac{1}{4}$ of an SD. So there is much more overlap of weights than there is of wing lengths. So the comment is correct. [3]

(ii) Females' weights will vary according to whether or not they have recently laid eggs. Males' weights will not have this element of variation. [1]

(iii) The vertical 'striping' in the data corresponds to the wing lengths being recorded to the nearest millimetre. [1]

(iv) The data cloud looks broadly elliptical. So it is appropriate to carry out the standard test on the pmcc. [2]

(v) There is almost no chance (probability less than 1 in 10^6) that so strong a correlation would be obtained in a sample of this size if there were no underlying correlation in the population. [2]

(vi) It seems very likely that there would be strong evidence for a correlation in the male blackbirds too. However, it is not certain. [1] It would *not* have been sensible to work with males and females as a single sample. Combining two separate samples can often lead to spurious or misleading results. [1]

Chapter 18

Review exercise (page 422)

1 (i)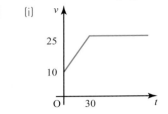

(ii) 30 s
(iii) 2275 m
(iv) $v = 10 + 0.5t; 0 \leqslant t \leqslant 30$
$v = 25; t \geqslant 30$

2 (i) $3\,\mathrm{m\,s^{-1}}$
(ii) $2.25\,\mathrm{m\,s^{-1}}$

3 (i) 200 m
(ii) 30 s
(iii) 150 m

4 (i) $1.875\,\mathrm{m\,s^{-2}}$
(ii) $14.6\,\mathrm{m\,s^{-1}}$

5 (i) $3\,\mathrm{m\,s^{-1}}$
(ii) 50 m
(iii) $9.6\,\mathrm{m\,s^{-2}}$
(iv) 35 m
(v) 5.25 m

6 78.4 m

7 (i) 4.11 s
(ii) $-20.2\,\mathrm{m\,s^{-1}}$

8 16 s

9 (i) $3.5\,\mathrm{m\,s^{-1}}, -3\,\mathrm{m\,s^{-2}}$
(ii) $21\frac{1}{3}$ m

10 (i) $v = 3t^2 - 8t + 6$;
$s = t^3 - 4t^2 + 6t + 4$
(ii) $v = -10t + 12$;
$s = -5t^2 + 12t - 2$

Discussion point (page 426)

$y = 0$ when $t = 2.4$ and then the projectile hits the ground.

Exercise 18.1 (page 433)

1 (i) $\mathbf{v} = 4t\mathbf{i} + 8\mathbf{j}$
(ii)

(0, 0); (2, 8); (8, 16);
(18, 24); (32, 32)
(iii) $8\mathbf{j}$; $4\mathbf{i} + 8\mathbf{j}$; $8\mathbf{i} + 8\mathbf{j}$;
$12\mathbf{i} + 8\mathbf{j}$; $16\mathbf{i} + 8\mathbf{j}$
(iv) $21.5\,\mathrm{m\,s^{-1}}$

2 $\mathbf{v} = \begin{pmatrix} -4 \\ -5 \end{pmatrix}$; $\mathbf{a} = \begin{pmatrix} 0 \\ 0 \end{pmatrix}$

3 $4.47\,\mathrm{m\,s^{-2}}$; $-153°$

4 (i) $\frac{1}{20}t^2\mathbf{i} + \frac{1}{30}t^3\mathbf{j}$

(ii) $5\mathbf{i} + 33\frac{1}{3}\mathbf{j}$

5 $\mathbf{v} = 2t^2\mathbf{i} + \left(6t - t^2\right)\mathbf{j}$;
$\mathbf{r} = \frac{2}{3}t^3\mathbf{i} + \left(3t^2 - \frac{1}{3}t^3\right)\mathbf{j}$

6 15.3°

7 $8.11\,\mathrm{m\,s^{-1}}$

8 (i) initial velocity = $3.54\mathbf{i} - 3.54\mathbf{j}$
(ii) $\mathbf{v} = 8.54\mathbf{i} + 11.46\mathbf{j}$;
$\mathbf{r} = 52.0\mathbf{i} + 14.6\mathbf{j}$

9 (i) $\mathbf{v} = \begin{pmatrix} 15 \\ 16 - 10t \end{pmatrix}$;
$\mathbf{a} = \begin{pmatrix} 0 \\ -10 \end{pmatrix}$
(ii) 1.6 s
(iii) $22.8\,\mathrm{m\,s^{-1}}$
(iv) $y = 2 + \frac{16}{15}x - \frac{1}{45}x^2$

10 (i)

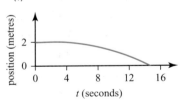

(0, 2); (4, 2); (8, 1.6);
(12, 0.8); (16, −0.4)
(ii) $\mathbf{v} = 20\mathbf{i} + (1 - 10t)\mathbf{j}$;
$20\mathbf{i} - \mathbf{j}$
(iii) $-10\mathbf{j}$
(iv) $y = 2 + \frac{1}{20}x - \frac{1}{80}x^2$

11 (i)

(ii) 5 s
(iii) $33.5\,\mathrm{m\,s^{-1}}$, $63.4°$
(or $116.6°$)
(iv) $\mathbf{a} = (12 - 6t)\mathbf{i} + (9 - 6t)\mathbf{j}$
is never 0

12 (i) $16.9\,\mathrm{m\,s^{-1}}$
(ii)
$$\mathbf{r} = \begin{pmatrix} 2 - 3t + t^2 \\ -5 + 2t - 0.5t^2 \\ 8 + 2t^2 \end{pmatrix}, \begin{pmatrix} 6 \\ -5 \\ 40 \end{pmatrix}$$

13 (i) A: $v\sin 35°\mathbf{i} + v\cos 35°\mathbf{j}$
B: $-8.66\mathbf{i} + 5\mathbf{j}$
(ii) A: $vt\sin 35°\mathbf{i} + vt\cos 35°\mathbf{j}$;
B: $(5 - 8.66t)\mathbf{i} + 5t\mathbf{j}$
(iii) $6.1\,\mathrm{km\,h^{-1}}$
(iv) $0.4111\ldots$ hours =
24.7 min

14 (i) $2t^2\mathbf{i} + 4t\mathbf{j}$
(ii) $\frac{2}{3}t^3\mathbf{i} + 2t^2\mathbf{j}$
(iii) $\frac{1}{6}t^4\mathbf{i} + \frac{2}{3}t^3\mathbf{j}$
(iv) speed $9.61\,\mathrm{m\,s^{-1}}$; $56.3°$
to $\mathbf{i}$ direction.

15 (i) $20\mathbf{i}+10\mathbf{j}$
(ii) $T = 5$ s; 125 m

16 (i) $16\,\mathrm{cm}$
(ii) $20\,\mathrm{cm}$
(iii) $0\,\mathrm{cm\,s^{-1}}, 0\,\mathrm{cm\,s^{-1}}$;
$20\,\mathrm{cm\,s^{-1}}, -16\,\mathrm{cm\,s^{-1}}$
(iv) $t = 0, \mathbf{r}_A = \mathbf{r}_B = 16\mathbf{j}$
$t = 2, \mathbf{r}_A = \mathbf{r}_B = 20\mathbf{i}$
(v) All components are zero,
so model B is better.

17 (i)

No it doesn't, maximum
$y = 9.48$ when $t = 2.67$
(ii)
$\mathbf{v} = 8t\mathbf{i} + \left(8 - 4t + \frac{3}{8}t^2\right)\mathbf{j}$;
$\mathbf{a} = 8\mathbf{i} + \left(-4 + \frac{3}{4}t\right)\mathbf{j}$;
$0 \leqslant t \leqslant 8$
$\mathbf{v} = 64\mathbf{i}$; $\mathbf{a} = 0\mathbf{i} + 0\mathbf{j}$;
$8 < t \leqslant 20$

(iii) $1200\mathbf{i} + (-600 + 112.5t)\mathbf{j}$
(iv) Because of air resistance
(v) The motorcycle leaves
the ground

18 (i) $2\mathbf{i} + \mathbf{j} + \mathbf{k}$
(ii) $(2t - 5)\mathbf{i} + (t + 1)\mathbf{j} + t\mathbf{k}$
(iii) 1.5 s; 3.5 m

19 $x = 6 - 4y + y^2$

20 $\mathbf{r} = (2 - \sin \omega t)\mathbf{i} + 2\cos \omega t\,\mathbf{j}$
$+ (\sin \omega t - 3)\mathbf{k}$;
$\mathbf{v} = -\omega\cos \omega t\,\mathbf{i} - 2\omega\sin \omega t\,\mathbf{j}$
$+ \omega\cos \omega t\,\mathbf{k}$

Chapter 19

Opening activity (page 438)

There are a number of forces
acting on the car which cancel
each other out, resulting in no
motion. In order for that to be
possible the cable must make
small angles with the horizontal
so that the vertical components
of the tension cancel out the
weight of the cable car. In that
case the tensions in the cable
will be greater than the weight
of the car.

Review exercise (page 444)

1 (i) $0.\dot{3}\,\mathrm{m\,s^{-2}}$
(ii) $500\,\mathrm{m\,s^{-2}}$

2 $65\,\mathrm{N}$

3 (i) $3\,\mathrm{m\,s^{-2}}$; $2\frac{2}{3}$ s
(ii) $1440\,\mathrm{N}$

4 (i) $29.5\,\mathrm{N}$
(ii) $19.5\,\mathrm{N}$
(iii) $24.5\,\mathrm{N}$

5 (i)
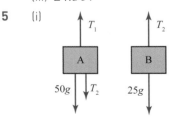
(ii) A: $50g + T_2 - T_1 = 100$
B: $25g - T_2 = 50$
(iii) $585\,\mathrm{N}$, $195\,\mathrm{N}$

6 (i)
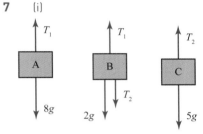
(ii) A: $T = 5a$;
B: $4g - T = 4a$
(iii) $4.35\,\mathrm{m\,s^{-2}}$

7 (i)
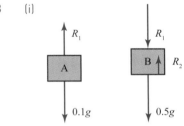
(ii) A: $8g - T_1 = 8a$;
B: $T_1 - T_2 - 2g = 2a$
C: $T_2 - 5g = 5a$
(iii) $0.65\,\mathrm{m\,s^{-2}}$, $73.2\,\mathrm{N}$, $52.3\,\mathrm{N}$

8 (i)

(ii) A: $R_1 - 0.98 = 0.08$;
B: $R_2 - R_1 - 4.9 = 0.4$
(iii) $R_1 = 1.06\,\mathrm{N}$,
$R_2 = 6.36\,\mathrm{N}$

9 (i)

300 N ⟵ caravan ⟶ T

250 N ⟵ car ⟶ 2250 N
T

(ii) $1\,\mathrm{m\,s^{-2}}$
(iii) $1100\,\mathrm{N}$

10 (i) $18000 - 900 = 90000a$
$\Rightarrow a = 0.19$
(ii) $4000\,\mathrm{N}$
(iii) $8000\,\mathrm{N}$
(iv) $-0.21\,\mathrm{m\,s^{-2}}$
(v) $-4000\,\mathrm{N}$ (thrust)

Discussion point (page 449)

Draw a vertical line to represent the weight, $10g = 98\,\text{N}$. Then add the line of the force T_2 at $45°$ to the horizontal (note the length of this vector is unknown), and then the line of the force T_1 at $30°$ to the horizontal ($60°$ to the vertical). C is the point at which these lines meet.

Discussion point (page 450)

The angles in the triangle are $180° - \alpha$, $180° - \beta$ and $180° - \gamma$. The sine rule holds and $\sin(180° - \alpha) = \sin\alpha$, $\sin(180° - \beta) = \sin\beta$ and $\sin(180° - \gamma) = \sin\gamma$.

Exercise 19.1 (page 452)

1   There is no force to the right to balance the force towards the left.

2   (i) $30\,\text{N}$, $36.9°$; $65\,\text{N}$, $67.4°$
     (ii)

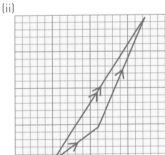

     (iii) $\begin{pmatrix} 49 \\ 78 \end{pmatrix}$; $92.1\,\text{N}$, $57.9°$

3   (i)

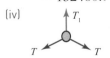

$20\,000g$

     (ii) $T\cos 40°$, $T\sin 40°$
     (iii) $T = \dfrac{20000 \times 9.8}{2 \times \sin 40°} = 152460.9\,\text{N}$
     (iv)

T_1

T T

     (v) $196\,000\,\text{N}$

(vi) Resolve vertically for whole system. It is the same as the weight of the girder.

4   (i) $78.4\,\text{N}$
     (ii)

R $15\,\text{N}$

$8g$

     (iii) $66.1\,\text{N}$, $8.60\,\text{N}$

5   $9.04\,\text{N}$

6   $5.39\,\text{N}$; $21.8°$

7   (i)

T_2

T_1

T_R T_2

$30\,\text{N}$

     (ii) Rod: $56.4\,\text{N}$, compression, Cable 1: $59.1\,\text{N}$, tension

8   $4.40\,\text{N}$, $3.26\,\text{N}$

9   (i) $15.04\,\text{kg}$
     (ii) Both read $10\,\text{kg}$
     (iii) Both read $7.64\,\text{kg}$
     (iv) Method A or C

10 (i)

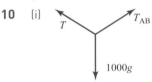

T T_{AB}

$1000g$

     (ii) A force to the right is required to balance the horizontal component of T.
     (iii)

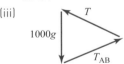

T

$1000g$

T_{AB}

     (iv) (a) $9800\,\text{N}$, $13859\,\text{N}$, $9800\,\text{N}$
           (b) $9800\,\text{N}$, $9800\,\text{N}$, $9800\,\text{N}$

11 (i) $\begin{pmatrix} 60\cos 75° \\ -60\sin 75° \end{pmatrix}$

         $\begin{pmatrix} 15.5 \\ -58.0 \end{pmatrix}$

     (ii) (a) $\begin{pmatrix} 56.1 \\ 61.2 \end{pmatrix}$

           (b) $83.0\,\text{N}$, $47.5°$
     (iii) $30.8\,\text{N}$, $-121°$

12 (i) Cable 1 ($5638\,\text{N}$, $2052\,\text{N}$); Cable 2: ($T_2\cos 30°$, $T_2\sin 30°$)
     (ii) $4104\,\text{N}$
     (iii) $9193\,\text{N}$

13 (i)

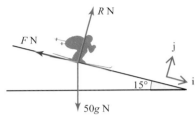

$R\,\text{N}$

$F\,\text{N}$

j

i

$15°$

$50g\,\text{N}$

     (ii) $-F\mathbf{i}$, $R\mathbf{j}$, $127\mathbf{i} - 473\mathbf{j}$
     (iii) $473\,\text{N}$, $127\,\text{N}$
     (iv) $254\,\text{N}$

14 (i)

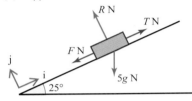

$R\,\text{N}$

$T\,\text{N}$

$F\,\text{N}$

j

i

$25°$

$5g\,\text{N}$

     (ii) $T\mathbf{i}$, $-F\mathbf{i}$, $R\mathbf{j}$, $-20.7\mathbf{i} - 44.4\mathbf{j}$
     (iii) $T = 29.4\,\text{N}$, $8.69\,\text{N}$
     (iv) $1.23\,\text{kg}$

15 (i) $58.0\mathbf{i} + 15.5\mathbf{j}$, $59.0\mathbf{i} - 10.4\mathbf{j}$
     (ii) (a) $117.04\,\text{N}$
           (b) $5.11\,\text{N}$
     (iii) $97\,\text{N}$ forwards
     (iv) $3\,\text{N}$

16 (i) $11.0\,\text{N}$, $63.4°$
     (ii) A circle centre A radius $1\,\text{m}$. No, 2 parallel forces and a third non-parallel force cannot form a triangle.

17 (i)   The wall can only push outwards.
     (ii) $T\sin 35° = R\sin\alpha$, $T\cos 35° + R\cos\alpha = 80g$
     (iii) $T = 562.92\ldots$, $R = 457\,\text{N}$
     (iv) $563\,\text{N}$ down along the rope
     (v) $563\,\text{N}$
     (vi) $1074\,\text{N}$

Discussion point (page 457)

Parallel to the slope, up the slope.

Discussion point (page 457)

Start with AB and BC. Then draw a line in the right direction for CD and another perpendicular line through A. These lines meet at D.

Discussion point (page 457)

(i)   The sledge accelerates up the hill.
(ii) The sledge is stationary or moving with constant speed (forces in equilibrium).
(iii) The acceleration is downhill.

Exercise 19.2 (page 461)

1 (i)

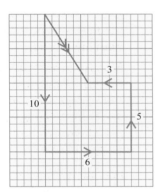

(iii) $3\mathbf{i} - 5\mathbf{j}$; 5.83 N, −59°

2 (i)

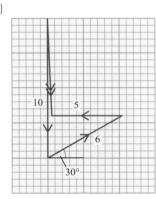

(iii) $0.196\mathbf{i} - 7\mathbf{j}$; 7.00 N, −88.4°

3 (i)

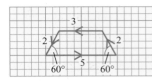

(ii) Equilibrium

4 (i)

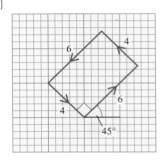

(ii) Equilibrium

5 (i)

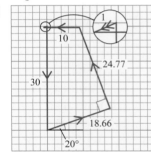

(iii) $-\mathbf{i}$; 1 N down incline

6 (i)

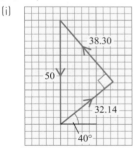

(ii) Equilibrium

7 5.73 N; −49.7°
8 (i) $6.04\mathbf{i} + 17.40\mathbf{j}$
   (ii) 18.4 N, 70.9°
9 $3.77\mathbf{i} + 2.60\mathbf{j}$; 4.58 N, 34.6°
10 $P\sqrt{3}$

Discussion point (page 462)

Down the slope.

Discussion point (page 464)

Sam and the sledge are a particle. There is no friction and the slope is straight. Friction would reduce

both accelerations so Sam would not travel as far on either leg of his journey.

Exercise 19.3 (page 466)

1 (i) $1.5\mathbf{i} - \mathbf{j}$
   (ii) $1.80\,\text{m s}^{-2}$
2 (i) $4\mathbf{i} + 11\mathbf{j}$
   (ii) $8\mathbf{i} + 8\mathbf{j}$, $2\mathbf{i} + 2\mathbf{j}$
3 (i)

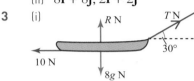

   (ii) 11.55 N
   (iii) $1\,\text{m s}^{-2}$
   (iv) 0.4 s
4 $0.37\,\text{m s}^{-2}$ down the plane, 71.2 N
5 (i)

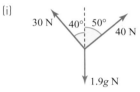

   (ii) 11.4 N, 30.1 N
   (iii) $16.9\,\text{m s}^{-2}$ at 69°
   (iv) The fish swings sideways as it moves up towards Jones.
6 (i)

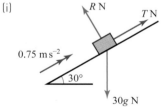

   (ii) $T\mathbf{i}$, $R\mathbf{j}$,
         $-30g\sin 30°\mathbf{i} - 30g\cos 30°\mathbf{j}$
   (iii) 169.5 N
   (iv) The crate slows down to a stop and then starts sliding down the slope.
7 $3.36\,\text{m s}^{-1}$, 0.1 N in compression
8 $0.71\,\text{m s}^{-2}$, 600 N
9 (i)

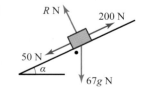

(ii) $13.2°$
(iii) $0.87\,\mathrm{m\,s^{-2}}$, $35.9\,\mathrm{m}$
(iv) $9.37\,\mathrm{m\,s^{-1}}$

10 (i)

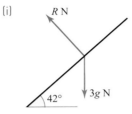

(ii) $6.56\,\mathrm{m\,s^{-2}}$
(iii) $1.75\,\mathrm{s}$
(iv) $13.0\,\mathrm{N}$

11 (i) Horizontal component of tension in the rope needs a balancing force.

(ii)

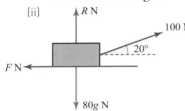

(iii) $94.0\,\mathrm{N}$, $750\,\mathrm{N}$
(iv) $128\,\mathrm{N}$
(v) $0.144\,\mathrm{m\,s^{-2}}$

12 (i)

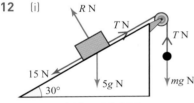

(ii) $4.03\,\mathrm{kg}$, $39.5\,\mathrm{N}$
(iii) $1.75\,\mathrm{m\,s^{-2}}$, $48.3\,\mathrm{N}$

13 $5.10\,\mathrm{m}$

14 $a = -0.121\,\mathrm{m\,s^{-2}}$
$\Rightarrow s = 413\,\mathrm{m}$

15 $319\,\mathrm{m}$

Chapter 20

Discussion point (page 473)
The tool shown on the left of Figure 20.8 works with one hand but has less leverage than the tool on the right.

Discussion point (page 477)
(i) $P + Q$ line of action parallel to P and Q and in same direction; distance from O is

$a + \dfrac{bQ}{(P + Q)}$ (between P and Q).

(ii) $P - Q$ line of action parallel to P and Q and in direction of larger; distance from O is

$a - \dfrac{bQ}{(P - Q)}$ (to the left of P for $P > Q$).

Discussion point (page 477)
You produce equal and opposite couples using friction between the other hand and the jar so that they turn in opposite directions. Pressing increases the normal reactions and hence the maximum friction possible.

Exercise 20.1 (page 478)
1 (i) $15\,\mathrm{Nm}$
(ii) $-22\,\mathrm{Nm}$
(iii) $18\,\mathrm{Nm}$
(iv) $-28\,\mathrm{Nm}$

2 (i) $2.1\,\mathrm{Nm}$
(ii) $6.16\,\mathrm{Nm}$
(iii) $-0.1\,\mathrm{Nm}$
(iv) $0.73\,\mathrm{Nm}$

3 $28.6\,\mathrm{N}$; $20.4\,\mathrm{N}$

4 (i) $1225\,\mathrm{N}$, $1225\,\mathrm{N}$
(ii) $1449\,\mathrm{N}$, $1785\,\mathrm{N}$.

5 $96.5\,\mathrm{N}$, $138.5\,\mathrm{N}$.

6 (i)

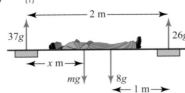

$55\,\mathrm{kg}$
(ii) $0.8\,\mathrm{m}$

7 (i) $P = 27.5g$, $Q = 147.5g$
(ii) $P = 2.5g$, $Q = 172.5g$
(iii) If child is less than $0.95\,\mathrm{m}$ from the adult, $P < 0$ so the bench tips unless A is anchored to the ground.
(iv) The bench tips if A is not anchored.

8 (i) $15g\,\mathrm{N}$, $30g\,\mathrm{N}$

(ii) $90g$, $5g$
(iii) 0
(iv) $\frac{2}{3}\,\mathrm{m}$

9 (i) $0.5g(30 - x)\,\mathrm{kN}$; $0.5g(20 + x)\,\mathrm{kN}$
(ii) Its centre of mass
(iii) Constant $15g\,\mathrm{kN}$ each

10 $6g$; $9\frac{1}{3}\,\mathrm{cm}$

11 $20\frac{5}{7}\,\mathrm{kg}$

12 (i) $35g\,\mathrm{N}$, $75g\,\mathrm{N}$
(ii) No
(iii) $36\,\mathrm{kg}$

13 $2\,\mathrm{N}$ upwards, $9\,\mathrm{m}$ to the left of A.

14 (i) $2262\,\mathrm{N}$, $7538\,\mathrm{N}$
(ii) 6
(iii) $784\,\mathrm{N}$

15 (i) $3600\,\mathrm{N}$
(ii) $0.0017\,\mathrm{m\,s^{-2}}$
(iii) $195\,\mathrm{N}$

16 Olwyn is correct. Laura's number is right but her units are wrong.

Meena correctly tried to find the component of the force perpendicular to the line from A but worked out $12\cos 60°$ instead of $12\sin 60°$.

Nat did not take account of the angle at which the force is acting.

Chapter 21

Activity 21.1 (page 484)
(i) (a) $v_y = 0$
(b) $y = 0$
(ii) These remain the same throughout.

Discussion point (page 485)
1 Yes for a parabolic path.
$u_y - gt = 0$ when $t = \dfrac{u_y}{g}$
and $u_y t - \frac{1}{2}gt^2 = 0$ when
$t = 2\dfrac{u_y}{g}$.

2 The balls and the bullet can be modelled as projectiles when there is no spin or wind and air resistance is negligible. Also a rocket with no power. The air affects the motion of the others.

Discussion point (page 485)

The only acceleration is in the vertical direction so any change in velocity happens vertically. This causes the projectile to reach a velocity of zero halfway through its flight.

Exercise 21.1 (page 485)

1 (i) (a)

(b) $u_x = 8.2$, $u_y = 5.7$

(c) $v_x = 8.2$,
$v_y = 5.7 - 9.8t$

(d) $x = 8.2t$,
$y = 5.7t - 4.9t^2$

(ii) (a)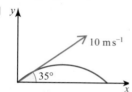

(b) $u_x = 2$, $u_y = 5$

(c) $v_x = 2$,
$v_y = 5 - 9.8t$

(d) $x = 2t$,
$y = 5t - 4.9t^2$

(iii) (a)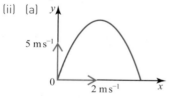

(b) $u_x = 4$, $u_y = 0$

(c) $v_x = 4$, $v_y = -9.8t$

(d) $x = 4t$, $y = -4.9t^2$

(iv) (a)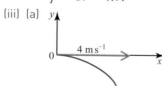

(b) $u_x = 9.74$,
$u_y = -2.25$

(c) $v_x = 9.74$,
$v_y = -2.25 - 9.8t$

(d) $x = 9.74t$,
$y = -2.25t - 4.9t^2$

(v) (a)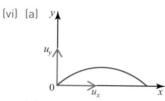

(b) $u_x = U\cos\alpha$,
$u_y = U\sin\alpha$

(c) $v_x = U\cos\alpha$,
$v_y = U\sin\alpha - gt$

(d) $x = Ut\cos\alpha$,
$y = Ut\sin\alpha - \frac{1}{2}gt^2$

(vi) (a)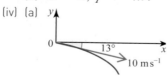

(b) u_x, u_y

(c) $v_x = u_x$,
$v_y = v_x - gt$

(d) $x = u_xt$,
$y = v_xt - \frac{1}{2}gt^2$

2 (i) (a) 1.5 s
(b) 11.025 m
(ii) (a) 0.51 s
(b) 1.28 m

3 (i) (a) 4 s
(b) 80 m
(ii) (a) 0.88 s
(b) 2.21 m

4 (i) $\dfrac{u_y^2}{2g}$

(ii) $\dfrac{2u_y}{g}$

(iii) $\dfrac{2u_xu_y}{g}$

Discussion point (page 489)

The particle's direction is a feature of its motion and so is determined by its velocity. In contrast the position vector is not a feature of the motion; a stationary particle has a position vector.

Exercise 21.2 (page 489)

1 (i) (a)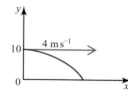

(b) $\mathbf{v} = \begin{pmatrix} 4 \\ -9.8t \end{pmatrix}$

(c) $\mathbf{r} = \begin{pmatrix} 4t \\ 10 - 4.9t^2 \end{pmatrix}$

(ii) (a)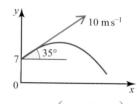

(b) $\mathbf{v} = \begin{pmatrix} 8.2 \\ 5.7 - 9.8t \end{pmatrix}$

(c)
$\mathbf{r} = \begin{pmatrix} 8.2t \\ 7 + 5.7t - 4.9t^2 \end{pmatrix}$

(iii) (a)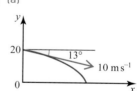

(b)
$\mathbf{v} = \begin{pmatrix} 9.74 \\ -2.25 - 9.8t \end{pmatrix}$

(c)
$\mathbf{r} = \begin{pmatrix} 9.74t \\ 20 - 2.25t - 4.9t^2 \end{pmatrix}$

(iv) (a)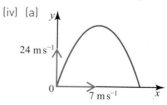

(b) $\mathbf{v} = \begin{pmatrix} 7 \\ 24 - 9.8t \end{pmatrix}$

(c) $\mathbf{r} = \begin{pmatrix} 7t \\ 24t - 4.9t^2 \end{pmatrix}$

(v) (a)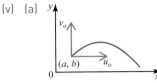

(b) $\mathbf{v} = \begin{pmatrix} u_0 \\ v_0 - gt \end{pmatrix}$

(c) $\mathbf{r} = \begin{pmatrix} a + u_0 t \\ b + v_0 t - \frac{1}{2} gt^2 \end{pmatrix}$

2 (i) (a) 1.5 s
    (b) 26.025 m
  (ii) (a) 0.31 s
    (b) 10.46 m

3 (i) 2.86 m
  (ii) 2.86 m
  (iii)

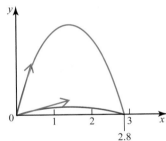

4 $T = \dfrac{u\sin\theta + \sqrt{u^2\sin^2\theta + 2gh}}{g}$;

$R = \dfrac{u^2 \sin\theta\cos\theta}{g}$

$R = \dfrac{u^2 \sin\theta\cos\theta}{g} + \dfrac{u\cos\theta}{g}\sqrt{u^2\sin^2\theta + 2gh}$

Discussion point (page 495)

They land together because u, s and a in the vertical direction are the same for both.

Exercise 21.3 (page 495)

1 (i) $17.3\,\text{m s}^{-1}$, $10\,\text{m s}^{-1}$
  (ii) $0, -9.8\,\text{m s}^{-2}$
  (iii) 35.3 m
  (iv) 1.02 s
  (v) 5.10 m

2 (i) 25.8 m
  (ii) 4.59 s
  (iii) 178.9 m

3 (i) $41.0\,\text{m s}^{-1}$, $28.7\,\text{m s}^{-1}$
  (ii)

t	0	1	2	3	4	5	6
x	0	41	82	123	164	205	246
y	0	24	38	42	36	21	−4.3

  (iii)

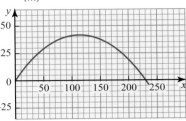

  (iv) 42.0 m, 239.7 m
  (v) The ball is a particle, no spin, no air resistance, so acceleration = g.

4 (i) $17.21, 8.03\,\text{m s}^{-1}$
  (ii) 1.64 s
  (iii) 28.2 m
  (iv) 0.82 s
  (v) 3.29 m
  (vi) 2.72 m, no

5 (i) $10.32, 14.74\,\text{m s}^{-1}$
  (ii) 2.91 s
  (iii) Into goal
  (iv) No

6 (i) 45.2 s
  (ii) 13.55 km
  (iii) $535\,\text{m s}^{-1}$
  (iv) 55.9°

7 (i) 3.2 m, vertical component of velocity is always less than $8\,\text{m s}^{-1}$
  (ii) 5.54 m
  (iii) 52.2°

8 (i) 1.74 s
  (ii) 3.50 m, hits Juliet's window
  (iii) $9.12\,\text{m s}^{-1}$

9 (i) 2.02 s
  (ii) No, height is 0.2 m
  (iii) $21.57\,\text{m s}^{-1}$

10 (i) 0.47 s
  (ii) 0.64 s
  (iii) $25.44\,\text{m s}^{-1}$
  (iv) $28.8\,\text{m s}^{-1}$

11 (i) Yes, the range is 70.4 m
  (ii) $32.7\,\text{m s}^{-1}$

12 $19.8\,\text{m s}^{-1}$; $17.7\,\text{m s}^{-1}$

13 (i) (a) 34.6 m
    (b) 39.4 m
    (c) 40 m
    (d) 39.4 m
    (e) 34.6 m
  (ii) $800\sin\alpha\cos\alpha = 800\cos(90° - \alpha)\sin(90° - \alpha)$
  (iii) 57.9°
  (iv) −30 cm; +31 cm. The lower angle is slightly more accurate.

14 (i) $26.06\,\text{m s}^{-1}$
  (ii) $27.35\,\text{m s}^{-1}$
  (iii) $26.15 < u < 26.88$

15 25.48 m

Discussion point (page 498)

$20\mathbf{i} + 30\mathbf{j}$; $(0, 6)$; $10\,\text{m s}^{-2}$

Exercise 21.4 (page 498)

1 (i) $y = \frac{5}{16} x^2$
  (ii) $y = 6 + 0.4x - 0.2x^2$
  (iii) $y = -14 + 17x - 5x^2$
  (iv) $y = 5.8 + 2.4x - 0.2x^2$
  (v) $y = 2x - \dfrac{gx^2}{2u^2}$

2 (i) $x = 40t$
  (ii) $t = \dfrac{x}{40}$;
$y = 30\left(\dfrac{x}{40}\right) - 5\left(\dfrac{x}{40}\right)^2$
  (iii)

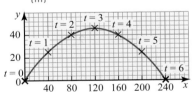

3 (i) $y = 0.75x - \dfrac{1}{320} x^2$
  (ii) Air resistance would reduce x.
  (iii)

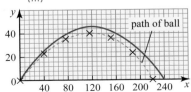

Yes; horizontal acceleration = $-0.5\,\text{m s}^{-2}$

4 (ii) $11.54\,\text{m s}^{-1}$; 30.96°

Discussion point (page 501)

The projectile is a particle. There is no air resistance or wind. The particle is projected from the origin.

Discussion point (page 501)

Substitute the x coordinate into the equation and see if it gives the y coordinate.

Exercise 21.5 (page 501)

1. (i) $\left(21.21t,\ 21.21t - 4.9t^2\right)$
    (ii) $t = \dfrac{x}{21.21..} \Rightarrow$
    $y = x - \dfrac{9.8}{900}x^2$
    (iii) $8.9\dot{1}$
    (iv) 29.4 or 62.4
2. (i) $y = 1.5 + x\tan 14°$
    $\qquad - \dfrac{4.9}{100}\left(1 + \tan^2 14°\right)x^2$
    (ii) 7.0 m
3. (i) $y = 2.5 + x\tan 5°$
    $\qquad - \dfrac{4.9}{625}\left(1 + \tan^2 5°\right)x^2$
    (ii) 24.2 m
    (iii) Yes, $y = 2.43$ m
4. (i) $y = 1 + 0.70...x$
    $\qquad - \dfrac{7.30...}{u^2}x^2$
    (ii) $u > 7.73\,\text{m s}^{-1}$
    (iii) $u < 8.41\,\text{m s}^{-1}$
5. (i) 1 m, (25, 0)
    (ii) 10 m
    (iii) 0.6 s
    (iv) $u_x = 25\,\text{m s}^{-1}$, 1 s
    (v) $25.7\,\text{m s}^{-1}$
6. $56.3°$, $18.43\,\text{m s}^{-1}$
7. (ii) 1.8, 0.6
    (iii) 22.5 m when $\tan\alpha = 1.2$
    (iv) 15.4 m
8. (ii) $39.2°$, $72.6°$

Chapter 22

Discussion point (page 508)

1. Downward slope would extend skid so u is an overestimate; opposite for upward slope.

2. u would be an underestimate if the motor cycle was not actually at rest when it hit the car.

3. If the deceleration is not uniform then u could either be an over- or and underestimate.

4. If the motorcycle were not a particle it would be effected by air resistance. Air resistance would reduce skid so u is an underestimate. Smaller μ would extend skid so u is an overestimate.

    The model gives a good indication that the motorcyclist was speeding, but it would be sensible to take further measurements to try to improve the model, e.g. slope of road, a more accurate estimation of μ.

Discussion point (page 509)

Friction is forwards when pedalling, backwards when freewheeling.

Discussion point (page 510)

Either the object is moving or it is on the point of moving.

Exercise 22.1 (page 513)

1.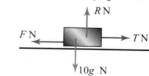
    (i) 0.102
    (ii) 0.051
2. (a) $0.61\,\text{m s}^{-2}$, $T = 45.9\,\text{N}$; $F = 36.8\,\text{N}$
    (b) $2.61\,\text{m s}^{-2}$, $T = 7.19\,\text{N}$; $F = 1.96\,\text{N}$
    (c) $0.31\,\text{m s}^{-2}$; $T_1 = 76.0\,\text{N}$, $T_2 = 40.4\,\text{N}$; $F = 29.4\,\text{N}$
    (d) $a = 0$; $T_1 = 9.8\,\text{N}$, $T_2 = 29.4\,\text{N}$; $F = 19.6\,\text{N}$

3. 4802 N
4. (i) 42.7 N
    (ii) 0.32
5. 0.82
6. (i) $(-)1.02\,\text{m s}^{-2}$
    (ii) 0.102 N
    (iii) 0.104
    (iv) 49 m (independent of mass)
7. (i) smoother contact
    (ii) 0.204
    (iii) 140 N
8. (i) $7.35\,\text{m s}^{-2}$
    (ii) $17.7\,\text{m s}^{-1}$
    (iii) 59.9 m
9. (i) 58.8 N
    (ii) 62.5 N
10. (i) 0.577
    (ii) $35.0°$
    (iii) 2.14
    (iv) $50.2°$
11. (i) $4.5\,\text{m s}^{-2}$
    (ii) $F = 2.30\,\text{N}$, $R = 18.4\,\text{N}$, $\mu = 0.125$
    (iii) $2.20\,\text{m s}^{-2}$, $4.20\,\text{m s}^{-1}$
12. (i)
    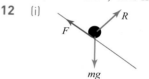
    (ii) $4.42\,\text{m s}^{-2}$
    (iii) $5.15\,\text{m s}^{-1}$
    (iv) 5.42 m
13. Greater than, equal to, less than $16.7°$, respectively.
14. (i)
    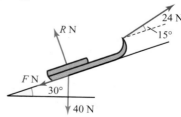
    (ii) $F = 3.18\,\text{N}$, $R = 28.4\,\text{N} \Rightarrow \mu = 0.11$
    (iii) $3.46\,\text{N}$, $4.05\,\text{m s}^{-2}$
15. (i) $1.4\,\text{m s}^{-2}$
    (ii) 1.20 s
    (iii) $0.92\,\text{m s}^{-1}$
16. $144.1 < T < 1810.9$

17 (i) (a) 37.94 N

   (b) 37.16 N

   (c) 37.52 N

   (ii) $\dfrac{40}{\cos\alpha + 0.4\sin\alpha}$

   (iii) 21.8°

18 Least force to move object up plane $= W(\sin\alpha + \mu\cos\alpha)$

Least force to prevent object sliding down plane $= W(\sin\alpha - \mu\cos\alpha)$

Multiple choice questions

1 B

2 C

3 D   The time, horizontal displacement and vertical displacement from the origin

4 C   Its initial speed is 7 m s⁻¹

Practice questions: Mechanics (page 518)

1 **Method 1**

Weight $= 0.5g$ [1]

Resolve horizontally and vertically

$-T_A\cos 60 + T_B\cos 30° = 0$

$T_A\sin 60 + T_B\sin 30° = 0.5g$

[1, 1]

Solve simultaneous equations

$T_A = 4.24\text{N}, \quad T_B = 2.45\text{N}$

[1, 1]

Method 2 – note the strings are perpendicular

Weight $= 0.5g$ [1]

Resolve along the strings [1]

$T_A = 0.5g\cos 30° = 4.24\text{N}$

$T_B = 0.5g\cos 60° = 2.45\text{N}$

[1, 1, 1]

Method 3 triangle of forces

Weight $-0.5g$ [1]

Diagram arrows forming a loop   [1]

and right angle marked

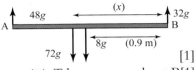

[1]

$T_A = 0.5g\cos 30° = 4.24\text{N}$ [1]

$T_B = 0.5g\sin 30° = 2.45\text{N}$ [1]

2 (i)   Contact force 72g [1]

   (ii) Downward forces   [1]
         Upward forces

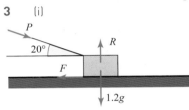

[1]

   (iii) Take moments about B[1]

$(72g)x + (8g) \times 0.9 = (48g)$
$\times 1.8$

$x = 1.10$ [1]

Centre of mass is 1.1 m above his feet.   [1]

3 (i)

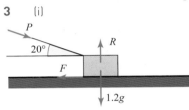

Weight and P shown [1]

normal reaction   [1]

Arrows showing friction opposing P [1]

   (ii) Resolve vertically   [1]
         $R = P\sin 20° + 1.2g$

$F_{\max} = \mu R = 0.4(P\sin 20° + 1.2g)$

[1]

On the point of sliding when   [1]

$P\cos 20° = F_{\max}$

$P\cos 20° = 0.4(P\sin 20° + 1.2g)$ [1]

$P(\cos 20° - 0.4\sin 20°) = 0.48g$ [1]

$P = \dfrac{0.48g}{\cos 20° - 0.4\sin 20°} = 5.86$[1]

4 (i)   Stationary when both components of $\mathbf{v} = 0$

$-0.4t + 8 = 0$

$t = 20$ [1]

When $t = 20$

$-0.03(20)^2 + 0.8(20) - 0.4 = 3.6 \neq 0$

So boat is never stationary.   [1]

   (ii) $\mathbf{r} = \displaystyle\int \mathbf{v}\,dt$

$= \begin{pmatrix} -0.01t^3 + 0.4t^2 - 0.4t \\ -0.2t^2 + 8t + 0 \end{pmatrix}$ [1, 1]

   (iii) When $t = 5$,

$\mathbf{r} = \begin{pmatrix} -0.01\times 5^3 + 0.4\times 5^2 - 0.4\times 5 \\ -0.2\times 5^2 + 8\times 5 \end{pmatrix}$

$= \begin{pmatrix} 6.75 \\ 35 \end{pmatrix}$ [1]

$\arctan\left(\dfrac{35}{6.75}\right) = 79.1$

Bearing $= 011°$ [1, 1]

   (iv) Travelling NE when components of velocity equal (and positive)   [1]

$-0.03t^2 + 0.8t - 0.4 = -0.4t + 8$

$-0.03t^2 + 1.2t - 8.4 = 0$

$t = 30.95, \quad 9.0$ [1]

$30.95 > 20$ so time 9.0s   [1]

5 (i)   In equilibrium for the bucket

$T = mg$ [1]

In equilibrium for the block resolve up the slope

$T = 1.5g\sin 40°$ [1]

$m = \dfrac{1.5g\sin 40°}{g}$

$= 0.964\text{kg}$ [1]

   (ii) Resolve downwards for the bucket N2L

$1.7g - T_1 = 1.7a$ [1]

Resolve up the slope for the block

$T_1 - 1.5g\sin 40 = 1.5a$ [1]

Add equations

$1.7g - 1.5g\sin 40 = 3.2a$

$a = 2.25$

Acceleration is 2.25 m s⁻²   [1]

   (iii) To find velocity when the bucket reaches the ground   [1]

$s = 0.8$

$u = 0$ $v^2 = u^2 + 2as$

$v = ?$ $v^2 = u^2 + 2as$

$a = 2.25$ $= 4.5 \times 0.8$

$= 3.6$

$t =$ $v = 1.90$ [1]

When tension removed
$$-1.5g\sin 40 = 1.5a$$
$$a = -g\sin 40 \qquad [1]$$

$s = ?$
$u = 1.90$ $v^2 = u^2 + 2as$: $v^2 = 3.6$
$v = 0$ $0 = 3.6 - 2 \times (g\sin 40)s$
$a = -g\sin 40$
$t =$ $s = \dfrac{3.6}{2g\sin 40} = 0.286$
$\qquad\qquad\qquad\qquad [1]$

Block travels another
28.6 cm. $\qquad\qquad [1]$

6 (i) Vertical motion

$s = 0$ $s = ut + \frac{1}{2}at^2$
$u = u\sin\alpha$ $0 = u\sin\alpha t - \frac{1}{2}gt^2$
$v =$
$a = -g$ $0 = t\left(u\sin\alpha - \frac{1}{2}gt\right)$ [1]
$t = ?$ $t = 0,\ \dfrac{u\sin\alpha}{\frac{1}{2}g} = \dfrac{2u\sin\alpha}{g}$

Horizontal motion
when $t = \dfrac{2u\sin\alpha}{g}$ [1]

$R = (u\cos\alpha)t = u\cos\alpha\left(\dfrac{2u\sin\alpha}{g}\right)$

$\quad = \dfrac{2u^2\sin\alpha\cos\alpha}{g}$ [1]

(ii) Negligible air
    resistance   [1]
(iii) $\sin 2\theta = 2\sin\theta\cos\theta$

$R = \dfrac{2u^2\sin\alpha\cos\alpha}{g}$

$\quad = \dfrac{u^2(2\sin\alpha\cos\alpha)}{g} = \dfrac{u^2\sin 2\alpha}{g}$ [1]

$\sin 2\alpha$ takes maximum
value of 1 when
$\quad 2\alpha = 90°$ [1]
$\quad R_{max} = \dfrac{u^2}{g}$ when
$\quad \alpha = 45°$ [1]

(iv) Using $\alpha = 30$ to find
time and range
For B2 time
$t = \dfrac{2\times 10\sin 30}{g} = 1.02$
$\qquad\qquad\qquad\qquad [1]$
For C2 range
$x = \dfrac{2\times 10^2\sin 30\cos 30}{g}$
$\quad = 8.84$ [1]

(v) Maximum range
$\dfrac{u^2}{g} = 7.1$ [1]
$u = \sqrt{7.1\times 9.8} = 8.34$ [1]

(vi) The measurements
give the maximum
range when the angle
is 40°. Changing the
value for u will still
have the maximum
range at 45°.   [1]

(vii)
$\mathbf{v} = \dfrac{d\mathbf{r}}{dt} = \begin{pmatrix} 10\cos\alpha - 3t \\ 10\sin\alpha - 9.8t \end{pmatrix}$ [1]

When $t = 0$
$\mathbf{v} = \begin{pmatrix} 10\cos\alpha - 0 \\ 10\sin\alpha - 0 \end{pmatrix}$ [1]

Speed $=$
$\sqrt{(10\cos\alpha)^2 + (10\sin\alpha)^2}$
$= \sqrt{100(\cos^2\alpha + \sin^2\alpha)}$
$= \sqrt{100\times 1} = 10$ [1]

(viii) To find T in new model
y component of $\mathbf{r} = 0$
$10t\sin\alpha - 4.9t^2 = 0$
$T = \dfrac{2u\sin\alpha}{g}$ which
is exactly the same
equation as in the
standard model in
part (ii)   [1]

(ix) For E5, we need
horizontal component
when $t = 1.44$
$x = 10\times 1.44\cos 45° - 1.5\times 1.44^2$
$\quad = 7.07$ [2]

(x) The vector model is
an improvement. It fits
all the measurements
better than the standard
model and it gives the
maximum range of
7.53 for 35° as
compared with the
measured value of 7.4 m
occurring at 40°.   [1]

Index

F

factor theorem 61–2, 160–61
factorials 152
Fibonacci sequence 42
fixed point iteration 297–302
force
    definition 439
    diagrams 441
    as a function of time 431–32
    moment 470, 472–73
    resolving 446–7
    vector quantity 439
fractions 158–62
frequency, definition 322
frequency chart 325
friction 438, 440, 507–16
fulcrum, definition 470
functions
    composite 80–9, 110–15
    domain 72–3
    graph 74
    implicit 193
    increasing and decreasing 98–9
    inverse 83–8
    language and notation 71, 72
    as mappings 72, 74
    order in a composite function 81–2

G

geometric sequences and series 47–52
Goldbach's conjecture 3
grade, measurement of angle 13
gradient
    of a curve at a point 97
    of a straight line 240
gradient function 190, 200–1
graphs
    of exponential functions 29
    from parametric equations 252–5
    of a function and its inverse 85–8
    of logarithmic functions 30
    of a mapping 74
    of the modulus function 90–1
    of proportional relationships 66
    of reciprocal functions 66, 138
    transformation 68–9, 75–7
    of trigonometric functions 15–16, 138
gravity 419–20, 439, 483
growth and decay, exponential 52, 67

H

hinge, reaction force 475
histogram 325
hypothesis testing 378–82, 386–9

I

image, definition 71
implication, symbols 3
implicit function, differentiation 193–6

indices 28
inequalities 59, 91–3
input, definition 71
integration
    by parts 229–33
    by recognition 212, 221
    by substitution 211–16, 227–9
    choice of method 234
    definite and indefinite 198–9, 213–14, 217, 231
    definition 198
    finding areas 202–10
    involving the natural logarithmic function 220–26
    as the limit of a sum 202
    notation 198
    of trigonometric functions 214–18
interquartile range 324
interval, bisection 294–5
irrational numbers 72
iteration 297

L

Lami's theorem 450
levers 476–7
limits, of a series 36
line segments, properties 240–42
logarithmic functions 30, 67, 185–8, 232
logarithms 30
lowest common multiple 160

M

mappings 71–2, 74
maximum and minimum points 98–9
mean absolute deviation 326
measure of central tendency 324
median 324
mils, measurement of angle 12, 14
modelling, vocabulary and assumptions 438
modulus function 89–94
moments, of forces 470, 472–73
motion
    of connected particles 443–4
    direction 414–15, 428–9
    equation of 442–43
    in one dimension 414–23
    parabolic 482
    resistance to 440
    in a straight line 420–22
    in two or three dimensions 423–36
    under gravity 419–20
    vocabulary and notation 415, 423
    see also projectile motion
multiplication and division
    algebraic 160–1
    of fractions 159–62
    with indices 28
    of logarithms 30
    of polynomials 61

    of surds 27
    with trigonometric functions 172
    of a vector by a scalar 268, 269

N

natural logarithmic function, in integration 220–26
Newton–Raphson method 304–7
Newton's law of cooling 278
Newton's laws of motion 441–2
    second law 431–2, 441, 456, 458, 462–68
Normal curve 362, 368
Normal distribution 362–76
    experimental data 369–70, 384–93
    mean 363, 364, 386–9
    notation 364, 368
    standard deviation 363, 364
    standardised form 364, 388
    tables 363, 367
    use in modelling 369–71
normal reaction 439–40
normal to a curve 97–8
null hypothesis 379
numbers, notation 72
numerical methods 292–304
    integration 308–15, 316–17

O

object, definition 71
opportunity sampling 322
outcomes see events
outliers, definition 323
output, definition 71

P

p-value 379
parallel lines, properties 240
parametric equations
    of a circle 257–8
    definition 251
    eliminating the parameter 255–6
    graphs 252–5
    in projectile motion 498
    trigonometric 256–7
    use in modelling 250–1
partial fractions 163–7, 223, 224
Pascal's triangle 151–52
path see trajectory
perpendicular lines, properties 240
point, coordinates 268
point of inflection 102–4, 106
point of intersection, coordinates 242, 245
polynomial equations, solving 61–2
polynomial functions 61, 64–5
population mean, estimating 385